Definitions, units and conversion factors

In the SI, units are given for both defining relations of the magnetization, $\boldsymbol{B}=\mu_0(\boldsymbol{H}+\boldsymbol{M})$ and $\boldsymbol{B}=\mu_0\boldsymbol{H}+\boldsymbol{M}$, respectively. $\mu_0=4\pi\cdot10^{-7}\,\mathrm{Vs\,A^{-1}\,m^{-1}}$, A: molar mass, ϱ: mass density, $\boldsymbol{P}$: magnetic moment, $\boldsymbol{M}$: magnetic moment per unit volume (magnetization, magnetic polarization).

Quantity	cgs/emu	SI	
$\boldsymbol{B}$	$\mathrm{G}=(\mathrm{erg\,cm^{-3}})^{1/2}$	$\mathrm{T}=\mathrm{Vs\,m^{-2}}$	
	$1\,\mathrm{G}\hat{=}$	$10^{-4}\,\mathrm{T}$	
$\boldsymbol{H}$	$\mathrm{Oe}=(\mathrm{erg\,cm^{-3}})^{1/2}$	$\mathrm{Am^{-1}}$	
	$1\,\mathrm{Oe}\hat{=}$	$10^3/4\pi\,\mathrm{Am^{-1}}$	
$\boldsymbol{M}$	$\boldsymbol{B}=\boldsymbol{H}+4\pi\boldsymbol{M}$	$\boldsymbol{B}=\mu_0(\boldsymbol{H}+\boldsymbol{M})$	$\boldsymbol{B}=\mu_0\boldsymbol{H}+\boldsymbol{M}$
	G	$\mathrm{Am^{-1}}$	T
	$1\,\mathrm{G}\hat{=}$	$10^3\,\mathrm{Am^{-1}}$	$4\pi\cdot10^{-4}\,\mathrm{T}$
$\boldsymbol{P}$	$\boldsymbol{P}=\boldsymbol{M}V$	$\boldsymbol{P}=\boldsymbol{M}V$	$\boldsymbol{P}=\boldsymbol{M}V$
	$\mathrm{G\,cm^3}$	$\mathrm{Am^2}$	$\mathrm{Vs\,m}$
	$1\,\mathrm{G\,cm^3}\hat{=}$	$10^{-3}\,\mathrm{Am^2}$	$4\pi\cdot10^{-10}\,\mathrm{Vs\,m}$
σ	$\sigma=M/\varrho$	$\sigma=M/\varrho$	$\sigma=M/\varrho$
	$\mathrm{G\,cm^3\,g^{-1}}$	$\mathrm{Am^2\,kg^{-1}}$	$\mathrm{Vs\,m\,kg^{-1}}$
	$1\,\mathrm{G\,cm^3\,g^{-1}}\hat{=}$	$1\,\mathrm{Am^2\,kg^{-1}}$	$4\pi\cdot10^{-7}\,\mathrm{Vs\,m\,kg^{-1}}$
σ_m	$\sigma_\mathrm{m}=\sigma A$	$\sigma_\mathrm{m}=\sigma A$	$\sigma_\mathrm{m}=\sigma A$
	$\mathrm{G\,cm^3\,mol^{-1}}$	$\mathrm{Am^2\,mol^{-1}}$	$\mathrm{Vs\,m\,mol^{-1}}$
	$1\,\mathrm{G\,cm^3\,mol^{-1}}\hat{=}$	$10^{-3}\,\mathrm{Am^2\,mol^{-1}}$	$4\pi\cdot10^{-10}\,\mathrm{Vs\,m\,mol^{-1}}$
χ	$\boldsymbol{P}=\chi\boldsymbol{H}$	$\boldsymbol{P}=\chi\boldsymbol{H}$	$\boldsymbol{P}=\chi\mu_0\boldsymbol{H}$
	$\mathrm{cm^3}$	$\mathrm{m^3}$	$\mathrm{m^3}$
	$1\,\mathrm{cm^3}\hat{=}$	$4\pi\cdot10^{-6}\,\mathrm{m^3}$	$4\pi\cdot10^{-6}\,\mathrm{m^3}$
χ_v	$\chi_v=\chi/V$	$\chi_v=\chi/V$	$\chi_v=\chi/V$
	$\mathrm{cm^3\,cm^{-3}}$	$\mathrm{m^3\,m^{-3}}$	$\mathrm{m^3\,m^{-3}}$
	$1\,\mathrm{cm^3\,cm^{-3}}\hat{=}$	$4\pi\,\mathrm{m^3\,m^{-3}}$	$4\pi\,\mathrm{m^3\,m^{-3}}$
χ_g	$\chi_\mathrm{g}=\chi_v/\varrho$	$\chi_\mathrm{g}=\chi_v/\varrho$	$\chi_\mathrm{g}=\chi_v/\varrho$
	$\mathrm{cm^3\,g^{-1}}$	$\mathrm{m^3\,kg^{-1}}$	$\mathrm{m^3\,kg^{-1}}$
	$1\,\mathrm{cm^3\,g^{-1}}\hat{=}$	$4\pi\cdot10^{-3}\,\mathrm{m^3\,kg^{-1}}$	$4\pi\cdot10^{-3}\,\mathrm{m^3\,kg^{-1}}$
χ_m	$\chi_\mathrm{m}=\chi_\mathrm{g}A$	$\chi_\mathrm{m}=\chi_\mathrm{g}A$	$\chi_\mathrm{m}=\chi_\mathrm{g}A$
	$\mathrm{cm^3\,mol^{-1}}$	$\mathrm{m^3\,mol^{-1}}$	$\mathrm{m^3\,mol^{-1}}$
	$1\,\mathrm{cm^3\,mol^{-1}}\hat{=}$	$4\pi\cdot10^{-6}\,\mathrm{m^3\,mol^{-1}}$	$4\pi\cdot10^{-6}\,\mathrm{m^3\,mol^{-1}}$

Experimental errors

In this volume, experimental errors are given in parentheses referring to the last decimal places. For example, 1.352 (12) stands for 1.352 ± 0.012, and 342.5 (21) stands for 342.5 ± 2.1.

Landolt-Börnstein / New Series

Landolt-Börnstein

Numerical Data and Functional Relationships
in Science and Technology

New Series
Editors in Chief: O. Madelung and W. Martienssen

Units and Fundamental Constants in Physics and Chemistry

Nuclear and Particle Physics (Group I)

Atomic and Molecular Physics (Group II)

Solid State Physics (Group III)

Macroscopic Properties of Matter (Group IV)

Geophysics (Group V)

Astronomy and Astrophysics (Group VI)

Biophysics (Group VII)

Landolt-Börnstein

Numerical Data and Functional Relationships in Science and Technology

New Series / Editors in Chief: O. Madelung and W. Martienssen

Group III: Solid State Physics

Volume 27

Magnetic Properties of Non-Metallic Inorganic Compounds Based on Transition Elements

Subvolume f 3

Perovskites II, Oxides with Corundum, Ilmenite and Amorphous Structures

Y. Endoh, K. Kakurai, A.K. Katori, M.S. Seehra, G. Srinivasan, H.P.J. Wijn

Edited by H.P.J. Wijn

Springer-Verlag

Berlin Heidelberg NewYork

London Paris Tokyo

Hong Kong Barcelona Budapest

ISSN 0942-7988 (Solid State Physics)

ISBN 3-540-57770-X Springer-Verlag Berlin Heidelberg New York
ISBN 0-387-57770-X Springer-Verlag New York Berlin Heidelberg

Library of Congress Cataloging in Publication Data
Zahlenwerte und Funktionen aus Naturwissenschaften und Technik, Neue Serie
Editors in Chief: O. Madelung, W. Martienssen
Vol. III/27f3: Edited by H.P.J. Wijn
At head of title: Landolt-Börnstein. Added t.p.: Numerical data and functional relationships in science and technology.
Tables chiefly in English.
Intended to supersede the Physikalisch-chemische Tabellen by H. Landolt and R. Börnstein of which the 6th ed.
began publication in 1950 under title: Zahlenwerte und Funktionen aus Physik, Chemie, Astronomie,
Geophysik und Technik.
Vols. published after v. 1 of group I have imprint: Berlin, New York, Springer-Verlag
Includes bibliographies.
1. Physics--Tables. 2. Chemistry--Tables. 3. Engineering--Tables.
I. Börnstein, R.(Richard), 1852-1913. II. Landolt, H. (Hans), 1831-1910.
Physikalisch-chemische Tabellen. IV. Title: Numerical data and functional relationships in science and
technology.
QC61.23 502`.12 62-53136

Production: PRODUserv Springer Produktions-Gesellschaft, Berlin
Typesetting: Macmillan India Ltd.
Printing: Druckhaus Langenscheidt KG, Berlin
Binding: Lüderitz & Bauer, Berlin

SPIN 10465197 63/3020 - 5 4 3 2 1 0 - Printed on acid-free paper

Editor

H. P. J. Wijn, Institut für Werkstoffe der Elektrotechnik der Rheinisch-Westfälischen Technischen Hochschule Aachen, Templergraben, 52056 Aachen, FRG

Authors

Y. Endoh, Neutron Scattering Laboratory, Department of Physics, Tohoku University, Aramaki, Aoba-ku, Sendai 980, Japan

K. Kakurai, Neutron Scattering Facility, Institute for Solid State Physics, University of Tokyo, Tokai mura, Ibaraki 319-11, Japan

A. K. Katori, Institute for Solid State Physics, University of Tokyo, Roppongi, Minato-ku, Tokyo 107, Japan

M. S. Seehra, Department of Physics, West Virginia University, Morgantown, West Virginia 26506 - 6315, USA

G. Srinivasan, Department of Physics, Oakland University, Rochester, Michigan 48309 - 4401, USA

H. P. J. Wijn, Institut für Werkstoffe der Elektrotechnik der Rheinisch-Westfälischen Technischen Hochschule Aachen, Templergraben, 52056 Aachen, FRG

Preface

The Landolt-Börnstein Volume 27 deals with the magnetic properties of non-metallic inorganic compounds based on transition elements, such as there are the pnictides, chalcogenides, oxides, halides, borates, silicates and phosphates. A preliminary survey of the contents of all subvolumes that have already appeared or have been planned is printed on the inside of the front cover.

The subvolume 27f consists of three parts and is mainly devoted to the magnetic properties of oxides with the perovskite and provskite-related structures. It covers the compounds with magnetic properties based on the constituting 3d, 4d, 5d elements or perovskite-type layered cuprates, a family of compounds well-known in the research of high-temperature superconductors.

In the present and last part 27f3 the magnetic properties of RMO_3-type perovskite (R = rare earth element, M = 3d element), of oxides with corundum and ilmenite structure, as well as some structurally amorphous oxides have been compiled. In many aspects this part is a supplement to the previous surveys of J.B. Goodenough and J.M. Longo in Landolt-Börnstein Vol. III/4a (publ. 1970) and Vol. III/12a (publ. 1978). However this subvolume presents in itself also the present state of knowledge of the magnetic properties of the considered compounds.

Many thanks are due to the authors for the agreeable co-operation, the Landolt-Börnstein editorial office in Darmstadt, especially Dr. W. Polzin, Frau R. Brangs and Frau I. Lenhart for the great help with the editorial work, and to Springer Verlag for their thoughtful help in the final preparation of this volume.

Aachen, June 1994

The Editor

Table of contents

Magnetic properties of non-metallic inorganic compounds based on transition elements
Subvolume III/27f3: Perovskites II, oxides with corundum, ilmenite and amorphous structures

5 Magnetic properties of oxides with perovskite, corundum, ilmenite and amorphous structures

(Chap. 5.1 ⋯ 5.3 see subvolume III/27f1; chap. 5.4 see subvolume III/27f2)

5.5 Perovskite-type oxides RMO_3 (R = rare earth element, M = 3d element or Al)

5.5.1 Introduction

5.5.1.1 General remarks

This chapter is a supplement to both Landolt-Börnstein, New Series, Group III,

- subvolume 4a "Magnetic Oxides and Related Compounds, Chapter 3, crystallographic and magnetic properties of perovskite and perovskite-related compounds" [70G1], abbreviated as III/4a and dealing with the literature up to the year of 1969.
- subvolume 12a "Magnetic and Other Properties of Oxides and Related Compounds, Chapter 2", with the same title as above [78N1], abbreviated III/12a, covering the literature that appeared in the period 1969–1975.

Because of the overwhelming amount of new data on all kinds of perovskites published since 1975, the present compilation is restricted to the substances RMO_3 having an orthorhombic structure, and where R stands for a rare earth element including Sc and Y or for a combination of rare earth elements, and M for a 3d element and/or Al, or a combination of such elements.

The Introduction of Goodenough and Longo on the various perovskite structures and on the magnetic order in these substances in III/4a remains of incalculable value. The present compilation adds only minor new information to the extensive data on the crystallographic properties already given in III/4a and III/12a. Major emphasis is on the recently published magnetic data. Reference is made to all orthorhombic perovskites RMO_3 for which magnetic data were already given in III/4a and III/12a. The sequence in the list of compounds has more or less been retained, i.e. the leading element for the classification is M, which means that the aluminates precede the Ti, V, Cr, Mn, Fe, Co and Ni oxides in this sequence. However if the compound contains a combination of Al and/or 3d elements, than it is entered in the compilation according to the leading element of the above given sequence of elements. Exceptions are made for small concentrations that can be considered as a doping of the main compound. Compounds with the same 3d elements have been arranged according to the place of the rare earth element in the periodic system of elements. In the case of more RE elements as the constituents of a compound, this compound can be found under the subchapter of the simple RE compound with the lowest atom number.

For each compound the data are divided into four parts as follows:

I Information about magnetic order from neutron diffraction, DTA, specific heat, thermal expansion and magnetic measurements. Information about crystallographic and electrical properties.

II Magnetic properties above the magnetic ordering temperature.

III Magnetic properties derived from resonance measurements including NMR, ME, AFMR and ESR.
IV Optical properties from reflectance and absorption measurements.

For semiconducting properties of RMO_3 compounds see LB III/17h, p. 226ff.

5.5.1.2 Magnetic configurations

The orthorhombic perovskites RMO_3 have the orthorhombically distorted perovskite crystal structure belonging to the space group D_{2h}^{16}-Pbnm. For the position of the R and M ions on the positions (4c) and (4b) in the unit crystal cell, see Fig. 1. The magnetic moments of the ions can order at low temperatures in various possible configurations with a symmetry compatible to the crystal symmetry as is described by Bertaut [63B1]. The magnetic classification system he introduced is now generally accepted. For the definition of the magnetic-moment configurations of the M and R ions see Table A. The different structures are described by the following linear combinations of the normalized spin vectors $\boldsymbol{S}_i$ (for both the M or the R ions): the ferromagnetic vector $\boldsymbol{F} = \boldsymbol{S}_1 + \boldsymbol{S}_2 + \boldsymbol{S}_3 + \boldsymbol{S}_4$, the antiferromagnetic vector $\boldsymbol{G} = \boldsymbol{S}_1 - \boldsymbol{S}_2 + \boldsymbol{S}_3 - \boldsymbol{S}_4$ and the two antiferromagnetic combinations with parallel nearest neighbors $\boldsymbol{C} = \boldsymbol{S}_1 + \boldsymbol{S}_2 - \boldsymbol{S}_3 - \boldsymbol{S}_4$ and $\boldsymbol{A} = \boldsymbol{S}_1 - \boldsymbol{S}_2 - \boldsymbol{S}_3 + \boldsymbol{S}_4$. See also Fig. 26 of chapter 3 in LB III/4a, p. 207. For the ferromagnetism basis vector $\boldsymbol{F}$ and the antiferromagnetism basisvectors $\boldsymbol{G}$, $\boldsymbol{C}$ and $\boldsymbol{A}$ holds that $\boldsymbol{F}$, $\boldsymbol{C}$ and $\boldsymbol{A}$ are very small compared to $\boldsymbol{G}$. The from a point of view of crystal symmetry possible magnetic structures Γ_i are given in Table B. The spin configurations Γ_2 and Γ_4 of the four inequivalent M^{3+} ions in the unit cell are shown in Fig. 2. For the configuration of the magnetic moments of the R sublattice see Table C. Under the influence of changing external parameters like temperature, magnetic field or pressure, transitions between the magnetic structures can take place [76B4]. These spin rotation transition (SRT) can have an abrupt character, a first order transition, or they can occur gradually with changing temperature, manifesting second order transitions at the beginning and at the end of the transition. In the latter case the intermediate phase between the phases Γ_4 and Γ_2 occurring in orthochromites and orthoferrites in a temperature interval $T_1 - T_2$ is designated as Γ_{24}. The weak ferromagnetic moment of the M^{3+} ions rotates gradually in the *ac* plane from the *c* direction into the *a* direction of the crystal.

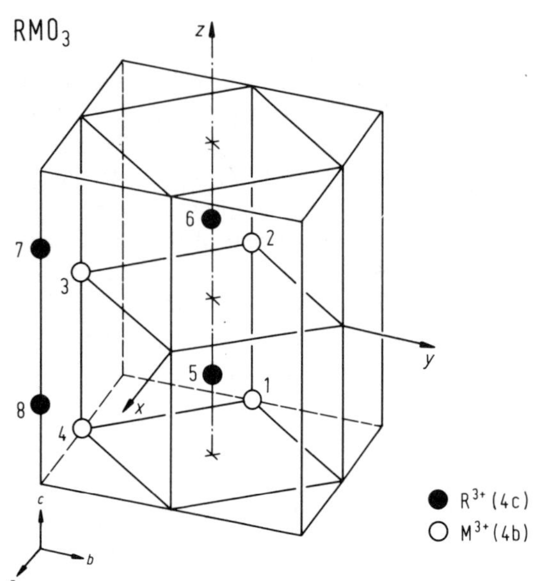

Fig. 1. RMO_3. Positions of the metal ions in the orthorhombic perovskite unit cell. The positions (4b) for the M^{3+} ions are (0, 1/2, 0), (0, 1/2, 1/2), (1/2, 0, 1/2), (1/2, 0, 0) and for the (4c) positions of the R^{3+} ions (1/2 − *x*, 1/2 + *y*, 1/4), (1/2 + *x*, 1/2 − *y*, 3/4), (− *x*, − *y*, 3/4), (*x*, *y*, 1/4). The parameters *x* and *y* are small deviations from zero, neglected in the figure [74Y1].

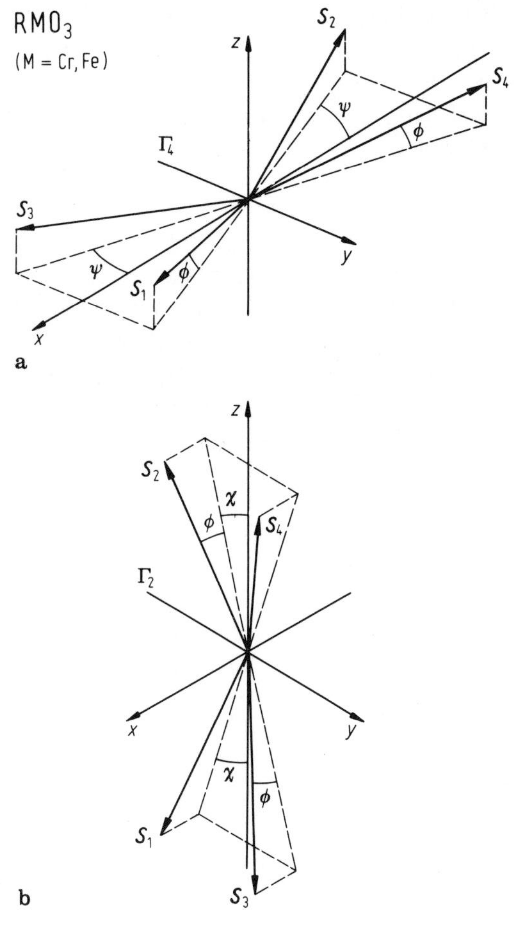

Fig. 2. RMO_3, M = Cr, Fe. Spin configurations Γ_2 and Γ_4 of the four M^{3+} ions in the unit cell, showing the "overt" canting angles ϕ leading to weak ferromagnetism, the "hidden" canting angles χ and ψ [82L1].

Table A. RMO_3. Definition of the possible magnetic-moment configurations according to [63B1], for the four R or the four M ions on the different sites of the unit cell shown in Fig. 1. The definition applies to both types of ions.

Configuration designation	Site 1 (8)	2 (7)	3 (6)	4 (5)
F	+	+	+	+
G	+	−	+	−
A	+	−	−	+
C	+	+	−	−

Table B. RMO_3. Irreducible representations of the group Pbnm expressed as magnetic phases Γ_i of possible magnetic structures for the components of the magnetic moments of the M and R ions on the (4b) and (4c) positions of the D_{2h}^{16} space group. The indices x, y, z refer to the coordinate axes.

Space group	Magnetic phase	Site (4b) M			Site (4c) R		
Pbnm	Γ_1	A_x	G_y	C_z	–	–	C_z
Pbn′m′	Γ_2	F_x	C_y	G_z	F_x	C_y	–
Pb′nm′	Γ_3	C_x	F_y	A_z	C_x	F_y	–
Pb′n′m	Γ_4	G_x	A_y	F_z	–	–	F_z
Pb′n′m′	Γ_5	–	–	–	G_x	A_y	–
Pb′nm	Γ_6	–	–	–	–	–	A_z
Pbn′m	Γ_7	–	–	–	–	–	G_z
Pbnm′	Γ_8	–	–	–	A_x	G_y	–

Table C. Rare earth moment directions in RMO_3 as derived from neutron diffraction experiments at low temperature [83T2].

RMO_3	R-moment configuration cooperative order	R-moment configuration induced order [a])	R-moment direction
$TbAlO_3$	G_xA_y	–	34° to [100]
$TbTiO_3$	–	F_xC_y	36° to [100]
$TbCrO_3$	complex		
	A_xG_y or G_xA_y	F_xC_y	–
$TbFeO_3$	A_xG_y	F_xC_y	38° to [100]
$TbCoO_3$	A_xG_y	–	–
$DyAlO_3$	G_xA_y	–	33° to [010]
$DyTiO_3$	–	C_xF_y	31° to [010]
$DyCrO_3$	G_xA_y	–	29° to [010]
$DyFeO_3$	G_xA_y	–	30° to [010]
$DyCoO_3$	G_xA_y	–	–
$HoAlO_3$	–	C_xF_y or C_yF_x	–
$HoTiO_3$	–	F_xC_y	24° to [010]
$HoCrO_3$	–	F_xC_y	26° to [010]
$HoFeO_3$	–	F_xC_y	27° to [010]
$HoCoO_3$	G_xA_y	F_xC_y	–
$ErAlO_3$	C_z	–	[001]
$ErTiO_3$	–	F_z	[001]
$ErCrO_3$	–	C_z	[001]
$ErFeO_3$	–	C_z	[001]
$ErCoO_3$	–	–	–
$TmAlO_3$	–	–	–
$TmTiO_3$	–	F_z	[001]
$TmCrO_3$	–	F_z	[001]
$TmFeO_3$	–	–	–
$TmCoO_3$	–	–	–

[a]) Induced either by the coupling to the M sublattice or to an applied magnetic field.

5.5.1.3 List of frequently used symbols and abbreviations

Symbols

Symbol	Unit	Property
a,b,c	Å	lattice parameters
B	G	magnetic induction
C_m	$cm^3\,K\,mol^{-1}$	molar Curie constant
C_p	$J\,K^{-1}\,mol^{-1}$	molar heat capacity at constant pressure
D	K	crystal field splitting parameter
E	$dyn\,cm^{-2}$	Young's modulus
E_a	eV	activation energy of electrical conductivity
ΔE_Q	$mm\,s^{-1}$	electric quadrupole splitting
g	–	spectroscopic splitting factor
H	Oe, $A\,m^{-1}$	magnetic field (sometimes given as $\mu_0 H$ in tesla (T))
h		ac magnetic field amplitude
H_c, H_{cr}		critical magnetic field
H_D		Dzyaloshinsky (spin-canting) field
H_{ex}		Heisenberg exchange field
H_{hyp}		effective hyperfine field
H_{int}		internal magnetic field
H_μ		local muon field
H_0	$kJ\,mol^{-1}$	molar enthalpy
J	–	magnetic quantum number
J	cm^{-1}	exchange coupling constant
J_{nn}/k	K	isotropic exchange constant for nearest neighbor interactions
k	$Å^{-1}$	wave vector
k	$J\,K^{-1}$	Boltzmann constant
K_2, K_4	$erg\,cm^{-3}$	second and fourth order magnetic anisotropy constants; $E_a = K_2 \sin^2\theta + K_4 \sin^4\theta$, where E_a is the magnetocrystalline anisotropy energy and θ the angle between the z-axis and the weak ferromagnetic moment
$\Delta l/l$	–	thermal expansion
M	G	magnetic moment per unit volume (= magnetization)
M_s		spontaneous magnetization
$n^{a,b,c}$	–	indices of refraction along the main crystallographic directions
Δn	–	birefringence
p	bar	hydrostatic pressure
p_{eff}	μ_B	effective magnetic moment
p_{FU}	μ_B	magnetic moment per formula unit
$p_{M^{3+}}$	μ_B	magnetic moment per M^{3+} ion
p_s	μ_B	spontaneous magnetic moment per formula unit
S		electron spin quantum number

Symbol	Unit	Property
S_0	$J\,mol^{-1}\,K^{-1}$	molar entropy
T	°C, K	temperature
T_c, T_{cr}		critical temperature (for phase transitions)
T_C		Curie temperature
T_{comp}		compensation temperature for zero magnetization
T_m		melting point
T_M		Morin temperature
T_N, T_{N1}, T_{N2}		Néel temperature, antiferromagnetic ordering temperatures
T_R		spin reorientation temperature
T_s		crystallographic transition temperature
v_S	$cm\,s^{-1}$	sound velocity
v_W	$cm\,s^{-1}$	domain wall velocity
V	$Å^3$	unit cell volume
V_{zz}	$V\,m^{-2}$	electric field gradient along z direction
x,y,z	–	coordinate axes
α,β	deg	canting angle of magnetic moment with respect to crystallographic axes (α is also rhombohedral angle)
α_{ij}	arb. units	magneto-electric susceptibility matrix elements
γ	$erg\,cm^{-2}$	domain wall energy density
Γ_i	–	spin configurations; see subsection 5.5.1.2, Table B
δ	$mm\,s^{-1}$	isomer shift
δ	–	equatorial Kerr rotation
$\varepsilon', \varepsilon''$	–	real and imaginary part of the dielectric constant
η	–	asymmetry parameter
θ	deg	angle between weak magnetic moment and c axis; angle between H_{hyp} and principal axis of EFG tensor
θ_F	$deg\,cm^{-1}$	Faraday rotation
Θ_D	K	Debye temperature
Θ_p	K	paramagnetic Curie temperature
λ	$W\,m^{-1}\,K^{-1}$	thermal conductivity
$\lambda^{a,b,c}$	–	linear magnetostriction along the main crystallographic axes
μ_B	$J\,T^{-1}$, $erg\,G^{-1}$	Bohr magneton
μ_w	$cm\,s^{-1}\,Oe^{-1}$	domain wall mobility
ν	Hz	frequency
ν_Q		quadrupole frequency
ν_{res}		NMR or AFMR frequency
$\bar{\nu}$	cm^{-1}	wavenumber

Symbol	Unit	Property
ρ	Ω cm	electrical resistivity
σ	emu g^{-1} = G cm^3 g^{-1}, A m^2 kg^{-1}	magnetic moment per unit mass (specific magnetization)
σ_m	G cm^3 mol^{-1}	magnetic moment per mole (molar magnetization)
$\sigma_{m,s}$	G cm^3 mol^{-1}	spontaneous molar magnetization
σ_s	G cm^3 g^{-1}	spontaneous specific (or mass) magnetization
$\sigma_s^{a,b,c}$	G cm^3 g^{-1}	id. along the main crystallographic directions
ϕ	deg, rad	"overt" canting angle, Fig. 2
χ	deg, rad	"hidden" canting angle, Fig. 2
χ', χ''	–	real, imaginary part of magnetic susceptibility
χ_g	emu g^{-1} = cm^3 g^{-1}	magnetic mass susceptibility
$\chi_g^{a,b,c}$	cm^3 g^{-1}	id. in main crystallographic directions
χ_m	emu mol^{-1} = cm^3 mol^{-1}	molar magnetic susceptibility
ψ	deg	"hidden" canting angle, Fig. 2
$\hbar\omega$	cm^{-1}, eV	photon energy, magnon energy
$\parallel$, $\perp$	–	parallel, perpendicular to a crystallographic axis

Abbreviations

ac	alternating current
A, C, G	antiferromagnetic vectors (see subsection 5.5.1.2)
AF	antiferromagnetism
AFMR	antiferromagnetic resonance
calc	calculated
cr	critical
C	cubic
DM	Dzyaloshinsky-Moriya antisymmetric exchange field
eff	effective
EFG	electric field gradient
EPR (ESR)	electron paramagnetic (spin) resonance
F	ferromagnetism, ferromagnetic vector (see subsection 5.5.1.2)
FU	formula unit
HT	high temperature
INS	inelastic neutron scattering
LT	low temperature
M	metal

ME	Mössbauer effect
NES	neutron elastic scattering
NMR	nuclear magnetic resonance
O	orthorhombic
P	paramagnetism
R, RE	rare earth element
Rh	rhombohedral
RT	room temperature
SMS	spontaneous magnetization and susceptibility measurements
SR	spin rotation
SRT	spin rotation transition
TIP	temperature independent paramagnetism
WF	weak ferromagnetism
ZFC	zero field cooled

5.5.2 $RAlO_3$ compounds

$RAlO_3$

I At RT the crystallographic structures are rhombohedral (D_{3d}^6) for La to Nd and orthorhombic (D_{2h}^{16}) for Sm to Lu. At higher temperatures phase transitions occur, see Table 1 and Fig. 3. Survey of magnetic characteristics Table 2.

II Magnetic susceptibility below 4K, see Fig. 4. Curie-Weiss constants Table 3.

III ^{170}Yb Mössbauer data for Yb^{3+}-host lattice magnetic interactions, see [78B4, 80B1].

IV For fluorescence properties see [88S1].

Table 1. $RAlO_3$. Crystallographic phase transitions at temperature T_s, and melting points T_m. Rh: rhombohedral, O: orthorhombic, C: cubic [84C1].

$RAlO_3$	T_s Rh ⇄ C [°C]	T_s O ⇄ Rh [°C]	T_s Rh ⇄ C [°C]	T_m [°C]
$LaAlO_3$	500			2110
$PrAlO_3$	1330			2080
$NdAlO_3$	1550			2090
$SmAlO_3$		770	1950	2040
$EuAlO_3$		1330		2020
$GdAlO_3$		1700		2000

Table 2. $RAlO_3$. Néel temperatures T_N and directions of the local moments, from [80B1].

$RAlO_3$ R	$4f^n$ n	Ground Multiplet	T_N [K]	Local moment directions $T < T_N$
Eu^{3+}	6	$^7F_0$	—	—
Gd^{3+}	7	$^8S_{7/2}$	3.87	*ab* plane, along + and − *a* axis
Tb^{3+}	8	$^7F_6$	3.95	*ab* plane, along + 35° and − 35° from *a* axis
Dy^{3+}	9	$^6H_{15/2}$	3.5	*ab* plane, along + 33° and − 33° from *b* axis
HO^{3+}	10	$^5I_8$	0.16	*ab* plane, along + 31° and − 31° from *b* axis
Er^{3+}	11	$^4I_{15/2}$	0.6	along + and − *c* axis
Tm^{3+}	12	$^3H_6$	—	—
Yb^{3+}	13	$^2F_{7/2}$	0.8	*ab* plane, along ~ + 30° and − 30° from *a* axis (single ion EPR data)

Table 3. $RAlO_3$. Curie-Weiss constants. Paramagnetic Curie temperature Θ_p and effective paramagnetic moment p_{eff} expressed in Bohr magnetons μ_B per R ion for powder samples. The calculated quantity $p_{eff,calc} = g\sqrt{J(J+1)}\,\mu_B$ [80M1].

Compounds	Temperature range [K]	p_{eff} [μ_B/R ion]	Θ_p [K]	$p_{eff,calc}$ [μ_B]
$PrAlO_3$	< 325	3.58	− 53	3.58
	> 325	3.75	− 90	
$NdAlO_3$	< 325	3.58	− 67	3.62
	325 ··· 550	3.87	− 132	
	> 550	4.14	− 237	
$GdAlO_3$	< 370	7.78	− 4	7.94
	370 ··· 500	8.52	− 62	
$TbAlO_3$	< 350	10.20	− 2	9.72
	350 ··· 625	10.76	− 40	
	> 625	11.10	− 82	
$DyAlO_3$	< 310	10.18	− 1	10.63
	310 ··· 603	10.76	− 40	
	> 625	11.08	− 79	
$HoAlO_3$	< 350	10.57	− 4	10.6
	350 ··· 650	11.26	− 44	
	> 650	11.75	− 109	
$YbAlO_3$	< 700	4.69	− 99	4.54
	> 700	4.50	− 37	

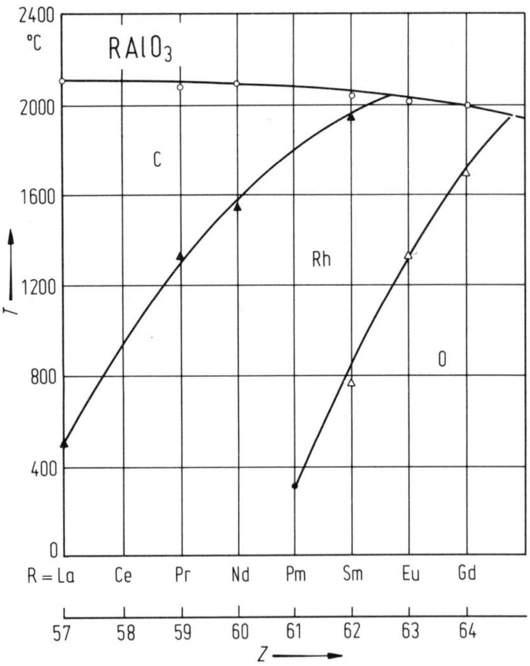

Fig. 3. $RAlO_3$. Phase transition temperatures and melting points. C = cubic, Rh = rhombohedral, O = orthorhombic [84C1].

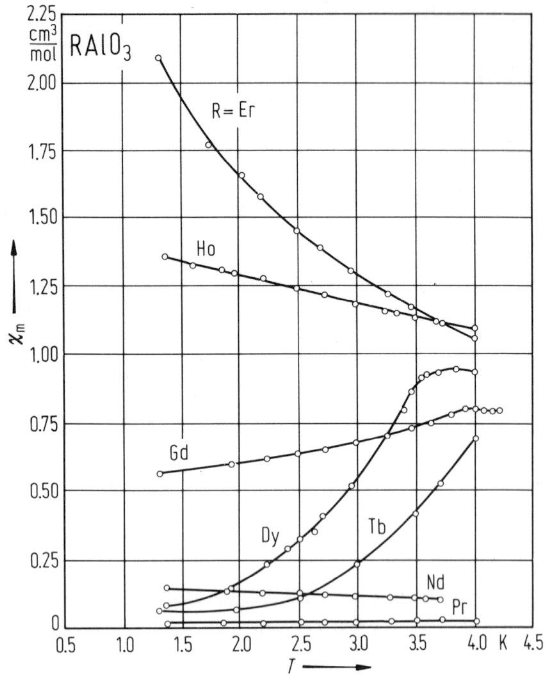

Fig. 4. $RAlO_3$. Molar magnetic susceptibility χ_m vs. temperature T [68C1].

$LaAlO_3$

I Calculated (p, T) phase diagrams [84M1].

III NMR of ^{27}Al in $LaAlO_3$ [83V4].

$PrAlO_3$ (see also LB III/4a, p. 225 and III/12a, p. 439).

II Curie-Weiss constants Table 3.

$NdAlO_3$ (see also LB III/4a, p. 225 and III/12a, p. 439).

I Lattice parameters Fig. 5 (see also LB III/4a, p. 161 and III/12a, p. 387).

II $80 < T < 300$ K: $\Theta_p = -45$ K, $p_{eff} = 3.62\ \mu_B$/FU [70S1].

IV Optical spectra: Raman, fluorescence and optical absorptions in the visible and near IR region [73F1].

$SmAlO_3$

I Lattice parameters Fig. 6 (see also LB III/4a, p. 161).
Specific heat: $T_N = 1.3$ K, $\Theta_D = 170$ K [68C2].

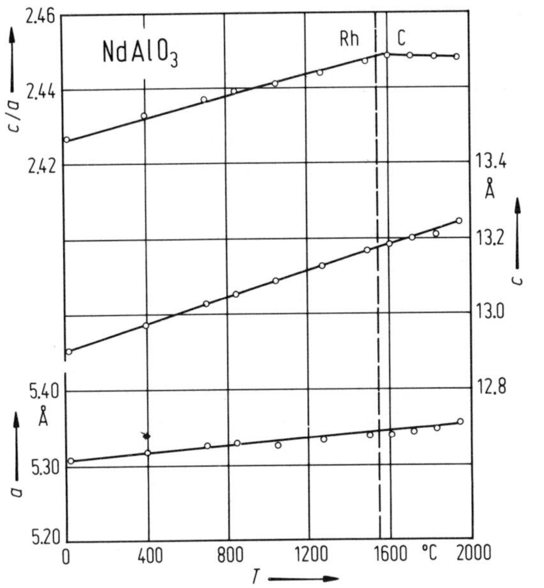

Fig. 5. NdAlO$_3$. Lattice parameters a and c vs. temperature T. Rh = rhombohedral, C = cubic [84C1].

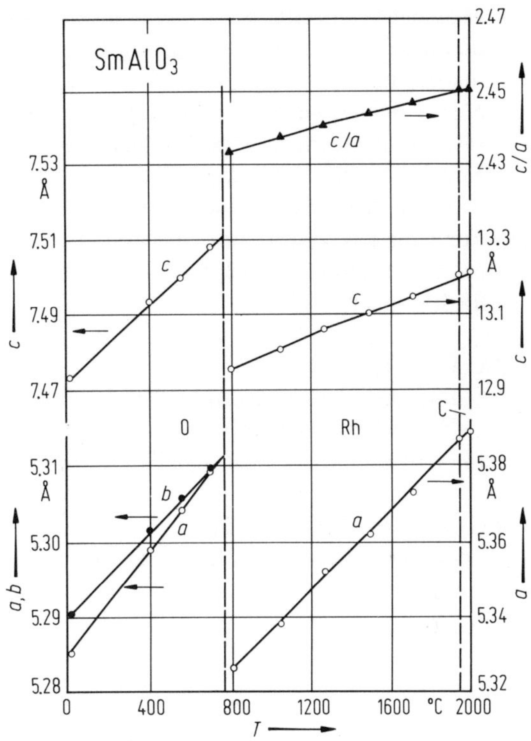

Fig. 6. SmAlO$_3$. Similar to Fig. 5 [84C1]. O = orthorhombic.

EuAlO$_3$ (see LB III/4a, p. 225 and III/12a, p. 439).
IV ESR spectra of Gd^{3+} in EuAlO$_3$ [82A1].

EuAl$_{0.2}$Fe$_{0.8}$O$_3$, see EuFeO$_3$.

GdAlO$_3$ (see also LB III/4a, p. 225 and III/12a, p. 440).

I In the past there has been much ambiguity regarding the type of magnetic ordering. However, Mössbauer spectra [76C1], magnetic susceptibility [76C2] and magnetoelectric susceptibility [77H1] data show that the ordering mode is predominantly antiferromagnetic, G_x, along the orthorhombic a axis. For the canting angle α see Fig. 7. From magnetoelectric susceptibility data, Figs. 8, 9, a magnetic ordering temperature $T_N = 3.876(2)$ K is derived [77H1]. Neutron diffraction analysis [82Q1] reveals for $T/T_N < 0.83$ a simple two sublattice antiferromagnet with the antiferromagnetic direction along the a axis. The magnetic moment value for the Gd^{3+} ion is $(p_{Gd^{3+}}) = 6.1\ \mu_B$. For magnetic phase diagram in the presence of an externally applied magnetic field see [77R1] and Fig. 10. The spin-flop multicritical phase point in the three dimensional space with coordinates given by the fields $H_{\parallel}$ and $H_{\perp}$ and the temperature T is: $T_M = 3.1242$ K and the external field $H_M = 12.494$ kOe. This point is bicritical in the $(H_{\parallel}\text{-}T)$ plane and tetracritical in the $(H_{\perp}\text{-}T)$ plane [77R1]. Magnetostriction vs. applied field Fig. 11.

II Curie-Weiss constants Table 3 and Fig. 12.

III Mössbauer spectra at $T > T_N$ of the 86.5 keV transition of ^{155}Gd reveal [80C1] a large quadrupole ground state splitting $\Delta E_Q = 2.88(5)\,\mathrm{mms^{-1}}$, from which follows an EFG at the Gd nucleus of $V_{zz} = -10.0(2)\cdot 10^{21}\,\mathrm{V\,m^{-2}}$ with an asymmetry parameter $\eta = 0.52(3)$. Isomer shift $\delta = +0.50(1)\,\mathrm{mms^{-1}}$ with respect to $SmPd_3$ above and below T_N (see also [78C1]). The saturation magnetic hyperfine field at $T = 0\mathrm{K}$ found by extrapolation is $\mu_0 H_{hyp}(0) = 29.1(2)$ T. For its temperature dependence see Fig. 13. Nearest neighbor exchange constant $J_{nn}/k = -0.070(2)$ K, corresponding to an exchange field $\mu_0 H_{ex} = -2.18(6)$ T [80C1].

IV For light absorption and emission characteristics of Cr^{3+} in $GdAlO_3$ see [83M1]. The Cr–Gd exchange coupling constant is found to be 2.1 and 1.46 $\mathrm{cm^{-1}}$ at zero and at high applied magnetic field, respectively.

$GdAl_xCr_{1-x}O_3$

I Spontaneous magnetization Fig. 14. It is concluded that the effective exchange field acting on Gd^{3+} decreases strongly with increasing Al^{3+} ion concentration, in contrast to what is found for the corresponding ferrites [87K1]. For the H-T phase diagram see Fig. 15.

$GdAl_{0.1}Fe_{0.9}O_3$; $GdGa_{0.1}Fe_{0.9}O_3$

I Spontaneous magnetization Fig. 16.

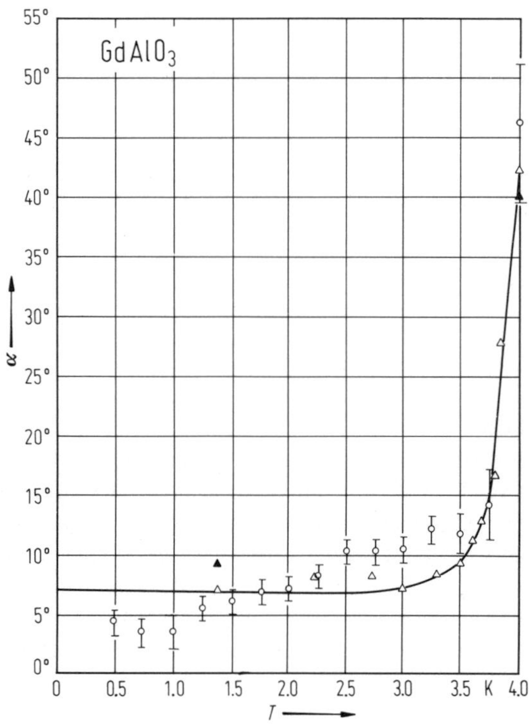

Fig. 7. $GdAlO_3$. Temperature dependence of the sublattice magnetization canting angle α measured from the a axis. Triangles data from ME, circles derived from susceptibility measurements [80C1].

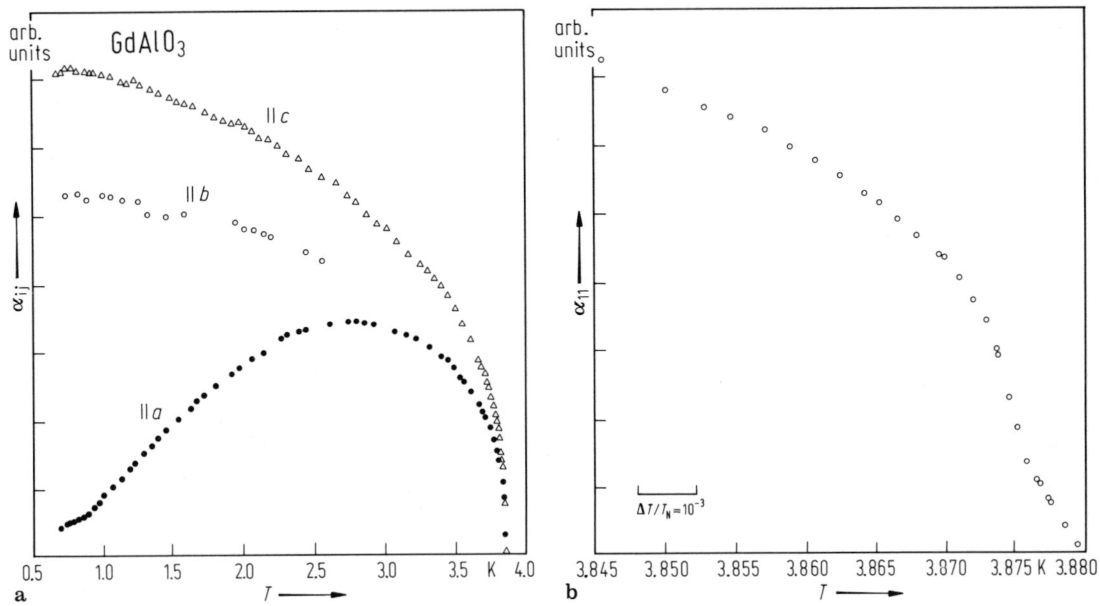

Fig. 8. $GdAlO_3$. (a) Magnetoelectric susceptibility matrix elements along the crystallographic axes vs. temperature T [77C1] and (b) the susceptibility along the a axis near the Néel temperature T_N [77H1]. $T_N = 3.88$ K.

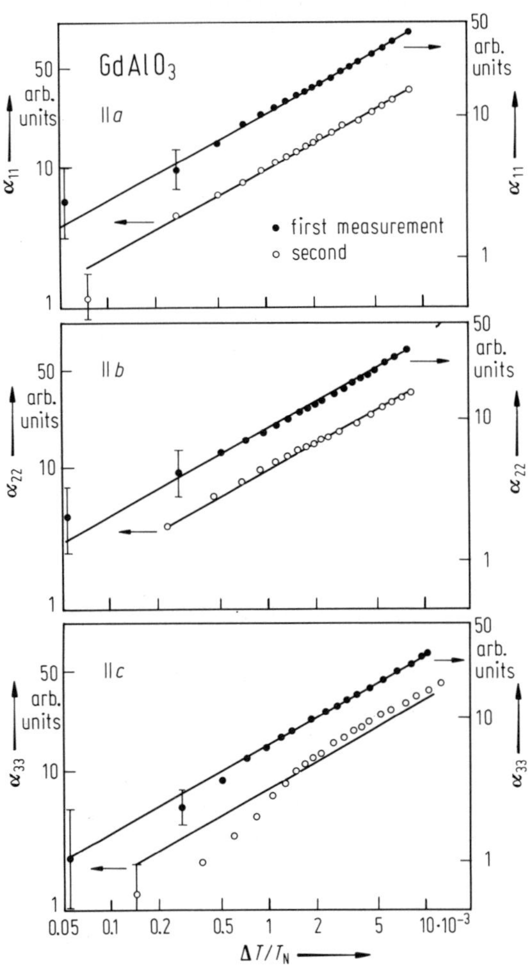

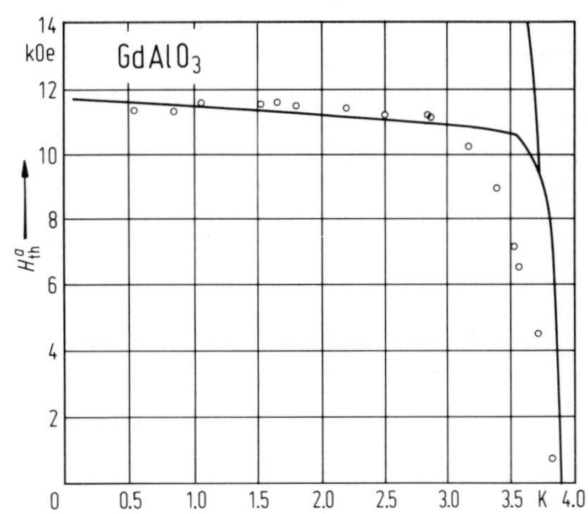

Fig. 10. $GdAlO_3$. Magnetic phase diagram derived from magneto-electric susceptibilities [77C1].

◀

Fig. 9. $GdAlO_3$. Log-log plots of the magnetoelectric susceptibility matrix elements α_{ii} in arbitrary units vs. relative temperature difference $(T_N - T)/T_N$. Two sets of measurements are given. The straight lines are theoretical fits of the form $\alpha = D(1 - T/T_N)^{\beta}$, where $\beta = 0.31$ (1) for all crystallographic directions [77H1]. $T_N = 3.88$ K.

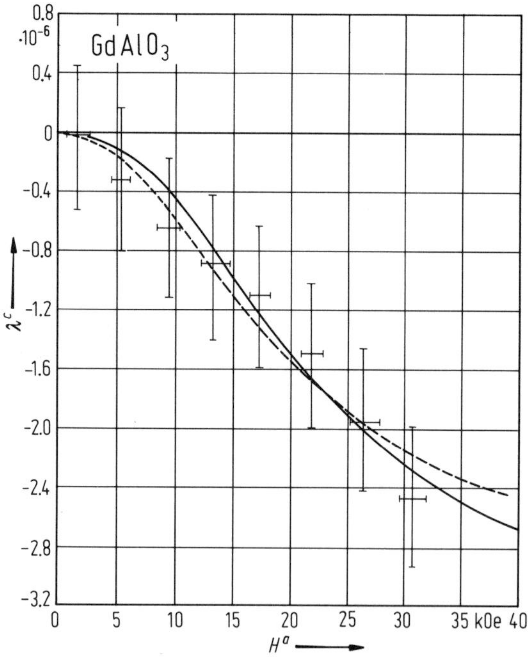

Fig. 11. $GdAlO_3$. Magnetostriction λ^c along the c axis vs. the magnitude of an applied magnetic field directed parallel to the a axis at 4.02 K. The curves result from single-(solid curve) and two-ion (dashed) based calculations [89K1].

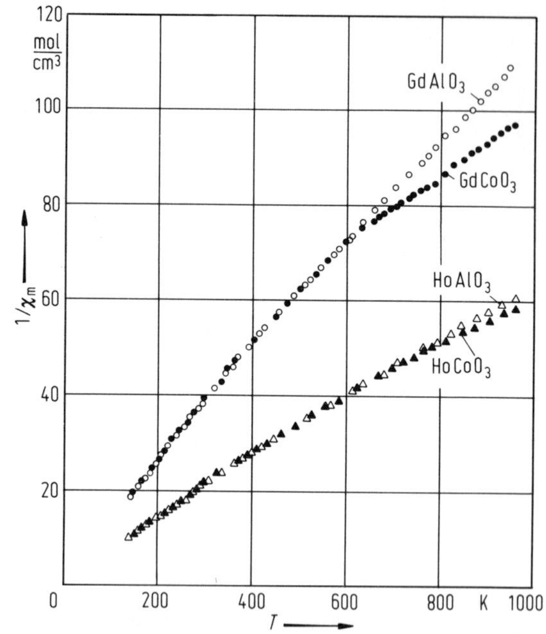

Fig. 12. RMO_3. Inverse molar magnetic susceptibility χ_m^{-1} vs. temperature T [80M1].

For Fig. 14, see next page.

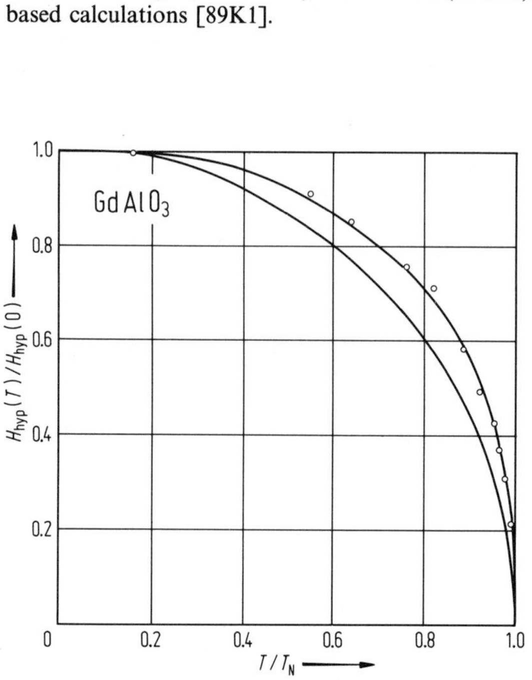

Fig. 13. $GdAlO_3$. Temperature dependence of the reduced magnetic hyperfine field H_{hyp} $(T)/H_{hyp}(0)$. Measuring points derived from ME; T_N: Néel temperature. Upper curve calculated sublattice magnetization from susceptibility measurements, lower curve Brillouin function for $S = 7/2$ [80C1].

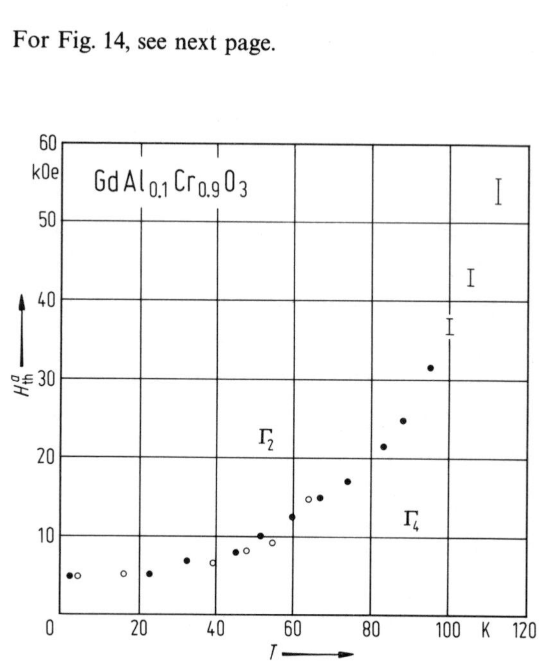

Fig. 15. $GdAl_{0.1}Cr_{0.9}O_3$. H–T phase diagram. The threshold field H_{th} for a phase transition $\Gamma_4 \rightarrow \Gamma_2$ depends on temperature. Open and closed circles: magnetic field applied in ab and ac plane, respectively [87K1].

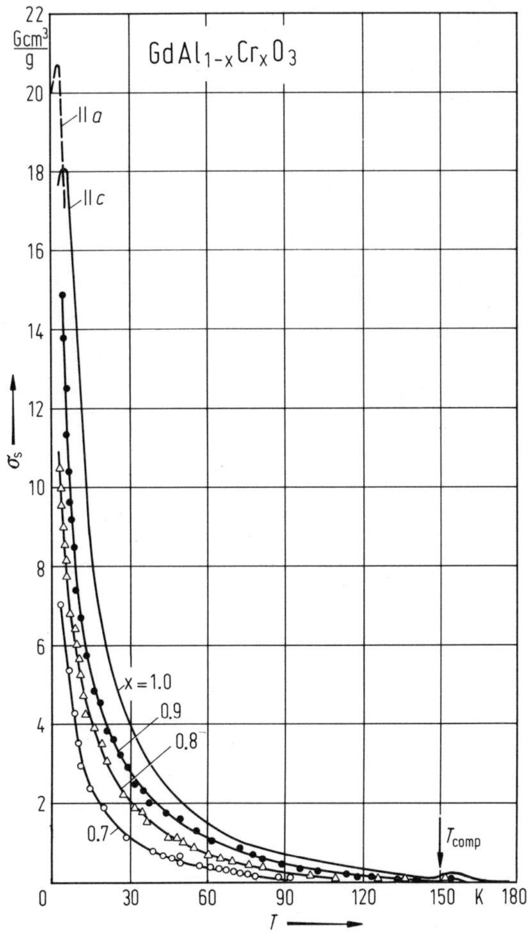

Fig. 14. $GdAl_{1-x}Cr_xO_3$. Single crystals. Spontaneous mass magnetization σ_S vs. temperature T. T_{comp} is the compensation point where $\sigma_S = 0$. For x = 0.7, T_{comp} = 110 K, for x = 1.0, T_{comp} = 143 K [87K1]; for x = 1.0 see [85B1].

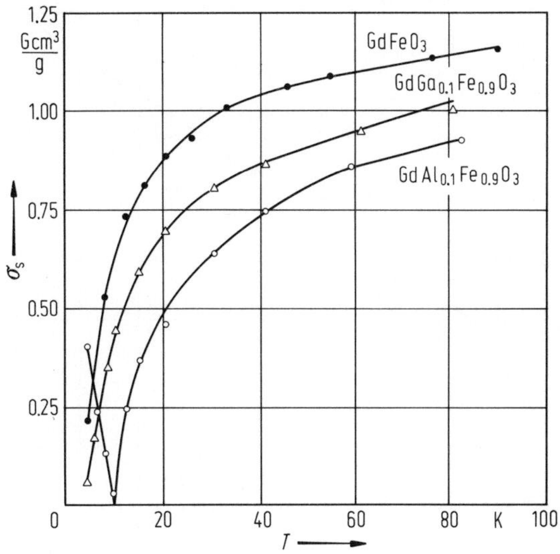

Fig. 16. $GdAl_{0.1}Fe_{0.9}O_3$; $GdGa_{0.1}Fe_{0.9}O_3$; $GdFeO_3$. Spontaneous mass magnetization σ_S vs. temperature T [75D1].

$TbAlO_3$ (see also LB III/4a, p. 225 and III/12a, p. 440)

I Magnetostriction vs. applied field Fig. 17.

II Curie-Weiss constants Table 3.

IV For the frequency and temperature dependence of the Faraday effect, see [87V1].

$TbAl_xFe_{1-x}O_3$

I Magnetization Figs. 18, 19. H-T phase diagram Fig. 20. The spontaneous SR $\Gamma_4(G_xF_z) \rightarrow \Gamma_2(G_zF_x)$ of $TbFeO_3$ at T_R = 6.5 K is not observed in the Al-substituted compounds [84D1].

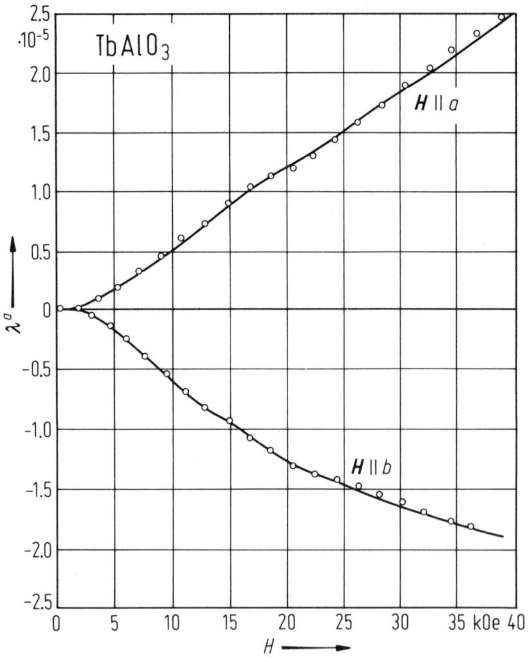

Fig. 17. $TbAlO_3$. Magnetostriction λ^a along the a axis at 4.2 K for magnetic fields parallel to this axis and parallel to the b axis. The curves are calculated [89K1].

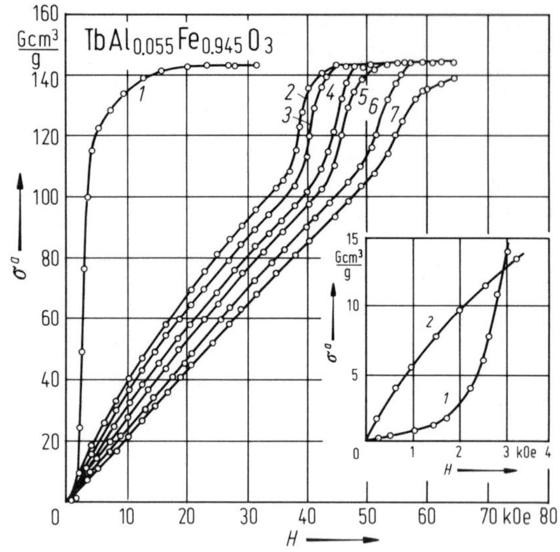

Fig. 18. $TbAl_{0.055}Fe_{0.945}O_3$. Magnetic isotherms at various temperatures for magnetic fields parallel to the a axis. Curves *2* to *7* apply to $T = 1.8, 4.2, 6.8, 8.2, 11$ and 12.6 K, respectively. Curve *1* applies to $TbFeO_3$ at 1.8 K [84D1].

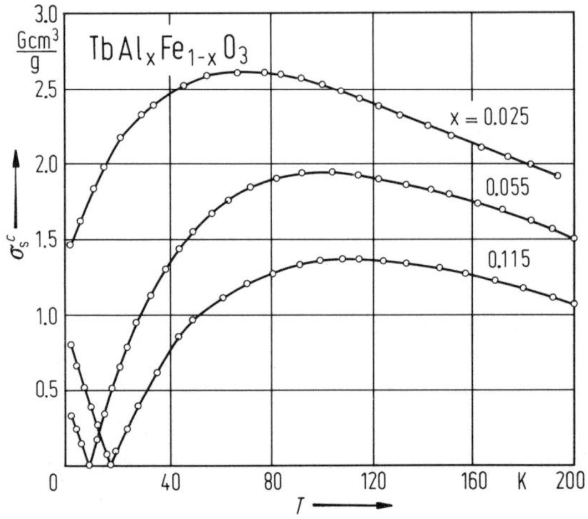

Fig. 19. $TbAl_xFe_{1-x}O_3$. Spontaneous mass magnetization σ_s^c along the c axis [84D1].

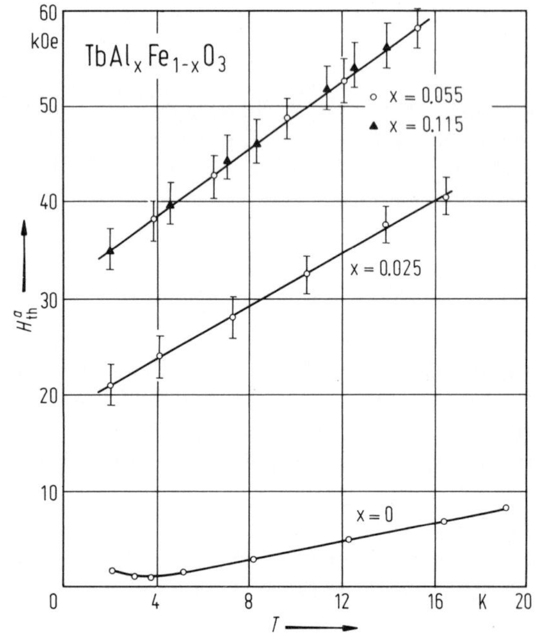

Fig. 20. $TbAl_xFe_{1-x}O_3$. Threshold field H_{th}^a for the first order SR transition $\Gamma_4\,(G_xF_z) \rightarrow \Gamma_2\,(G_zF_x)$ of the moments of the Fe^{3+} ions in an externally applied magnetic field parallel to the a axis [84D1].

$DyAlO_3$ (see also LB III/4a, p. 226 and III/12a, p. 440)

I The metamagnetic transitions at low temperature are related to the ground state of Dy^{3+} [90K1]. The magnetic structure consists of two pairs of antiferromagnetically coupled sublattices not interacting between themselves. The ferromagnetic order within a sublattice is governed by an exchange energy $J_{11}/k = 1.6\ (4)$ K. The antiferromagnetic ordering is determined by $J_{12}/k = J_{34}/k = -3.1\ (4)$ K [90K2]. Magnetostriction vs. applied field Fig. 21.

II Curie-Weiss constants Table 3. Paramagnetic susceptibility Fig. 22.

$DyAl_xFe_{1-x}O_3$

I Magnetic vacancies (e.g. Al ions) on the Fe-sites stabilize the low temperature antiferromagnetic phase Γ_1, see Fig. 23. The Morin temperature T_M increases with x [80Z1]. For magnetostriction see Fig. 24 and for the magnetic phase diagram see Fig. 25.

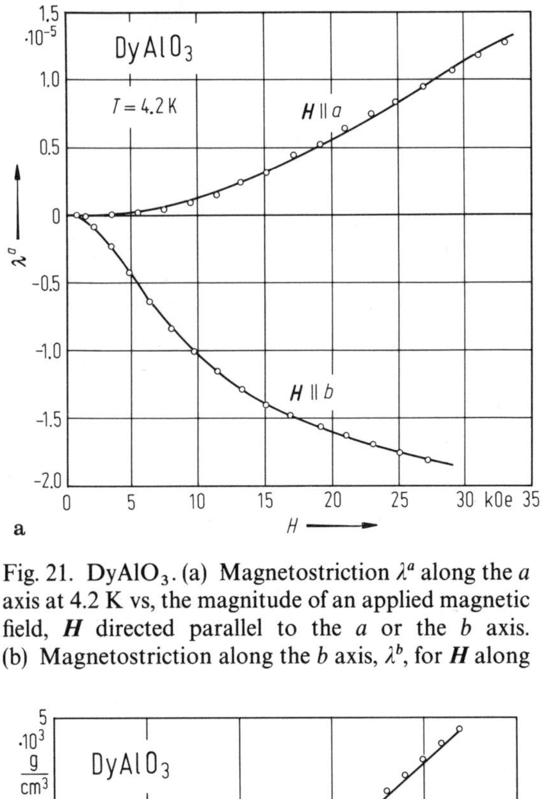

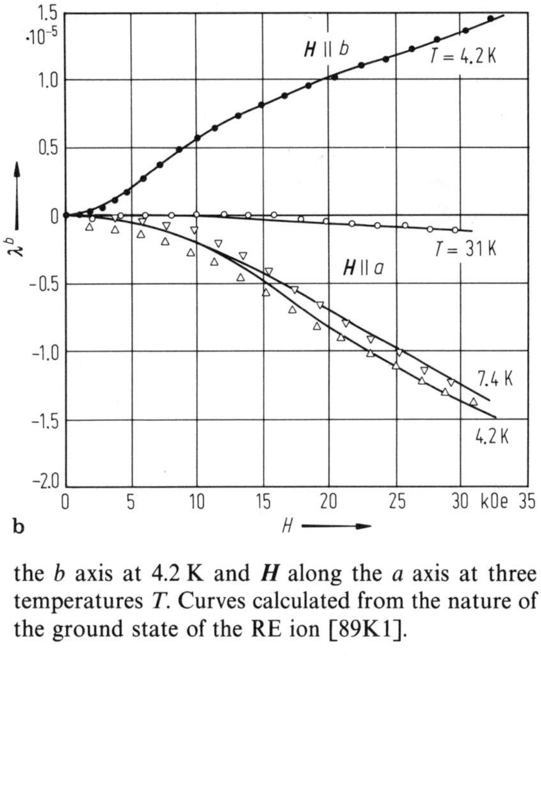

Fig. 21. $DyAlO_3$. (a) Magnetostriction λ^a along the a axis at 4.2 K vs, the magnitude of an applied magnetic field, $\boldsymbol{H}$ directed parallel to the a or the b axis. (b) Magnetostriction along the b axis, λ^b, for $\boldsymbol{H}$ along the b axis at 4.2 K and $\boldsymbol{H}$ along the a axis at three temperatures T. Curves calculated from the nature of the ground state of the RE ion [89K1].

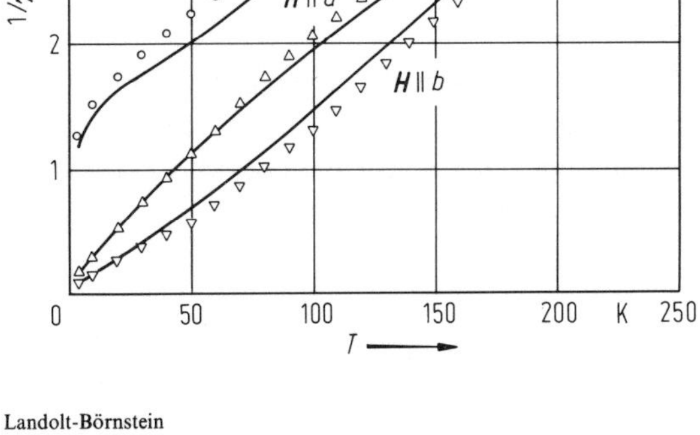

Fig. 22. $DyAlO_3$. The inverse magnetic mass susceptibility, χ_g^{-1} vs. T for three crystallographic directions. The solid curves are results calculated from crystal field parameters and the paramagnetic Curie temperatures $\Theta_p^a = -2.5$ K and $\Theta_p^b = -1.5$ K [90K1].

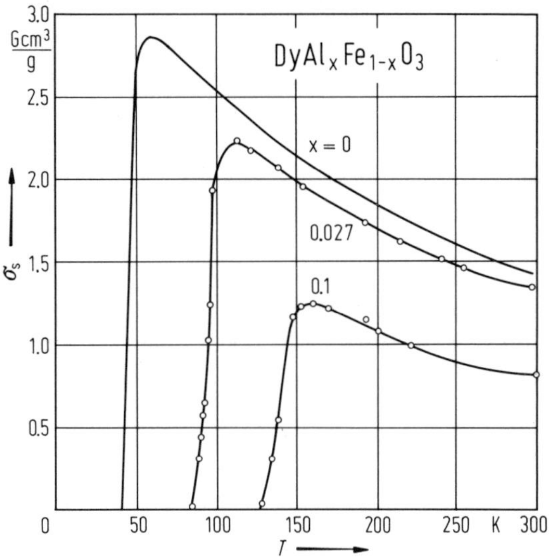

Fig. 23. $DyAl_xFe_{1-x}O_3$. Spontaneous mass magnetization σ_S vs. temperature T [80Z1].

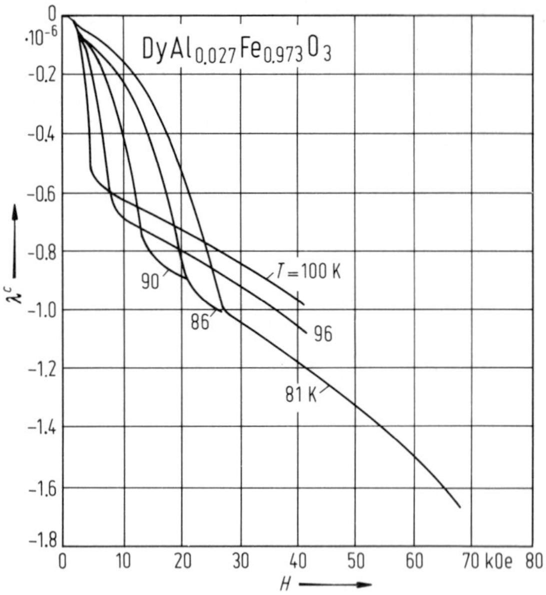

Fig. 24. $DyAl_{0.027}Fe_{0.973}O_3$. Linear magnetostriction λ^c in the crystallographic c direction vs. applied magnetic field H in this direction for various temperatures T. The magnetostriction due to rotation of the iron sublattice is found by extrapolation to be $0.8 \cdot 10^{-5}$ [77K1].

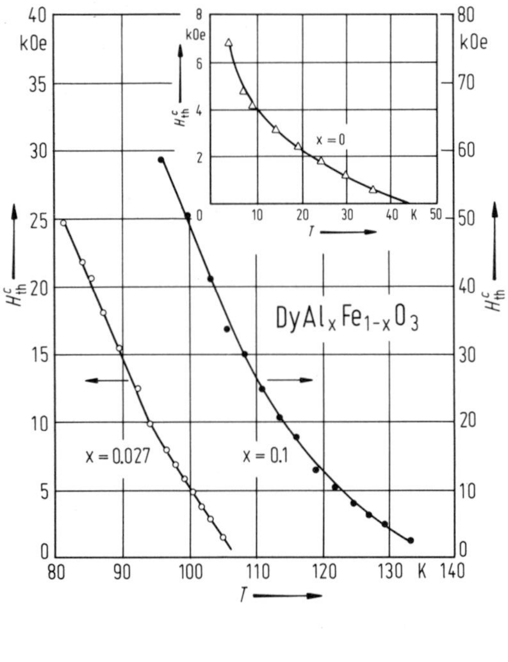

Fig. 25. $DyAl_xFe_{1-x}O_3$, single crystals. Magnetic phase diagram. Threshold field H^c_{th} in the direction of the c axis for the phase transition $\Gamma_1 \rightarrow \Gamma_4$, i.e. for the spin reorientation transition $G_y \rightarrow G_xF_z$ [77K1].

$HoAlO_3$

I X-ray lattice parameter $a = 5.182$ Å, $b = 5.324$ Å, $c = 7.37$ Å, private communication Kouchkovski in [77H2]. Neutron diffraction [77H2] shows an antiferromagnetic structure G_xA_y below $T_N = 0.16$ K. The magnetic moment as derived from susceptibility measurements at 4.2 K is $p_{Ho^{3+}} = 7.13\ (5)\ \mu_B$ [80H1]. For electronic and nuclear polarizations see Fig. 26.

II Paramagnetic properties Table 3, Figs. 12 and 27.

$HoAl_{0.03}Fe_{0.97}O_3$; $HoAl_{0.05}Fe_{0.95}O_3$

I Weak ferromagnetic moments in *a* and *c* direction Fig. 28. At high *T* an ordinary $c \rightarrow a$ reorientation occurs upon cooling (i.e. a phase transition $\Gamma_4 \rightarrow \Gamma_2$). At lower temperature an increased contribution of the magnetic vacancies to the effective anisotropy constant in the *bc* plane leads via an orientational transition to the antiferromagnetic state Γ_1 [80Z1]. An analysis showing the complexity of the $\Gamma_4 \rightarrow \Gamma_2$ transition via intermediate Γ_{24} phases was possible, Fig. 29.

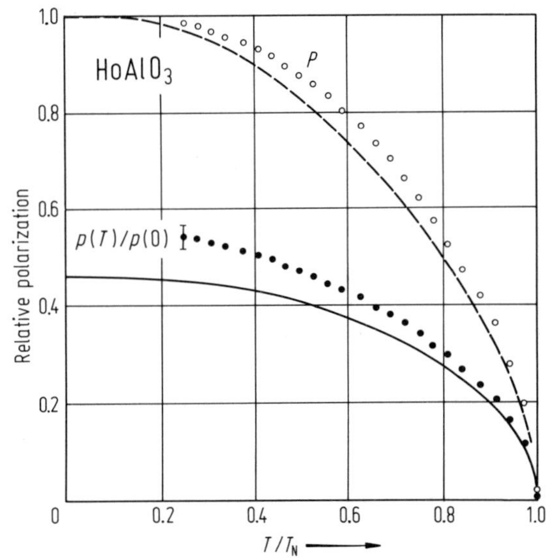

Fig. 26. $HoAlO_3$. Temperature dependence of the relative electronic and the relative nuclear polarizations of the Ho^{3+} ions $p(T)/p(0)$ and P, respectively. $T_N = 0.16$ K. Experimental points derived from susceptibility data, curves calculated [80H1].

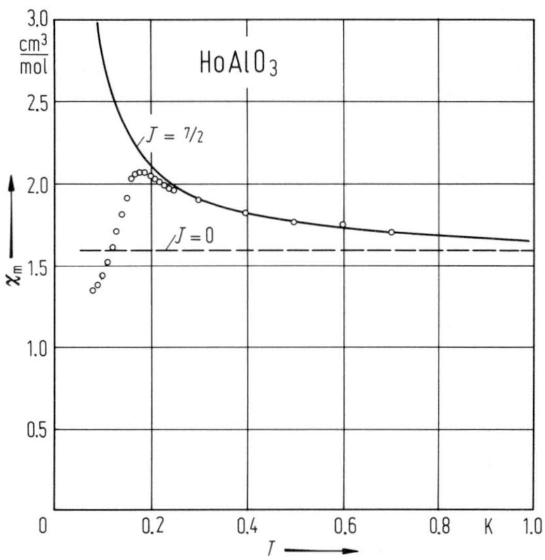

Fig. 27. $HoAlO_3$, powder. Molar magnetic susceptibility χ_m vs. temperature T. Drawn curve: calculated for $J = 7/2$ without interactions, dashed curve $J = 0$. The experimental points are adjusted at high temperature to the calculated curve [80H1].

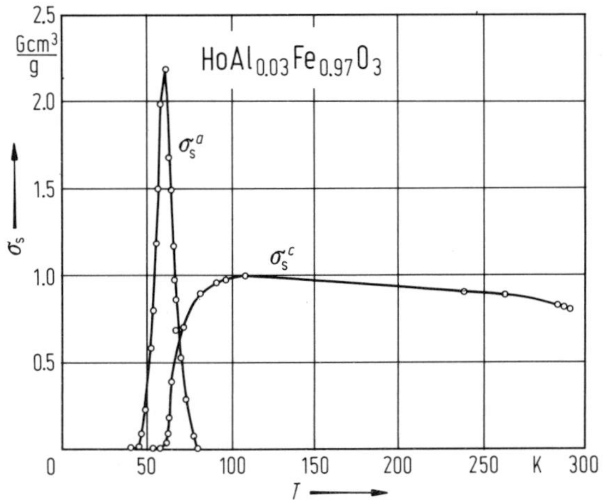

Fig. 28. $HoAl_{0.03}Fe_{0.97}O_3$. Specific spontaneous magnetization σ_s vs. temperature T for the weak ferromagnetism in the *a* and *c* direction [80Z1].

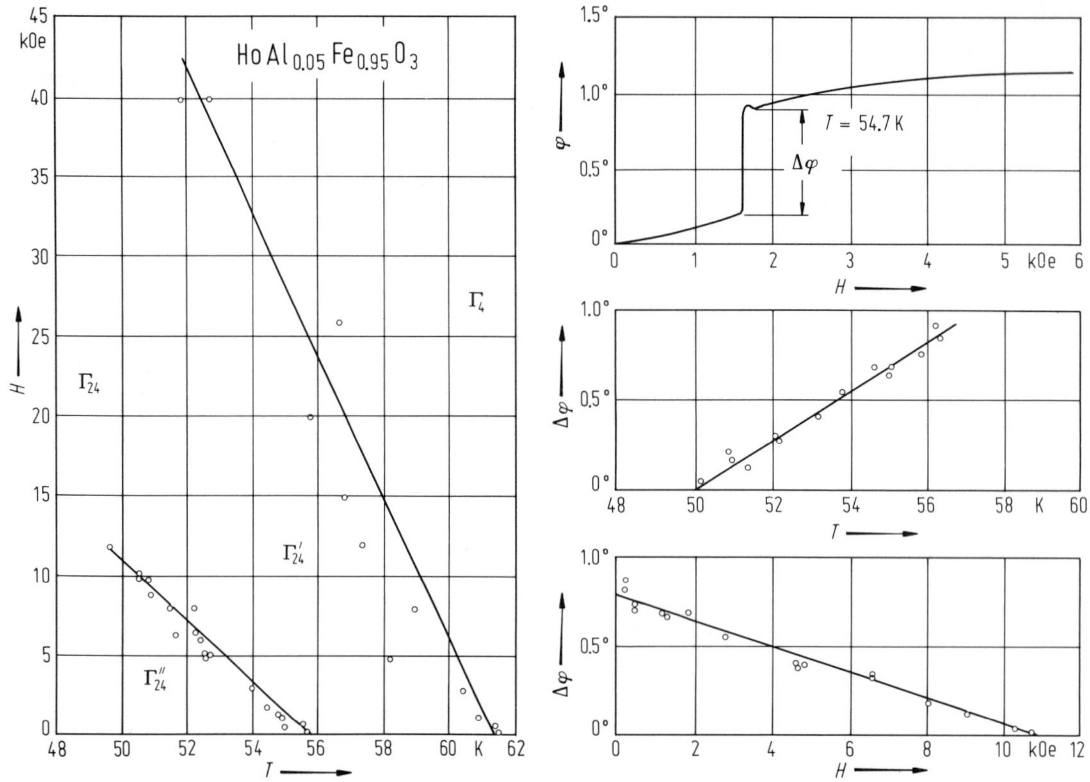

Fig. 29. $HoAl_{0.05}Fe_{0.95}O_3$. Magnetic phase diagram, parameters temperature T and externally applied magnetic field H, parallel to the c axis of the crystal plate. φ is the Faraday rotation angle. The boundary of the Γ_4 phase space represents a second order phase transition. The line separating the phases Γ''_{24} and Γ'_{24} represents a first order transition and exists only under certain combinations of H and T for which a jump $\Delta\varphi$ in the Faraday rotation angle φ is found [87P1].

$ErAlO_3$

I Lattice parameters Fig. 30 (see also LB III/12a, p. 387).
Neutron diffraction: antiferromagnetic, $T_N \approx 0.6$ K [71S1].

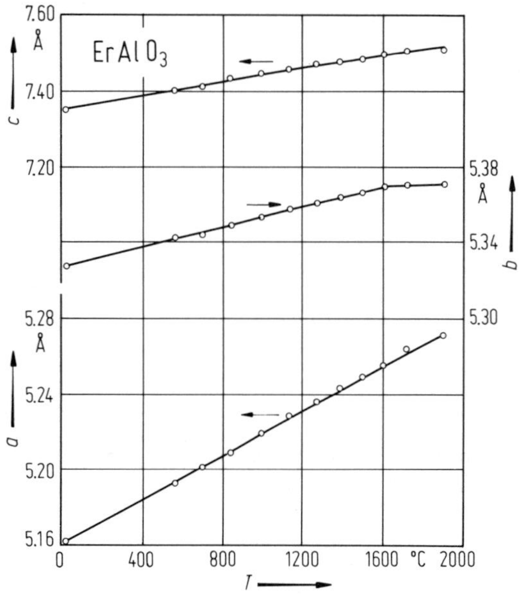

Fig. 30. $ErAlO_3$. Lattice parameters a, b and c vs. temperature T for the orthorhombic structure [84C1].

$TmAlO_3$

II Paramagnetic susceptibility Fig. 32.

III Mössbauer absorption measurements using ^{169}Tm in the range 4.2 K to 300 K show a symmetric quadrupole hyperfine doublet with no evidence of any magnetic interaction [84H1]. For the quadrupole hyperfine splitting see Fig. 31.

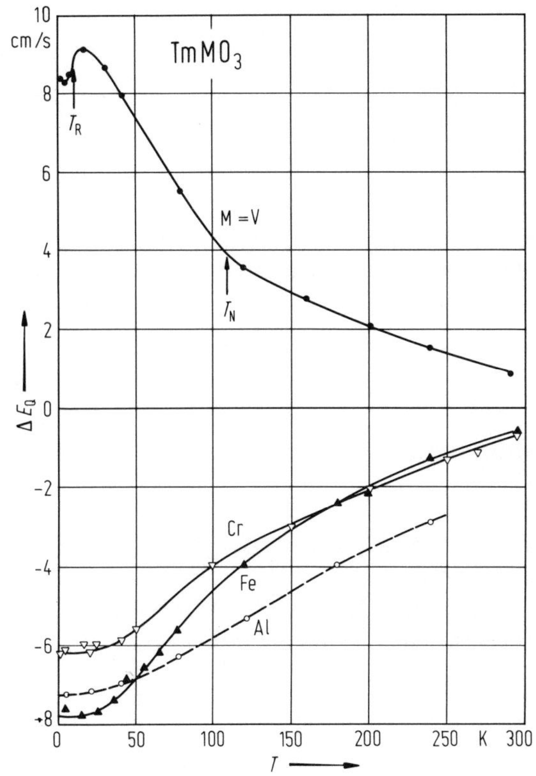

Fig. 31. $TmMO_3$. Thermal variation of the quadrupole hyperfine splitting ΔE_Q for ^{169}Tm Mössbauer spectra [84H1]. For $TmAlO_3$ the sign of ΔE_Q is experimentally not accessible.

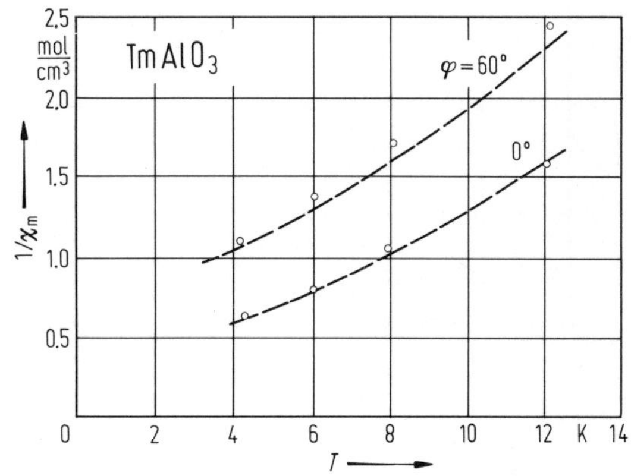

Fig. 32. $TmAlO_3$. Reciprocal molar magnetic susceptibility χ_m^{-1} vs. temperature T. Applied magnetic field in the ab plane making an angle φ with the b axis. Dashed curves calculated [90A1].

$YbAlO_3$

II Curie-Weiss constants Table 3. Low temperature magnetic susceptibility Fig. 33. Estimated critical temperature for A_xG_y-type ordering $T_N = 0.78$ K; paramagnetic moment $p_{eff} = 3.4(5)\ \mu_B$/FU [81R1].

$YAl_{1-x}Fe_xO_3$ (see LB III/12a, p. 461)

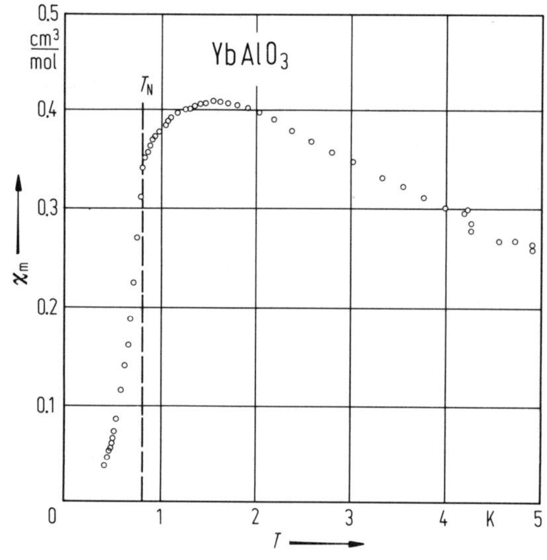

Fig. 33. $YbAlO_3$. Low temperature molar magnetic susceptibility χ_m vs. temperature T [81R1].

5.5.3 $RScO_3$ compounds

$CeScO_3$

I Orthorhombic, Pnma, cell with lattice constants in Å: $a = 5.787(3)$, $b = 8.047(8)$ and $c = 5.626(4)$ and $V = 262.0$ Å^3. No magnetic order down to 4.2 K [81G1].

II Magnetic susceptibility Fig. 34. Curie-Weiss law in the temperature range 100 ··· 250 K gives $C_m = 0.70(3)$ cm^{3}K mol^{-1} and $\Theta_p = -50(3)$ K, [81G1]. Magnetization curve Fig. 35.

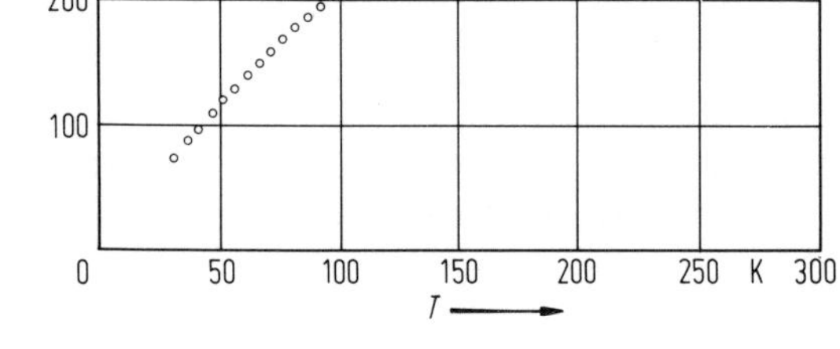

Fig. 34. $CeScO_3$. Inverse molar magnetic susceptibility χ_m^{-1} vs. T [81G1].

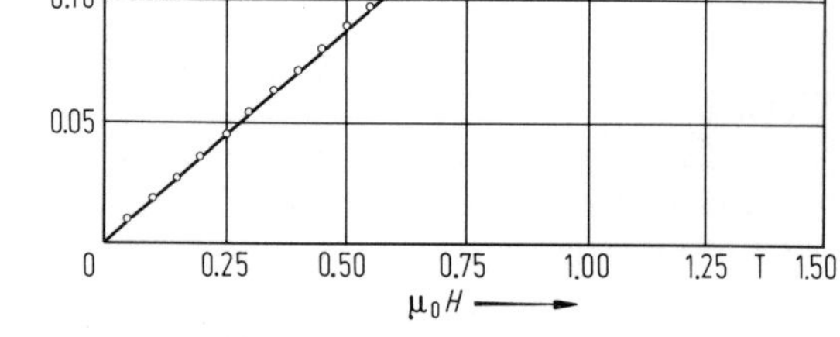

Fig. 35. $CeScO_3$. Magnetic moment per formula unit p_{FU}, vs. applied magnetic field H. The curve is calculated for a $J = 1/2$ Brillouin function with $\langle g_{eff} \rangle$, averaged for a powder, equal to 2.07 [81G1].

$PrScO_3$

I Orthorhombic, Pbnm cell; $a = 5.609(4)$ Å, $b = 5.777(3)$ Å and $c = 8.031(9)$ Å [81M1].

II Magnetic susceptibility and Curie-Weiss constants Fig. 36 and Table 4

$NdScO_3$

I Orthorhombic, Pbnm cell; $a = 5.586(4)$ Å, $b = 5.772(3)$ Å and $c = 8.017(2)$ Å [81M1].

II Magnetic susceptibility and Curie-Weiss constants Fig. 36 and Table 4.

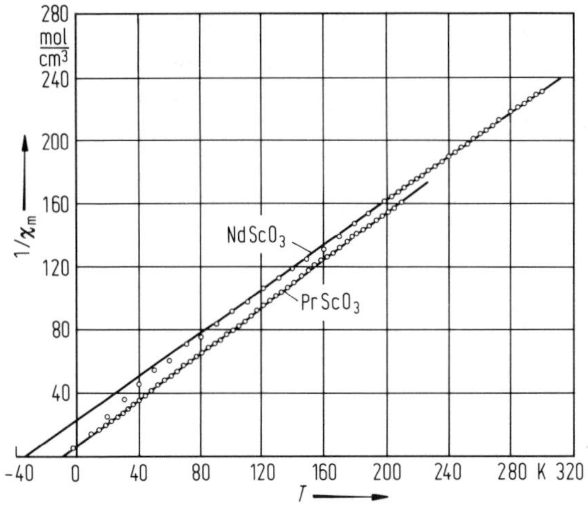

Fig. 36. $NdScO_3$, $PrScO_3$. Reciprocal molar susceptibility χ_m^{-1} vs. temperature. Applied fields 1.3 and 1.4 T, respectively, [81M1].

Table 4. $NdScO_3$, $PrScO_3$, $NdTiO_3$ and $PrTiO_3$. Curie-Weiss constants C_m and Θ_p as derived from the curves in Figs. 36 and 60 [81M1].

Compound	C_m [$cm^3\,K\,mol^{-1}$]	Θ_p [K]
$PrTiO_3$	1.73(3)	− 49(3)
$PrScO_3$	1.34(3)	− 8(2)
Pr^{3+} (free ion)	1.60	
$NdTiO_3$	1.77(3)	− 35(3)
$NdScO_3$	1.51(3)	− 32(3)
Nd^{3+}	1.62	

$EuScO_3$

III Mössbauer parameters [84S3].

T [K]	δ*) [mms^{-1}]	ΔE_Q [mms^{-1}]	η
297.6	0.76(7)	− 7.19(31)	0.54(21)
77.6	0.88(6)	− 9.33(29)	0.63(15)

*) With respect to $^{151}Sm(SmF_3)$.

$GdScO_3$

II $1.2 < T < 80$ K: $C_m = 11.0\ cm^3\,K\,mol^{-1}$, $\Theta_p \approx -6$ K [56B1].

$TbSc_xFe_{1-x}O_3$

I Spontaneous magnetic moment of weak ferromagnetism Fig. 37.

$DySc_xFe_{1-x}O_3$

I Spontaneous mass magnetization Fig. 38 and magnetic threshold field Fig. 39.

$HoScO_3$

I High pressure synthesis [78C2].

$HoSc_xFe_{1-x}O_3$

I Spontaneous magnetic moment of weak ferromagnetism Fig. 37.

$ErScO_3$; $YScO_3$

I High pressure synthesis [78C2].

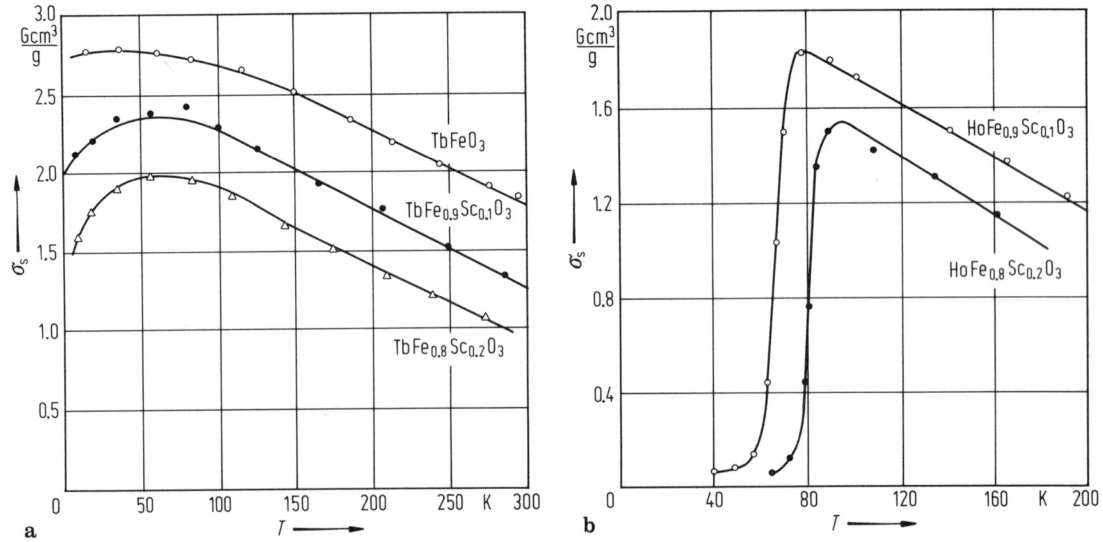

Fig. 37. $TbSc_xFe_{1-x}O_3$; $HoSc_xFe_{1-x}O_3$. Temperature dependence of spontaneous mass magnetization σ_S [89G1].

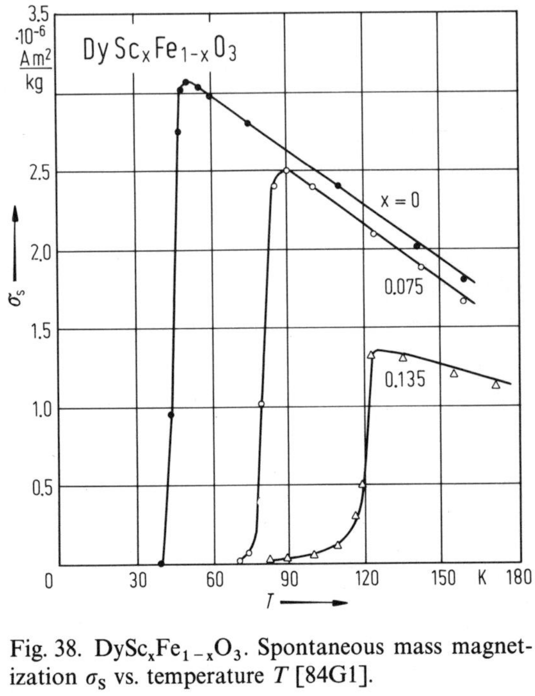

Fig. 38. $DySc_xFe_{1-x}O_3$. Spontaneous mass magnetization σ_S vs. temperature T [84G1].

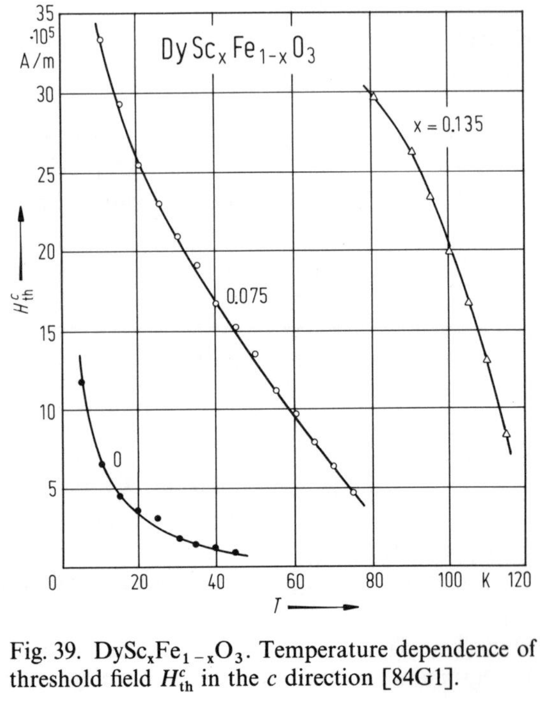

Fig. 39. $DySc_xFe_{1-x}O_3$. Temperature dependence of threshold field H^c_{th} in the c direction [84G1].

5.5.4 $RTiO_3$ compounds

$RTiO_3$

Lattice parameter	Table 5	Fig. 40
Electrical resistivity	Tables 6, 7	Fig. 41
Magnetic ordering	Table 7	Fig. 42
Paramagnetic properties	Table 8	Fig. 43

Table 5. $RTiO_3$. Lattice data at RT derived from X-ray spectra on the basis of the orthorhombic centric space group Pbnm [79M1, 80T1]. Position parameters, interatomic distances and bond angles are given in detail in [79M1] for R = La, Nd, Sm, Gd and Y.

R	a [Å]	b [Å]	c [Å]	V [Å^3]
La[a])[b])	5.601(2)	5.590(2)	7.906(4)	247.5(2)
Nd[a])	5.495(3)	5.589(3)	7.779(4)	238.9(2)
Sm[a])	5.454(2)	5.660(2)	7.722(4)	238.4(2)
Gd[a])[b])	5.393(2)	5.691(2)	7.664(3)	235.2
Tb	5.355(2)	5.665(2)	7.647(4)	232.0
Dy	5.355(2)	5.681(2)	7.655(3)	232.9
Ho	5.347(2)	5.672(5)	7.621(5)	230.9
Er	5.295(2)	5.649(2)	7.588(2)	226.9
Tm	5.293(2)	5.648(2)	7.586(3)	226.8
Yb	5.288(3)	5.638(3)	7.585(4)	226.1
Lu	5.282(2)	5.617(1)	7.585(3)	225.0
Y[a])[b])	5.316(2)	5.679(2)	7.611(3)	229.8(2)

[a]) Data from a single crystal; all other data are from powders.
[b]) See also Table 10 or/and 12.

Table 6. $RTiO_3$. Resistivity ρ at 290 K and activation energy of conductivity E_a [78B1].

R	ρ [Ω cm]	E_a [eV]
La	$8\cdot10^{-4}$	–
Ce	0.01	–
Pr	0.02	–
Nd	0.15	0.06
Sm	4.2	0.15
Gd	28	0.19
Tb	110	0.20
Ho	90	0.20
Er	140	0.24
Yb	150	0.24
Y	120	0.23

Table 7. $RTiO_3$. Survey of electrical and magnetic properties [85G1].

Electrical properties

La	Ce	Pr	Nd	Sm	Gd	Y
M or SM ($T > 125$ K) SC(0.01 eV) ($T < 125$ K)	M or SM ($T > 100$ K) SC(0.001 eV) ($T < 60$ K)	SC(0.03 eV) (50 K $< T <$ 500 K)	SC(0.03 eV) (50 K $< T <$ 500 K)	SC(0.15 eV)	SC(0.19 eV)	SC(0.22 eV)

Magnetic properties

R	La	Ce	Pr	Nd	Sm	Gd	Tb	Dy	Ho	Y	Er	Tm	Yb
T_N, T_C [K]	125	116	90	< 4	?	34	49	64	56	29	41	58	41
Paramagnetism													
Ti	TIP	TIP	C–W	C–W	?					C–W			–
						Fi	Fi	Fi	Fi		Fi	Fi	
R	–	NON C–W	C–W	C–W	?					–			–
Ordered													
Ti	AF(G)	AF(G)	AF	–	?	F_z	F_x	F_y	F_y	F_z	F_z	F_z	F_z
R	–	F, CF	CF	–	?	F_z	C_yF_x	C_xF_y	C_xF_y	–	F_z	F_z	F_z

Symbols: M, metal; SM, semimetal; SC, semiconductor (activation energy in eV); TIP, temperature independent paramagnetism; C–W, Curie–Weiss; Fi, ferrimagnetic, i.e., hyperbolic χ^{-1} vs. T; AF(G), antiferromagnetic type G; F, ferromagnetic; CF, non-collinear structure with both AF and F components. Symbols G, F, C as defined in [63B1] (see Table A).

Table 8. $RTiO_3$. Paramagnetic parameters derived from the high-temperature susceptibility curves of Fig. 43 according to the eq. $\chi_m^{-1} = (T - \Theta_p)/C_m - \xi/(T - \Theta')$, based on the molecular field model for a two-sublattice ferrimagnet [80T1].

R	T_C [K] ± 2	C_m [cm^3 K mol^{-1}]	$C_{m, calc}$ [cm^3 K mol^{-1}]	Θ_p [K]	ξ [mol K cm^{-3}]	Θ' [K]
Gd	34	8.12(3)	8.25	– 4.7(9)	52(4)	41(1)
Tb	49	9.69(8)	12.19	– 1(2)	40(4)	48(2)
Dy	64	11.9(1)	14.54	16(1)	36(4)	52(2)
Ho	56	13.40(4)	14.44	– 5.4(7)	26(3)	52(1)
Er	41	6.1(4)	11.85	– 9(1)	18(2)	40(1)
Tm	58	6.1(2)	7.52	– 19.6(7)	55(7)	59(1)
Lu[a])	30	–	–	–	–	–
Y[b])	29	0.36	0.37	33	–	–

[a]) T_C = 38 K in [75J2]

[b]) [75J2].

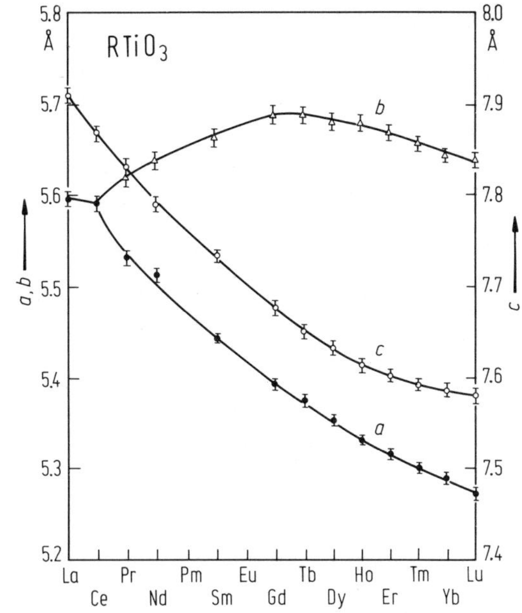

Fig. 40. $RTiO_3$. Room temperature X-ray unit cell parameters *a*, *b* and *c*, obtained from powder spectra [77B1].

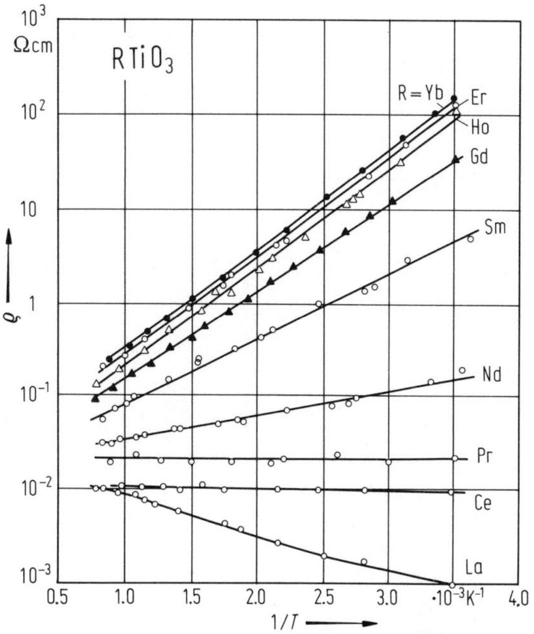

Fig. 41. $RTiO_3$. Temperature dependence of the electrical resistivity ρ for sintered samples. The stoichiometry of the compounds is such that the oxygen contents correspond to $3 + x$, where $0.01 < x < 0.03$. Only for $LaTiO_3$ an x value of 0.05 was found [78B1, see also 76G1, 75B2].

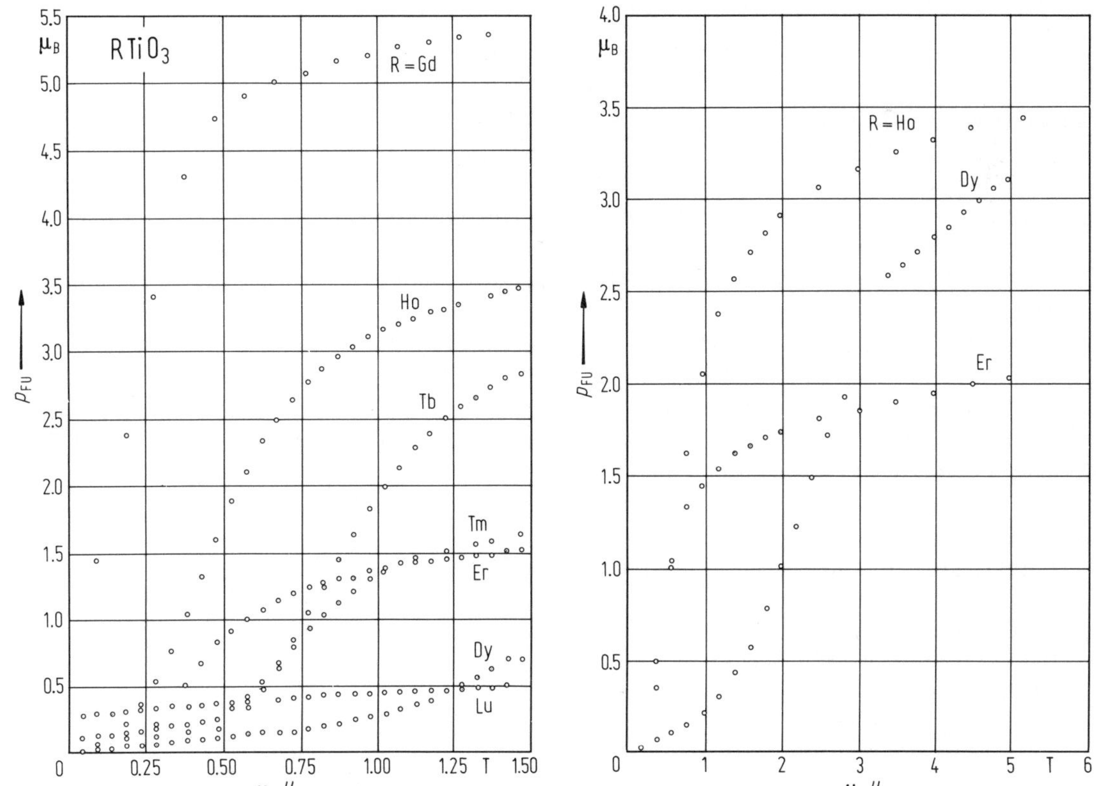

Fig. 42. $RTiO_3$. Magnetic moment per formula unit p_{FU} vs. applied magnetic field *H* at 4.2 K [80T1].

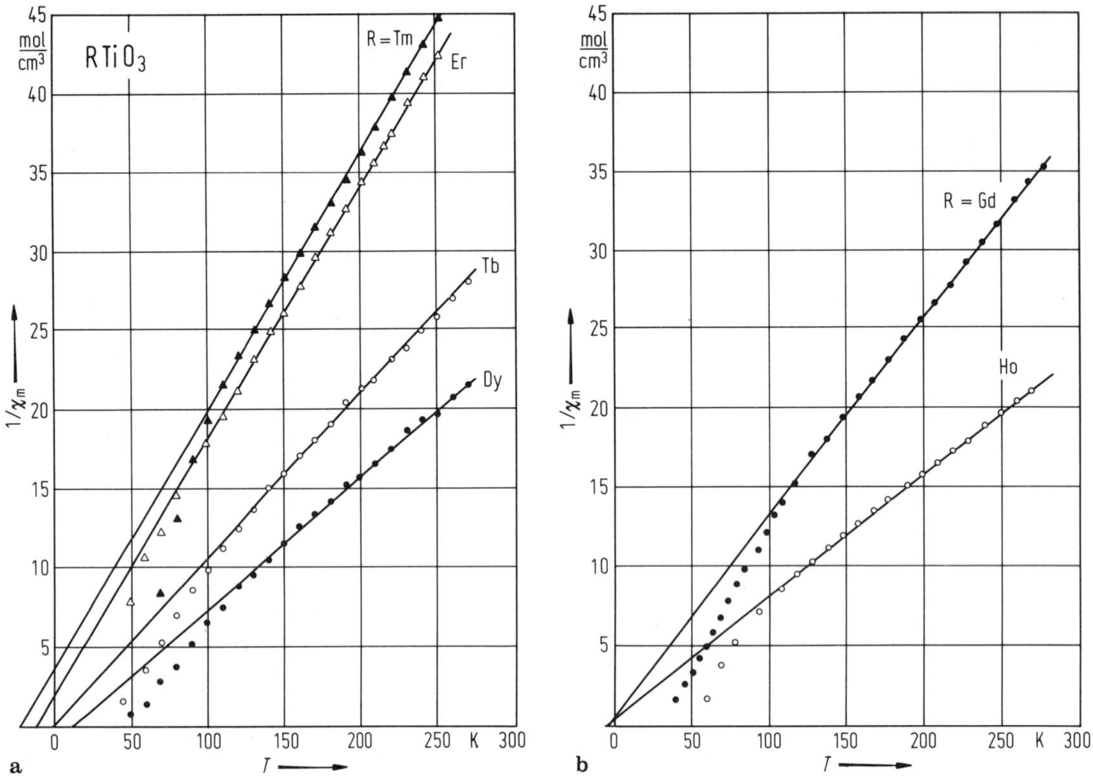

Fig. 43. $RTiO_3$. Inverse molar magnetic susceptibility χ_m^{-1} vs. temperature T for polycrystalline samples [80T1].

$LaTiO_3$ (see also LB III/4a, p. 226)

I Orthorhombic, Pbnm cell, Table 5. Electrical resistivity Figs. 41, 44 and Table 6. Magnetization at 4.2 K Fig. 45. Below 125 K there is evidence for weak ferromagnetism with a magnetization of $7 \cdot 10^{-3}$ μ_B per formula unit at 4.2 K. Powder neutron diffraction at $T = 152$ and 10 K reveal a G-type antiferromagnet with $p_{Ti^{3+}} = 0.45(5)\,\mu_B$ at 10 K. The small ferromagnetic component mentioned above can be present, since this would correspond to a canting angle of less than 1 deg [83G1]. No magnetic order was found by [82Z1].

II Magnetic susceptibility Figs. 44, 46. Temperature independent susceptibility $\chi_{TIP} = 560(10) \cdot 10^{-6}$ $cm^3\,mol^{-1}$ [78B1]. In [81M2] is found $\chi_{TIP} = 800(30) \cdot 10^{-6}$ $cm^3\,mol^{-1}$. Below 130 K, the magnetization depends on H according to $\sigma = \sigma_0 + \chi \cdot H$, where σ_0 is the spontaneous magnetization. At 77 K is found $\sigma_0 = 7.1 \cdot 10^{-2}$ Gcm^3/g ($= 0.003$ μ_B/FU) [78B1, see also 82G1].

III Magnetic Franck-Condon effect [83A1].

$LaTi_{1-x}Cr_xO_3$

I Preparation, structure, electrical conductivity [92D1].

$LaTi_{1-x}Co_xO_3$

I Electrical conductivity [79R1]

II Magnetic susceptibility Fig. 47 and Table 9

$La_{1-x}Gd_xTiO_3$

I Lattice constants Table 10. Magnetic phase diagram as derived from magnetization curves, Fig. 48.

$x \le 0.5$	Anomaly in magnetization-temperature data associated with magnetic ordering within Ti^{3+} sublattice
$0.7 \le x \le 1$	Two sublattice ferrimagnets
$0.2 \le x \le 0.4$	Spin glass behavior [82G1].
	Low temperature anomalies (LTA) in the magnetization-temperature curve, see Table 11.

II Paramagnetic properties Table 11 and Fig. 49.

$La_{1-x}Y_xTiO_3$

I Lattice constants Table 12. Magnetization vs. temperature, see Fig. 50.

II Paramagnetic susceptibilities Table 13, Fig. 51. Magnetization-temperature data suggest a transition from negative $Ti^{3+} - Ti^{3+}$ exchange interactions at the La-rich end to positive exchange interaction at the Y-rich end [82G2].

III Resistivity Fig. 52.

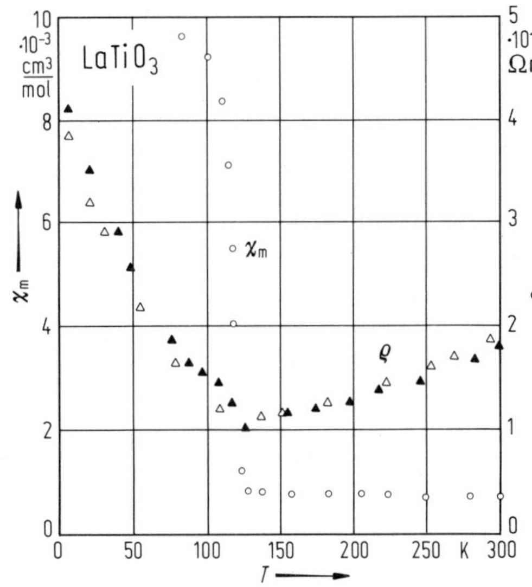

Fig. 44. $LaTiO_3$. Temperature dependence of the electrical resistivity ρ of a 0.2 mm thick crystal and molar magnetic susceptibility χ_m of a polycrystalline sample vs. temperature T. For ρ two sets of measurements are given [81M2].

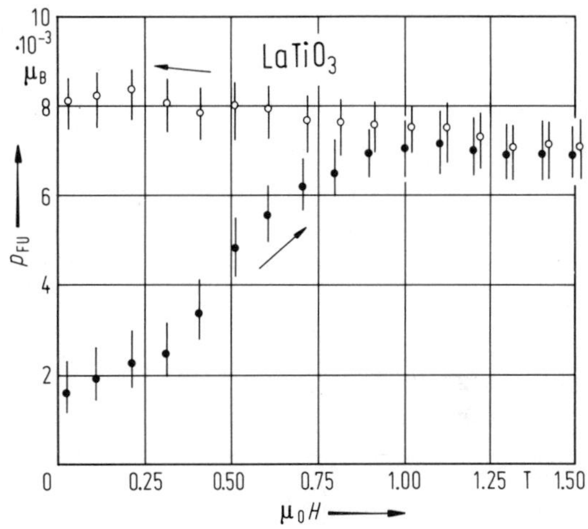

Fig. 45. $LaTiO_3$. Magnetic moment per formula unit p_{FU} vs. applied magnetic field H at 4.2 K. The arrows indicate the field sequence [81M2].

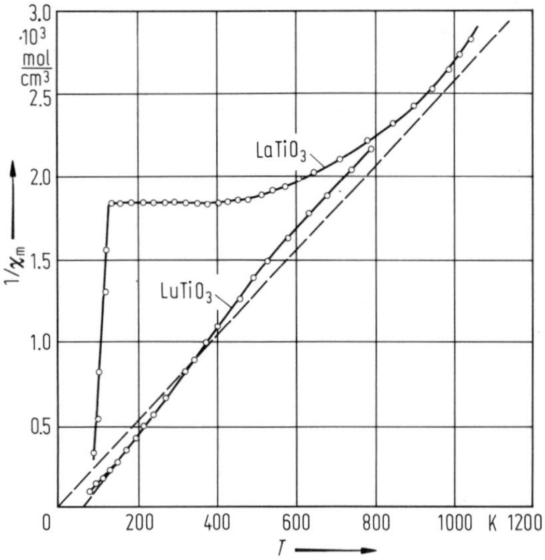

Fig. 46. $LaTiO_3$; $LuTiO_3$. Inverse molar magnetic susceptibility χ_m^{-1} vs. temperature T. The dashed curve is calculated for the Ti^{3+} ion with $S = 1/2$ [78B1].

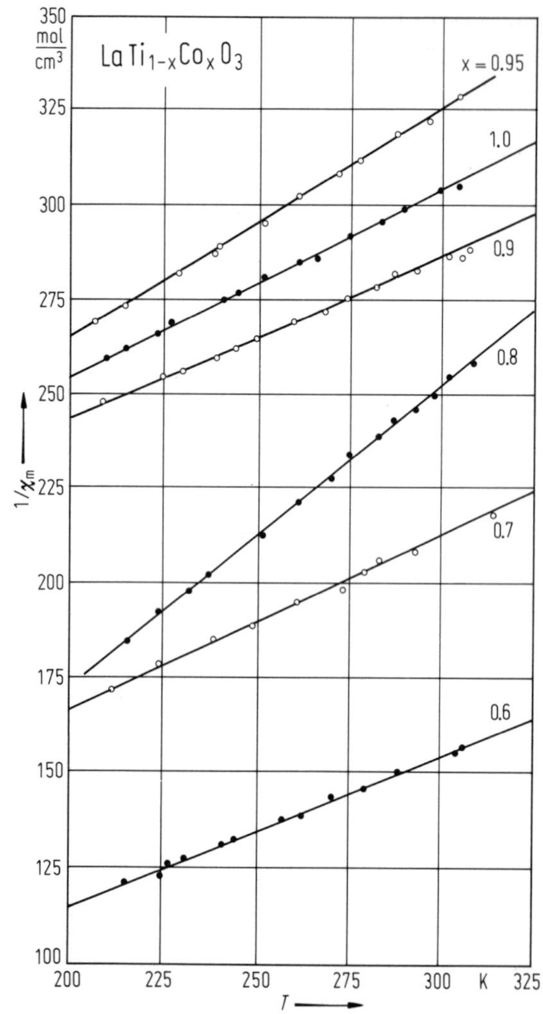

Fig. 47. $LaTi_{1-x}Co_xO_3$. Inverse molar magnetic susceptibility χ_m^{-1} vs. temperature T, see also Table 9 [79R1].

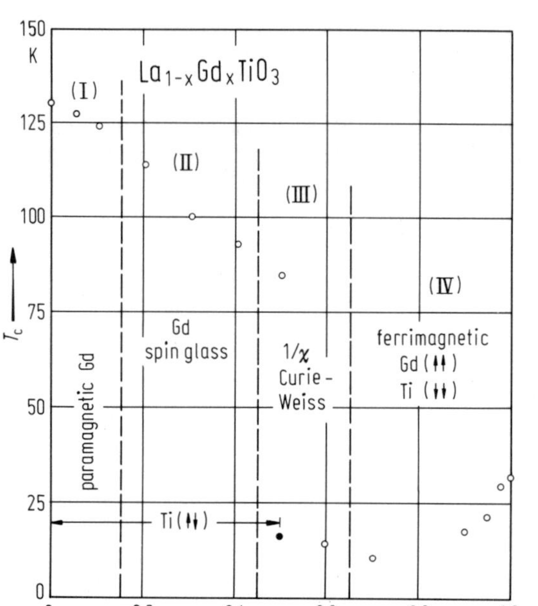

Fig. 48. $La_{1-x}Gd_xTiO_3$. Magnetic phase diagram as derived from magnetization measurements. Magnetic transition point T_c vs. composition parameter x. The arrows represent the magnetic ordering of the ion moments [82G1].

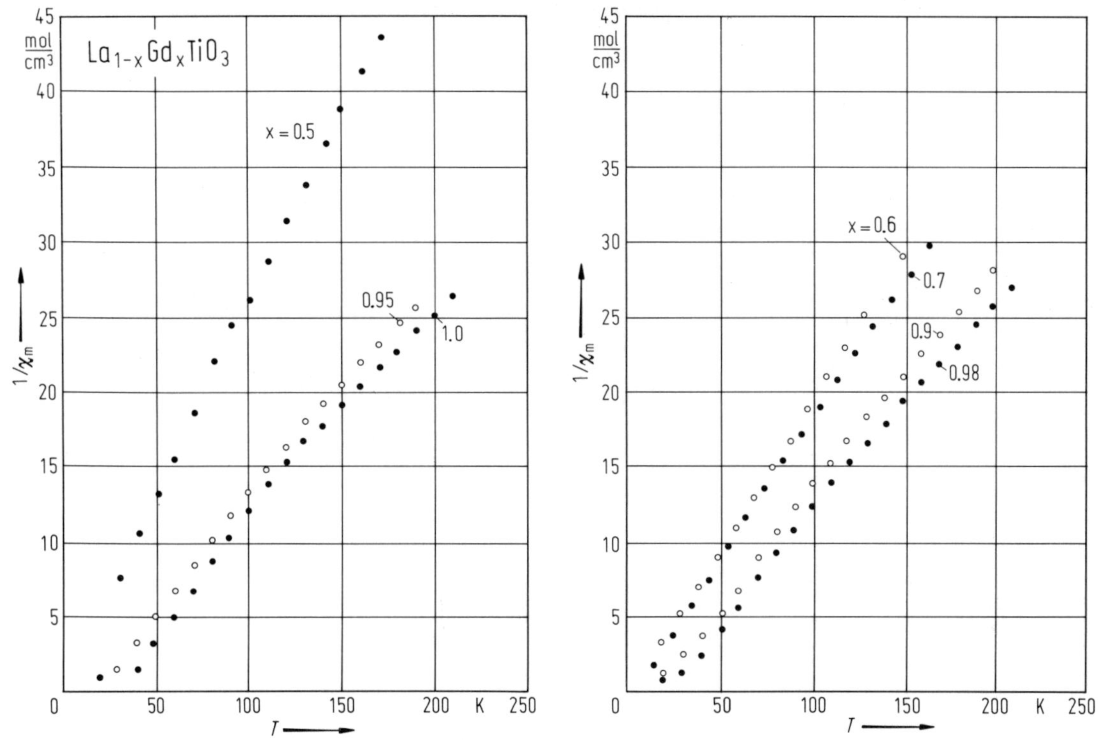

Fig. 49. $La_{1-x}Gd_xTiO_3$. Inverse molar magnetic susceptibility χ_m^{-1} vs. temperature T. Magnetic measuring field about 10 kOe [82G1].

For Fig. 50, see next page.

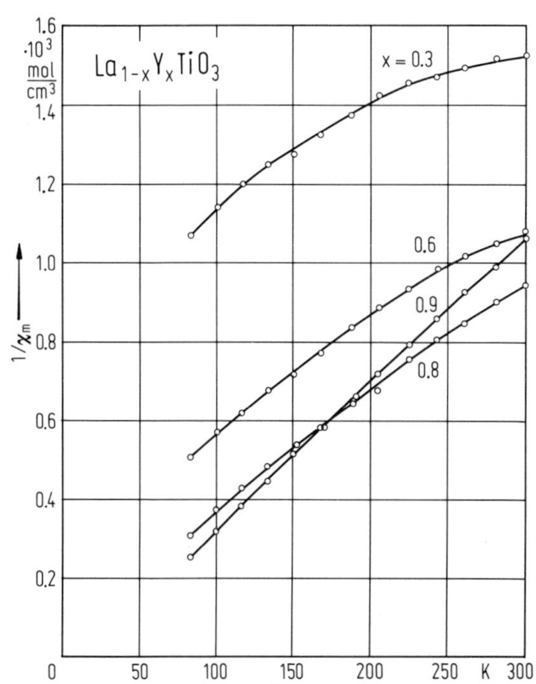

Fig. 51. $La_{1-x}Y_xTiO_3$. Inverse molar magnetic susceptibility χ_m^{-1} vs. temperature T [82G2].

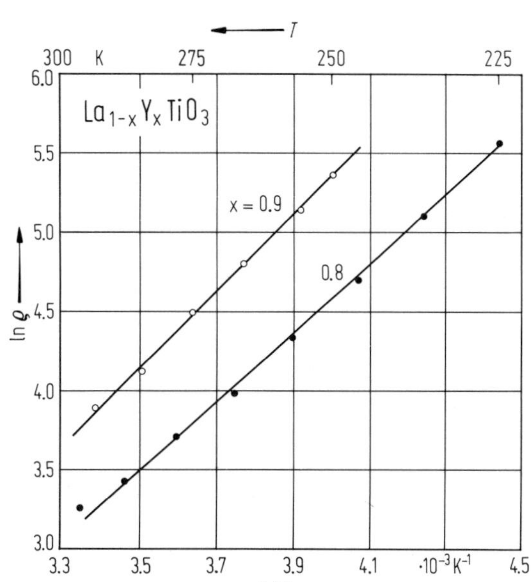

Fig. 52. $La_{0.2}Y_{0.8}TiO_3$; $La_{0.1}Y_{0.9}TiO_3$. Natural logarithm of the electrical resistivity ρ (in Ω cm) vs. reciprocal temperature T^{-1}. Activation energies are 0.187(4) eV and 0.221(4) eV, respectively [82G2].

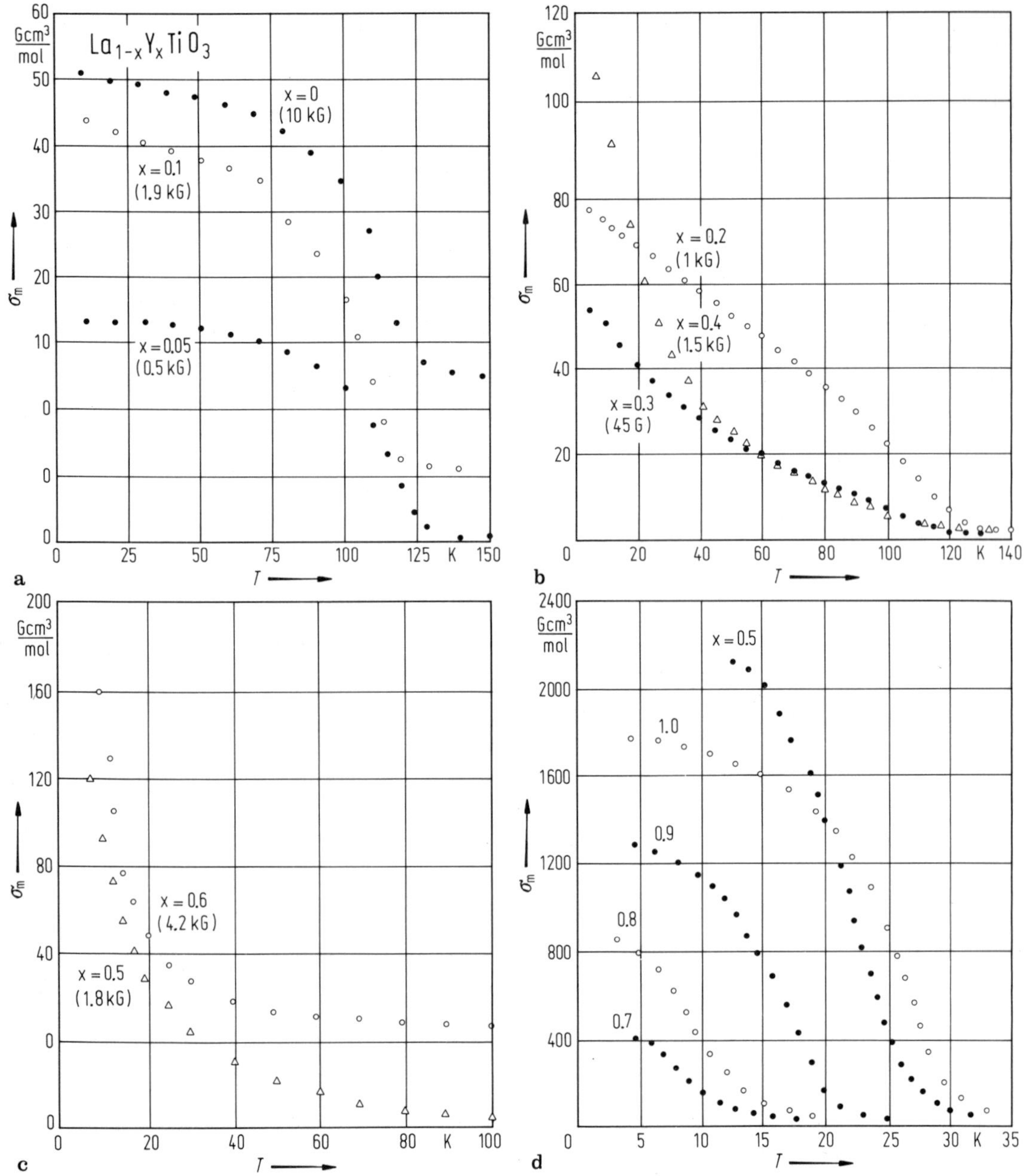

Fig. 50. $La_{1-x}Y_xTiO_3$. Molar magnetization σ_m in a magnetic field (as indicated in brackets) as a function of temperature T. The data of Fig. 50d are obtained for low magnetic fields [82G2, see also 78G1].

Table 9. $LaTi_{1-x}Co_xO_3$. Paramagnetic Curie temperatures Θ_p and effective magnetic moments p_{eff} per FU as derived from the data of Fig. 47. The irregularity may be caused by the influence of Ti^{4+} content on the spin state of cobalt ions [79R1].

x	p_{eff} [μ_B/FU]	Θ_p [K]
1	4.01	− 310
0.95	3.65	− 237
0.90	4.36	− 375
0.80	3.16	− 14
0.70	4.11	− 148
0.60	4.67	− 100

Table 10. $La_{1-x}Gd_xTiO_3$. RT lattice parameters and cell volume V of the orthorhombic unit cell (Pbnm) as derived from powder X-ray diffraction [82G1].

x	a [Å]	b [Å]	c [Å]	V [$Å^3$]
1 ($GdTiO_3$)	5.409(5)	5.712(4)	7.681(7)	237.3(6)
0.98	5.409(5)	5.700(5)	7.683(6)	236.9(6)
0.95	5.414(5)	5.693(5)	7.694(8)	237.1(7)
0.90	5.429(5)	5.681(6)	7.705(8)	237.6(7)
0.80	5.450(5)	5.677(5)	7.733(9)	239.3(7)
0.70	5.475(6)	5.664(4)	7.762(5)	240.7(6)
0.60	5.497(6)	5.656(4)	7.789(6)	242.2(6)
0.50	5.516(6)	5.634(4)	7.820(6)	243.0(6)
0.40	5.532(11)	5.654(11)	7.830(12)	244.9(13)
0.30	5.561(9)	5.640(6)	7.864(16)	246.6(12)
0.20	5.593(10)	5.627(6)	7.885(11)	248.1(11)
0.10	5.622(4)	5.619(6)	7.908(11)	249.8(8)
0.05	5.621(4)	5.609(5)	7.917(12)	249.6(8)
0 ($LaTiO_3$)	5.633(11)	5.614(5)	7.940(6)	251.1(9)

Table 11. $La_{1-x}Gd_xTiO_3$. Critical temperatures for magnetic ordering, T_c, low-temperature anomalies LTA, indicating spin-glass behavior, Néel hyperbola parameters Θ_p, C_m, ξ, Θ' (definition Table 8) and magnetic moment of the Ti^{3+} ions, $p_{Ti^{3+}}$ [82G1].

x	T_c [K]	LTA [K]	$p_{Ti^{3+}}$ [μ_B]	Θ_p [K]	C_m [$cm^3 K mol^{-1}$]	ξ [$mol K cm^{-3}$]	Θ' [K]
1 ($GdTiO_3$)	33(2)		1.1(2)	− 13(5)	7.9(1)	140(40)	30(5)
0.98	30(2)		1.0(2)	− 4(4)	7.77(9)	80(40)	30(10)
0.95	22(2)		1.0(2)	− 6(5)	7.4(1)	90(40)	15(8)
0.90	18(2)		1.0(2)	− 3(3)	7.07(9)	60(30)	14(10)
0.70	12(2)			− 3(1)	5.60(4)	20(10)	8(7)
0.60	14(2)			3.7(2)	5.0(1)		
0.50	85(7)	16(2)		− 5.2(2)	4.01(1)		
0.40	93(5)	16(2)					
0.30	100(4)	14(2)					
0.20	114(3)	13(2)					
0.10	124(2)						
0.05	127(2)						
0 ($LaTiO_3$)	130(2)						

Table 12. $La_{1-x}Y_xTiO_3$. RT lattice parameters and cell volume V of the orthorhombic unit cell (Pbnm) as derived from powder X-ray diffraction [82G2].

x	a [Å]	b [Å]	c [Å]	V [$Å^3$]
1 ($YTiO_3$)	5.340(5)	5.690(4)	7.611(5)	231.2(5)
0.9	5.366(3)	5.677(3)	7.651(5)	233.1(4)
0.8	5.395(3)	5.672(3)	7.684(5)	235.1(4)
0.7	5.429(5)	5.677(6)	7.725(15)	238.1(9)
0.6	5.462(10)	5.671(6)	7.736(7)	239.6(9)
0.5	5.477(10)	5.667(1)	7.769(8)	241.1(7)
0.4	5.517(11)	5.649(11)	7.815(1)	243.6(12)
0.3	5.589(11)	5.626(6)	7.879(15)	247.7(12)
0.2	5.566(11)	5.623(5)	7.930(16)	248.2(12)
0.1	5.593(11)	5.618(6)	7.937(8)	249.4(10)
0.05	5.626(4)	5.620(6)	7.913(14)	250.2(9)
0 ($LaTiO_3$)	5.633(10)	5.614(9)	7.940(6)	251.1(9)

Table 13. $La_{1-x}Y_xTiO_3$. Critical temperature for magnetic ordering T_c, and paramagnetic properties Θ_p and C_m. The latter derived from the curves of Fig. 51, allowing for a Curie-Weiss fit with a temperature-independent (TIP) contribution as found for $CeTiO_3$ [82G2].

x	T_c [K]	Θ_p [K]	C_m [$cm^3\,K\,mol^{-1}$]	χ_{TIP} ($\cdot 10^6$) [$cm^3\,mol^{-1}$]
1 ($YTiO_3$) [a]	29(2)	33	0.35	0
0.9	19(2)	19	0.27	0
0.8	14(2)	1	0.27	180
0.7	11(2)			
0.6	43 ··· 57	− 20	0.17	380
0.5	94 ··· 106			
0.4	112(4)			
0.3	123(4)	− 13	0.04	510
0.2	125(4)			
0.1	115(2)			
0.05	127(2)			
0 ($LaTiO_3$)	130(2)			800

[a]) Data from [75J2].

$CeTiO_3$ (see also LB III/4a, p. 226)

I Orthorhombic, Pbnm cell; a = 5.570 Å, b = 5.584 Å and c = 7.865 Å [79M1]. Electrical resistivity Figs. 41, 53; from 90 ··· 300 K metallic, below 60 K semiconducting with $E_a = 0.001$ eV [81M2]. Magnetization Figs. 54 and 55. Evidence for magnetic ordering below $T_N = 116(2)$ K. Saturation at 4.2 K is 0.60(1) μ_B per formula unit at $\mu_0 H = 1.5$ T [81M2] increasing to 0.9 μ_B at 4.8 T [83G1]. Neutron diffraction data suggest [83G1, 85G1] that for 80 K < T < 116 K the Ti^{3+} sublattice orders with a moment of 0.4 μ_B in a G-type configuration, either G_xF_z or G_zF_x, which induces a moment on Ce^{3+} (F_x, respectively F_z) of the same magnitude 0.4 μ_B. For T < 80 K the Ce^{3+} moments reorient into a F_xC_y configuration with a moment of 1.4 μ_B per ion. No magnetic order was found by [82Z1].

II Magnetic susceptibility Fig. 56. A Curie-Weiss behavior with $C_m = 0.55(3)$ $cm^3K\,mol^{-1}$, $\Theta_p = -5(3)$ K combined with a temperature independent paramagnetism $\chi_{TIP} = 860(10) \cdot 10^{-6}$ $cm^3\,mol^{-1}$ is found [81M2]. See also Fig. 57 and Table 14.

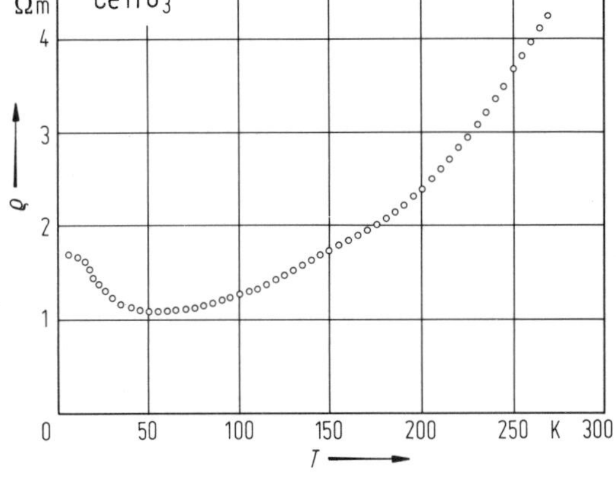

Fig. 53. $CeTiO_3$. Resistivity ρ vs. temperature T [81M2].

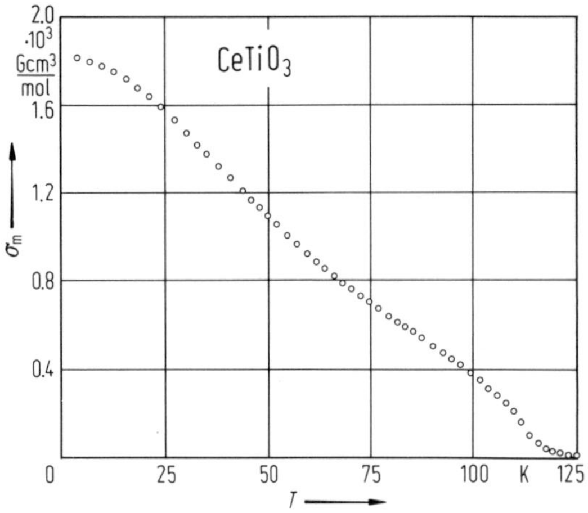

Fig. 54. $CeTiO_3$. Molar magnetization σ_m vs. temperature T measured in a magnetic field of 0.0045 T [81M2].

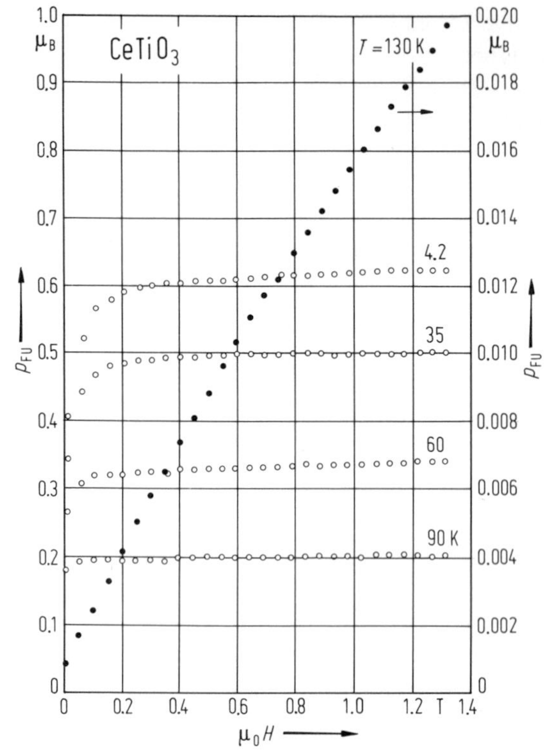

Fig. 55. $CeTiO_3$. Magnetic moment per formula unit p_{FU} vs. applied magnetic field H [81M2].

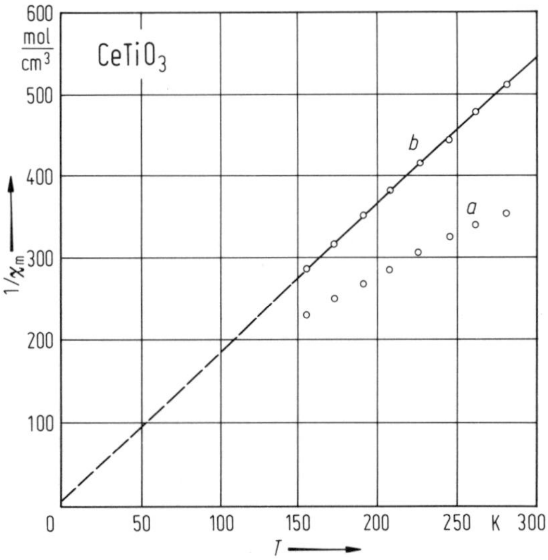

Fig. 56. $CeTiO_3$. Inverse magnetic molar susceptibility χ_m^{-1} vs. temperature T. Curve a: experimental data, b: data corrected for a temperature independent contribution $\chi_{TIP} = 860 \cdot 10^{-6}$ cm^3 mol^{-1} [81M2].

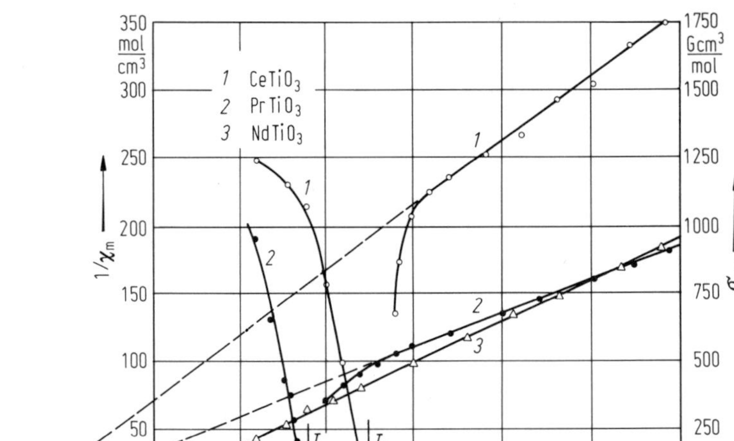

Fig. 57. $CeTiO_3$, $PrTiO_3$ and $NdTiO_3$, curves *1*, *2* and *3*, respectively. Polycrystalline sintered samples. Inverse molar magnetic susceptibility χ_m^{-1} and spontaneous molar magnetization. $\sigma_{m,s}$ vs. temperature T. For Curie-Weiss constants, see Table 14 [75B1].

Table 14. $CeTiO_3$, $PrTiO_3$, $NdTiO_3$. Magnetic properties derived from the curves of Fig. 57 [75B1].

Compound	χ_m at 290K [10^{-3} cm^3 mol^{-1}]	C_m [cm^3 K mol^{-1}]	Θ_p [K]	p_{eff} [μ_B/FU]	T_N [K]	p_{FU}[a]) [μ_B]
$CeTiO_3$	286.0	1.02	− 72	2.86	124	0.22
$PrTiO_3$	545.0	1.92	− 65	3.93	89	0.17
$NdTiO_3$	535.0	1.59	− 7	3.58	–	–

[a]) Magnetization per formula unit derived from the spontaneous magnetic moment at 60 K.

$PrTiO_3$ (see also LB III/4a, p. 226)

I Electrical resistivity Fig. 58, also Table 6. Semiconductor with $E_a = 0.04$ eV above and 0.03 eV below $T = 75$ K [81M1], see also [80T1, 80T2, 85G1]. Magnetization curves, Fig. 59, indicate a magnetic ordering at about $T_N = 96(2)$ K. At 4.2 K a saturation magnetization of 0.50 μ_B/FU is found [81M1]. Neutron diffraction at 7 K indicates a magnetic structure similar to that of $CeTiO_3$ at low temperature [84G2, 85G1] with $p_{Pr^{3+}} = 1.35(15)\,\mu_B$, making an angle of 45°(5) with the a or b axis.

II Magnetic susceptibility Figs. 57, 60, Curie-Weiss constants Tables 4 and 14.

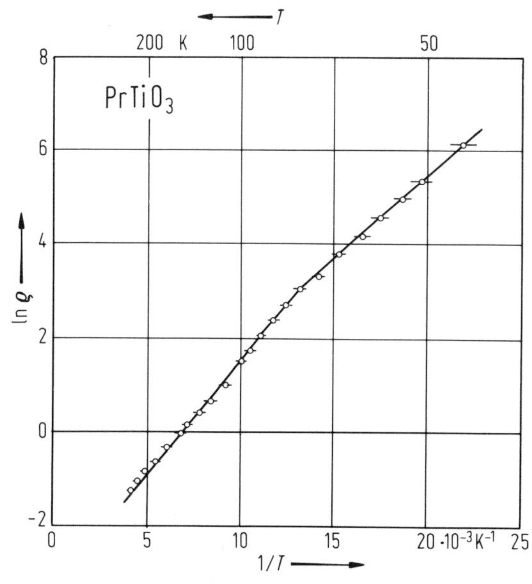

Fig. 58. $PrTiO_3$. Natural logarithm of the resistivity ρ (in Ω cm) vs. (reciprocal) temperature T. Activation energies are 0.03 and 0.04 eV below and above T = 75 K, respectively [81M1].

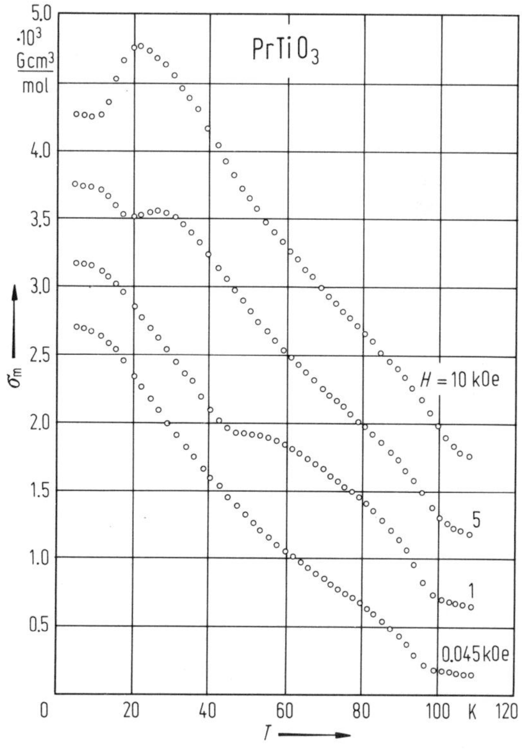

Fig. 59. $PrTiO_3$. Molar magnetization σ_m vs. temperature T for various magnitudes of the applied field H. The successive curves have been shifted upwards by an amount of 0.5 units [81M1].

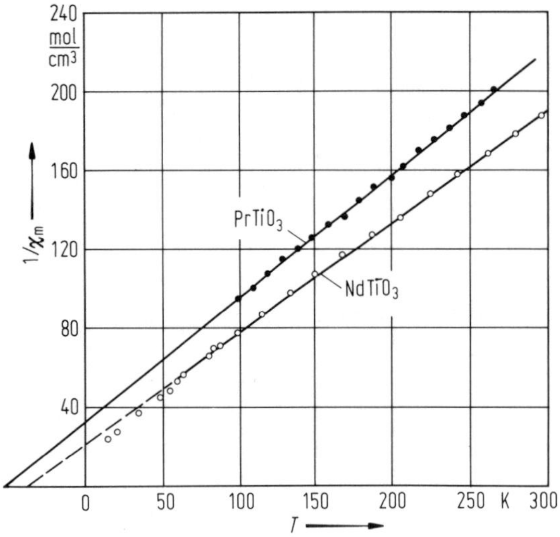

Fig. 60. $PrTiO_3$; $NdTiO_3$. Reciprocal molar magnetic susceptibility χ_m^{-1} vs. temperature T. Applied field 1.08 T [81M1].

$NdTiO_3$ (see also LB III/4a, p. 226)

I Unit cell dimensions Table 5. Electrical resistivity Figs. 41, 61, Table 6. Semiconductor with E_a = 0.03 eV [81M1], see also [80T1, 80T2, 85G1]. No sign of magnetic order down to 4.2 K. Magnetization curve at 4.2 K, Fig. 62, can be explained with an effective g-factor $g_{eff} = 2.32$.

II Magnetic susceptibility Figs. 57, 60. Curie-Weiss constants Tables 4 and 14.

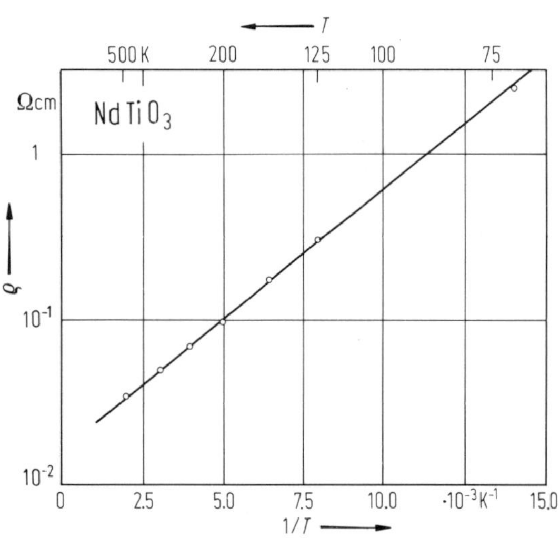

Fig. 61. $NdTiO_3$. Resistivity ρ vs. temperature T [81M1].

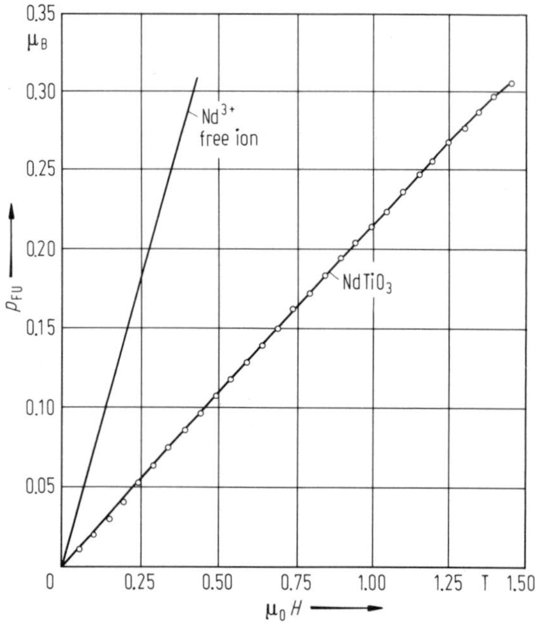

Fig. 62. $NdTiO_3$. Magnetic moment per formula unit p_{FU} vs. applied field H at 4.2 K. Curve through measuring points is a Brillouin function for $J = 1/2$ and $g_{eff} = 2.32$. Also shown is the free-ion Brillouin function for Nd^{3+}, $J = 9/2$ and $g = 8/11$ [81M1].

$SmTiO_3$ (see also LB III/4a, p. 226)

I Unit cell dimensions Table 5.

$GdTiO_3$ (see also LB III/4a, p. 226)

I Unit cell dimensions Tables 5, 10. Electrical resistivity Table 6, Fig. 41. Ti–Ti interaction is ferromagnetic and the strongest coupling of the order of 10^2 T. The Gd–Ti coupling is antiferromagnetic, but somewhat smaller, of the order of 10 T. The R–R interaction is weak. The result is a ferrimagnetic structure [85G1, 80T1]. For Curie point see Tables 8, 11. Magnetization curve at 4.2 K, Fig. 42. Saturation at $\mu_0 H = 6.0$ T with $p_s = 6.0\ \mu_B$ per formula unit, as predicted by a two-sublattice model [80T1]. Also measured at high field $5.8(2)\ \mu_B$/FU[82G1].

II Magnetic susceptibility, see Figs. 43, 49. For an analysis of the curve of Fig. 43b see Table 8. See also Table 11.

$TbTiO_3$

I Unit cell dimensions Table 5. Electrical resistivity Table 6. Curie point Table 8. Neutron powder diffraction spectra at 10 K can be explained by a canted spin structure. A ferromagnetic x component for the Tb^{3+} moments and antiferromagnetic y and z components, configuration F_xC_y. For a good refinement of the spectrum, the Ti^{3+} moments have to be chosen collinear and antiparallel to the F-component of the Tb^{3+} configuration. Moreover, $p_{Tb^{3+}} = 8.1(4)\ \mu_B$, making an angle of $\pm$ 36(3)° with the [1 0 0] direction, and $p_{Ti^{3+}} = 0.5(5)\ \mu_B$ along [1 0 0] [81T1, 85G1]. The same F_xC_y configuration for the moments of the Tb ions at 4.2 K is found by [82Z1] with magnetic moment projections 4.4(5) and 3.5(5) μ_B, respectively, and the Ti^{4+} moments ordered ferromagnetically along the c axis (F_z) with a magnetic moment of $-$ 0.65(3) μ_B. Magnetization curve at 4.2 K, see Fig. 42. Temperature dependence of the magnetization of the Tb ions Fig. 63. This magnetization is caused by the isotropic $R^{3+} - Ti^{3+}$ coupling. Since the titanium moment arises from one d electron, the curve can best be described by a spin 1/2 Brillouin function [82T1].

II Magnetic susceptibility, see Fig. 43. For an analysis of the curve, see Table 8.

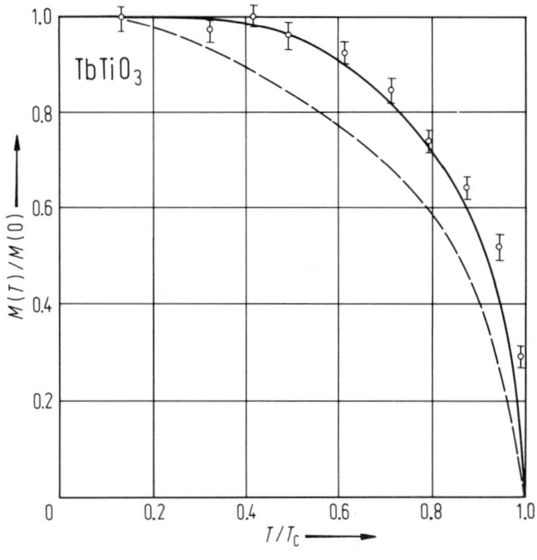

Fig. 63. $TbTiO_3$. Reduced magnetization $M(T)/M(0)$ vs. reduced temperature T/T_C as derived from neutron diffraction analysis. The curves apply to Brillouin functions calculated for $J = 6$ (dashed) and $J = 1/2$ (solid curve) [82T1].

$DyTiO_3$

I Unit cell dimensions Table 5. Curie point Table 8. Magnetization curve at 4.2 K Fig. 42. Neutron powder diffraction spectra can be explained by a spin configuration F_yC_x for the Dy^{3+} moments, i.e. the antiferromagnetic components are along the [1 0 0] direction, the ferromagnetic component along the [0 1 0] direction. The Ti moments align antiparallel to the F component of the Dy^{3+} moment. It is found $p_{Dy^{3+}} = 9.7(7)\,\mu_B$, making an angle of $\pm 31(5)°$ with the [0 1 0] direction, and $p_{Ti^{3+}} = -0.9(5)\,\mu_B$ [81T1, 85G1].

II Magnetic susceptibility, see Fig. 43. For an analysis of the curve, see Table 8.

$HoTiO_3$

I Unit cell dimensions Table 5. Electrical resistivity Table 6, Fig. 41. Curie point Table 8. According to neutron diffraction analysis [82Z1] the Ho^{3+} moments have a C_xF_y ordering (Γ_3 phase), the magnetic moments of the Ti ions order antiparallel to the F component of the Ho^{3+} moments. The x and y components of Ho^{3+} moment are 2.4(5) and 6.1(5) μ_B, respectively, according to [82Z1], whereas $p_{Ti^{3+}} = -0.7(3)\,\mu_B$. According to [80T2]: $p_{Ho^{3+}} = 8.1(5)\,\mu_B$, $p_{Ti^{3+}} = -0.55(30)\,\mu_B$. Magnetization curves Figs. 64, 65; see also Fig. 42. Easy direction of magnetization along the b axis. Saturation magnetization at 4.2 K is 7.3(2) μ_B per FU [84G3]. These experiments seem to confirm the results of the neutron diffraction analysis of [82Z1], see above.

II Magnetic susceptibility Figs. 43, 66, Curie-Weiss constants Tables 8, 15.

$HoTi_{0.2}Fe_{0.6}Ni_{0.2}O_3$

I Spontaneous magnetization Fig. 67.

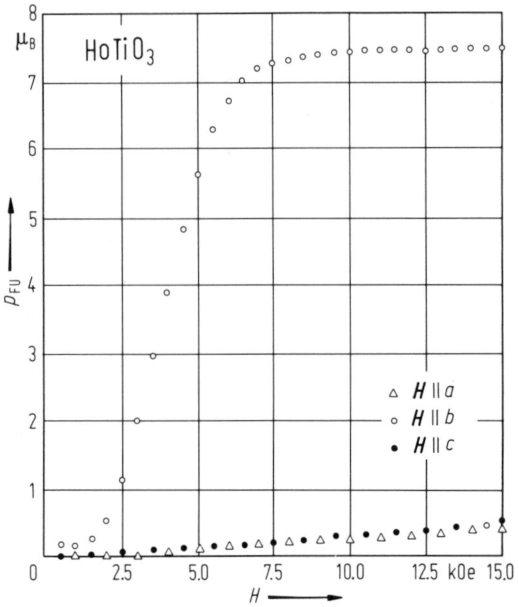

Fig. 64. $HoTiO_3$. Magnetic moment per formula unit p_{FU} of a single crystal as dependent on magnitude and direction of an applied magnetic field at 4.2 K [84G3].

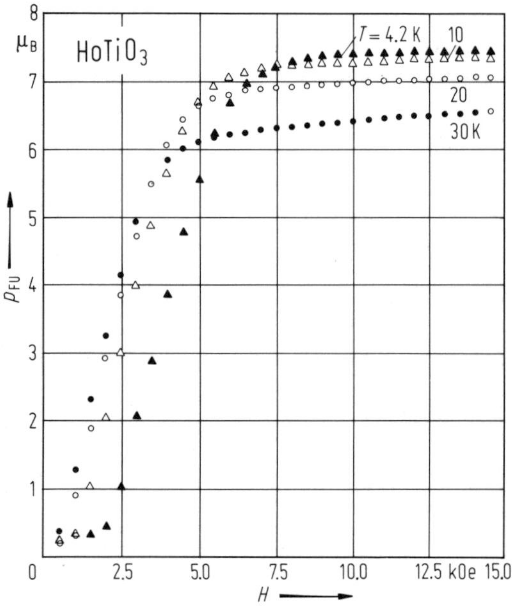

Fig. 65. $HoTiO_3$. The same as Fig. 64, but now for an applied magnetic field in the direction of the b axis at various temperatures [84G3].

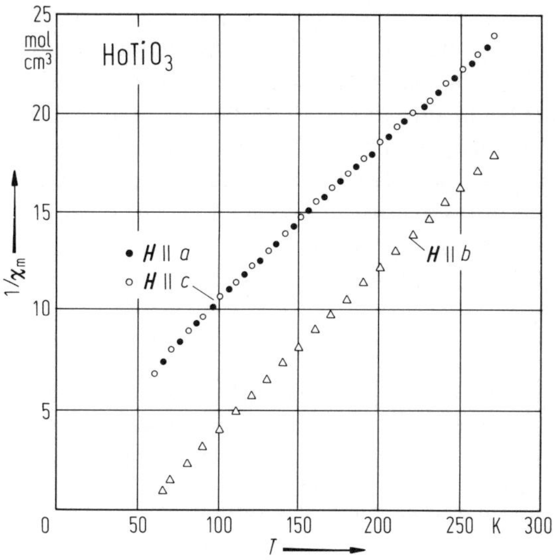

Fig. 66. $HoTiO_3$. Inverse molar magnetic susceptibility χ_m^{-1} vs. T measured along the three main crystal axes [84G3].

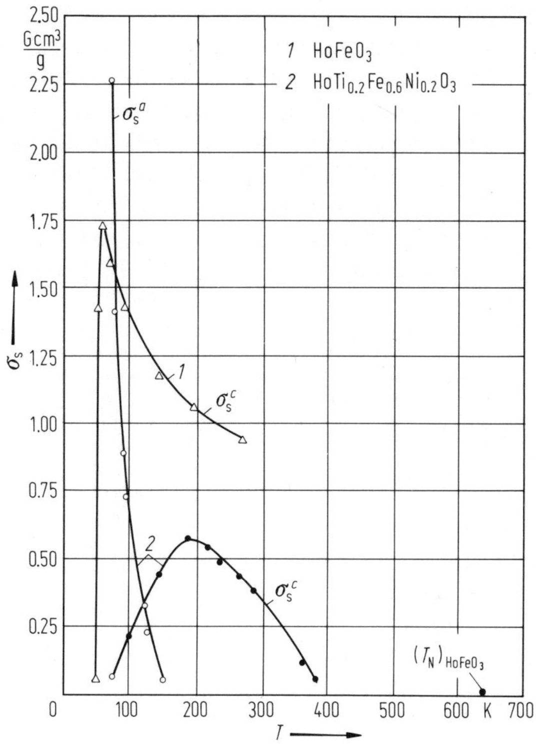

Fig. 67. $HoFeO_3$, $HoTi_{0.2}Fe_{0.6}Ni_{0.2}O_3$. Temperature dependence of magnitude and orientation of the spontaneous mass magnetization σ_S [81A1].

Table 15. $HoTiO_3$; $ErTiO_3$. Curie-Weiss constants derived from susceptibility data for the three main crystallographic directions. Theoretical values for Ho ($J = 8$) and Er ($J = 15/2$) are 14.1 and 11.5 $cm^3\,K\,mol^{-1}$, respectively [84G3].

	$HoTiO_3$		$ErTiO_3$	
	C_m [$cm^3\,K\,mol^{-1}$]	Θ_p [K]	C_m [$cm^3\,K\,mol^{-1}$]	Θ_p [K]
χ^a	13.3	− 44.1	11.2	− 20.9
χ^b	12.5	47.3	10.4	− 41.1
χ^c	13.0	− 41.3	10.7	27.9

$ErTiO_3$

I Unit cell dimensions Table 5. Electrical resistivity Table 6, Fig. 41. Curie point Table 8. Magnetization curves Figs. 42, 70, 71. Neutron diffraction on polycrystalline samples discloses a ferrimagnetic collinear structure, the Ti^{3+} moments being antiparallel to the Er^{3+} moments along the direction of the c axis. It was found: $p_{Er^{3+}} = 5.2(5)\,\mu_B$ [82Z1] and $8.5(5)\,\mu_B$ [80T2] whereas $p_{Ti^{3+}} = -0.64(30)\,\mu_B$ [82Z1] and $-0.7(3)\,\mu_B$ [80T2], respectively. Temperature dependence of the magnetization of the Er ions Fig. 68 and of the total magnetic moment Fig. 69. Saturation magnetization at 4.2 K is $6.9(2)\,\mu_B$ per FU in the c direction [84G3].

II Magnetic susceptibility Figs. 43, 72, and Curie-Weiss constants Tables 8 and 15.

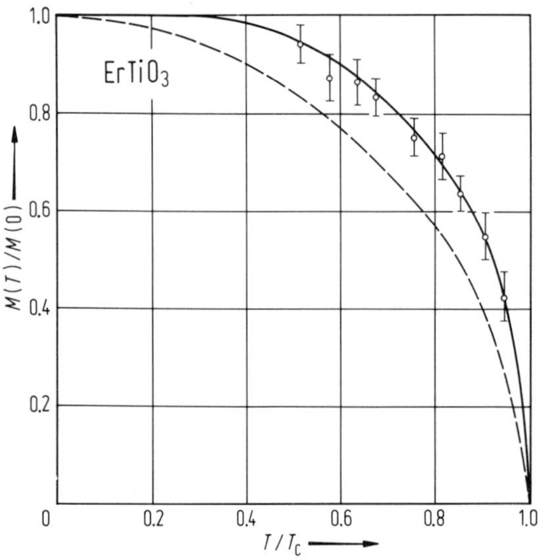

Fig. 68. $ErTiO_3$. Reduced magnetization $M(T)/M(0)$ vs. reduced temperature T/T_C as derived from neutron diffraction analyses. The curves apply to Brillouin functions calculated for $J = 1/2$ (solid) and $J = 15/2$ (dashed curve), respectively [82T1].

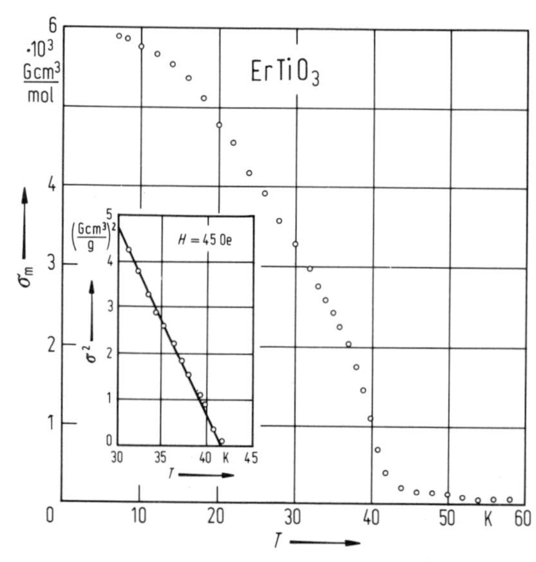

Fig. 69. $ErTiO_3$. Polycrystalline sample. Temperature dependence of the molar magnetization σ_m measured in a magnetic field of 5 kOe. Inset: σ^2 vs. T for $H = 45$ Oe, as a means to estimate the critical temperature T_c [80T1].

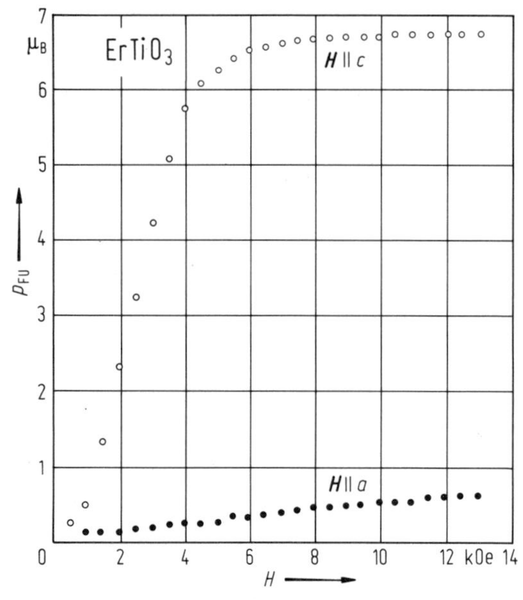

Fig. 70. $ErTiO_3$. Magnetic moment per formula unit p_{FU}, of a single crystal as dependent on magnitude and direction of an applied magnetic field at 4.2 K [84G3].

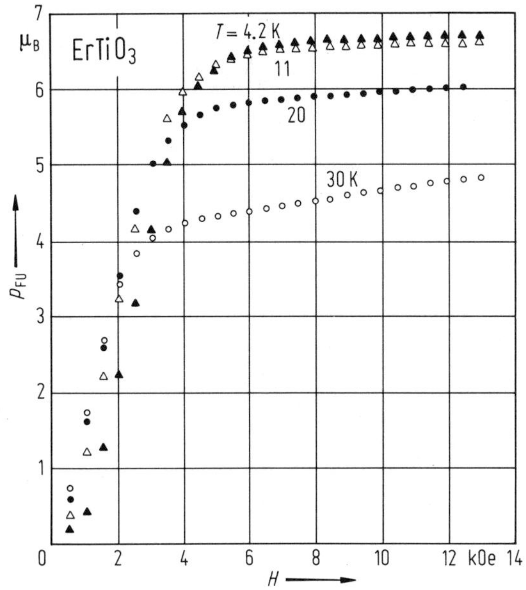

Fig. 71. $ErTiO_3$. The same as Fig. 70, but now for applied magnetic fields in the direction of the *c* axis at various temperatures [84G3].

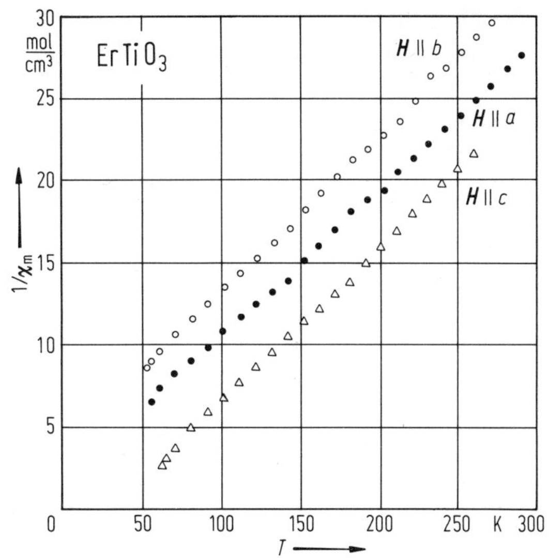

Fig. 72. $ErTiO_3$. Inverse molar magnetic susceptibility χ_m^{-1} vs. T measured along the three main crystal axes [84G3].

$TmTiO_3$

I Unit cell dimensions Table 5. Curie point Table 8. Magnetization curve at 4.2 K Fig. 42. Neutron diffraction on polycrystalline samples discloses a ferrimagnetic collinear structure, the Ti moments being antiparallel to the Tm moments along the *c* axis. It was found $p_{Tm^{3+}} = 6.0(4)\,\mu_B$ and $p_{Ti^{3+}} = -0.7(5)\,\mu_B$ [81T1].

II Magnetic susceptibility Fig. 43. Curie-Weiss constants Table 8.

$YbTiO_3$

I Unit cell dimensions Table 5. Electrical resistivity Table 6, Fig. 41. Magnetization measurements point to a magnetic transition at $T_c = 39(1)$ K and from a neutron diffraction spectrum analysis follows that $p_{Yb^{3+}} = 1.7(2)\,\mu_B$, considerably quenched as compared to its free ion value of 4.0 μ_B [81T1].

$LuTiO_3$

I Unit cell dimensions Table 5. Ordering temperature Table 8. Magnetization at 4.2 K Fig. 42.

II Magnetic susceptibility Fig. 46. $C_m = 0.312\ \mathrm{cm^3\,K\,mol^{-1}}$, $\Theta_p = 65$ K, and $p_{eff} = 1.59\ \mu_B$/FU [78B1].

$YTiO_3$

I Unit cell parameters Tables 5, 12. Electrical resistivity Table 6. Curie point Table 8. Low temperature neutron diffraction spectra of a powder sample can be explained by accepting ferromagnetic ordering of the magnetic moments of the Ti ions along the *c* axis, with $p_{Ti^{3+}} = 0.77(30)\,\mu_B$ [82Z1]. Magnetization measurements Figs. 73, 74, see also Fig. 50d. At 4.2 K is found: $p_{FU} = 0.84(1)\,\mu_B$, with both the *b* and *c* axis as directions of easy magnetization [81G2].

II Magnetic susceptibility [82G2]. Curie-Weiss constants Tables 8, 13.

$YTi_{x/2}Fe_{1-x}Ni_{x/2}O_3$

I Spontaneous magnetization Fig. 75.

$YTi_{x/2}Fe_{1-x}Co_{x/2}O_3$

I Magnetostriction Fig. 76.

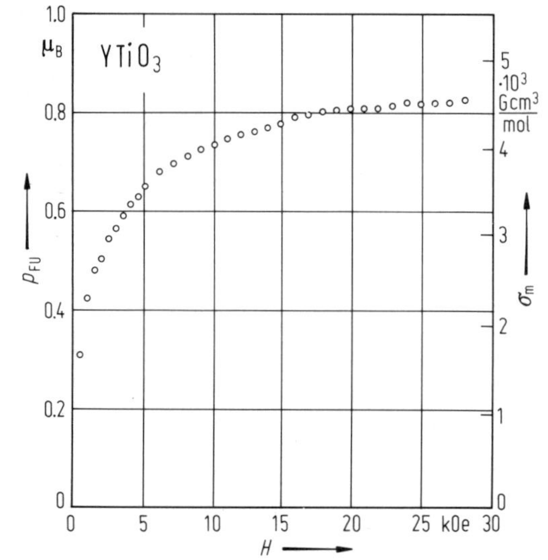

Fig. 73. $YTiO_3$, polycrystalline. Magnetic moment per FU, p_{FU} vs. applied magnetic field H at 4.2 K [81G2].

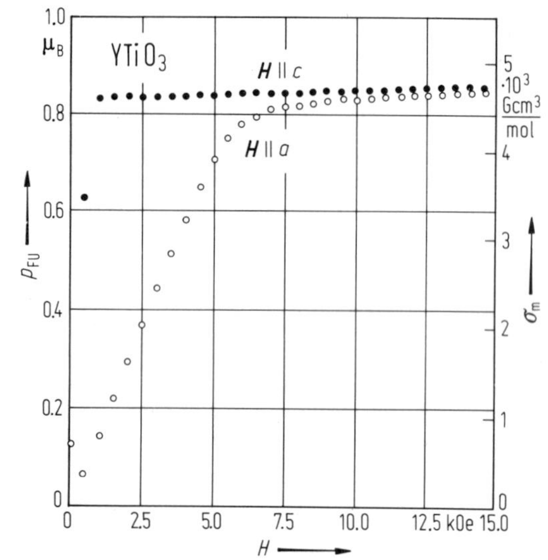

Fig. 74. $YTiO_3$. Magnetic moment per FU, p_{FU} vs. applied magnetic field H along the c and a axis as indicated [81G2].

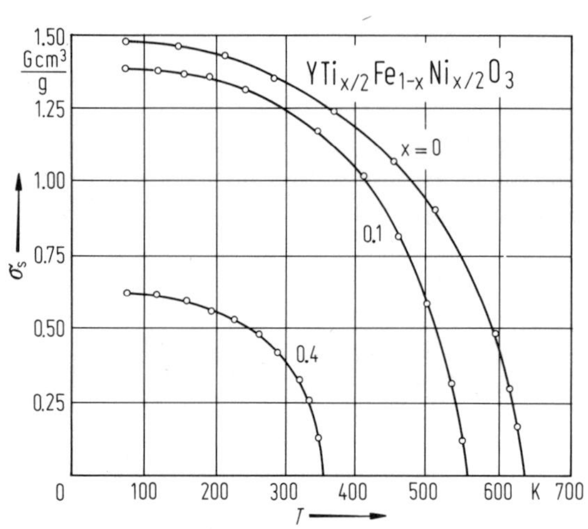

Fig. 75. $YTi_{x/2}Fe_{1-x}Ni_{x/2}O_3$. Spontaneous mass magnetization σ_S vs. temperature T [81A1].

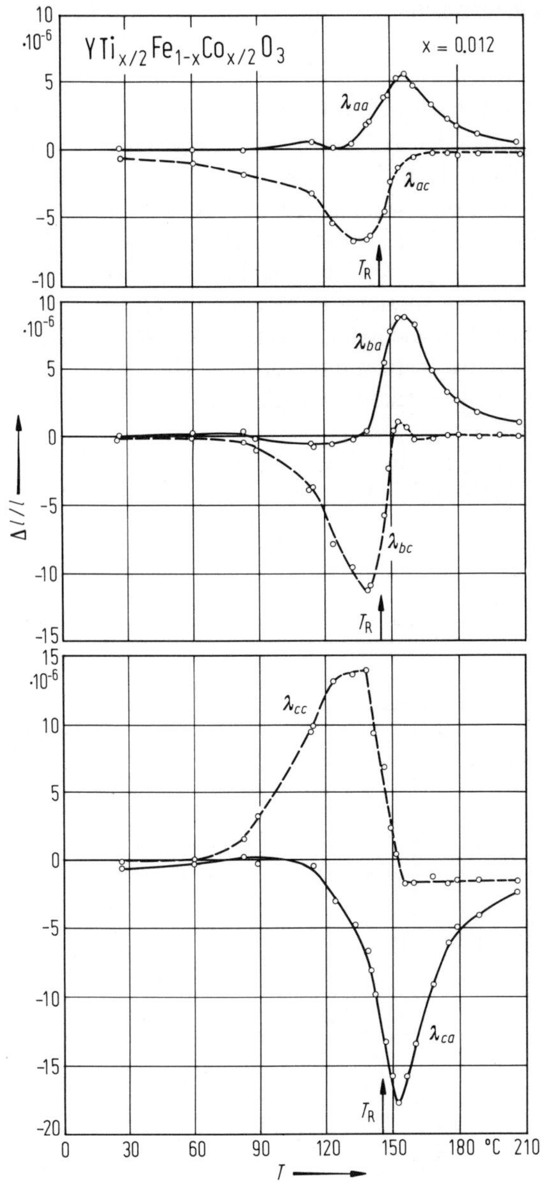

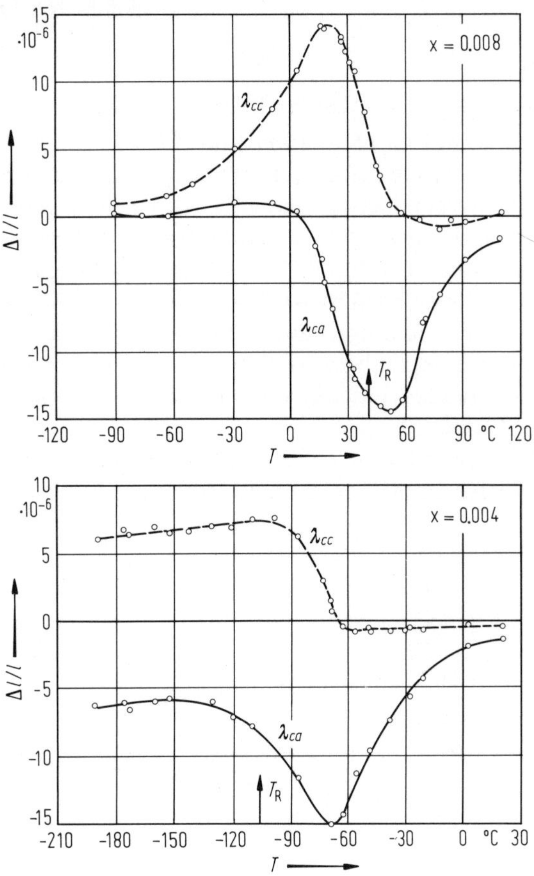

Fig. 76. $YTi_{x/2}Fe_{1-x}Co_{x/2}O_3$. Magnetostriction λ_{ij} measured along the crystallographic i axis for an external magnetic field H = 6.8 kOe applied along the crystallographic j axis. The reorientation temperature T_R is the temperature at which the rotation angle of the magnetization relative to the c axis is 45° [77A2].

5.5.5 RVO_3 compounds

RVO_3

I Unit cell parameters Table 16 and Fig. 77. The discontinuities for lanthanum and cerium vanadite point to a phase transition 0′ ($c/(a\sqrt{2}) < 1$) ⇆ 0 ($c/(a\sqrt{2}) > 1$) [76Z1]. The temperature dependence of the electrical resistivity is shown in Fig. 78, Table 17. The sintered compounds exhibit a semiconducting character with p-type conduction. There seem to be two activation energies, above and below a certain transition temperature T_1 [75B2]. Neutron diffraction spectra show ordering of the V^{3+} ions below the Néel temperature, T_{N1}, see Table 18. The C-type ordering is characterized by AF interaction in the (001) planes and F interaction between the planes. The G-type ordering is also characterized by AF interaction in the (0 0 1) planes, but additionally AF interaction between these planes. The ordering of the magnetic moments of the R^{3+} ions of some vanadites has been established, T_{N2}. For the spontaneous weak ferromagnetic moment at 77 K see Table 19.

II Paramagnetic properties Fig. 79 and Table 19. A temperature independent van Vleck contribution to the paramagnetic susceptibility becomes significant only at high temperatures [74B1].

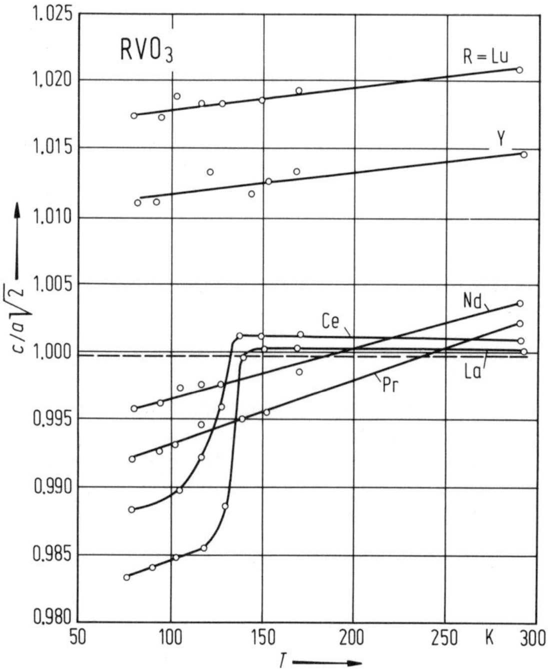

Fig. 77. RVO_3. Temperature dependence of the lattice parameter ratio $c/(a\sqrt{2})$ [76Z1].

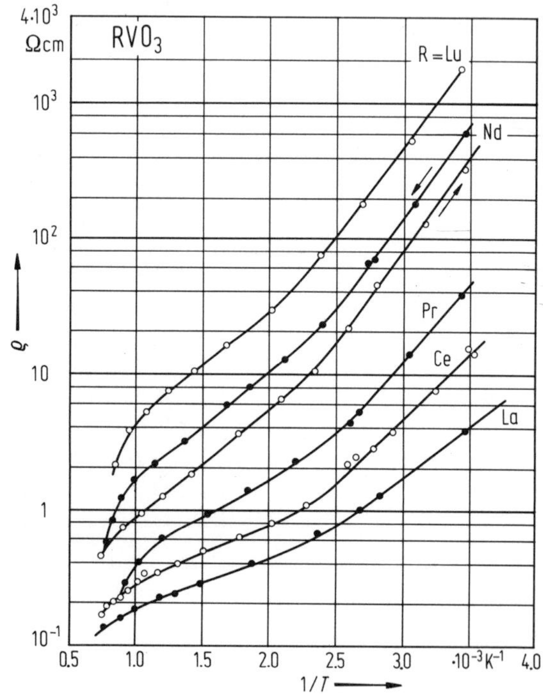

Fig. 78. RVO_3. Temperature dependence of the electrical resistivity ρ of sintered samples. For R = Nd the resistivity is measured during heating and during cooling as indicated by arrows [75B2].

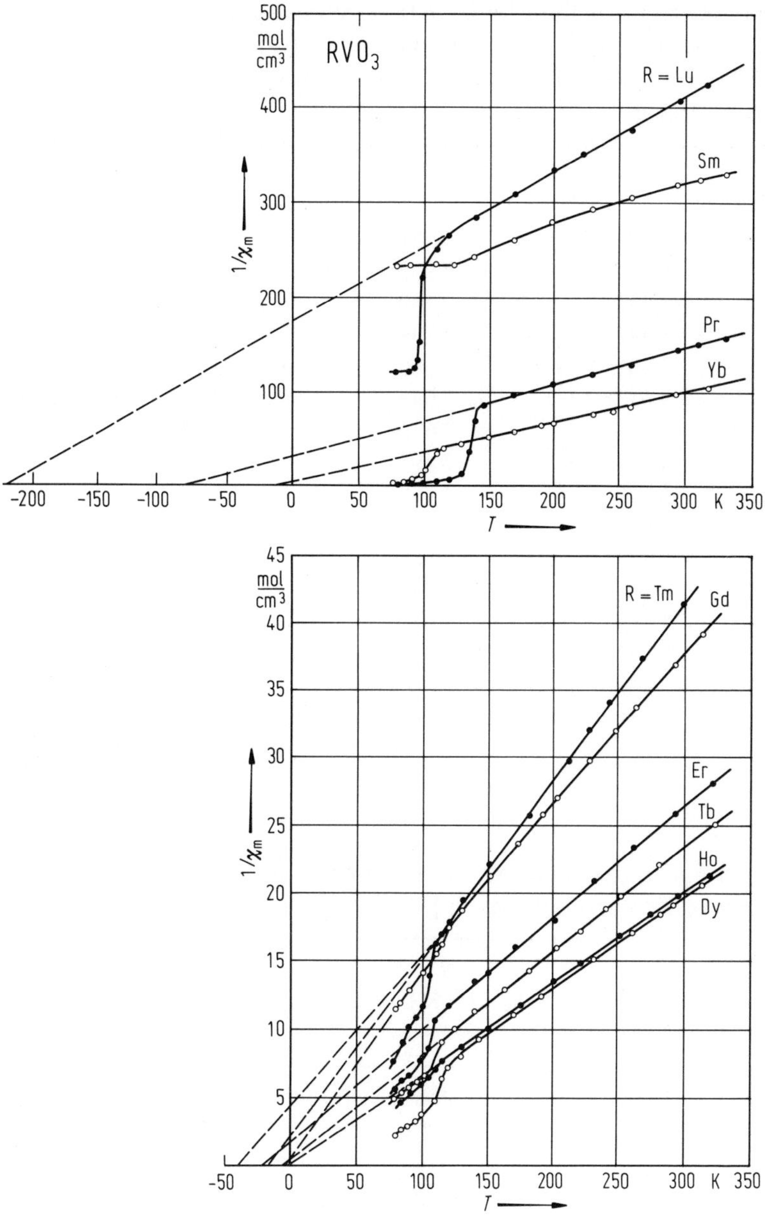

Fig. 79. RVO_3. Temperature dependence of the reciprocal molar susceptibility, χ_m^{-1}, for various constituting lanthanide ions [74B1].

Table 16. RVO_3. X-ray lattice parameters a, b and c in Å ($\pm$ 0.003Å) (and unit cell volume V in $Å^3$) at RT indexed on the basis of an orthorhombically distorted perovskite, space group D_{2h}^{16} – Pbnm. For the simple perovskite cell holds $a \approx b \approx a_0\sqrt{2}$, $c = 2a_0$ with $Z = 4$ FU/cell [74B1].

R	a	b	c	V
La	5.547	–	7.851	241.6
Ce[a])	5.514	5.557	7.808	239.2
Pr	5.474	5.572	7.760	236.7
Nd	5.451	5.582	7.738	235.5
Sm	5.395	5.601	7.683	232.2
Eu	5.364	5.617	7.645	230.3
Gd	5.344	5.623	7.632	229.3
Tb	5.319	5.621	7.605	227.4
Dy	5.297	5.608	7.590	225.5
Ho	5.278	5.601	7.573	223.9
Er	5.260	5.592	7.560	222.4
Tm	5.244	5.582	7.548	223.9
Yb	5.230	5.578	7.540	220.0
Lu	5.216	5.571	7.529	218.8

[a]) Ref. [78B2].

Table 17. RVO_3. Activation energy E_a of the conductivity for two temperature ranges corresponding to the straight parts of the curves of Fig. 78 [75B2].

R	E_a [eV]		T_1 [K]	T_2 [K]
	$T < T_1$	$T_1 < T < T_2$		
La	0.15	0.08	410	1190
Ce	0.18	0.08	570	1170
Pr	0.22	0.11	415	990
Nd	0.25	0.16	475	1060
Sm	0.28	0.17	450	1050
Gd	0.34	0.18	420	1080
Tb	0.32	0.18	430	1050
Dy	0.30	0.18	425	1010
Ho	0.33	0.20	445	1080
Tm	0.27	0.22	400	–
Yb	0.35	0.19	420	1020
Lu	0.26	0.15	455	1030
Er	0.33	0.20	520	1080

Table 18. RVO_3. Magnetic structures in the ordered state, magnetic moments of the constituting ions at 4.2 K as well as the ordering temperatures T_{N1} and T_{N2} of the magnetic sublattices of the V^{3+} and R^{3+} ions, respectively, as derived from neutron diffraction spectra. Spin rotation temperatures are indicated [76Z1].

R	Type of ordering		$p_{V^{3+}}$ [μ_B]	$p_{R^{3+}}$ [μ_B]	T_{N1} [K]	T_{N2} [K]
	V^{3+}	R^{3+}				
La	C, $p \perp c$	–	1.3	–	137	–
Ce	C, $p \perp c$	–	1.2	–	136	< 4.2
Pr	$C_xF_y \xrightarrow{50K} C_{xy}$	–	1.2	–	135	< 4.2
Nd	$C_xF_y \xrightarrow{15K} C_yF_x$	–	1.2	–	132	< 4.2
Sm*)	–	–	–	–	126	–
Eu*)	–	–	–	–	120	–
Gd*)	–	–	–	–	118	–
Tb	C_xF_y	A_xG_y	1.14	$x = 2.6$ $y = 1.7$	113	10.4
Dy	C_xF_y	–	1.2	–	112	< 4.2
Y	$G_xF_z \xrightarrow{77.7K} F_xC_y$	–	1.2	–	114	–
Ho	G_xF_z	C_xF_y	1.12	3.0	108	8.8
Er	G_xF_z	C_z	1.2	4.2	108	16
Tm	G_zE_x	C_z	1.2	3.0	106	13
Yb	G_zF_x	–	1.2	–	104	< 4.2
Lu	G_zF_x	–	1.18	–	101	–

*) The magnetic structure was not determined because of the strong resonant absorption of neutrons.

Table 19. RVO_3. Survey of magnetic properties of annealed samples. Molar susceptibility χ_m at RT, molar Curie-Weiss constant C_m, paramagnetic Curie temperature Θ_p, effective paramagnetic moment p_{eff}, derived from susceptibility curves. Néel temperature T_{N1} and weak spontaneous magnetic moment p_s per FU at $T = 77$ K [74B1]; see also [77S1].

R	χ_m at RT [$cm^3\,mol^{-1}$]	C_m [$cm^3\,K\,mol^{-1}$]	Θ_p [K]	p_{eff} [μ_B]	T_{N1} [b]) [K]	p_s at 77 K [μ_B/FU]
La[a])				4.2	128	
Ce[a])				3.7	145	
Pr	6825	2.56	− 85	4.55	135	0.30
Pr[a])				4.4	150	
Nd[a])				4.6	139	
Sm	3100			2.69	126	
Sm[a])					110	
Eu[a])					133	
Gd	27150	8.95	− 40	8.50	118	
Gd[a])				8.3	118	
Tb	43600	12.9	− 5	10.15	113	0.05
Tb[a])				10.2	112	

(continued)

Table 19, *continued*

R	χ_m at RT [$cm^3 mol^{-1}$]	C_m [$cm^3 K mol^{-1}$]	Θ_p [K]	p_{eff} [μ_B]	T_{N1} [b]) [K]	p_s at 77 K [μ_B/FU]
Dy	52025	15.1	0	11.10	112	0.24
Dy [a])				11.2	112	
Ho	51040	14.8	0	10.90	108	0.04
Ho [a])				11.8	110	
Er	38650	12.05	– 22	9.90	108	0.08
Er [a])				10.0	109	
Tm	24500	7.55	– 18	7.85	106	0.04
Tm [a])				8.8	88	
Yb	9900	3.07	– 20	5.00	104	0.14
Yb [a])				5.3	94	
Lu	2450	1.25	– 220	3.20	101	0.005
Lu [a])				2.8	105	
Y [a])				2.8	110	

[a]) [77S1].
[b]) For the reference [77S1] T_{N1} is quoted the temperature at which a weak ferromagnetism first appears.

$LaVO_3$ (see also LB III/4a, p. 226 and III/12a, p. 441)

I Crystallographic phase transitions Table 20, Fig. 80 [80B1, 80Z2, 91S1] and chemical stability at high temperatures [79N1]. A cubic to orthorhombic transformation, immediately followed by a magnetic transition into a strong diamagnetic state, occurs below about 135 K, Fig. 81. – See also Tables 18, 19. Resistivity Figs. 78, 82 and Table 17.

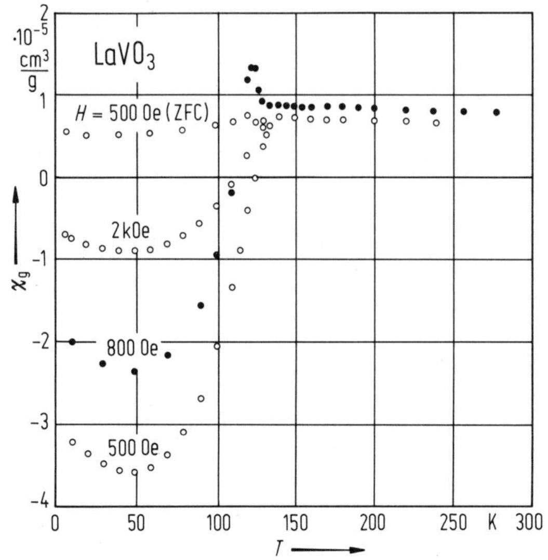

For Fig. 80, see next page.

Fig. 81. $LaVO_3$. Temperature dependence of the magnetic mass susceptibility χ_g of an arc melted (open circles) and a sintered (filled circles) sample. The data were taken while the sample was slowly cooled in a magnetic field H. The sample ZFC was quenched in zero field and measured in H = 500 Oe [91S1, see also 77S1].

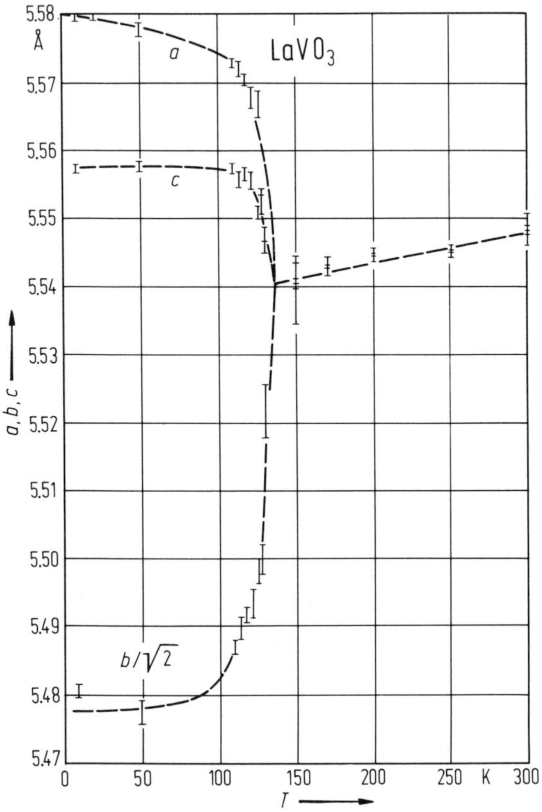

Fig. 80. $LaVO_3$. Lattice parameters a, c and $b/\sqrt{2}$ based on the orthorhombic Pnma structure as derived from powder X-ray diffraction spectra [91S1].

Table 20. $LaVO_3$. Lattice parameter a, rhombohedral angle α and cell volume V at various temperatures. X-ray diffraction peaks indexed on the basis of a simple cubic cell with $Z = 8$ FU [80Z2].

T [K]	a [Å]	α [deg]	V [Å^3]
293	7.851[c])	–	60.49
523	7.867	–	60.86
773	7.887	–	61.27
1023	7.904	–	61.72
1103	7.910	–	61.86
1113[a])	5.560	60.4	60.91
1123	5.562	60.4	60.98
1133	5.565	60.4	61.06
1153	5.567	60.4	61.11
1173	5.573	60.4	61.29
1273	5.580	60.4	61.65
1298[b])	3.950	–	61.63
1323	3.955	–	61.87
1373	3.960	–	62.1

[a]) Transformation of the doubled cubic lattice to a rhombohedral lattice R$\bar{3}$c.

[b]) Transformation from rhombohedral R$\bar{3}$c to a simple cubic perovskite-type lattice, space group Pm3m.

[c]) 7.846 Å [91S1].

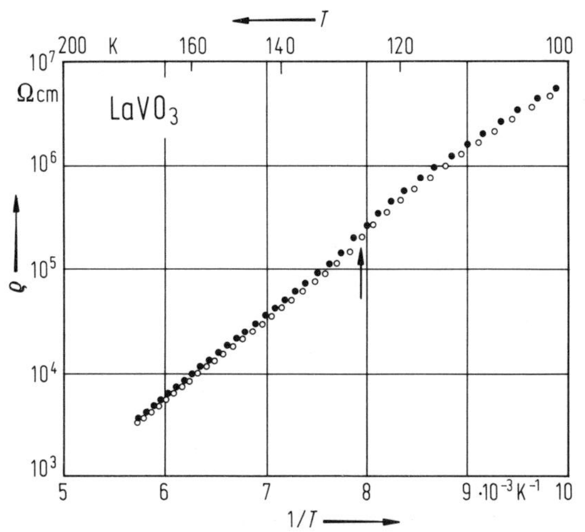

◀ Fig. 82. $LaVO_3$, sintered sample. Temperature dependence of the resistivity ρ in zero field (filled circles) and in a field of 1 kOe (open circles). The arrow points to a slight increase below 127 K [91S1].

$CeVO_3$ (see also LB III/4a, p. 226)

I Magnetic and thermodynamic properties Tables 18, 19, 20. The antiferromagnetic compound undergoes a crystallographic transition at a temperature T_s just below $T_{N1} = 135$ K [78B2]. Specific heat Fig. 83. Resistivity Table 17.

II Paramagnetic properties Table 21 and Fig. 83.

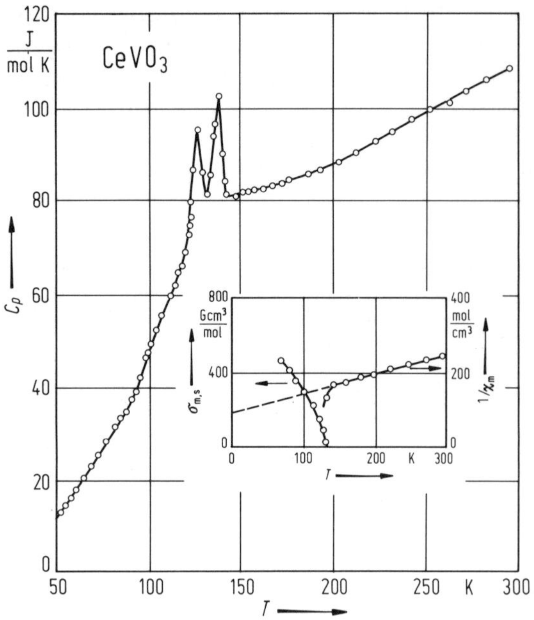

Fig. 83. $CeVO_3$. Specific heat C_p vs. temperature T. Inset: spontaneous molar magnetization $\sigma_{m,s}$ and inverse molar susceptibility χ_m^{-1} vs. temperature [78B2].

Table 21. $CeVO_3$, $GdVO_3$, $LuVO_3$. Néel temperature T_{N1}, crystallographic transition temperature T_s, paramagnetic Curie point Θ_p, and the specific heat C_p, the enthalpy ΔH_0, and entropy ΔS_0 at RT [78B2].

Compound	T_{N1} [K]	T_s [K]	Θ_p [K]	C_p [$J\,mol^{-1}\,K^{-1}$]	ΔH_0 [$kJ\,mol^{-1}$]	ΔS_0 [$J\,mol^{-1}\,K^{-1}$]
$CeVO_3$	135	124	−160	108.6	19.28	116.7
$GdVO_3$	113	111	−40	98.4	17.66	108.2
$LuVO_3$	96.8	85	−250	99.6	15.86	94.8

$PrVO_3$ (see LB III/4a, p. 226 and III/12a, p. 441)

$NdVO_3$ (see LB III/4a, p. 226 and III/12a, p. 441)

$SmVO_3$ (see LB III/4a, p. 226)

$EuVO_3$ (see LB III/12a, p. 441)

$GdVO_3$ (see also LB III/4a p. 227 and III/12a, p. 441)

I Magnetic and thermodynamic properties Tables 19, 21. Specific heat Fig. 84. Magnetization curve Fig. 84. For $T < T_N$ the magnetic susceptibility can be described by a model of free Gd^{3+} ions (ground state $^8S_{7/2}$) with $p_{Gd^{3+}} = 7.73\,\mu_B$, $\Theta_p = -5$ K, whereas $p_{Gd^{3+},\,theor} = 7.94\,\mu_B$ [74B1]. Resistivity Table 17.

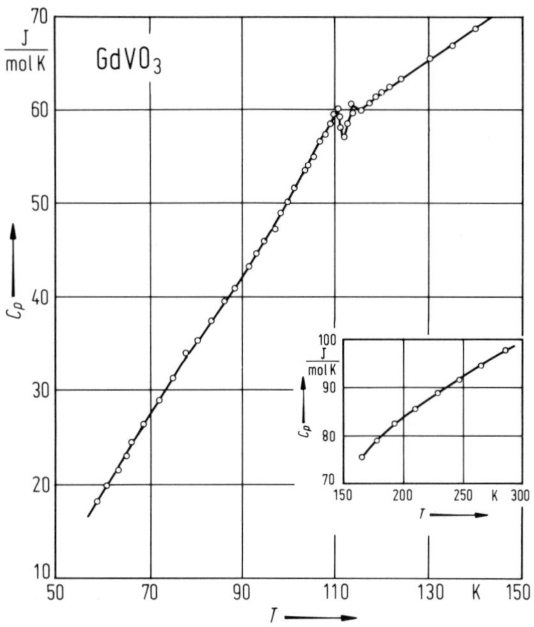

Fig. 84. $GdVO_3$. Specific heat C_p vs. temperature T [78B2].

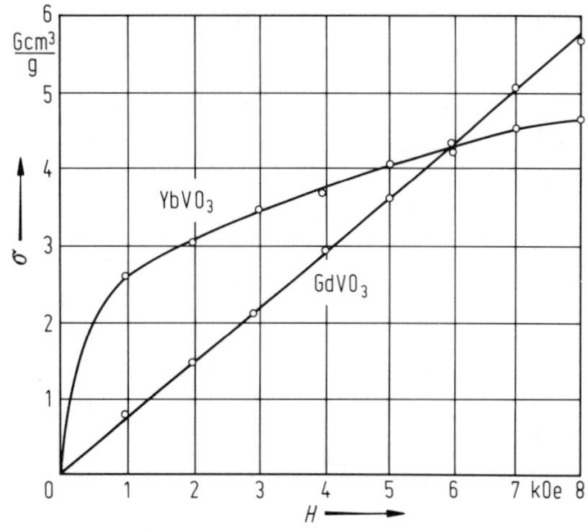

Fig. 85. $GdVO_3$; $YbVO_3$. Magnetization curves at 77 K [77S1].

$TbVO_3$ (see LB III/12a, p. 441)

$DyVO_3$ (see LB III/4a, p. 227; III/12a, p. 441)

$HoVO_3$ (see also LB III/12a, p. 441)

I The neutron diffraction results given in Table 18 have not been confirmed by [78B3]. The latter find the results of Table 22 with the magnetic structures C_xF_z and $G_xF_yA_z$ with a transition temperature $T_{N2} = 13(2)$K and $T_{N1} = 108(2)$K. Resistivity Table 17.

II Paramagnetic susceptibility Figs 79, 86 and Table 19.

Table 22. $HoVO_3$ Components (in μ_B) of the magnetic moments of the constituting ions at 4.2 and 60 K as derived from neutron diffraction patterns [78B3].

	$T = 4.2$K	$T = 60$K
Ho^{3+}	$p_x = 8.65(15)$	
	$p_z = 3.74(12)$	
V^{3+}	$p_x = 1.4(3)$	$p_x = 1.1(4)$
	$p_y = 0.04(40)$	$p_y = 0.2(10)$
	$p_z = 1.9(2)$	$p_z = 1.4(1)$

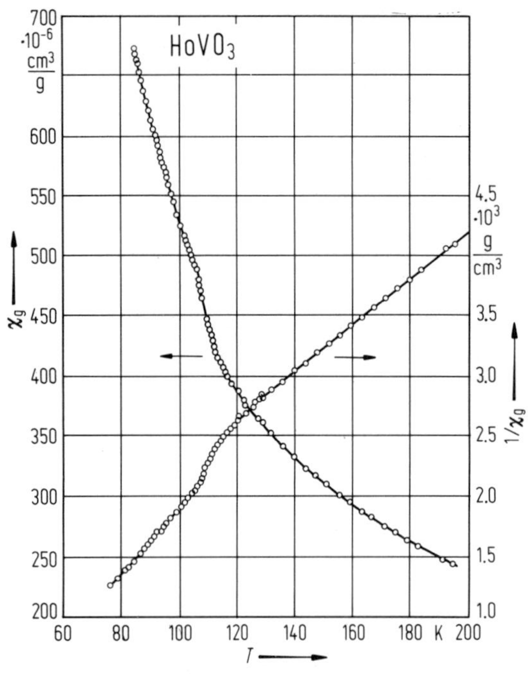

Fig. 86. $HoVO_3$. Magnetic mass susceptibility χ_g vs. temperature T [78B3].

$ErVO_3$ (see LB III/4a, p. 227 and III/12a, p. 441)

$TmVO_3$ (see also LB III/12a, p. 442)

I Resistivity Table 17. Magnetic structure Table 18. – See also Table 19.

III Mössbauer spectra [84H1]. Weak magnetic Tm^{3+} lattice from $T_{N1} = 106$ K down to about 15 K. Below 15 K much more strongly magnetic Tm^{3+} lattice, accompanied by a spin rotation. Magnetic hyperfine field of $^{169}Tm^{3+}$ see Fig. 87. At 15 K and 4 K the equivalent magnetic moments of Tm^{3+} are 0.3 and 2.1 μ_B, respectively. For the quadrupole hyperfine action see Fig. 31. In the high-temperature range the angle θ between H_{hyp} and the principle axis of the EFG tensor is estimated to be 60(15)°, at and below 7 K is found $\theta \approx 0$ [84H1].

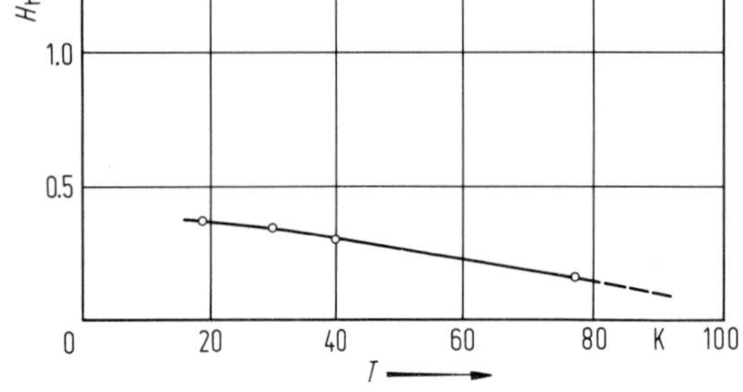

Fig. 87. $TmVO_3$. Magnetic hyperfine field H_{hyp} vs. temperature T derived from $^{169}Tm^{3+}$ Mössbauer spectra [86H1].

$YbVO_3$ (see also LB III/12a, p. 442)
I Magnetization curve Fig. 85. See also Tables 18, 19. Resistivity Table 17.
III ^{170}Yb Mössbauer data Table 23.

$YbV_{0.97}Fe_{0.03}O_3$
I $T_N = 94(1)$K, from ^{57}Fe Mössbauer spectra [74B2].

Table 23. $YbVO_3$. ^{170}Yb Mössbauer data at various temperatures T. δ isomer shift with respect to $^{170}Tm\ ^{11}B_{12}$, ΔE_Q quadrupole hyperfine splitting, H_{hyp} hyperfine field, θ angle between H_{hyp} and principal axis of the EFG tensor, η asymmetry parameter of the EFG [74B2].

T [K]	δ [mm/s]	ΔE_Q [mm/s]	H_{hyp} [kOe]	θ [deg]	η
4.2	0.03	3.70	738	0	0.16
10	0.07	3.70	712	0	0.16
15	0.13	3.65	648	0	0.17
23	0.20	3.54	520	0	0.30
30	0.13	3.56	465	2.7	0.22

$LuVO_3$ (see also LB III/12a, p. 442)
I No significant ordering of Lu^{3+} moments down to 4.2 K [75Z1]. See also Tables 17, 18, 19. Resistivity Table 17.
II Susceptibility Figs 79, 88 and Table 21.

YVO_3 (see also LB III/4a, p. 227; III/12a, p. 442)
I Magnetic structure Table 18, – See also Table 19.
II Susceptibility Fig. 88.

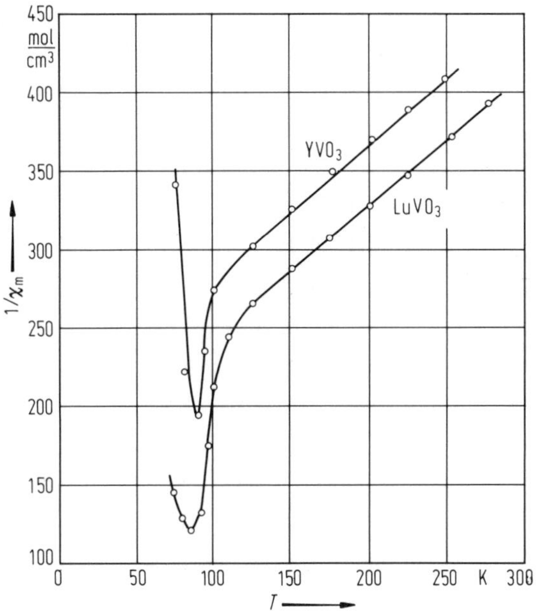

Fig. 88. $LuVO_3$; YVO_3. Temperature dependence of reciprocal molar susceptibility χ_m^{-1} [77S1].

5.5.6 $RCrO_3$ compounds

$RCrO_3$

I Structural data Figs. 89, 90 (see also LB III/4a, p. 163 and III/12a, p. 338); magnetic structures Fig. 91 and Table 24.

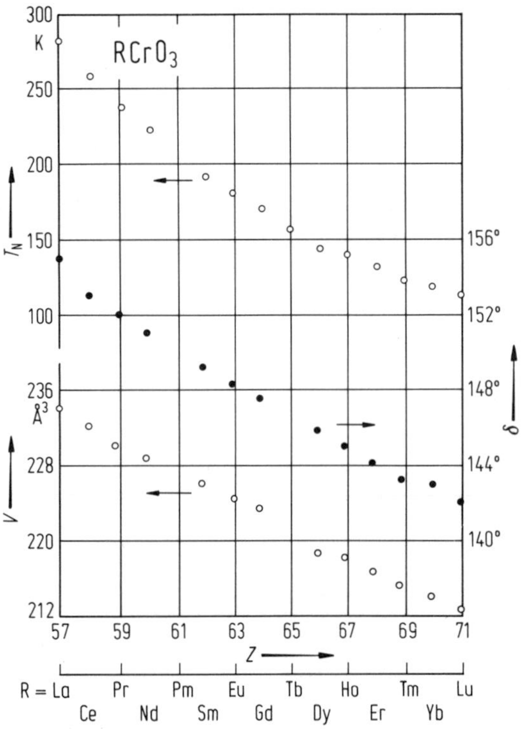

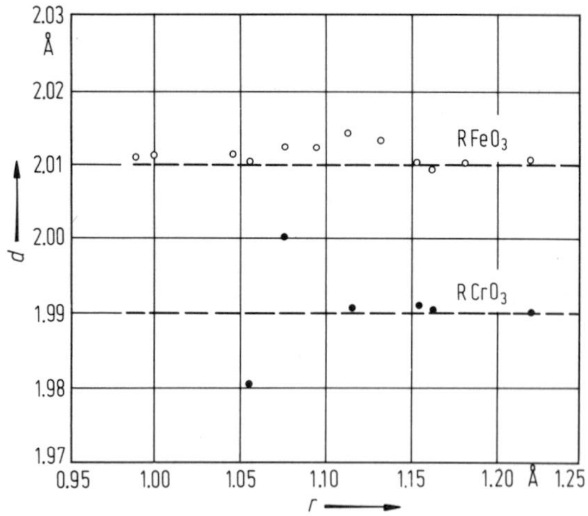

Fig. 90. $RCrO_3$; $RFeO_3$. Distances d for Cr–O and Fe–O [82K2]. r: ionic radius of R^{3+}.

◀

Fig. 89. $RCrO_3$. Angle δ of Cr–O–Cr configuration, cell volume V and Néel temperature T_N [82K2]. Z: atomic number of R element.

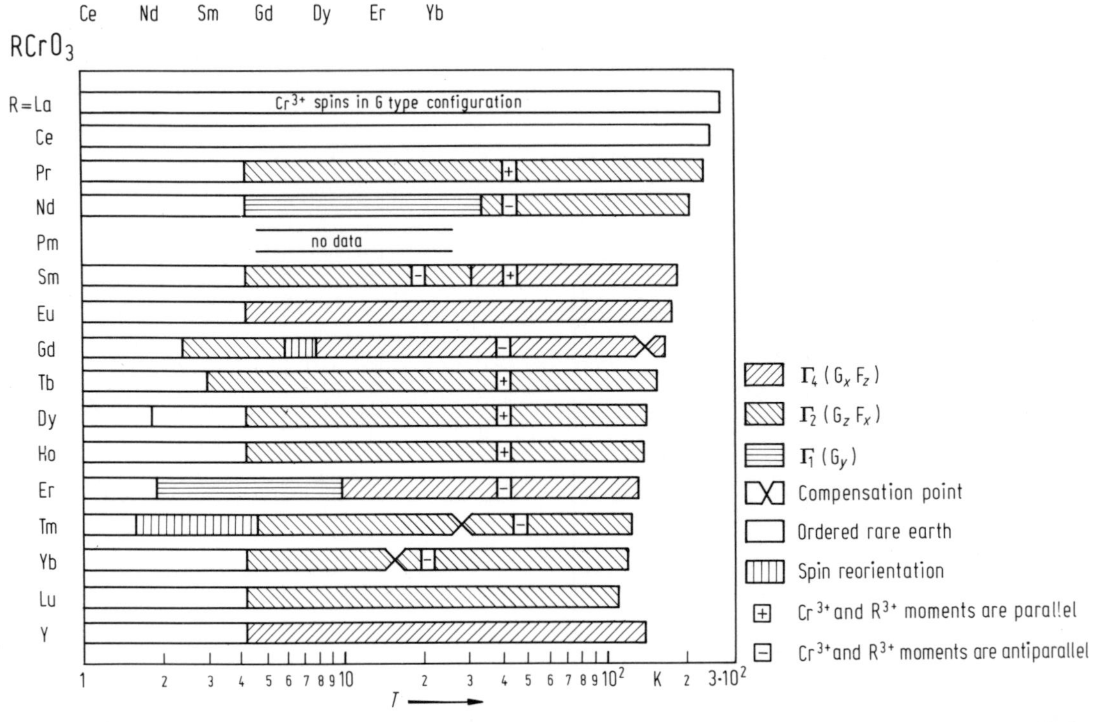

Fig. 91. $RCrO_3$. Survey of the spontaneous magnetic structures in the rare earth orthochromites [78H1]. See also [81G4].

Table 24. $RCrO_3$. Summary of magnetic structures obtained by various measuring techniques at 80 and 4.2 K [81S1]. NES: neutron elastic scattering. SMS: spontaneous magnetization and susceptibility measurements. The spin structures of Cr and R ions are written as [Cr; R]. [$G_{xy}60°$; C_z] for e.g. $NdCrO_3$ means the Cr^{3+} moment is 60° from the x axis. A hyphen in e.g. [G_x; −] indicates a paramagnetic state of the R ion. The sign of the configuration indicates the sign of the moments in site 1, e.g. $-C \equiv - - + +$ (see Table A). The symbol ± indicates that the direction of the magnetic component could not be determined. Angles are given in respect to the x axis.

R	T = 80 K			T = 4.2 K		
	SMS a)	NES a)	NES [81S1]	SMS a)	NES a)	NES [81S1]
Pr	[G_z, F_x; −]	[G; −]	[G; −]	[G_z, F_x; −C_y, −F_x]	[G_x; F_z]	[G_z, F_x; ±C_y, ±F_x]
Nd	[G_z; −]	[G_z; −]	[G; −]	[G_y; −C_z]	[$G_{xy}60°$; C_z]	[G_y; −C_z]
Ho	[G_z, F_x; −]	[$G_{xz}86°$; −]	[G_z; −]	[G_z, F_x; −C_y, −G_x]	[G_z; C_y, F_x]	[G_z; −C_y, −F_x]
Er	[G_x, F_z; −]	[G_x; −]	[G_x; −]	[G_y; −C_z]	[$G_{xy}47°$; C_z]	[G_y; −C_z]
	[G_z, F_x; −]			[G_z, F_x; −C_y, −F_x]		
Tm		[$G_{xz}62°$; −]	[G_z; −]		[G_x; F_z]	[G_z; −C_y, −F_x]
	[G_x, F_z; −]			[G_x, F_z; −F_z]		
Yb	[G_z, F_x; −F_x]	[$G_{xz}68°$; −]	[G_z; −]	[G_z, F_x; −F_x]	[$G_{xz}68°$; F_x]	[G_z; −F_x]
Lu	[G_z, F_x; −]	[$G_{xz}63°$; −]	[G_{xz}; −]	[G_z, F_x; −]	[$G_{xz}63°$; −]	[G_{xz}; −]

a) Various sources, mentioned in LB III/4a and III/12a.

$LaCrO_3$ (see also LB III/4a, p. 227)

I Crystallographic phase transitions are given in [87T1], sintering properties in [79N1, 81B2] and CO adsorption in [83F1]. Distorted perovskite-type structure with orthorhombic lattice parameters for a sintered sample $La_{1.00}Cr_{1.02}O_3$ at RT: a = 5.485 Å, b = 5.524 Å and c = 7.758 Å [82K2]. Electronic band structure [80H3], resistance under pressure Fig. 92, spontaneous magnetization Fig. 93 and Néel point T_N = 291 K. The latter changes with pressure according to dT_N/dp = 0.6 K/kbar [82K2].

II Magnetic susceptibility Fig. 94. Curie-Weiss constants (SI units): Θ_p = −430 K, C_m = $11.9 \cdot 10^{-6}$ $m^3 K mol^{-1}$ and p_{eff} = 2.76 μ_B/FU [82T3].

$LaCr_{1-x}Ni_xO_3$

I Probably ferrimagnetic [84V2], Néel points Fig. 95.

II Magnetic susceptibility Fig. 96.

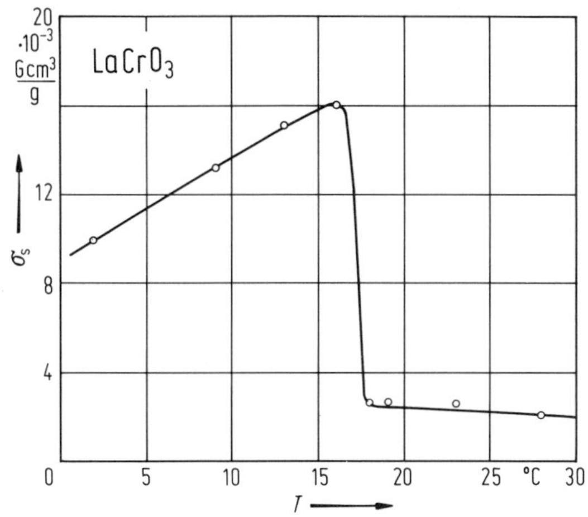

For Fig. 92, see next page.

Fig. 93. $LaCrO_3$. Spontaneous mass magnetization σ_S vs. temperature T [82K2].

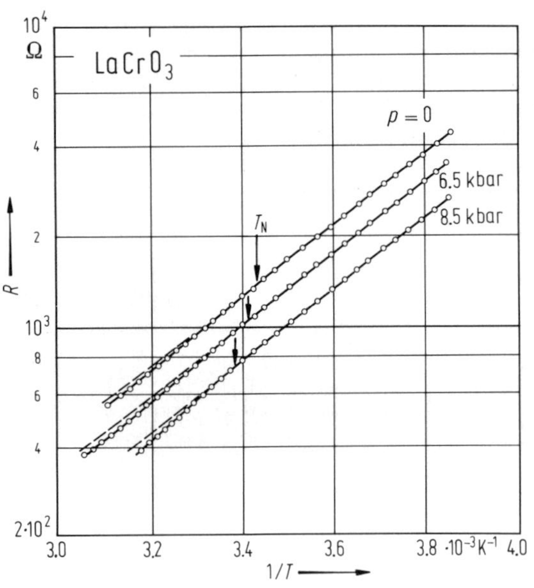

Fig. 92. $LaCrO_3$. Resistance under pressure p as a function of temperature [82K2].

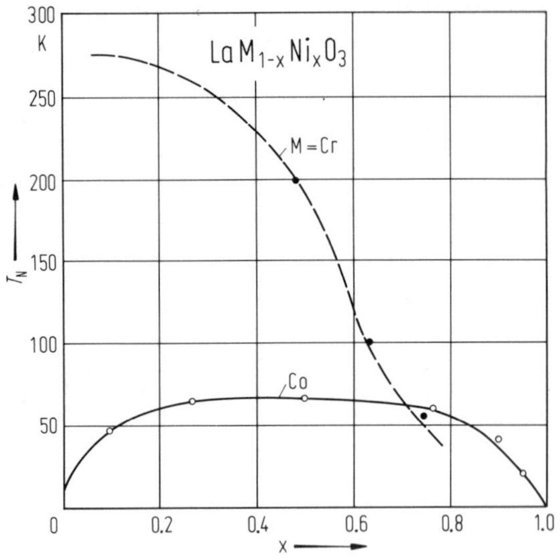

Fig. 95. $LaCr_{1-x}Ni_xO_3$; $LaCo_{1-x}Ni_xO_3$. Nèel temperature T_N vs. composition parameter x [84V2].

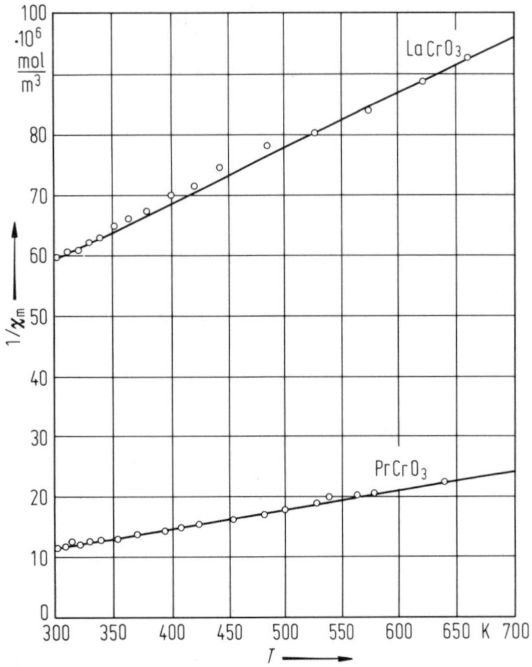

Fig. 94. $LaCrO_3$; $PrCrO_3$. Inverse molar magnetic susceptibility χ_m^{-1} (in SI units) vs. temperature T [82T3].

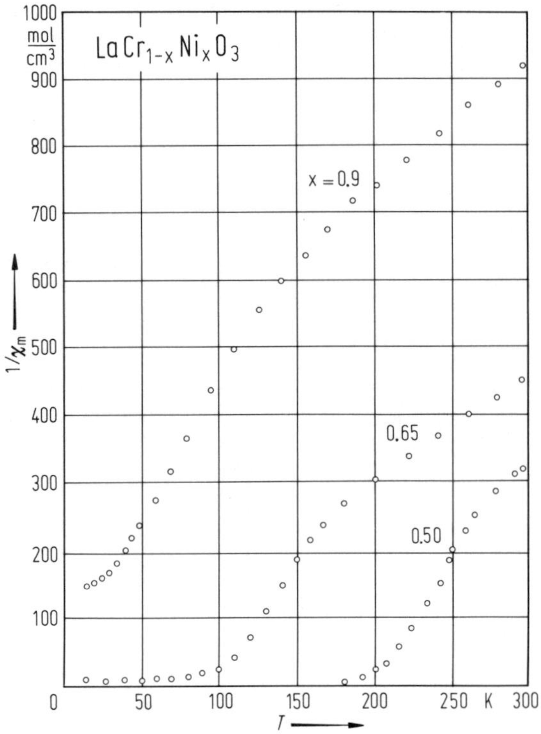

Fig. 96. $LaCr_{1-x}Ni_xO_3$. Inverse molar magnetic susceptibility χ_m^{-1} vs. temperature T [84V2].

$PrCrO_3$

I Axial antiferromagnetic structure of the Cr sublattice (G_z; $-C_y$, $-F_x$) with $p_{Pr^{3+}} = 0.6\ \mu_B$ from NES experiments at 9 K [81S1] in contrast to a value of 2.2 μ_B and 47° as the angle of the Pr^{3+} moment to the y axis in the xy plane as derived from SMS measurements [76G2]. Magnetization and susceptibility Figs. 97 ⋯ 99. $T_N = 237$ K. Molecular magnetic parameters Table 25.

II Magnetic susceptibility Fig. 94, Curie-Weiss constants (in SI units): $\Theta_p = -133$ K, $C_m = 33.0 \cdot 10^{-6}\ m^3\ K\ mol^{-1}$, $p_{eff} = 4.59\ \mu_B/FU$ [82T3].

IV Specific heat: No anomalies in the range $1.2 < T < 5$ K [68D1].
INS value of first exited Pr^{3+} level is 4.1 meV (48 K) an additional level is found at 15.3 meV (178 K) [77S2].

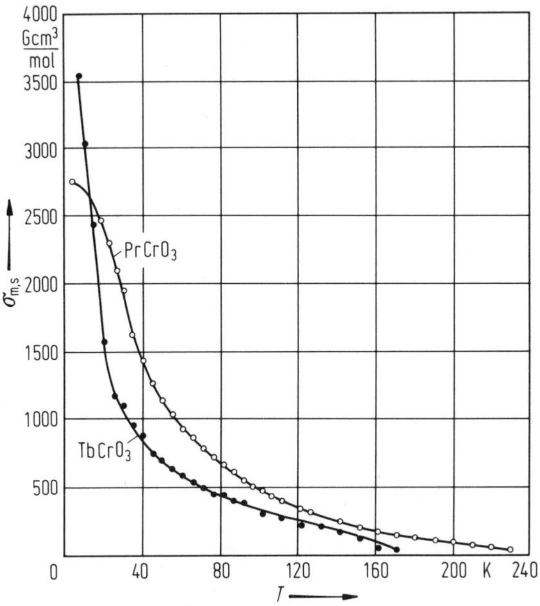

◄ Fig. 97. $PrCrO_3$; $TbCrO_3$, single crystals. Spontaneous molar magnetization $\sigma_{m,s}$ vs. temperature T. $\sigma_{m,s} \parallel a$ axis. Theoretical best fit curve using the data of Table 25 [76G2].

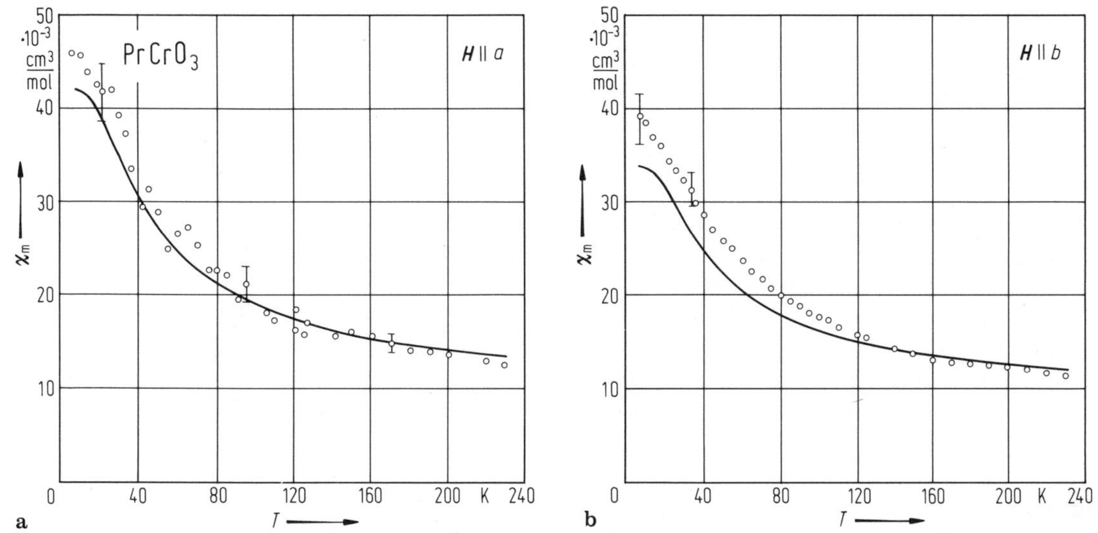

Fig. 98. $PrCrO_3$. Molar magnetic susceptibility in high magnetic fields, measured in the direction of the a axis (a) and b axis (b). Theoretical best fit curves using the data given in Table 25 [76G2].

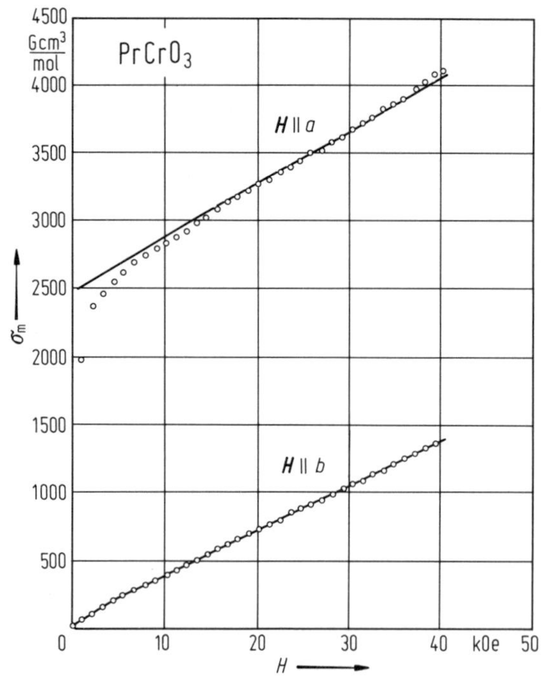

Fig. 99. $PrCrO_3$. Magnetization curves in two crystallographic directions at 4.2 K. Theoretical best fit curves using the data given in Table 25 [76G2].

Table 25. $PrCrO_3$, $TbCrO_3$, $HoCrO_3$. Magnetic parameters derived from magnetization data represented in Figs. 97–99. *D* crystal field splitting; Δ_0 Cr^{3+}-induced Zeeman splitting at the rare-earth ion site extrapolated to 0 K; g_{eff}, ϕ magnitude of effective *g* factor and its angle with the *b* axis; α canting angle of Cr sublattice moments relative to *b* axis [76G2]

	D [K]	Δ_0 [K]	g_{eff}	ϕ [deg]	α [mrad]
$TbCrO_3$	14.0(15)	1.8(1)	17.9(4)	43.0(15)	< 5
$PrCrO_3$	55(5)	17.8(5)	4.4(3)	47(3)	– 18(1)
$HoCrO_3$[a])	4(2)	14.7(2)	17.3(3)	25(1)	56(12)

[a]) [72H1]; see also Table 26.

$NdCrO_3$ (see also LB III/4a, p. 227 and III/12a, p. 442).

I Magnetic structure (G_y; $-$ C_z) at 4.2 K and magnetic moment $p_{Nd^{3+}} = 1.93\ \mu_B$ as derived from NES [81S1]. Magnetization Fig. 103.

II Temperature dependence of the susceptibility Figs. 100 and 101. Curie-Weiss constants (in SI units): $\Theta_p = -292$ K, $C_m = 32.5 \cdot 10^{-6}\ m^3\ K\ mol^{-1}$, $p_{eff} = 4.57\ \mu_B$/FU [82T3].

IV Ground multiplet levels of the Nd^{3+} ion as determined by INS: weak line at 2.4 meV (28 K) corresponding to transitions in the ground state doublet and a second doublet at 10.9 meV (127 K) [77S2].

$NdCr_xFe_{1-x}O_3$

I Magnetization at various temperatures Figs. 102 and 103.

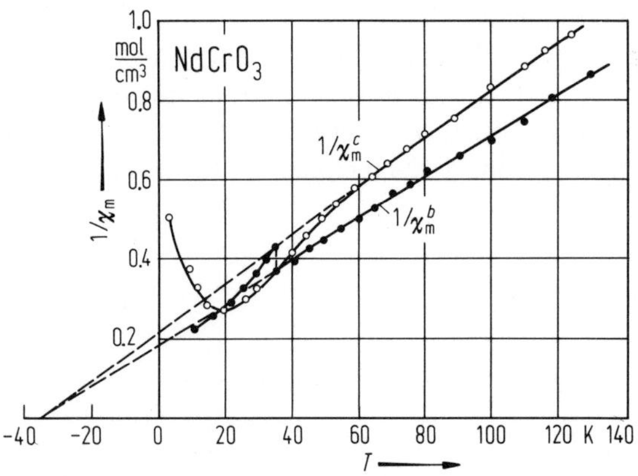

Fig. 100. $NdCrO_3$. The inverse molar magnetic susceptibility χ_m^{-1} vs. temperature for two crystallographic directions [76G2, 80G2].

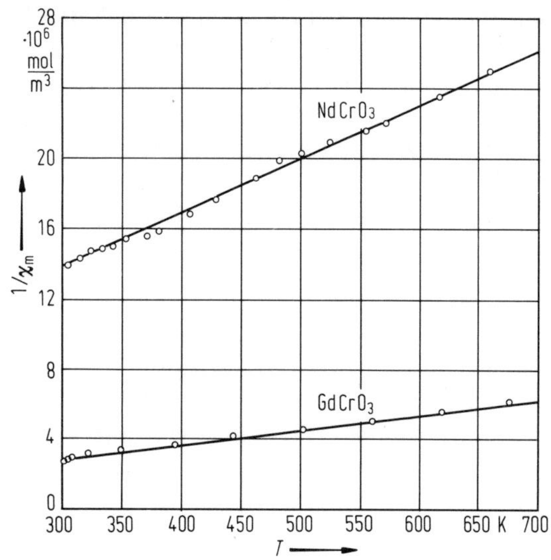

Fig. 101. $NdCrO_3$; $GdCrO_3$. Inverse molar magnetic susceptibility χ_m^{-1} (in SI units) vs. temperature T [82T3].

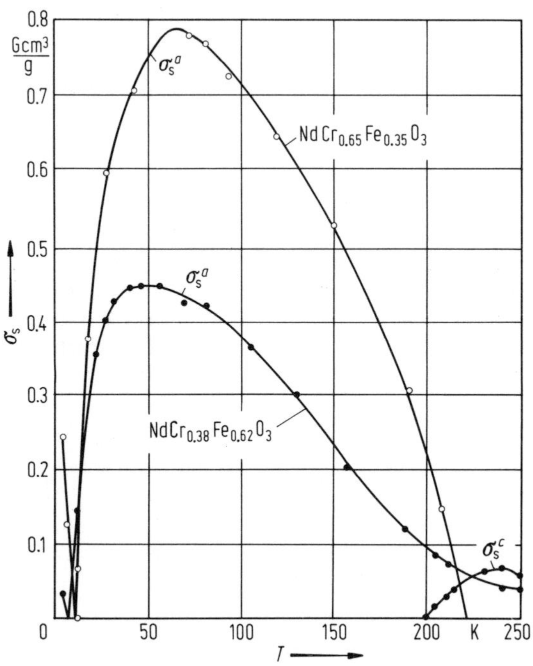

Fig. 102. $NdCr_xFe_{1-x}O_3$, single crystals. Temperature dependence of the spontaneous mass magnetization σ_S. For x = 0.38 the canted spin structure below T_N is $(G_{xz}F_{zx})$, which upon cooling changes into (G_zF_x). T_{comp} = 7 K. For x = 0.65 the spin configuration is Γ_2 (G_z,F_x) [78K2].

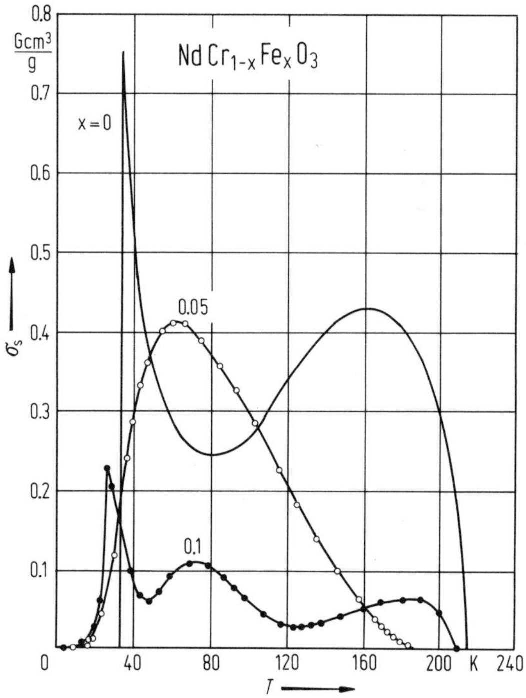

Fig. 103. $NdCr_{1-x}Fe_xO_3$. Spontaneous mass magnetisation σ_S vs. temperature T [91M1].

$SmCrO_3$ (see also LB III/4a, p. 227 and III/12a, p. 442)

I Acoustic velocity data Fig. 104. It is concluded from these data that the spin-reorientation phase transition $\Gamma_4 \rightleftarrows \Gamma_2$ in the temperature range $32.4\ K < T < 32.9\ K$ is continuous involving two second-order phase transitions. Specific heat data Fig. 105. It was concluded that the uniaxial anisotropy field for rotation of the moments of the Cr^{3+} ions in the $a\,c$ plane is: $H_K = 2.1(2)$ kOe [77G1].

II Magnetic susceptibility Fig. 106.

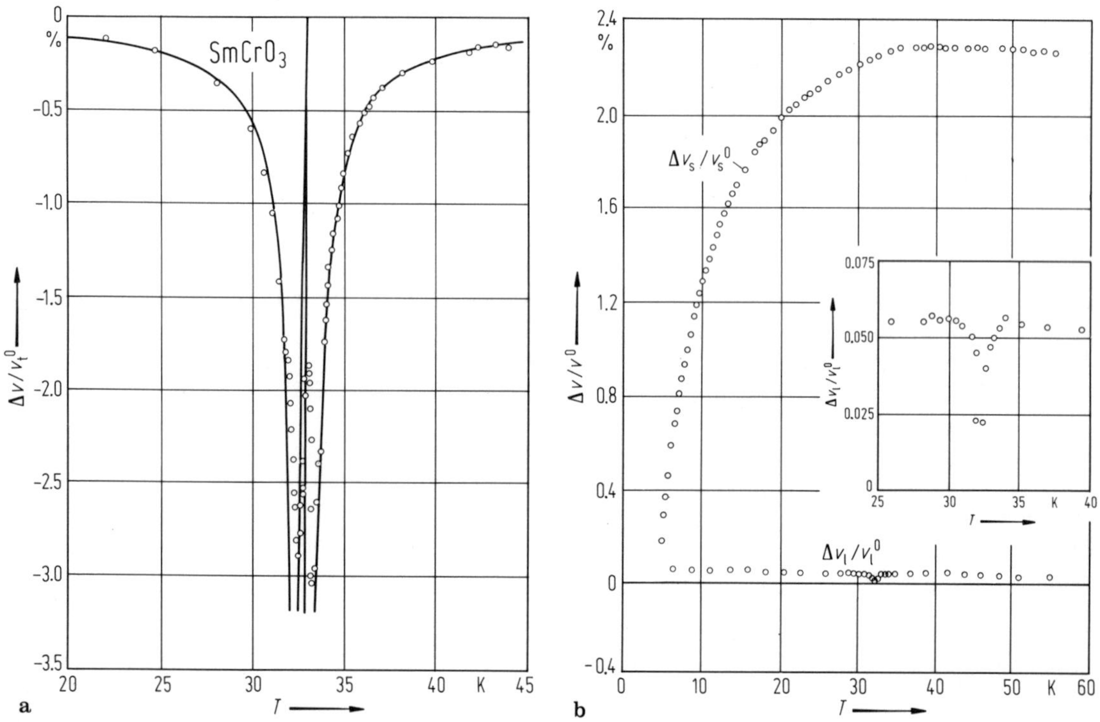

Fig. 104. $SmCrO_3$, single crystal. Temperature dependence of velocity change of a 30 MHz sound wave propagating along the c axis. (a) Shear wave polarized $\| a$, $\Delta v/v_t^0$; $v_t^0 = 3.55(4) \cdot 10^5\ cm\,s^{-1}$ at RT. (b) Shear wave polarized $\| b$, $\Delta v_s/v_t^0$; longitudinal wave $\Delta v_l/v_l^0$, see also inset; $v_l^0 = 7.00(7) \cdot 10^5\ cm\,s^{-1}$ at RT. The solid line in (a) is a theoretical best fit resulting from a conventional free-energy model calculation. Similar results for the case of the presence of a magnetic field are given also in [77G1].

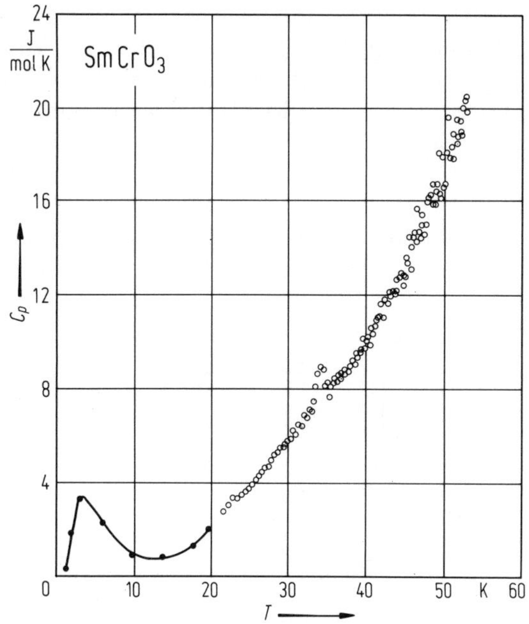

Fig. 105. $SmCrO_3$. Specific heat C_p vs. temperature T [77G1]. The data for $T < 20$ K are from [68C2].

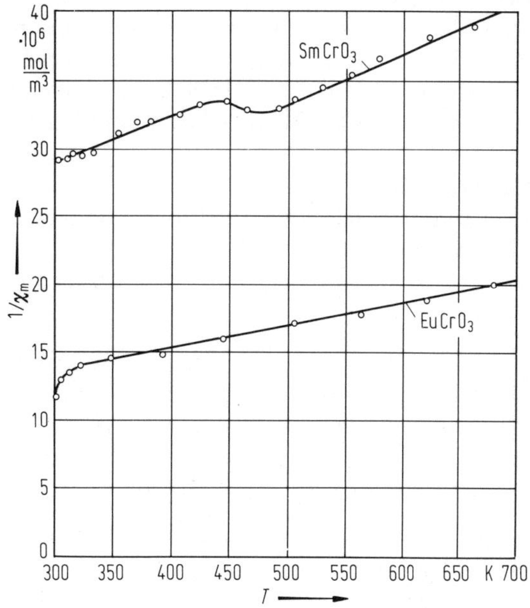

Fig. 106. $SmCrO_3$; $EuCrO_3$. Inverse molar magnetic susceptibility χ_m^{-1} (in SI units) vs. temperature T [82T3].

$EuCrO_3$ (see also LB III/4a, p. 227 and III/12a, p. 442)
II Magnetic susceptibility Fig. 106.
III ^{151}Eu Mössbauer parameters [84S3, see also 81G6]:

T [K]	δ*) [mm s^{-1}]	ΔE_Q [mm s^{-1}]	η
295.5	0.68(4)	− 5.87(31)	0.51(19)
77.4	0.72(4)	− 6.05(25)	0.56(21)

*) with respect to the source ^{151}Sm (SmF_3).

AFMR: Fig. 107

IV Optical absorption spectra [82G3]. Optical pumping with an energy above a certain threshold level influences the AFMR spectrum for some time after switching off the pumping power [80G1, 81G3, 82G3]. The pumping can induce a long-lived excitation of the Eu^{3+} ions, influencing their magnetic state. The Cr^{3+} ions retain their magnetic ordered Γ_4 (G_xF_z) structure, while the Eu^{3+} ions become magnetized and consequently antiferromagnetically ordered as a result of 4f-3d-4f exchange interaction. Fig. 108 shows the influence of optical pumping on the magnetization. Dielectric constant Fig. 109. The change in the polarization properties is related to the thermal excitation of the Jahn-Teller state of Eu^{3+} ions ($^7F_1$) with an activation energy of about $6 \cdot 10^3$ K [84S1]. For the influence of optical pumping on electric dipole ordering, see [81G2, 86G1].

$EuCr_{1-x}Fe_xO_3$

III Collapse of magnetic hyperfine structure in ^{57}Fe Mössbauer spectra at T_N = 185, 234, 330, 415, 497, 572 and 622 K for x = 0.4, 0.5, 0.6, 0.7, 0.8, 0.9 and 1.0, respectively. At 295 K the ^{151}Eu spectrum for x = 0.5 is fitted with a coupling constant $\Delta = -7.23(8)\,\mathrm{mm\,s^{-1}}$, an asymmetry parameter $\eta = 0.71(2)$ and a linewidth $2.34(2)\,\mathrm{mm\,s^{-1}}$ [83G2].

$Eu_{0.3}Er_{0.7}CrO_3$

I Saturation magnetization Fig. 151. Magnetization curves at low T, Fig. 110. (H-T) magnetic phase diagram Fig. 152.

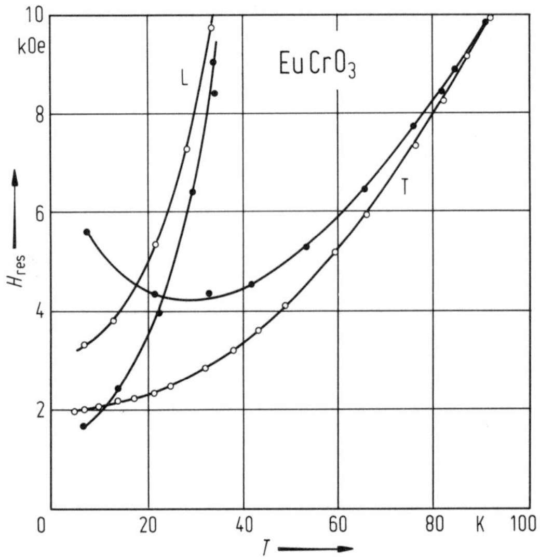

Fig. 107. $EuCrO_3$. The resonance field H_{res} parallel to the crystallographic c axis, vs. temperature T for the transverse (T) and the longitudinal (L) AFMR modes at ν = 210 GHz. Open circles before pumping, closed circles after laser pumping at 532 nm [81G3].

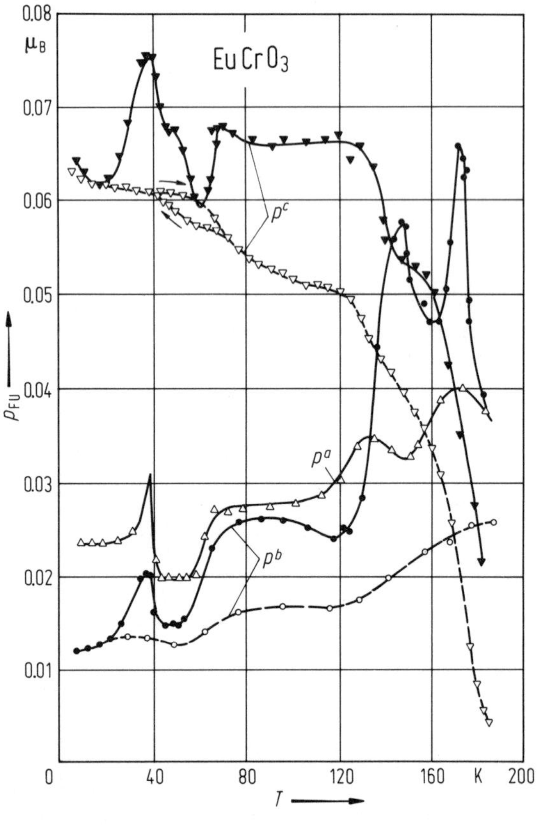

Fig. 108. $EuCrO_3$. Temperature dependence of the magnetic moment p_{FU} in a magnetic field of 5.7 kOe applied in the indicated crystallographic directions. Dashed curves before optical pumping, drawn curves after pulsed 532 nm wave length laser pumping [81G3].

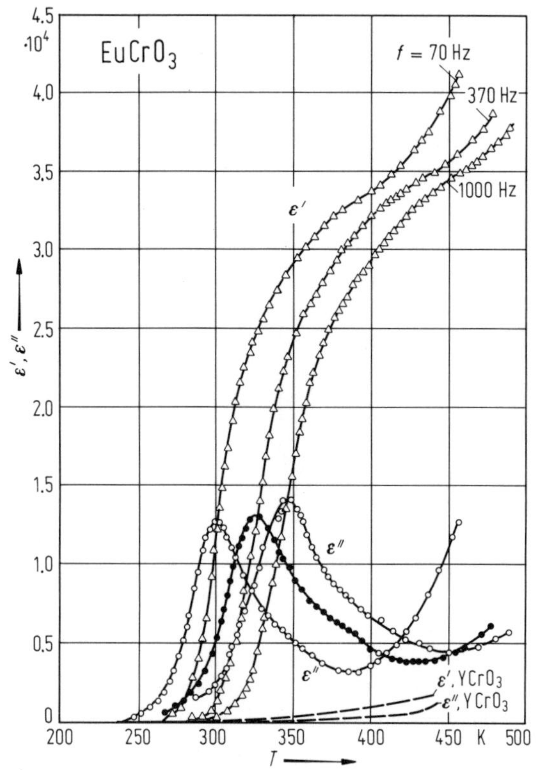

Fig. 109. $EuCrO_3$. Real and imaginary part, ε′ and ε″ respectively vs. temperature T for various frequencies f of a low amplitude electric field. For comparison the dielectric constants of $YCrO_3$, measured at 370 Hz are given [84S1].

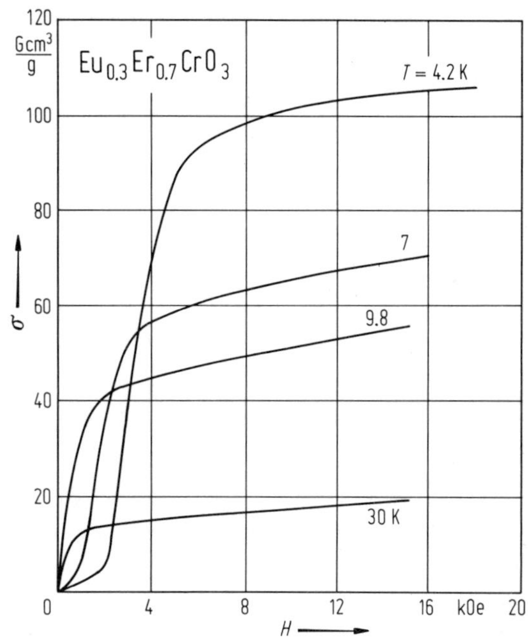

Fig. 110. $Eu_{0.3}Er_{0.7}CrO_3$. Magnetization curves at various low temperatures T [91A1].

$GdCrO_3$ (see also LB III/4a, p. 227 and III/12a, p. 444).

I, Weak ferromagnet. Magnetization Fig. 113; see also Fig. 14. [78W1] concludes from an analysis III of the magnetization properties that a $\Gamma_4(G_xA_yF_z;\ F_z) \rightarrow \Gamma_2(F_xC_yG_z;\ F_xC_y)$ spin rotation transition takes gradually place in the temperature region of 4 ··· 10 K. However NMR measurements show a transition temperature of 7 K, Fig. 111. The LT phase has a symmetric quadrupole splitting $\nu_Q = 1100$ kHz, the HT phase an asymmetric, 510 kHz between the upper and central line and 440 kHz between the central and lower line. The temperature for Γ_{42} $(G_{xz}F_{zx}) \rightarrow \Gamma_2$ (G_zF_x) spin transitions induced by a magnetic field have been derived from magnetostriction measurements, Fig. 112. For an analysis of the magnetic interactions see [85B1] and [78W1].

II Magnetic susceptibility Fig. 101. Curie-Weiss constants (SI units): $\Theta_p = -20$ K, $C_m = 111.1 \cdot 10^{-6}\,m^3\,K\,mol^{-1}$, $p_{eff} = 8.39\ \mu_B$/FU [82T3].

$GdCr_{1-x}Mn_xO_3$

I Spontaneous magnetization Fig. 113. A spin-reorientation temperature is not found [87K2].

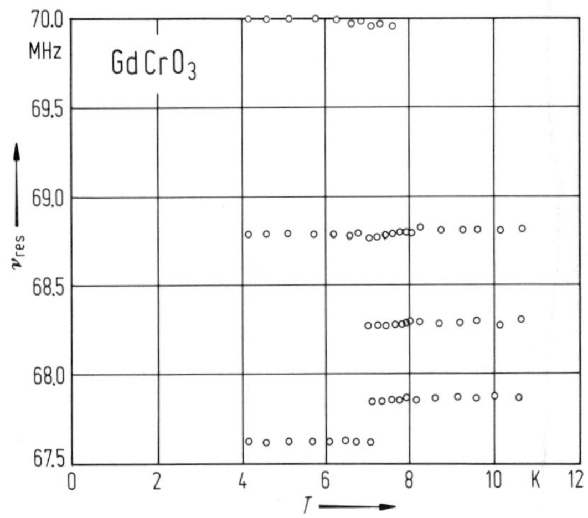

Fig. 111. $GdCrO_3$. Single crystals. ^{53}Cr NMR frequencies ν_{res} vs. temperature T, showing the three lines for the spin $J = 3/2$ [83K1].

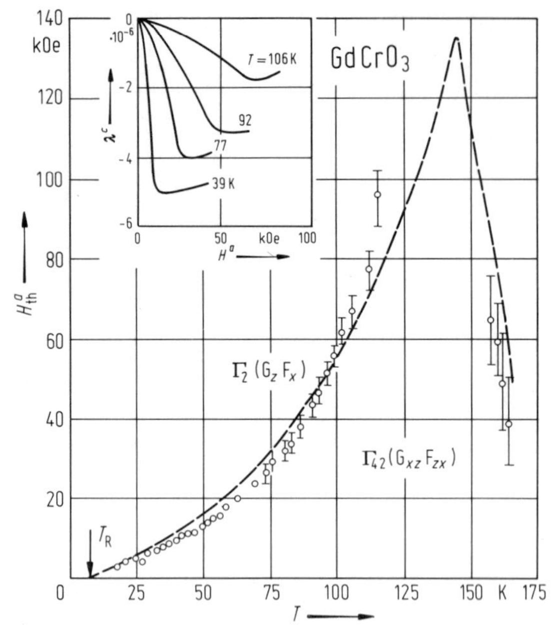

Fig. 112. $GdCrO_3$. Threshold field H^a_{th} in the crystallographic a direction for the spin transition $\Gamma_{42} \rightarrow \Gamma_2$ as a function of temperature. Dashed curve: calculated on the basis of a set of magnetic interaction parameters. Inset: typical magnetostriction curves for λ^c in the c direction as a consequence of a magnetic field H^a in the a direction [85B1].

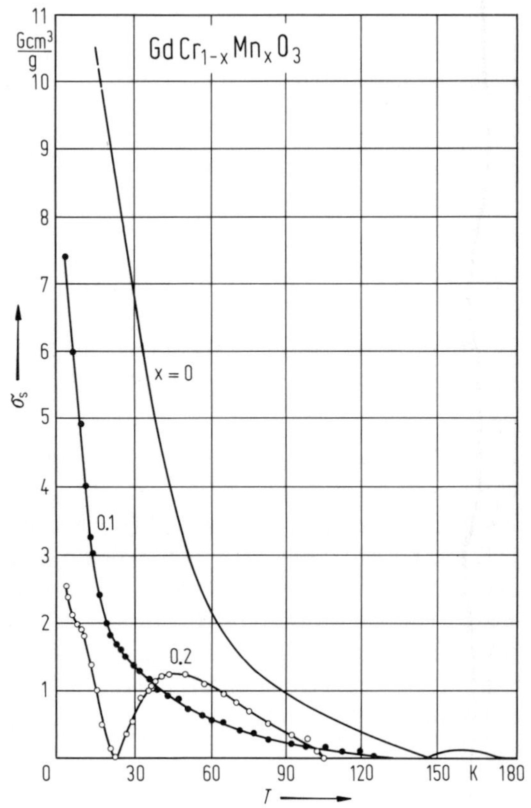

◀ Fig. 113. $GdCr_{1-x}Mn_xO_3$. Spontaneous mass magnetization σ_S vs. temperature T [87K2].

$TbCrO_3$ (see also LB III/4a, p. 227)

I Magnetization curves Figs. 97, 114, 115. Magnetic structure below $T_{N1} \approx 153$ K is Γ_2 (F_xG_z). Below $T_{N2} = 3$ K onset of antiferromagnetic ordering of Tb^{3+} moments. An orientational phase transition Γ_2 $(G_zF_x) \rightarrow \Gamma_4$ (G_xF_z) can be induced by a magnetic field in the c direction just below T_{N1}, see Fig. 116.

III AFMR: Fig. 117. An analysis of the data shows that the magnetic moment of the ground quasidoublet of Tb^{3+} is $p_{Tb^{3+}} = 7.7\ \mu_B$, oriented in the *ab* plane under an angle of 35.5° with the *a* axis. Exchange splitting of the quasidoublet is 1.6 K and splitting of the doublet in the crystal field $\leq 6 \cdots 7$ K [89Z1], see also [76G2].

IV For optical absorption spectra, see [85K1].

$TbCr_{1-x}Fe_xO_3$ (see LB III/12a, p. 456)

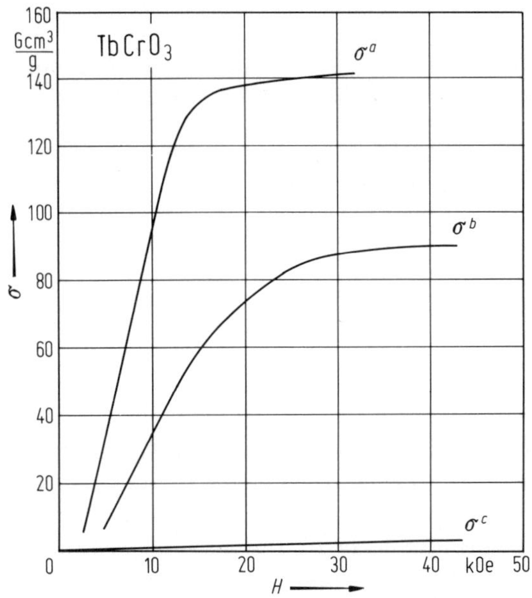

Fig. 114. $TbCrO_3$. Magnetization σ vs. applied magnetic field H for three crystallographic directions at T = 1.65 K [89Z1].

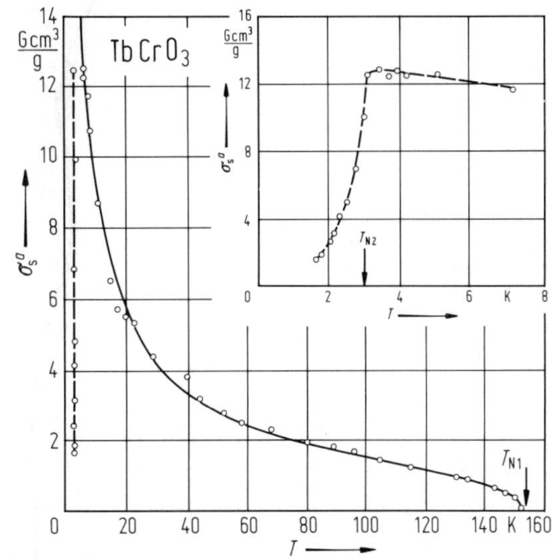

Fig. 115. $TbCrO_3$. Temperature dependence of the spontaneous mass magnetization σ_s^a along the *a* axis. The magnetic moments measured along the other axes vanish after extrapolation to zero field strength. The drawn curve is the result of an analysis on the basis of magnetic interactions. The inset shows the enlarged temperature region where the Tb^{3+} subsystem orders antiferromagnetically [89Z1].

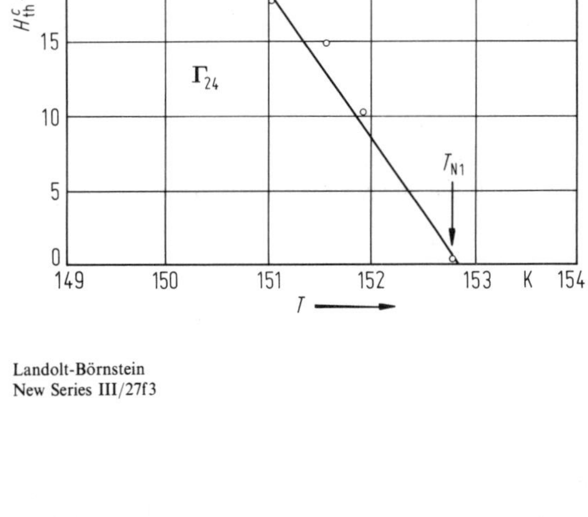

Fig. 116. $TbCrO_3$. Magnetic (H–T) phase diagram for a magnetic field applied along the *c* axis of the crystal [89Z1].

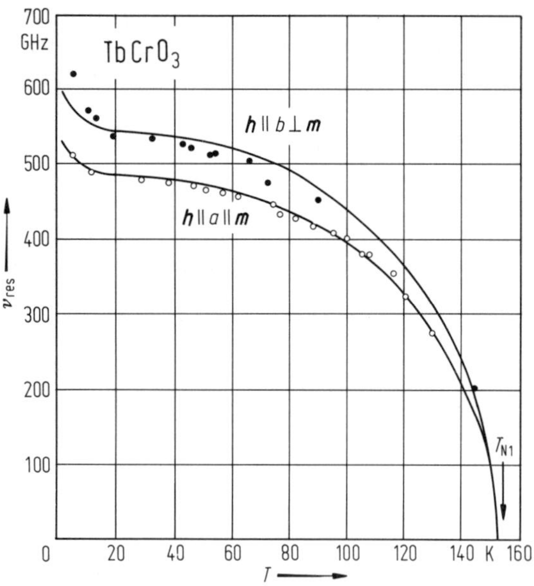

Fig. 117. $TbCrO_3$. Plane parallel *c*-cut crystal. Antiferromagnetic resonance mode frequencies (ν_{res}). Quasiferromagnetic; $\boldsymbol{h} \| b \perp \boldsymbol{m}$; quasiantiferromagnetic mode: $\boldsymbol{h} \| a \| \boldsymbol{m}$ [89Z1]. *a* and *b* crystallographic directions, ***h*** magnetic vector of radiation, ***m*** magnetization. The curve is the result of a model calculation.

$DyCrO_3$ (see also LB III/4a, p. 227 and III/12a, p. 444)

I The Cr system is in a weakly ferromagnetic state Γ_2 (G_zF_x) below T_{N1} = 140 K. Magnetization curves Figs. 118, 119. Saturation magnetization at 4.2 K is 88 and 175 $G\,cm^3\,g^{-1}$ for the *a* and *b* crystallographic directions, respectively. The magnetic specific susceptibility in high fields along the *c* axis is $\chi_g^c = 1.8 \cdot 10^{-4}\,cm^3\,g^{-1}$ [87K3]. The Ising axis of the Dy^{3+} ions in the *ab* plane makes an angle $\pm\alpha = \pm 63°$ with the *a* axis with $p_{Dy^{3+}} = 9.1\,\mu_B$. At T_{N2} = 2 K the Dy^{3+} moments order in the G_xA_y (Γ_5) mode. Magnetic phase diagram Fig. 121. A sufficiently strong magnetic field in the *z* direction causes a spin reorientation $G_{zx}F_{xz} \rightarrow G_xF_z$. A detailed analysis of the magnetic interactions is given. The splitting of the ground doublet of the Dy^{3+} ion in the Γ_2 phase is caused by a magnetic field of 3.5 kOe [87K3].

II For the susceptibilities, see Fig. 120.

$DyCr_{0.92}Al_{0.08}O_3$

I Magnetization curves Figs. 122, 123, magnetic phase diagrams Fig. 124. Weak ferromagnetism parallel to *a* axis, magnetic structure Γ_2 (G_zF_x). For $T < 18$ K a reorientation to an antiferromagnetic Γ_1 (G_y) state takes place [88Z1]. In the region $1.5 < T < 18$ K a strong magnetic field parallel to the *a* axis can induce a spin reorientation Γ_1 (G_y) $\rightarrow \Gamma_2$ (G_zF_x).

$DyCr_{1-x}Mn_xO_3$

I Magnetization curve Fig. 125. Below T_N the compounds exhibit a G_xF_z magnetic structure. Below T_M = 9.5 and 23 K for x = 0.065 and 0.083, respectively, the weak ferromagnetism disappears and a G_y antiferromagnetic phase occurs [86K1].

$DyCr_xFe_{1-x}O_3$

I Magnetization curves Figs. 126, 127. Reorientation transition from the weak ferrimagnetic high-temperature Γ_4 to the low temperature Γ_1 configuration. This Morin transformation can be very complex [84A1].

$Dy_xEr_{1-x}CrO_3$

I Spontaneous magnetization curve for x = 0.3 in Fig. 128. For $T < T_N = 127$ K weak ferromagnetism Γ_2 (G_zF_x). At $T_R = 14$ K spin reorientation from the *a* to the *c* axis ($\Gamma_2 \rightleftarrows \Gamma_4$), and at $T_M = 5.9$ K antiferromagnetism occurs (first order Morin-type transition $\Gamma_4 \rightleftarrows \Gamma_1$). For x = 0.2 was found $T_R = 17$ K and $T_M = 7.8$ K [88K1]. ($H-T$) phase diagrams based on magnetization and magnetostriction measurements Fig. 129.

$Dy_{1-x}Y_xCrO_3$

I Spontaneous magnetization Fig. 130. For x = 0.89 and 0.79 the spontaneous spin-reorientation transitions $\Gamma_4 \rightarrow \Gamma_2$ ($G_xF_z \rightarrow G_zF_x$) occur in the temperature ranges 3 ··· 1.7 K and 15 ··· 8 K, respectively. Threshold fields for spin-reorientation transition in the *ac* crystal plane are given in Fig. 131.

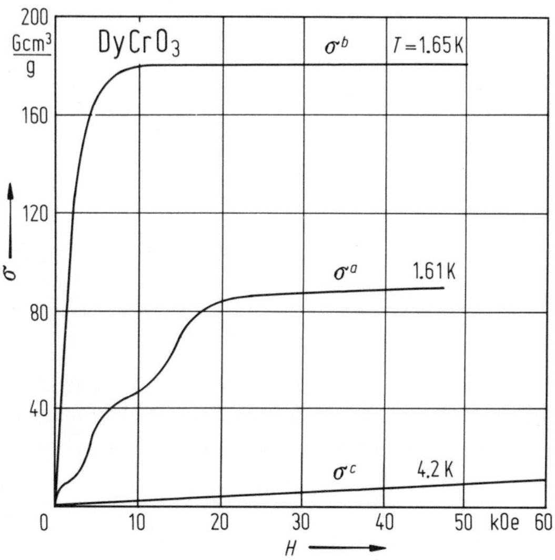

Fig. 118. $DyCrO_3$. Specific magnetization σ vs. applied magnetic field H at low temperature for the three main crystallographic directions [87K3].

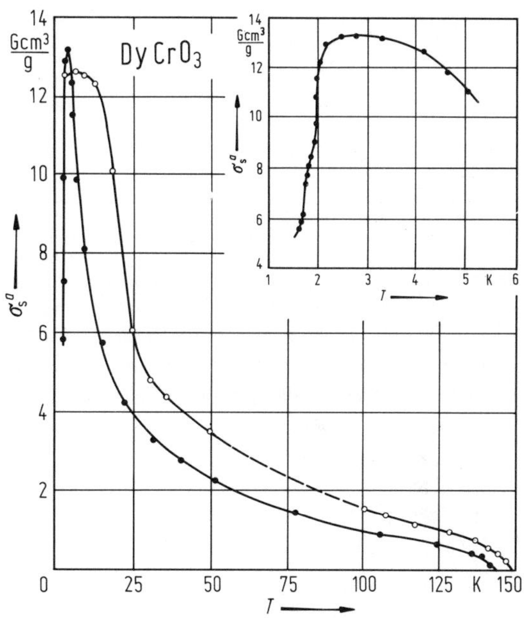

Fig. 119. $DyCrO_3$. Spontaneous mass magnetization σ_s^a along the *a* axis in the Γ_2 phase as a function of temperature *T*. Full circles [87K3], open circles [74T1].

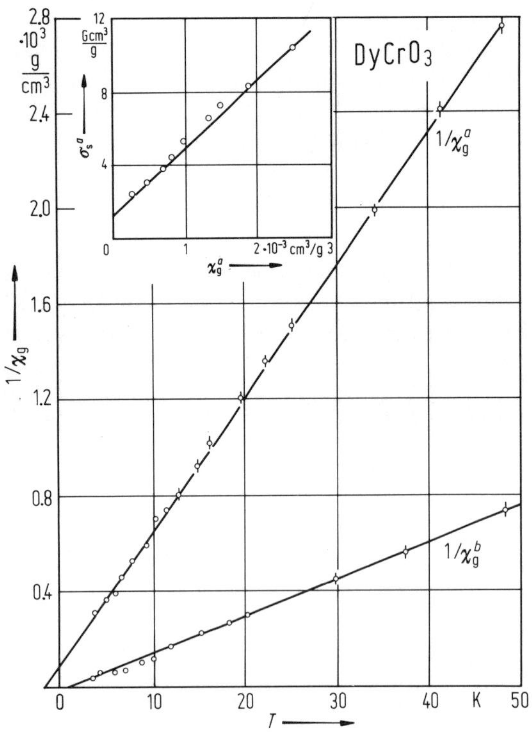

Fig. 120. $DyCrO_3$. Temperature dependence of the reciprocal mass susceptibility χ_g^{-1} measured along the crystallographic *a* and *b* directions. Inset shows the spontaneous mass magnetization σ_s^a in the *a* direction vs. the corresponding susceptibility χ_g^a [87K3].

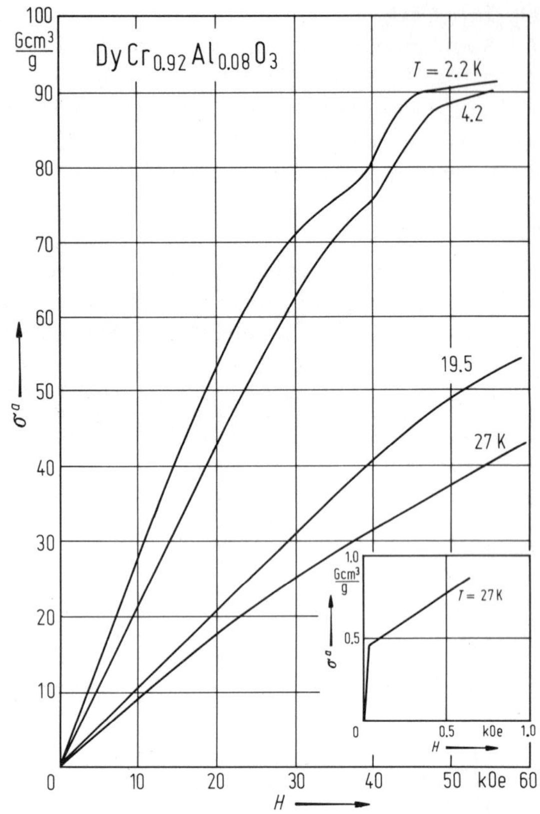

Fig. 122. $DyCr_{0.92}Al_{0.08}O_3$. Mass magnetization curves at various temperatures *T* for applied fields *H* in the direction of the crystallographic *a* axis [88Z1].

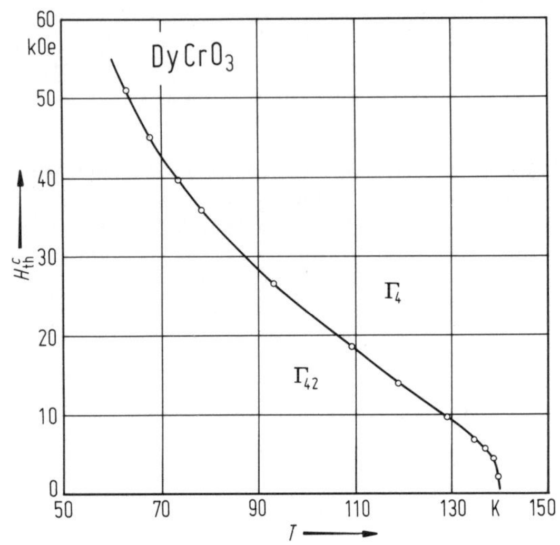

Fig. 121. $DyCrO_3$. The magnetic threshold field H_{th}^c for the phase transition $\Gamma_{42} \rightarrow \Gamma_4$ as a function of temperature [87K3].

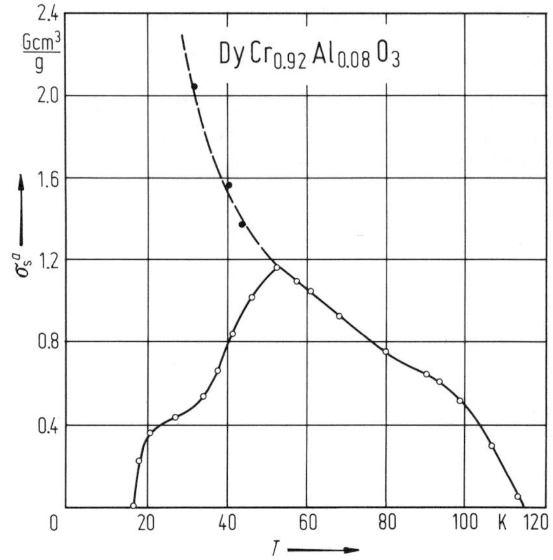

Fig. 123. $DyCr_{0.92}Al_{0.08}O_3$. The spontaneous mass magnetization σ_s^a along the crystallographic *a* axis obtained by extrapolation of the isotherms of Fig. 122. Broken curve: field induced magnetic moment $T_{N1} = 112$ K [88Z1].

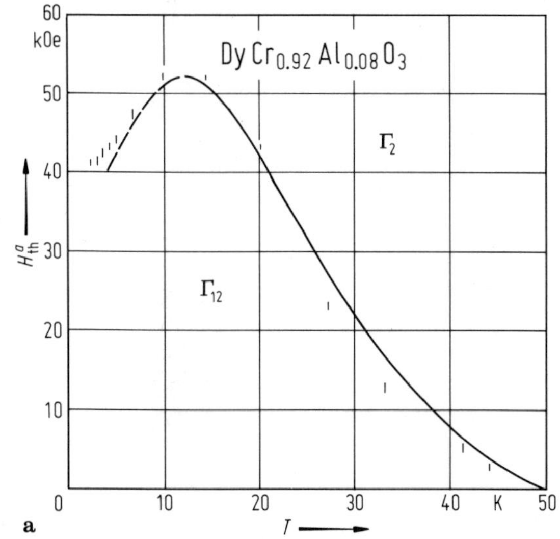

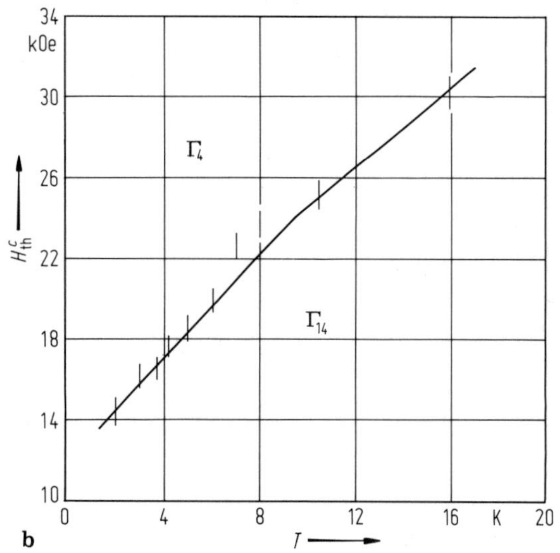

Fig. 124. $DyCr_{0.92}Al_{0.08}O_3$. Magnetic phase diagrams. (a) Magnetic field parallel *a* axis, (b) magnetic field parallel *c* axis. Bars represent experimental results. Curves calculated for a first order phase transition (dashed) and a second order phase transition (fully drawn) [88Z1].

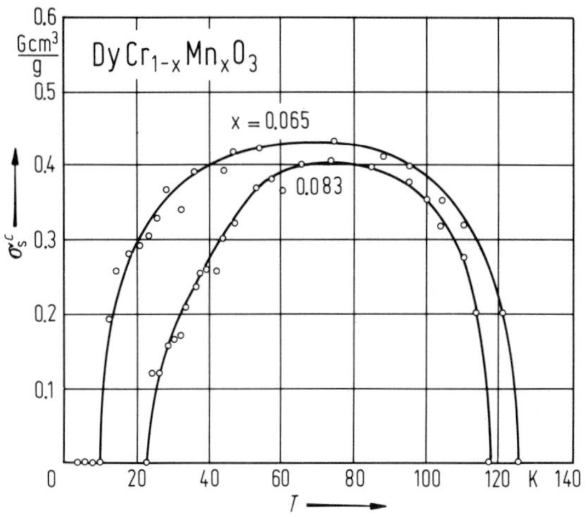

Fig. 125. $DyCr_{1-x}Mn_xO_3$. Spontaneous mass magnetization, σ_s^c, in the crystallographic *c* direction as a function of temperature [86K1].

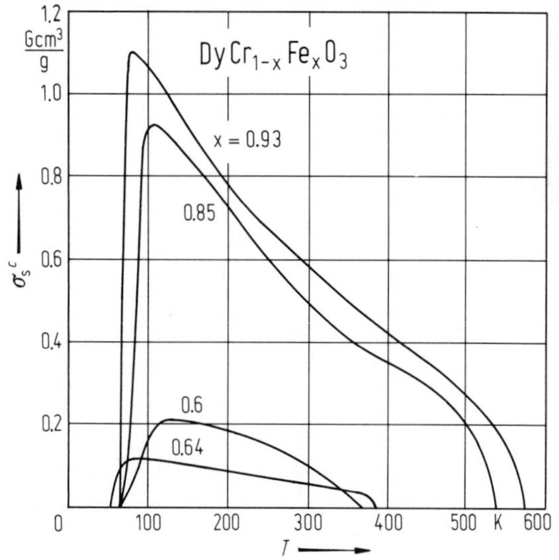

Fig. 126. $DyCr_{1-x}Fe_xO_3$. Spontaneous mass magnetization, σ_s^c, along the *c* axis vs. temperature [84A1].

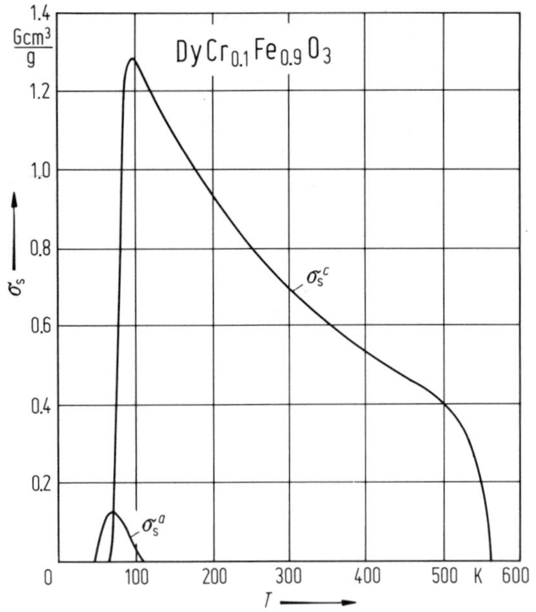

Fig. 127. $DyCr_{0.1}Fe_{0.9}O_3$. Spontaneous mass magnetizations, σ_s^c and σ_s^a, along the c and a axis, respectively, vs. temperature [84A1].

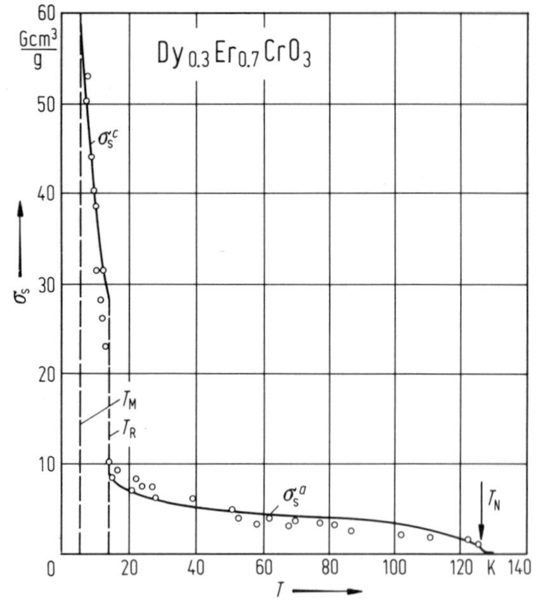

Fig. 128. $Dy_{0.3}Er_{0.7}CrO_3$. Spontaneous mass magnetizations, σ_s^a and σ_s^c, along the direction of the a and c axis, respectively, vs. temperature T. Curve calculated on the basis of a one-doublet approximation for the rare earth ions [88K1].

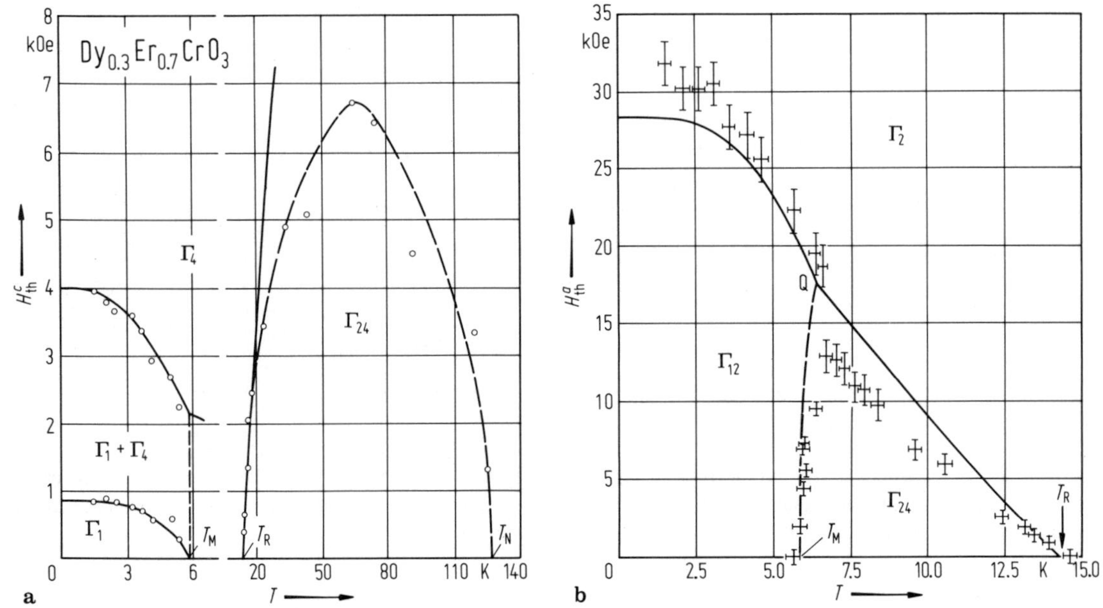

Fig. 129. $Dy_{0.3}Er_{0.7}CrO_3$. Magnetic structure phase diagram for applied magnetic fields in (a) the crystallographic c direction. Circles and dashed line experimental results. Drawn lines result from calculations based on anisotropy of the exchange splitting of the ground doublet of Dy^{3+} and Er^{3+}. The vertical dashed line represents T_M. (b) For $\boldsymbol{H}$ in the crystallographic a direction. Drawn and dashed curves are calculated second order and first order transitions, respectively. Q is a bicritical point [88K1].

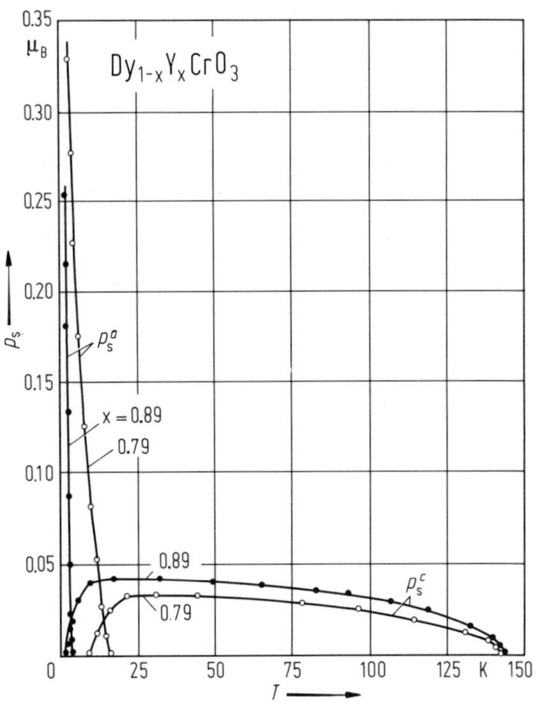

Fig. 130. $Dy_{1-x}Y_xCrO_3$. Spontaneous magnetic moments per FU in the crystallographic *a* and *c* directions, p_s^a and p_s^c, respectively, as a function of temperature *T* [87B3].

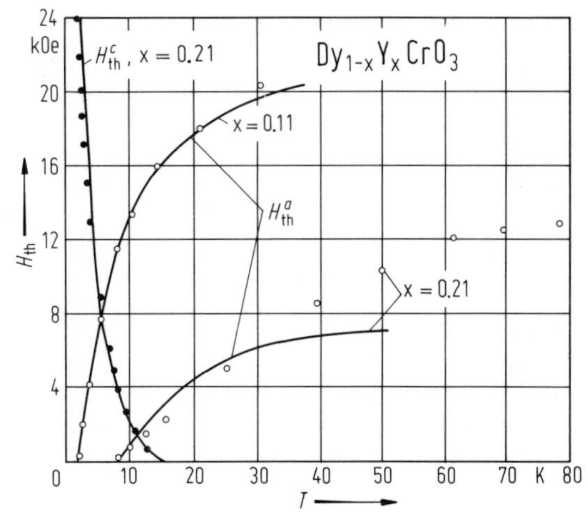

Fig. 131. $Dy_{1-x}Y_xCrO_3$. Threshold fields H_{th}^a and H_{th}^c in the crystallographic *a* and *c* direction, respectively, as a function of temperature, curves calculated [87B3].

$HoCrO_3$ (see also LB III/4a, p. 229 and III/12a, p. 444)

I Below $T_N = 140$ K the weak ferromagnetism is a consequence of the Γ_2 (G_zF_x) magnetic structure of the Cr^{3+} sublattice. The compatible Ho^{3+} sublattice structure is C_yF_x [72H1]. Magnetization curves Figs. 132 ··· 134. Neutron diffraction analysis confirms temperature independent canting angles α and β (see Fig. 135): $\alpha = 3.2(5)°$, $\beta = 25(1)°$ with $p_{Cr^{3+}} = 2.94\ \mu_B$ and $p_{Ho^{3+}} = 7.75\ \mu_B$. Threshold field for a phase transition $\Gamma_2 \rightarrow \Gamma_4$, caused by a rearrangement of the directions of the Cr^{3+} spins from Γ_2 (G_zF_x) to Γ_4 (G_xF_z), as derived from optical absorption spectra, Fig. 136. Magnetic and crystal field parameters Table 26. Magnetic phase diagram Fig. 138.

IV Optical absorption spectra [76C3]. Temperature dependence of the ground state splitting of Ho^{3+}, Fig. 137 and for the influence of a magnetic field on this splitting, see [76C3, 82K3].

$Ho_{0.07}Y_{0.93}CrO_3$

I Spontaneous magnetization Fig. 132. Magnetic structure Figs. 138, 139.

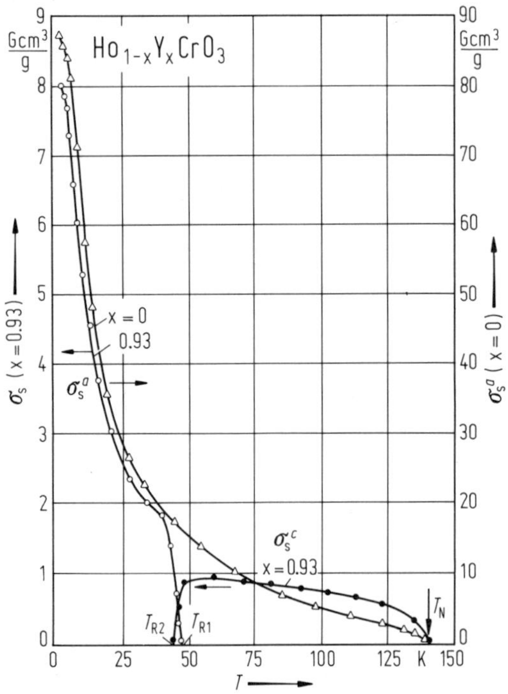

Fig. 132. $HoCrO_3$; $Ho_{0.07}Y_{0.93}CrO_3$. Temperature dependence of the spontaneous mass magnetizations σ_s^a and σ_s^c along the crystallographic a and c axis, respectively, as derived from magnetization isotherms along these axes [87Z1].

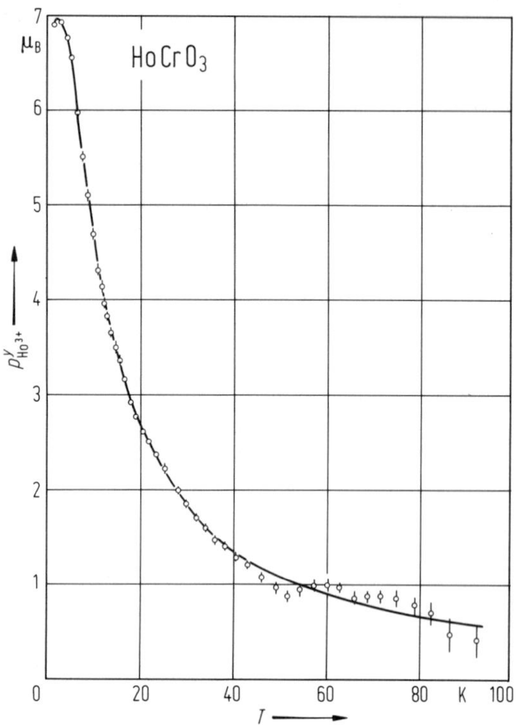

Fig. 134. $HoCrO_3$. The antiferromagnetic component of the Ho^{3+} moment $p^y_{Ho^{3+}}$ in the y direction as derived from neutron diffraction spectra. Curve best theoretical fit [77S3].

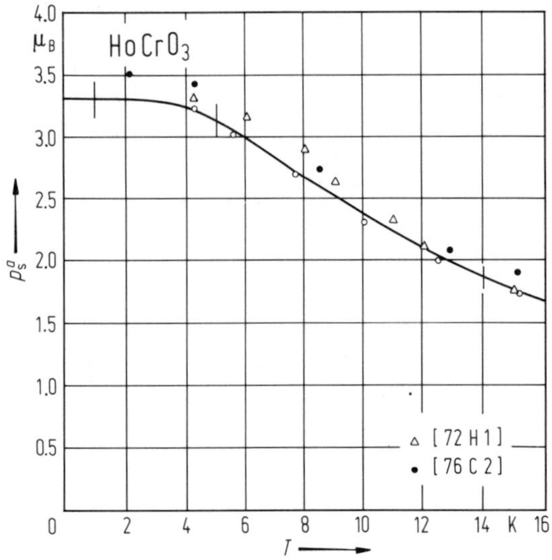

Fig. 133. $HoCrO_3$. Low temperature spontaneous magnetic moment p_s^a per FU along the a axis derived from magnetic measurements. The curve and open circles are the result of calculations on the basis of optical absorption data [76C2].

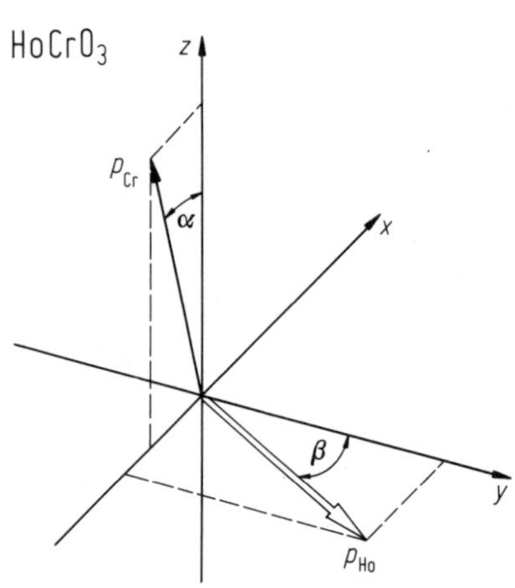

Fig. 135. $HoCrO_3$. Definition of canting angles α and β [77S3].

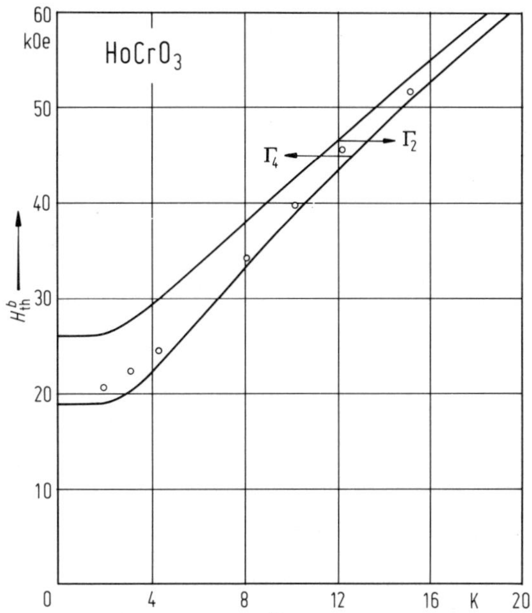

Fig. 136. $HoCrO_3$. Threshold field H^b_{th} in the direction of the b axis vs. temperature T for the phase transition $\Gamma_2 \leftrightarrows \Gamma_4$, as derived from optical measurements. The curves represent the stability boundaries for both phases and have been calculated using the data of [76C3, see also 82K3] in Table 26. A field $H = 60$ kOe along the c axis did not induce a $\Gamma_2 \rightarrow \Gamma_4$ reorientation [87Z1]. See also Fig. 138 for similar results obtained from magnetization measurements.

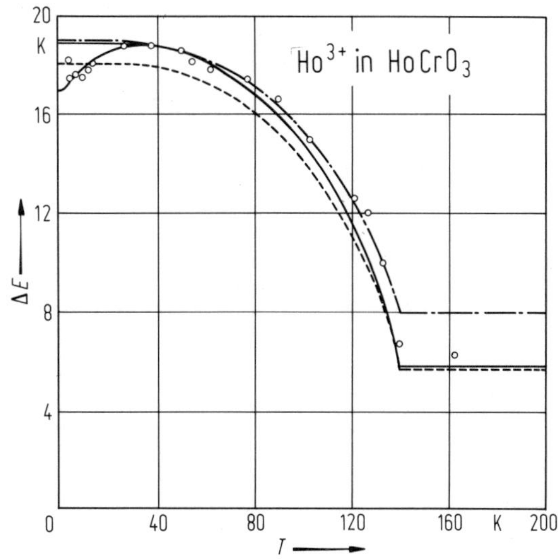

Fig. 137. $HoCrO_3$. Temperature dependence of the ground state splitting ΔE of Ho^{3+} as deduced from optical absorption measurements. Curves calculated according to various sets of doublet splitting parameters [76C3]; see also [77S3] for similar results derived from elastic neutron diffraction.

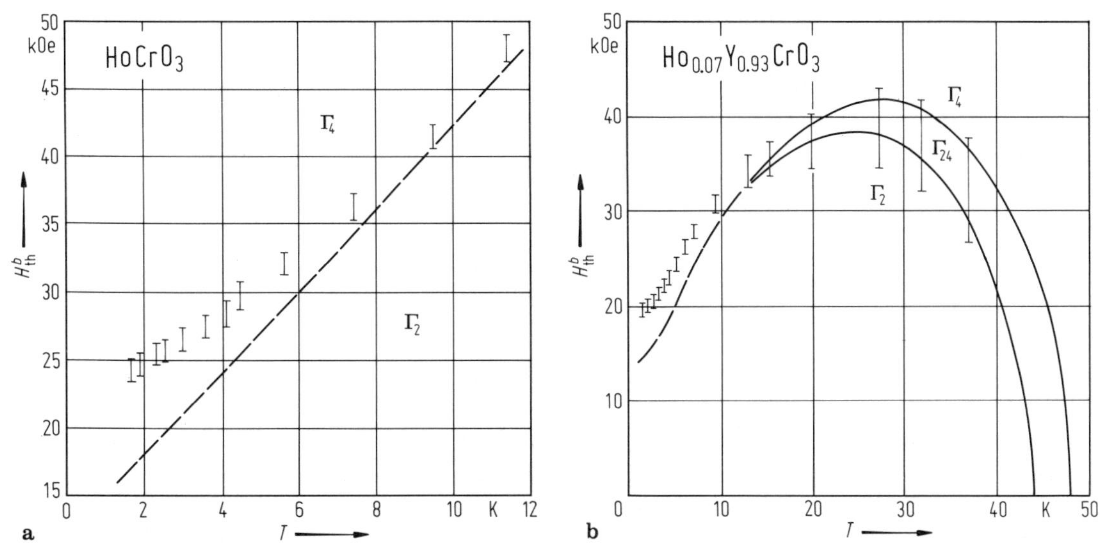

Fig. 138. $HoCrO_3$; $Ho_{0.07}Y_{0.93}CrO_3$. Magnetic phase diagrams derived from magnetization and magnetostriction measurements. Curves calculated on the basis of a simple one-doublet model. Continuous curves represent second order phase transitions, the dashed curves apply to first order transitions [87Z1].

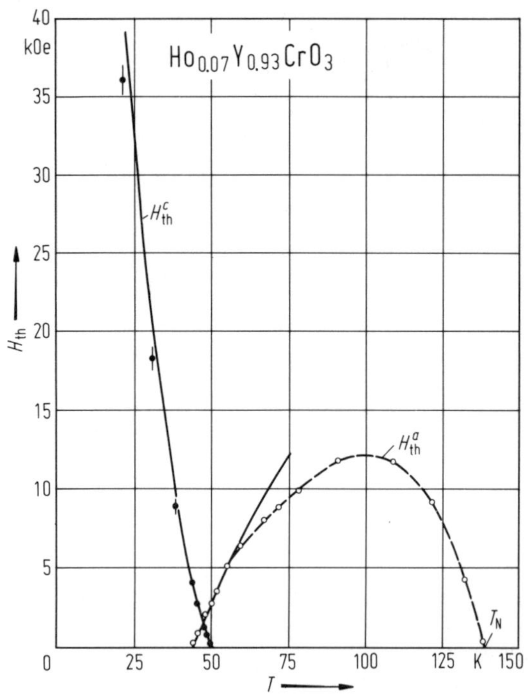

Fig. 139. $Ho_{0.07}Y_{0.93}CrO_3$. Threshold fields H_{th}^a for transitions $\Gamma_4 \rightarrow \Gamma_2$ and H_{th}^c for transitions $\Gamma_2 \rightarrow \Gamma_4$ of the magnetic phases for magnetic fields applied along the *a* and *c* axis of the crystal, respectively. Curves calculated [87Z1].

Table 26. $HoCrO_3$. Magnetic and crystal field parameters of the ground "accidental" doublet of Ho^{3+}. *D* crystal field splitting; Δ_0 magnetic splitting at $T = 0$ K (see Table 25); ΔE total splitting at $T = 0$ K, $g_{x,y}$ components of *g* vector.

	Elastic neutron diffraction [77S3]	Single crystal magnetization [72H1]	Powder magnetization [66B1] Specific heat [70P1] Susceptibility [from 77S3]	Optics [76C3]	Inelastic neutron scattering [77S4]
D [K]	4.1(3)	4(2)		5.7(3)	5.3(8)
Δ_0 [K]	15.5(5)	14.7(2)		16.1(3)	17.8(12)
ΔE [K]	16.9(2)	15.2(2)	16	17.1(3)	18.6(12)
g_x	6.8(3)	7.3(2)	6.8	7.0(3)	
g_y	14.3(5)	15.7(3)	14.0	15.0(3)	
g	15.8(6)	17.3(3)	15.4	16.6(3)	

$ErCrO_3$ (see also LB III/4a, p. 229 and III/12a, p. 445)

I Below $T_N \approx 133$ K the weak ferromagnet has the symmetry Γ_4 ($G_xA_yF_z$; F_z); at $T_R \approx 9.4(5)$K, [75C1], a spin reorientation occurs to the antiferromagnetic phase Γ_1 ($A_xG_yC_z$; C_z).

Magnetic phase diagram Figs. 140 ··· 142. See also Fig. 152. Spontaneous magnetization Figs. 143, 144. High-field susceptibilities Figs. 145, 146. For the susceptibility behaviour in the neighborhood of $T_N = 132.9$ K, see [81M3].

II Paramagnetic susceptibility Fig. 147.

III NMR data Fig. 148. For NMR spectra of ^{53}Cr at $T = 1.7$ K in the presence of an external magnetic field [88K2]. Domain walls have a Bloch-type character with the rotation of the sublattice magnetization in the (*ac*) plane [80S1].

IV Optical absorption spectra with and without an applied magnetic field [75C1, 75H1, 77K2, 79T1, 85K1]. Ground doublet splitting of Er^{3+}, Fig. 149. The magnetic phase transition $\Gamma_1 \rightarrow \Gamma_4$ can be photo-induced [81K1, 82T4, 83T1, 87K4].

For a survey of various parameters of $ErCrO_3$, see Table 27.

$ErCr_{0.95}Mn_{0.05}O_3$

I Spontaneous magnetization Fig. 150.

$Er_{0.7}Lu_{0.3}CrO_3$

I Saturation magnetization Fig. 151.

$Er_{1-x}Y_xCrO_3$

I Saturation magnetization Fig. 151. Morin temperature $T_M = 7.5$ K and 6.3 K for x = 0.2 and 0.3, respectively [91A1]. (*H-T*) phase diagram Fig. 152.

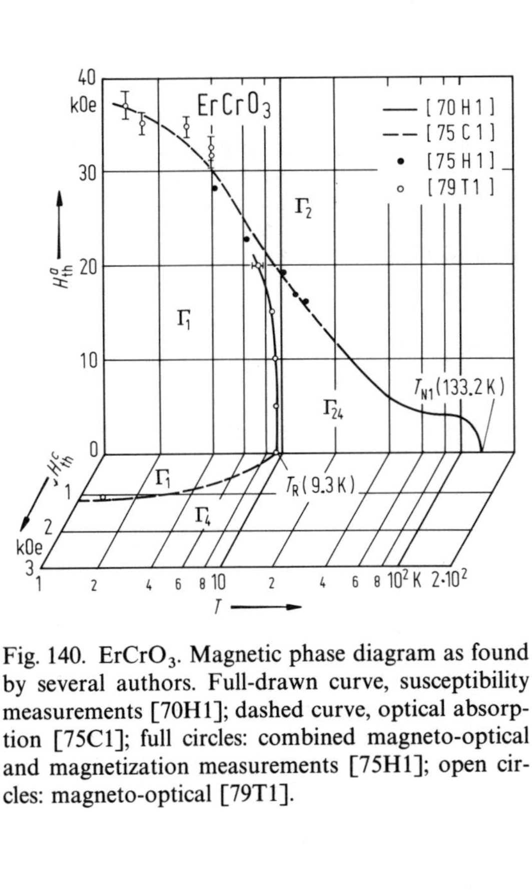

Fig. 140. $ErCrO_3$. Magnetic phase diagram as found by several authors. Full-drawn curve, susceptibility measurements [70H1]; dashed curve, optical absorption [75C1]; full circles: combined magneto-optical and magnetization measurements [75H1]; open circles: magneto-optical [79T1].

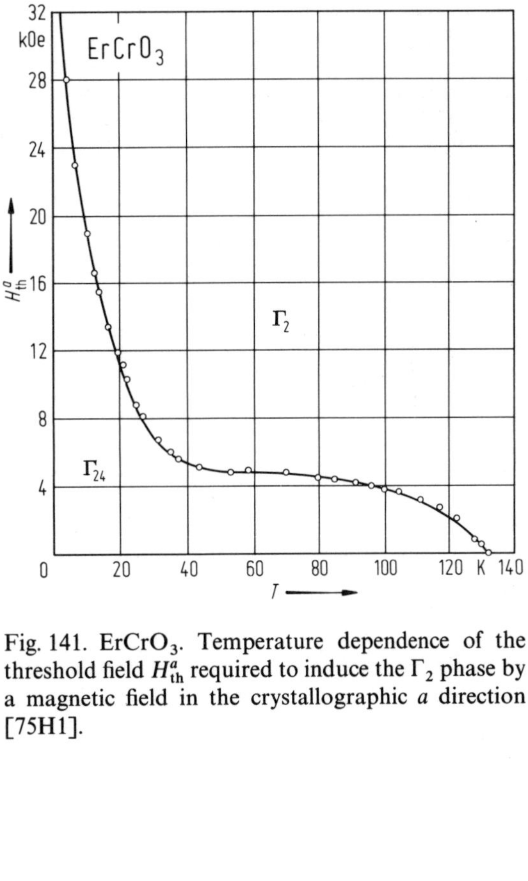

Fig. 141. $ErCrO_3$. Temperature dependence of the threshold field H^a_{th} required to induce the Γ_2 phase by a magnetic field in the crystallographic *a* direction [75H1].

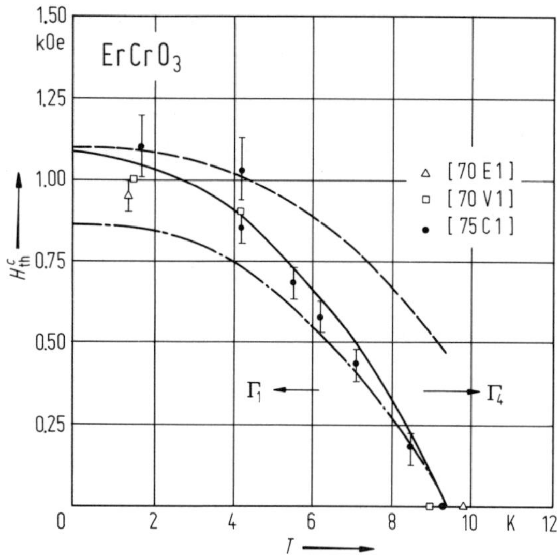

Fig. 142. $ErCrO_3$. Γ_1 phase. Temperature dependence of the threshold field H^c_{th} required to induce the Γ_4 phase by a magnetic field in the crystallographic c direction as derived from optical absorption properties. The dashed curves represent several calculated fits [75C1]. Also shown earlier results based on magnetization data.

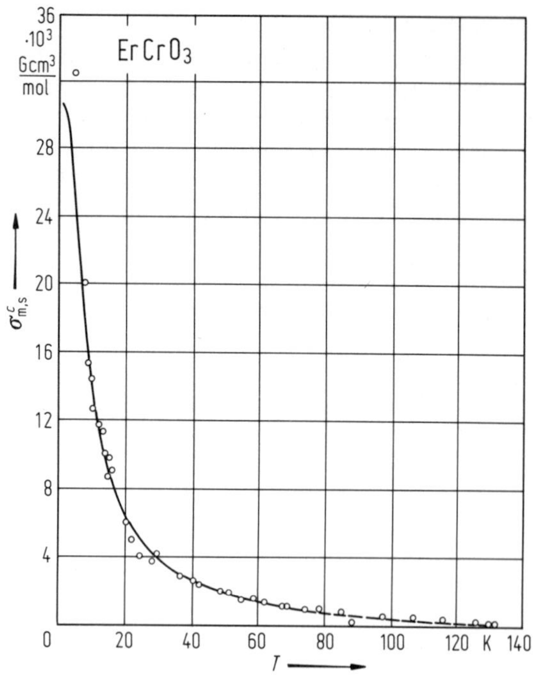

Fig. 143. $ErCrO_3$. Γ_4 phase. Spontaneous molar magnetization $\sigma^c_{m,s}$ in the crystallographic c direction as a function of temperature T. The data for $T < T_R$ were obtained by extrapolation from high magnetic fields parallel to the c axis. The drawn curve is a calculated fit, the dashed curve an extrapolation [75H1].

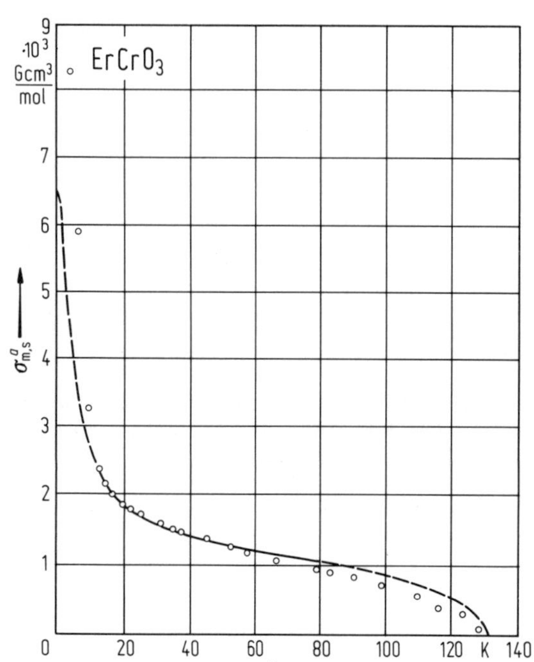

Fig. 144. $ErCrO_3$, Γ_2 phase. Temperature dependence of the spontaneous molar magnetization, $\sigma^a_{m,s}$ obtained by extrapolating high field magnetization curves to zero field strength. Drawn curve: calculated fit; dashed curve: extrapolation [75H1].

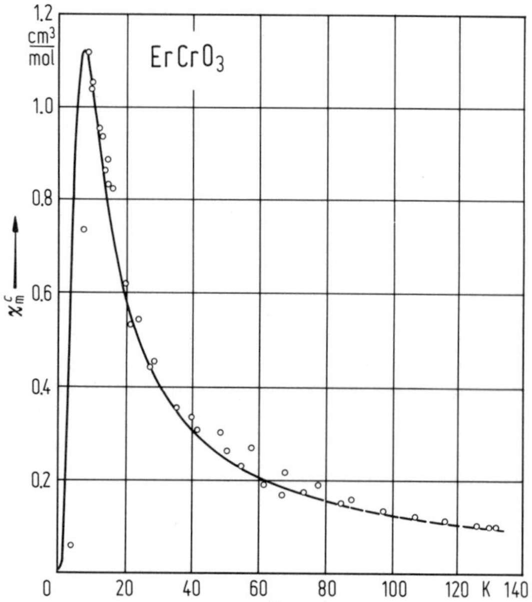

Fig. 145. $ErCrO_3$, Γ_4 phase. Molar susceptibility, χ^c_m, measured as the slope of the magnetization curve at high field strengths in the crystallographic c direction. For $T < T_R$ the curves were extrapolated to zero field strength. Curve: calculated best fit [75H1], see also [75C1].

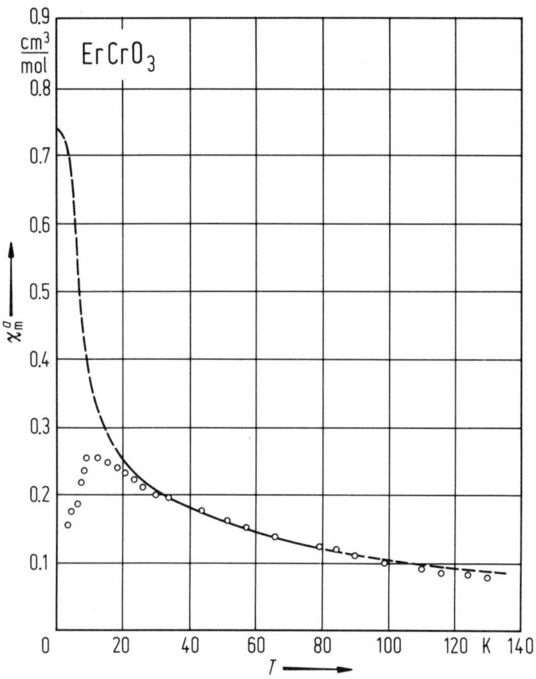

Fig. 146. $ErCrO_3$. Γ_2 phase. Similar to Fig. 145 but now for applied fields in the crystallographic a direction [75H1]. Dashed line extrapolation of calculated curve.

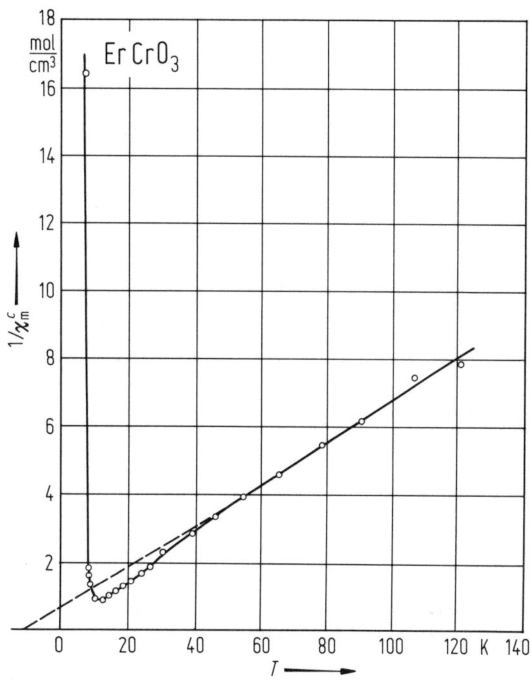

Fig. 147. $ErCrO_3$. Inverse molar susceptibility, $1/\chi_m^c$, measured along the crystallographic c direction as a function of temperature T, data from Fig. 145 [80G2].

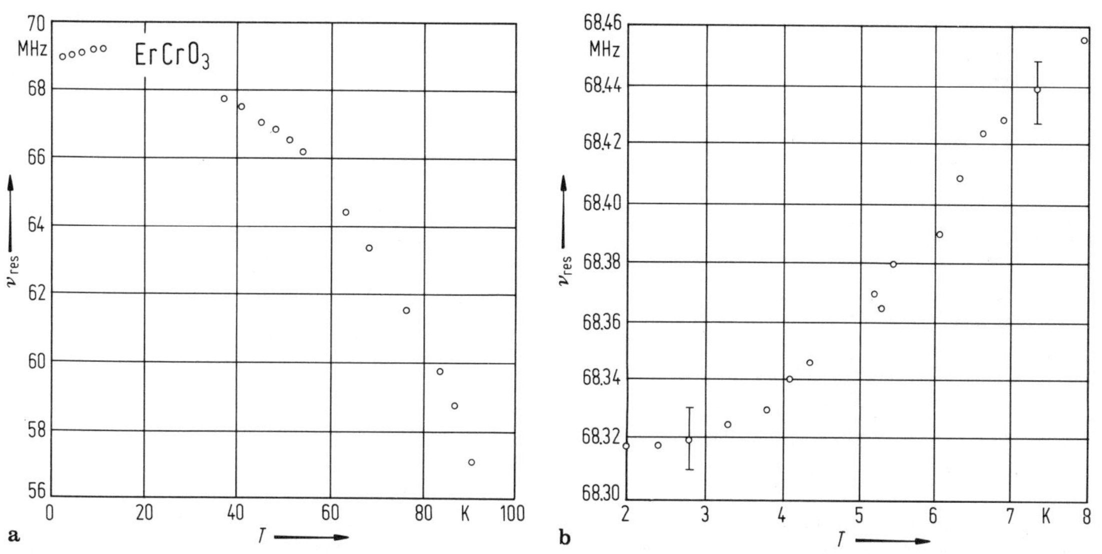

Fig. 148. $ErCrO_3$. Temperature dependence of the ^{53}Cr NMR frequency ν_{res}. (a) [79K1], (b) [78K3].

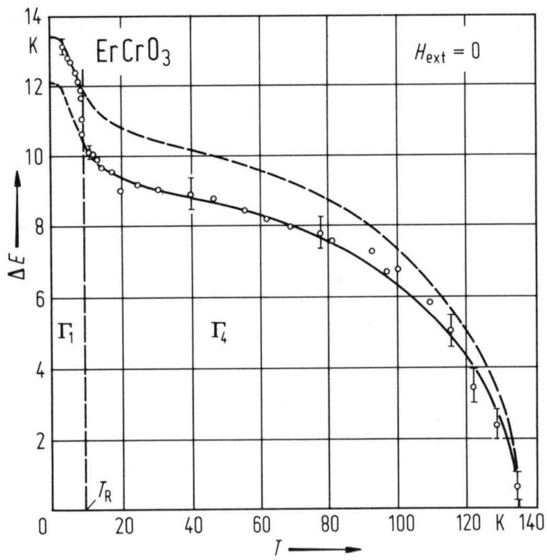

Fig. 149. $ErCrO_3$. Ground-doublet splitting, ΔE, in Γ_1 and Γ_4 phases as a function of temperature. Drawn curves calculated fits [75C1], for similar results [75H1].

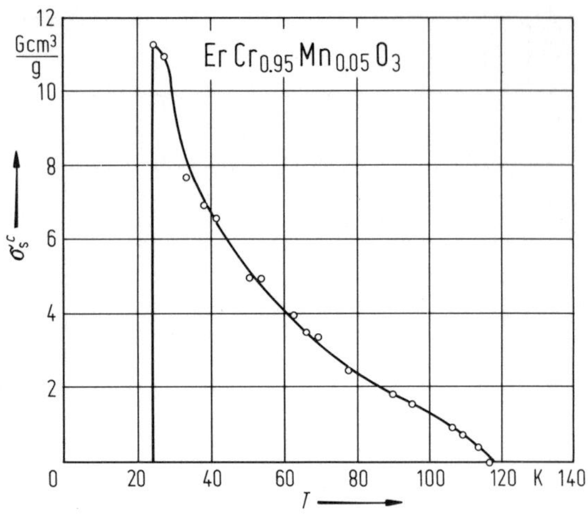

Fig. 150. $ErCr_{0.95}Mn_{0.05}O_3$. Spontaneous mass magnetization in the crystallographic c direction σ_s^c, vs. temperature T. At $T_R = 23$ K the $G_xF_z \rightarrow G_y$ Morin-type transformation takes place [87K2].

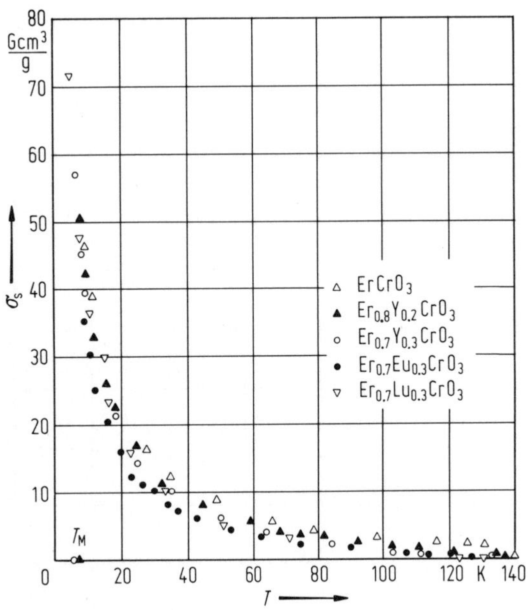

Fig. 151. $ErCrO_3$ with various substitutions of non-magnetic RE ions. Spontaneous mass magnetization σ_s vs. temperature T [91A1].

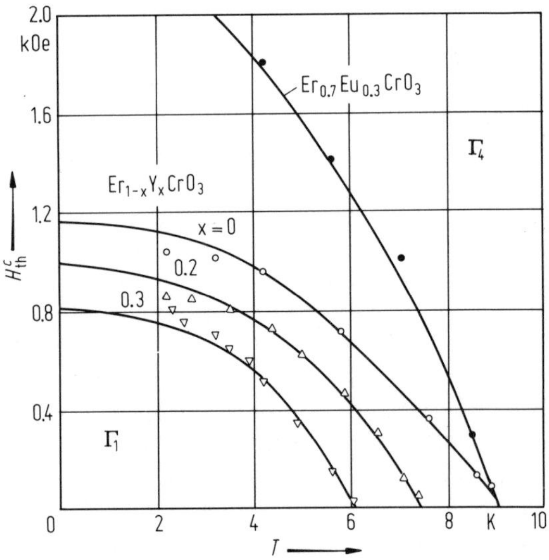

Fig. 152. $ErCrO_3$, diluted with non-magnetic RE ions. Magnetic phase diagram for the orientational transitions of the Morin type $\Gamma_4(G_xF_z) \rightarrow \Gamma_1(G_y)$ for magnetic fields applied along the c axis. The continuous curves are calculated [91A1].

Table 27. $ErCrO_3$. Magnetic and crystal field parameters derived from optical absorption spectroscopy studies. T_R spin reorientation temperature; $g_{z,x}$ components of g-factor of Er^{3+} along the axes; α_0 canting angle of Cr^{3+} spin system; ΔE_α ground-doublet splitting in Γ_4 phase; ΔE_β ground-doublet splitting in Γ_1 phase; ΔK difference in anisotropy energy of Γ_4 and Γ_1 phase at T_R; $T_R \Delta S$ latent heat associated with SR; $E_{II \to V}$ crystal field splittings between Kramers doublet and elevated doublets.

Property	Unit	References [75H1]	[75C1]	[72C1]
T_R	K	9.3(5)	9.4(5)	9.4(5)
g_z		11.3(3)	11.3(4)	11.1(9)
g_x		5.9(6)		
α_0	mrad	− 60(20)		
$\Delta E_\alpha(T)$	K	10.3(2) at 9.3 K	10.4(2) at 10 K	10.2(2) at 9.7 K
$\Delta E_\beta(T)$	K	11.6(2) at 9.3 K	13.2(1) at 4.2 K	11.6(2) at 9.0 K
$\Delta K/N\mathrm{k}$	K/FU	− 0.57(15)	− 0.18	− 0.45(8)
$T_R \cdot \Delta S/N\mathrm{k}$	K/FU	0.25(5)	0.19	0.15(5)
E_{II}	K	66(1)	65.7(3)	66(1)
E_{III}	K	164(3)	162.7(8)	162(2)
E_{IV}	K		258.9(4)	
E_V	K	248(4)	303.0(4)	

$TmCrO_3$ (see also LB III/12a, p. 446)

I, III Spontaneous magnetization and field dependence of magnetization Figs. 153 ··· 155. The magnetic structure below $T_{N1} = 124$ K is (G_z; $-C_y$, $-F_x$), [81S1]. At low temperature no evidence of magnetic ordering within the Tm^{3+} sublattice is found by ME [84H1]. However, [87K5], NMR spectra at low temperature showed the coexistence of two phases Γ_4 and Γ_2, Fig. 156. This is also confirmed by properties of the optical absorption spectra [85K2]. For Mössbauer data see Figs. 157, 31. The direction of H_{hyp} rotates as a function of temperature and consequently $\boldsymbol{p}_{Tm^{3+}}$ changes its direction. At 4.2 K this moment is oriented parallel to the crystallographic c axis: ($G_xF_z(Cr^{3+})$; F_z (Tm^{3+})) configuration. With increasing T there is a progressive rotation about the b axis, and at 80 K the ferromagnetic component of the chromium sublattice, and consequently also the Tm moments make an angle of 62° with the c axis. The canting angle α of the Cr^{3+} moments is independent of T: $\alpha = 7.5$ (15)° [84H1]. Evidence for $\Gamma_2 \leftrightarrow \Gamma_4$ spin-flip transitions in an external magnetic field in the high-temperature range is found in NMR spectra, whereas at low temperatures similar experiments confirm the two-phase configuration [87K5].

IV Detailed optical investigations with and without applied magnetic fields [79W1, 85K2].

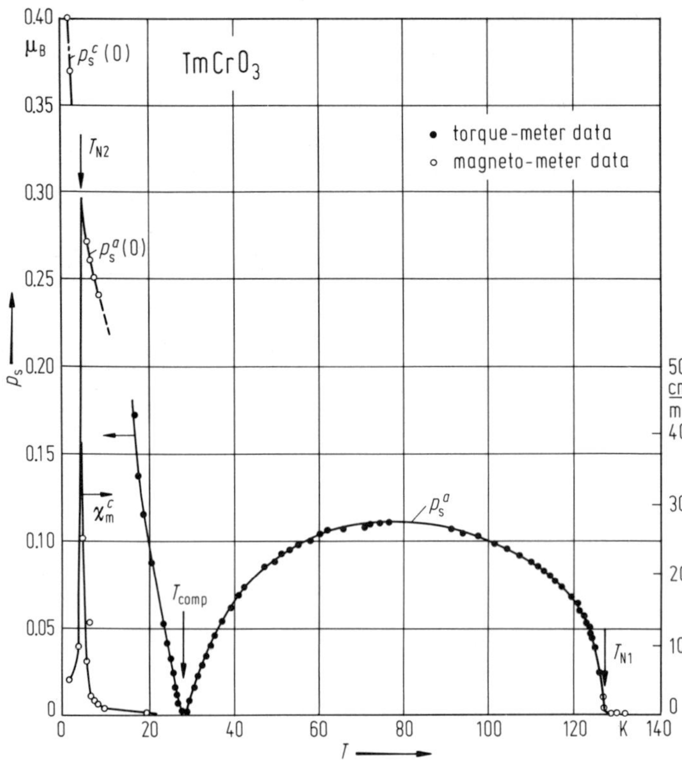

Fig. 153. $TmCrO_3$. Spontaneous magnetic moments per FU, p_s vs. temperature T along the crystallographic a and c direction, p_s^a, p_s^c, respectively. Also shown is the molar magnetic susceptibility, χ_m^c in the c direction. $T_{comp} = 29$ K, $T_{N2} = 5$ K [77T1]. $p_s^c(0)$, $p_s^a(0)$ indicate remanent magnetizations.

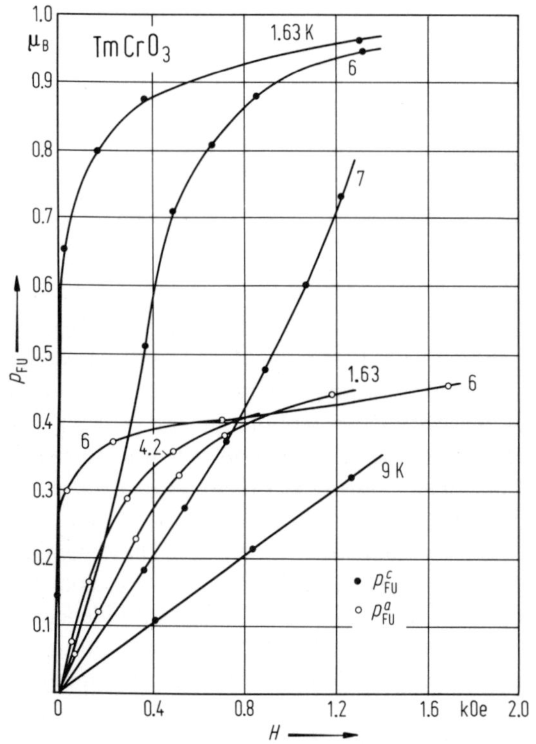

Fig. 154. $TmCrO_3$. Magnetic moment per FU, p_{FU}^a and p_{FU}^c vs. applied magnetic field in the crystallographic a and c direction at low temperatures T [77T1].

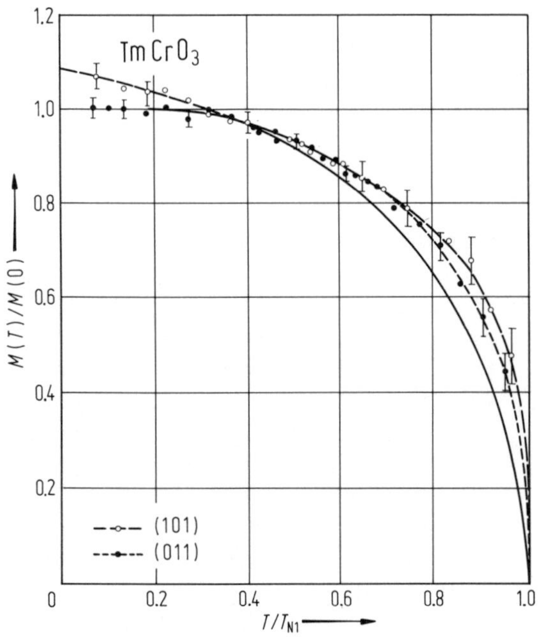

Fig. 155. $TmCrO_3$. Reduced magnetization $M(T)/M(0)$ vs. reduced temperature T/T_{N1} as derived from neutron elastic scattering spectra ($T_{N1} = 124$ K) for the (011) and (101) reflections. Full curve: Brillouin curve for $J = 3/2$. Dashed curves are guides to the eye. The (011) magnetization is scaled to unity at $T/T_{N1} = 0.3$ [81S1].

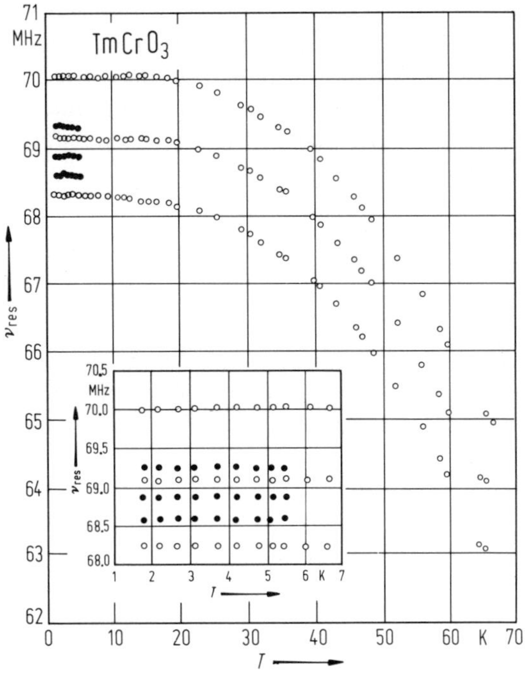

Fig. 156. $TmCrO_3$. ^{53}Cr NMR frequency ν_{res} vs. temperature T. Inset: low temperature part. Below $T \approx 5.6$ K a two-phase system Γ_2, Γ_4 is found [87K5].

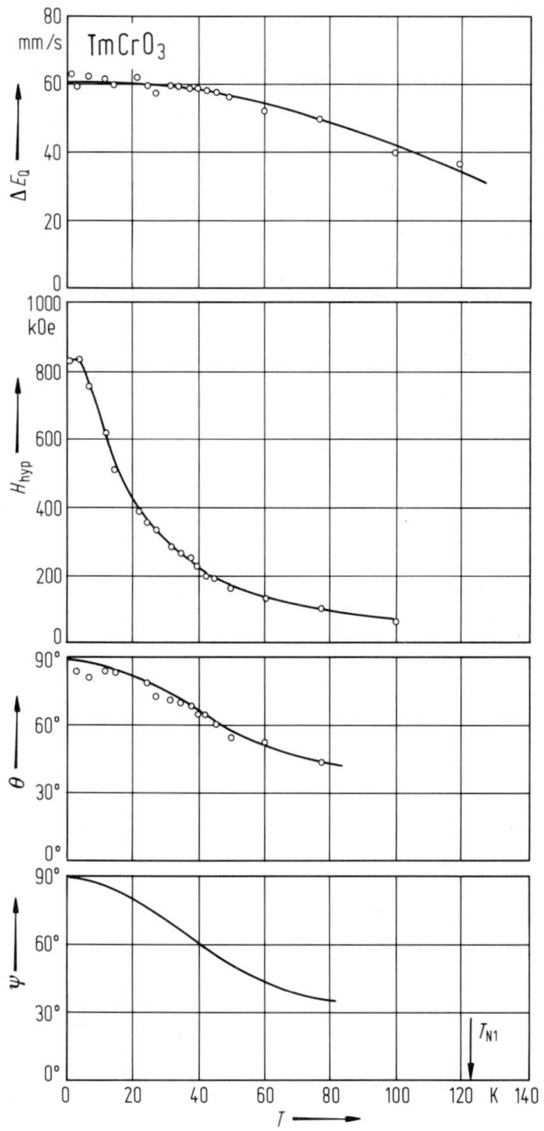

Fig. 157. $TmCrO_3$. Mössbauer data. ΔE_Q, quadrupole hyperfine splitting (which has a negative sign); H_{hyp}, effective hyperfine field; θ, angle between H_{hyp} and the main direction of the EFG tensor; ψ, angle between $\boldsymbol{p}_{Tm^{3+}}$ and direction of crystallographic a axis [84H1].

$YbCrO_3$ (see also LB III/4a, p. 229 and III/12a, p. 447)

I Below $T_N = 118$ K the magnetic structure Γ_2 (G_zF_x) is retained at all temperatures. The slightly canted Cr^{3+} moments are approximately antiferromagnetically oriented along the c axis with a net moment along the a axis. The latter being antiparallel to the net moment of the canted Yb^{3+} ions, configuration Γ_2 ($F_xC_yG_z$; F_xC_y) [71S2], see Fig. 158.

Field dependence of magnetization and spontaneous magnetization curves Figs. 159, 160. An analysis of the exchange fields is given by [71S2] and [88H1] on the basis of magnetization and magnetooptical data, respectively. The effective isotropic exchange field between Cr^{3+} ions is 97 kOe; the DM fields are 49 kOe and 65 kOe for the Cr^{3+} ions and for the Yb^{3+} ions, respectively.

II Magnetic susceptibilities Fig. 161.

III Mössbauer data Table 28.

IV Optical absorption spectra in high magnetic fields [80K3, 80K4, 84K1, 86H1, 86M1, 88H1].

$Yb_xY_{1-x}CrO_3$

I Spontaneous weak ferromagnetic moments Fig. 160. Spontaneous and magnetic field induced spin rotation transition Fig. 162. A reorientation of the weak magnetization from the *c* direction to the *a* direction occurs during cooling of the compound x = 0.1 in the temperature interval 96 ··· 74 K, i.e. a Γ_2 (G_zF_x) → Γ_4 (G_xF_z) phase transition. A field induced SRT, $\Gamma_4 \to \Gamma_2$, is found above T_2 = 74 K for $\boldsymbol{H}\|a$ and an SRT $\Gamma_2 \to \Gamma_4$ below T_1 = 96 K for $\boldsymbol{H}\|c$ [90V1].

$YbCr_{0.97}Fe_{0.03}O_3$

III Mössbauer data. T_N = 118 (1) K; ΔE_Q = 0.26 (2) mm s^{-1}. [74B2].

Table 28. $YbCrO_3$. Mössbauer data at low temperatures *T*. Two types of fits used for the spectra at 4.2 K and 1.3 K. δ isomer displacement relative to the $^{170}Tm^{11}B_{12}$ source; *G* Lorentz width (natural width 2.0 mm s^{-1}); $\alpha = 3\,eV_{zz}\,Q/[4I(2I-1)]$, quadrupole interaction; H_{hyp} hyperfine field; φ angle between H_{hyp} and principal axis of EFG; η asymmetry parameter of EFG [74B2].

T [K]	δ [mm/s]	G [mm/s]	α [mm/s]	H_{hyp} [kOe]	φ [deg]	η
25	0.18	2.58	3.76	0	–	0.11
4.2	0.12	2.58	3.83	76	0	0.11
	0.1	2.58	3.80	95	37	0.11
1.3	0.07	2.58	3.81	60	0	0.11
	0.1	2.58	3.80	106	55	0.11

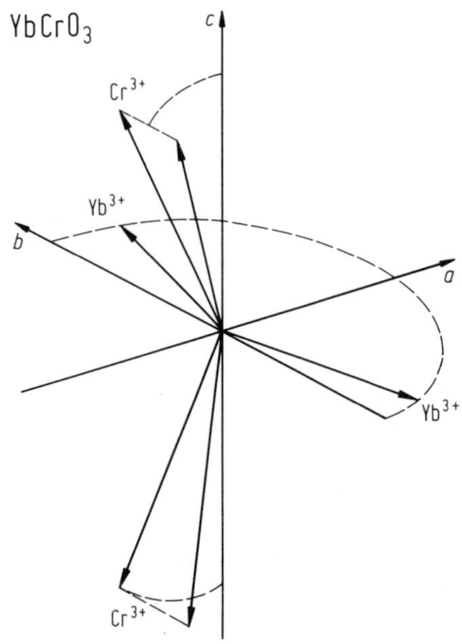

Fig. 158. $YbCrO_3$. Arrangement of the magnetic moments of the ions, magnetic configuration Γ_2 ($F_xC_yG_z$; F_xC_y).

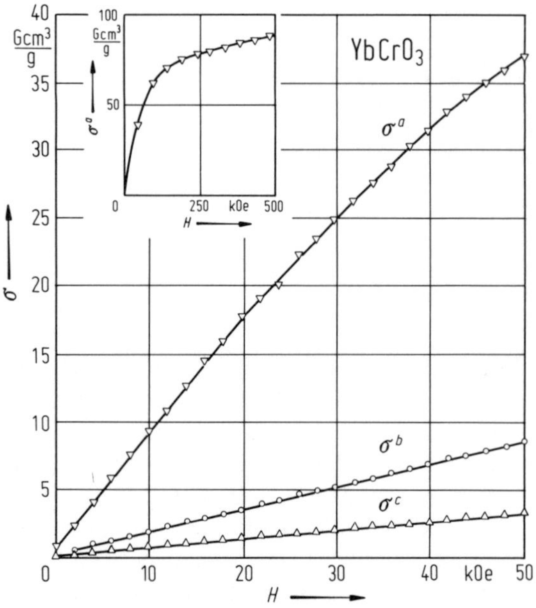

Fig. 159. $YbCrO_3$. Mass magnetization σ vs. applied magnetic field *H* at 1.6 K for the three main crystallographic directions. Curves calculated for optimized set of magnetic parameters. Inset the same at *T* = 1.3 K for $\boldsymbol{H} \parallel a$ axis [90V1, 86M1].

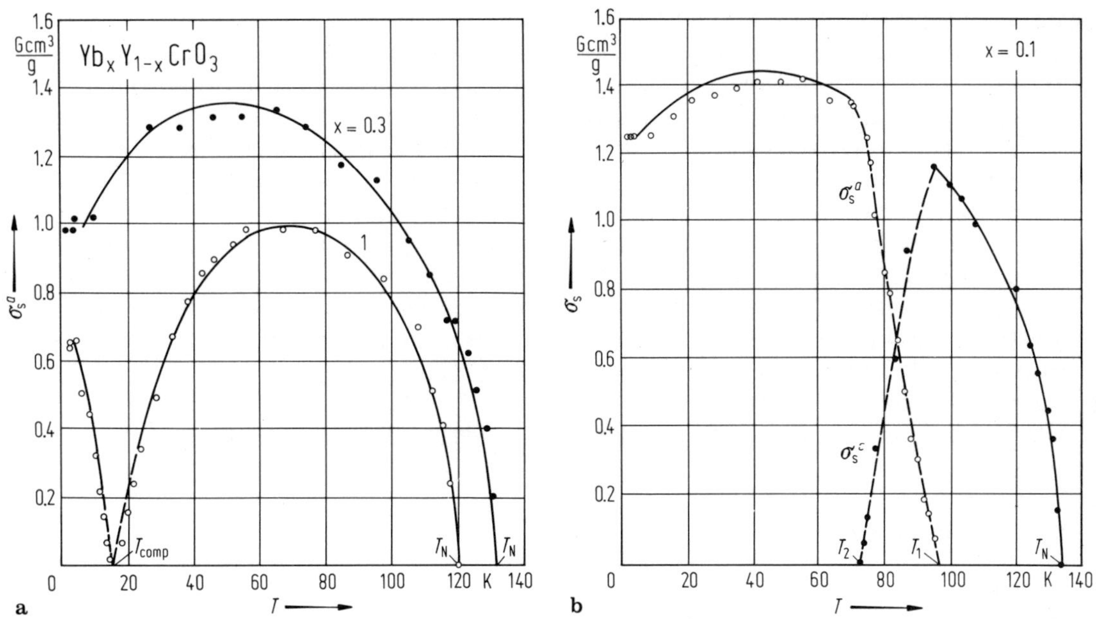

Fig. 160. $Yb_xY_{1-x}CrO_3$. Spontaneous weak ferromagnetic moment σ_s vs. temperature T measured along the crystallographic a and c axes. The curves are results of calculations of best fits of various sets of magnetic parameters [90V1].

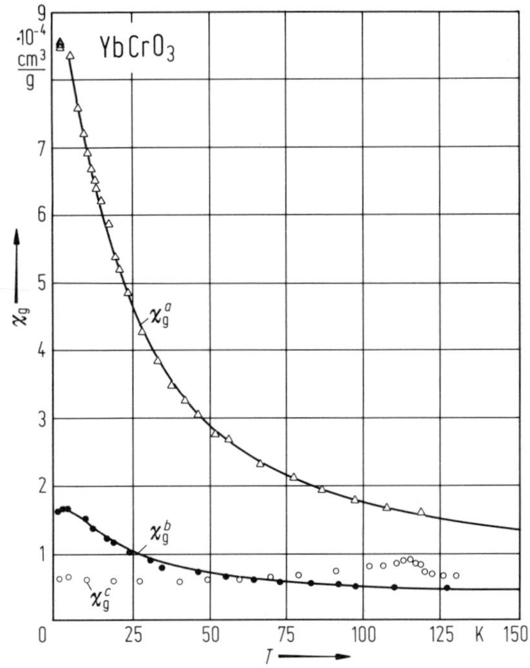

Fig. 161. $YbCrO_3$. Temperature dependence of the specific susceptibilities χ_g^a, χ_g^b and χ_g^c in the three main crystallographic directions [90V1].

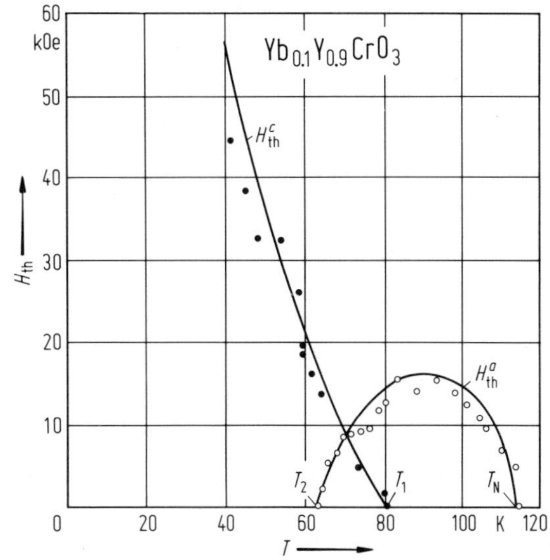

Fig. 162. $Yb_{0.1}Y_{0.9}CrO_3$. Magnetic (H–T) phase diagram, see text. Curves calculated for an optimally fitted set of magnetic parameters [90V1].

$LuCrO_3$ (see also LB III/4a, p. 229 and III/12a, p. 447)

I Magnetization measurements show that down to 4.2 K the weak ferromagnetic moment is oriented along the *a* axis with an experimental error of $\pm 3°$, magnetic configuration (G_zF_x); $T_N = 111$ (1) K. Spontaneous magnetization extrapolated to 0 K is $\sigma_{m,s}(0) = 360$ (20) G cm³ mol⁻¹ [76H1]. There exists still much uncertainty as regards the magnetic structure, Table 24. A magnetic field applied along the antiferromagnetic *c* axis can align the magnetization parallel to the *c* axis, Fig. 163. The system appears to undergo two second-order phase transitions. The critical field required for the G_xF_z phase at 4.2 K is $H_{cr} = 3.6$ (4) kOe. An analysis of the magnetization curves gives the following properties: Heisenberg exchange field $H_{ex} = 990$ (100) kOe; Dzyaloshinsky-Moriya antisymmetric exchange field $H_D = 42$ (4) kOe; quadratic anisotropy field $H_K = 0.084$ (10) kOe; quartic anisotropy field $H'_K = 0$; canting angle of the Cr sublattice moments with respect to the AF axis $\alpha = 21.3$ (10)° [76H1].

II Susceptibility curves Fig. 164.

III NMR in domain walls [84J1]. ESR spectra at $T = 140 \cdots 520$ K, see [83D1].

$LuCr_{1-x}Fe_xO_3$

I Spontaneous magnetization curves Fig. 165.

$Lu_xY_{1-x}CrO_3$

I Threshold field H_{th} for the Γ_4 (G_xF_z) → Γ_2 (G_zF_x) phase transition Fig. 166. Orientation of weak magnetization with respect to the *c* axis, Fig. 167. For partial disordering of the spin system during the phase transition, see [85M1].

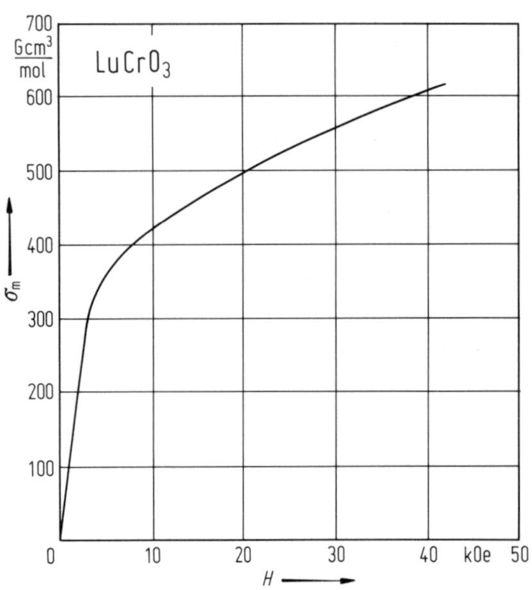

Fig. 163. $LuCrO_3$. Molar magnetization σ_m vs. applied field H along the *c* axis at $T = 4.2$ K [76H1].

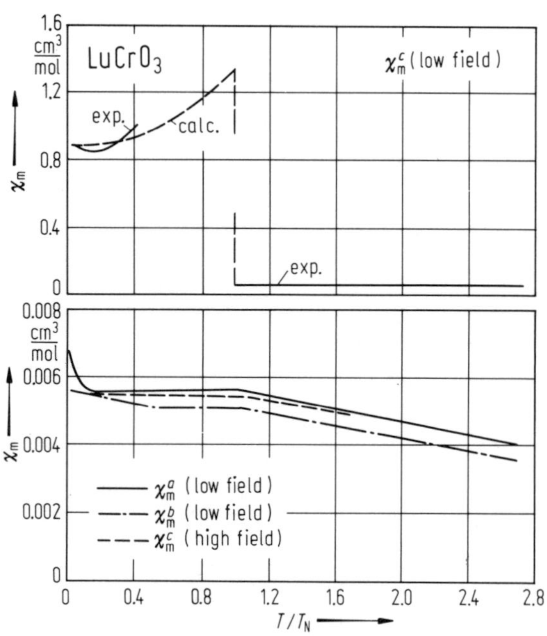

Fig. 164. $LuCrO_3$. Molar magnetic susceptibility χ_m (in various crystallographic directions indicated by suffices *a*, *b* or *c*) vs. reduced temperature T/T_N, where $T_N = 111$ K. The slight rise at low *T* in χ^a may be due to isolated paramagnetic impurities [76H1].

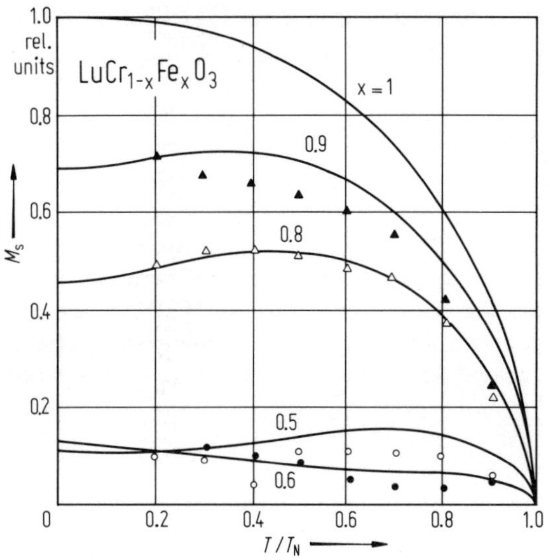

Fig. 165. $LuCr_{1-x}Fe_xO_3$. Temperature dependence of the spontaneous magnetization. The magnetization in relative units, the temperature in units relative to the Néel point of each substance. Curves calculated [83S1].

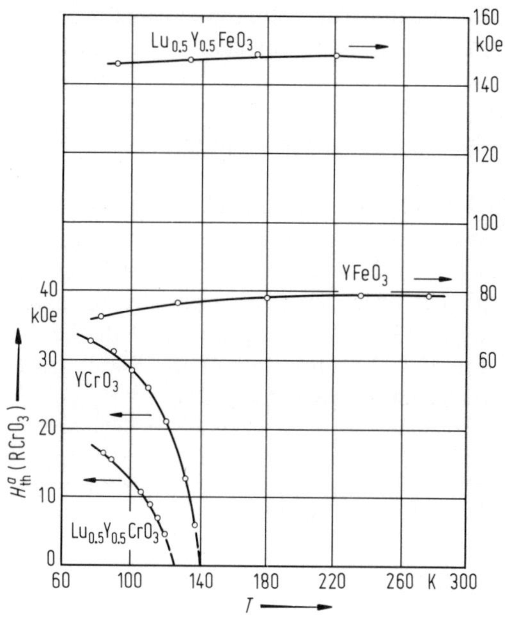

Fig. 166 $Lu_{0.5}Y_{0.5}CrO_3$; $Lu_{0.5}Y_{0.5}FeO_3$; $YCrO_3$; $YFeO_3$. Temperature dependence of threshold field H^a_{th} applied along the a axis for spin reorientation Γ_4 $(G_xF_z) \rightarrow \Gamma_2$ (G_zF_x), as derived from magnetostriction data [81K3].

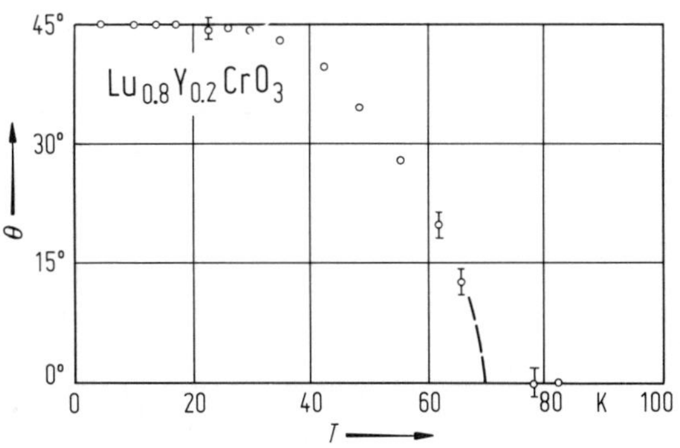

Fig. 167. $Lu_{0.8}Y_{0.2}CrO_3$. Temperature dependence of the angle θ between direction of weak magnetic moment and crystallographic c axis in the ac plane [83K2].

$YCrO_3$ (see also LB III/4a, p. 229 and III/12a, p. 447)

I Spontaneous magnetization Fig. 168. Below $T_N = 141$ K the weak ferromagnetism is oriented along the c axis, spin configuration Γ_4 (G_xF_z). Threshold field along the a axis for spin rotation, see Fig. 166. This phase transition is accompanied by a monoclinic lattice distortion [81K2].

III ^{53}Cr NMR at 4.2 K reveals the "overt" canting angle $\phi = 0.81$ (5)° leading to weak ferromagnetism and the "hidden" canting angle $\psi = 1.25$ (5)° [82L1] defined in Fig. 2. NMR quadrupole splitting 500 kHz independent of T. Reduced magnetization vs temperature, Fig. 169. Exchange integral from spin wave theory $J/k = 18.2$ (10) K, from molecular field theory 20.7 K [79K1]. NMR spectrum at 4.2 K in the presence of a magnetic field along the a axis shows 6 lines due to the AF structure and quadrupole interactions. The field dependence of the central lines can be described by the parameters derived from magnetization measurements [76Z2]: $H_{ex} = 2300$ kOe, $H_D = 61$ kOe, $H_K = 0.775$ kOe. The position of the satellite lines can be fitted with the parameters of Table 29. The EFG is approximately parallel to the axes of the oxygen octahedra. NMR in domain wall, see [84J1]. For ESR between $T = 140 \cdots 520$ K see [83D1].

IV For magnon- and phonon-assisted exciton transitions in magnetic fields, see [78Z1] and [75W1]. For the dielectric constant Fig. 109. Infrared spectra of Sr-doped $YCrO_3$ [88P1].

$YCr_{1-x}Mn_xO_3$

I Spontaneous magnetization Fig. 168. For x = 0.2 is found T_N = 93 K.

$YCr_xFe_{1-x}O_3$ (see also LB III/12a, p. 447, 461)

I Weak ferromagnets. Spontaneous magnetization Figs. 170, 171. Néel temperature Fig. 172. For x = 0.15 (T_N = 510 K) and x = 0.65 (T_N = 240 K) the high temperature magnetic configuration is (G_xF_z) which changes gradually on cooling into the (G_zF_x) configuration [77K3], see also Figs. 173, 175. A magnetic field in the *c* direction can also induce this configuration change Fig. 174. The rotation of the Cr^{3+} and Fe^{3+} magnetic moments is coherent in the whole reorientation region [78K1].

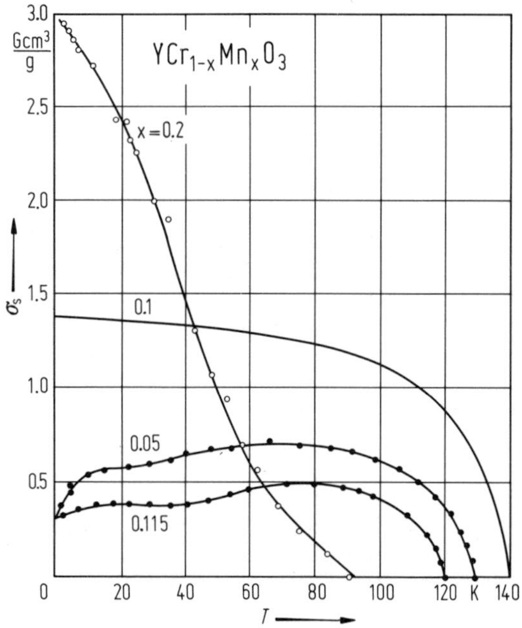

Fig. 168. $YCr_{1-x}Mn_xO_3$. Spontaneous mass magnetization σ_s vs. temperature T [87K2].

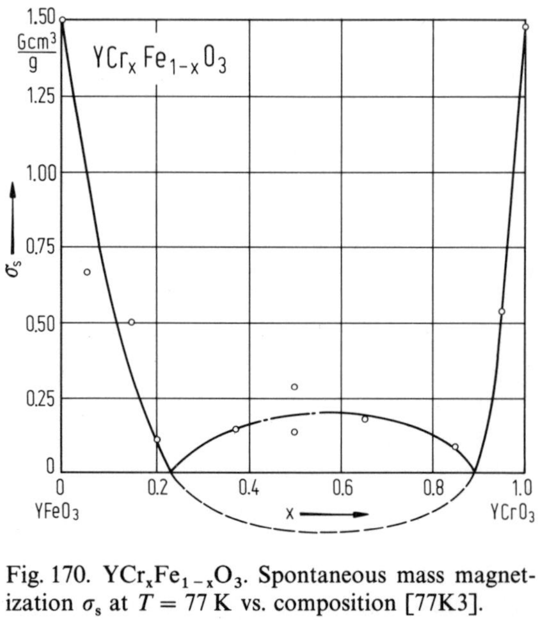

Fig. 170. $YCr_xFe_{1-x}O_3$. Spontaneous mass magnetization σ_s at T = 77 K vs. composition [77K3].

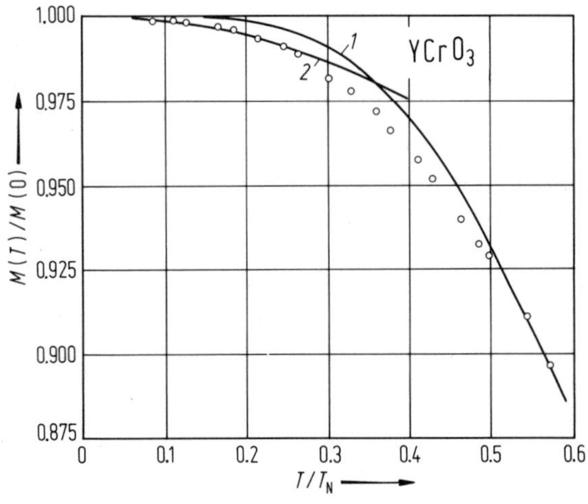

Fig. 169. $YCrO_3$. Temperature dependence of the reduced magnetization $M(T)/M(0)$ as derived from NMR. T_N = 141 K. Curves calculated *1*: molecular field theory, *2*: spin wave theory [79K1].

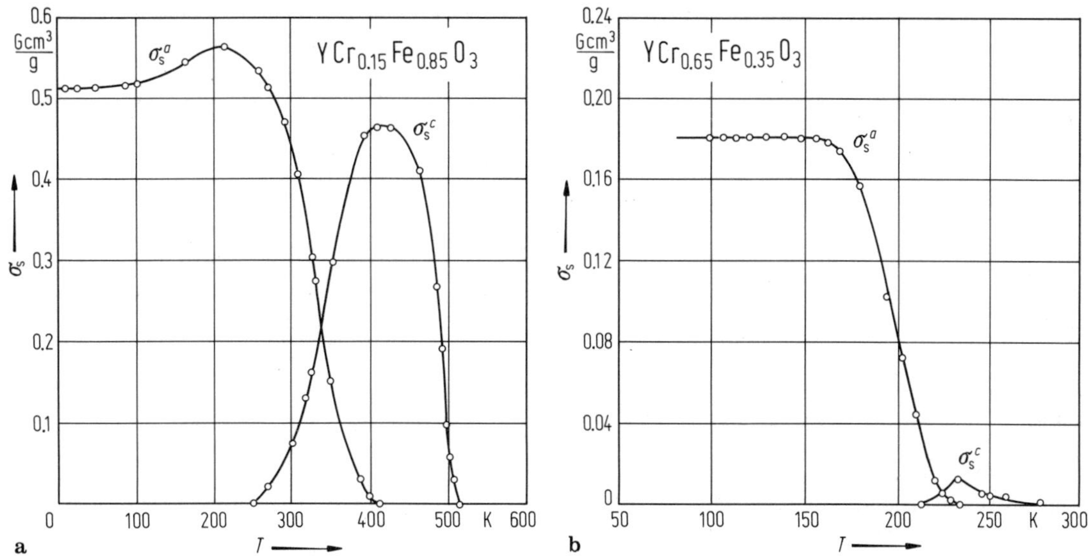

Fig. 171. $YCr_xFe_{1-x}O_3$. Temperature dependence of the spontaneous mass magnetization in crystallographic *a* and *c* direction, σ_s^a and σ_s^c, respectively, for x = 0.15 and x = 0.65 [77K3].

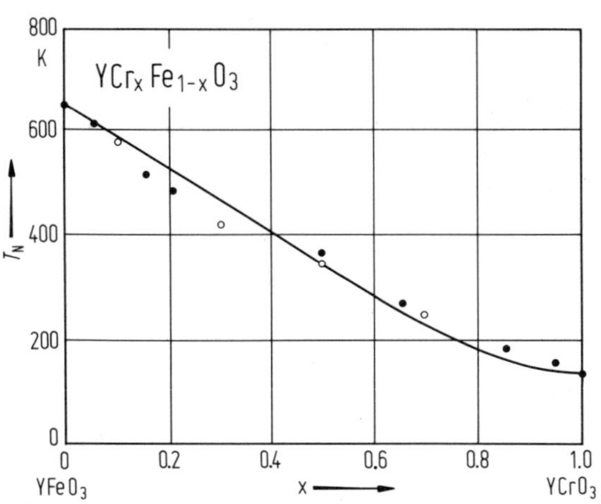

Fig. 172. $YCr_xFe_{1-x}O_3$. Néel temperature T_N vs. composition parameter x, for single crystals (●) and polycrystals (○). The curve is a calculated fit with a superexchange J/k (FeCr) = J/k (CrFe) = 6.7 K, independent of x [77K3].

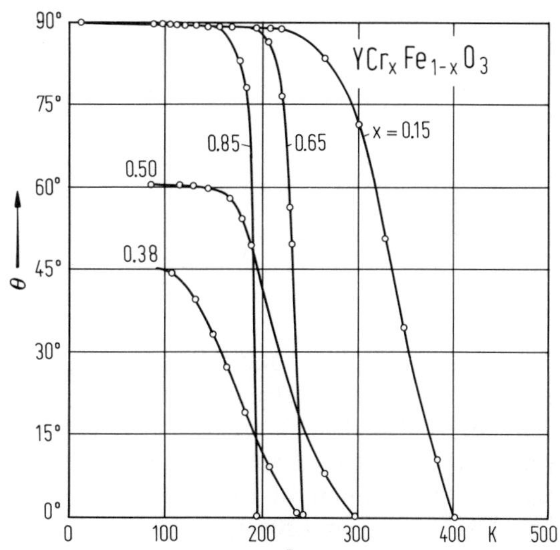

Fig. 173. $YCr_xFe_{1-x}O_3$. Temperature dependence of the angle θ between the direction of the weak ferromagnetism and the crystallographic *c* axis in the *ac* plane [77K3].

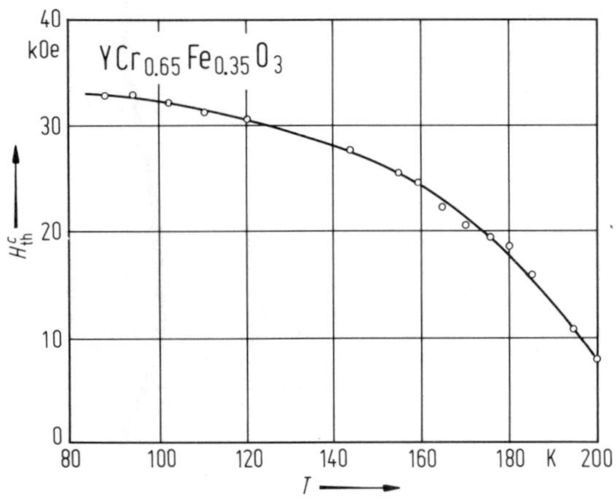

Fig. 174. $YCr_{0.65}Fe_{0.35}O_3$. Temperature dependence of the threshold field H_{th}^{c} in the crystallographic c direction for the spin-reorientation $G_zF_x \rightarrow G_xF_z$ [78K1].

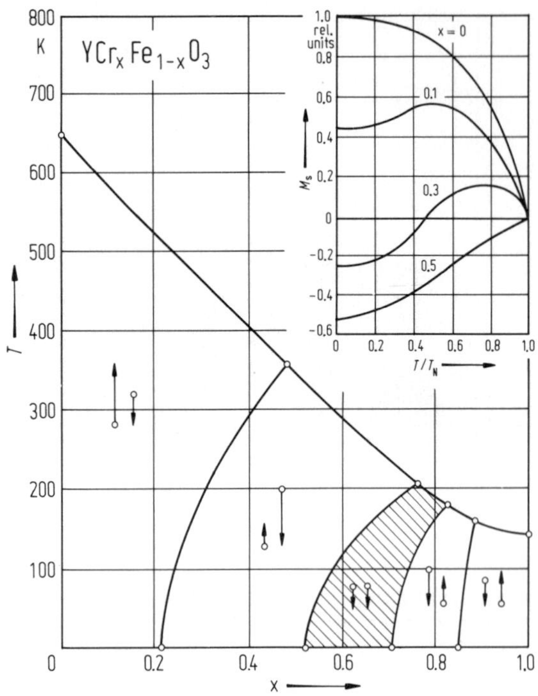

Fig. 175. $YCr_xFe_{1-x}O_3$. Magnetic phase diagram calculated on the basis of the molecular field theory, with exchange integrals and Dzyaloshinsky vectors derived from various magnetic properties of the compounds. The arrows represent orientations and relative magnitudes of the weak magnetic moments p_{Fe} and p_{Cr} averaged over the structure. The shaded region is the region of parallel orientation of the average magnetic moments of the Fe and Cr sublattices. Inset: typical temperature dependences of the magnetization M (relative scales), to be compared to Fig. 165 [83S1].

Table 29. $YCrO_3$ Polar and azimuthal angles ϕ, ε of the EFG with respect to z and x axis, respectively. Quadrupole frequency ν_Q and asymmetry parameter η derived from field dependence of the position of the sattelite lines in ^{53}Cr NMR spectrum [80N1]

Sub-lattice	ϕ [deg]	ε [deg]	ν_Q [MHz]	η
1	21.5	15.8	1.6	0
2	21.5	22	1.6	0

5.5.7 $RMnO_3$ compounds

(see also LB III/4a, p. 230–237 and III/12a, p. 447–451)

$LaMnO_3$ (see also LB III/4a, p. 230 and III/12a, p. 447)

I Chemical stability at high temperatures [79N1]. For detailed analysis of the influence of small lattice distortions on the magnetic properties, see [76S1, 77S5, 77S6, 80K5 and 81S2]. The simultaneous presence of an antiferromagnetic and a ferromagnetic phase is established by neutron diffraction patterns. A magnetic phase diagram is shown in Fig. 176. – For lattice parameters see also Fig. 180.

II For non-stoichiometric behavior see also [78K4] and [83B1].

$LaMn_{1-x}Ni_xO_3$

I Crystal structure and lattice parameters Table 30. For the problem of preparing homogeneous compounds see [79A1, 84V2]. Electrical resistivity Fig. 177. For x = 0.9 ferrimagnetic spin glass behaviour below 40 K, Fig. 178. For x = 0.5 ferromagnetism with $T_C = 275$ K [79A1].

II Paramagnetic properties Table 31 and Fig. 179.

$La_{1-x}Ca_xMnO_3$ (see also LB III/4a, p. 230 and III/12a, p. 448)

I Lattice parameters and magnetic properties Figs. 180, 181. Magnetic phase diagram Fig. 182. In the ferromagnetic state the samples are metallic conductors.

$La_{1-x}Ca_xMn^{3+}_{1-x}Ti^{4+}_xO_3$

I Lattice parameters and magnetic properties Fig. 183.

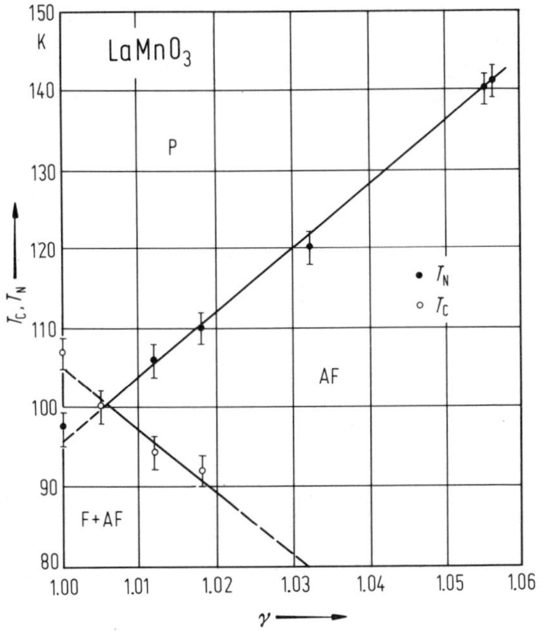

Fig. 176. $LaMnO_3$. Magnetic phase diagram showing the ferromagnetic and antiferromagnetic transition temperatures, T_C and T_N, respectively, for samples with different degrees of lattice distortion $\gamma = b\sqrt{2}/c$. The presence of the phases has been established by an analysis of neutron diffraction spectra [80K5].

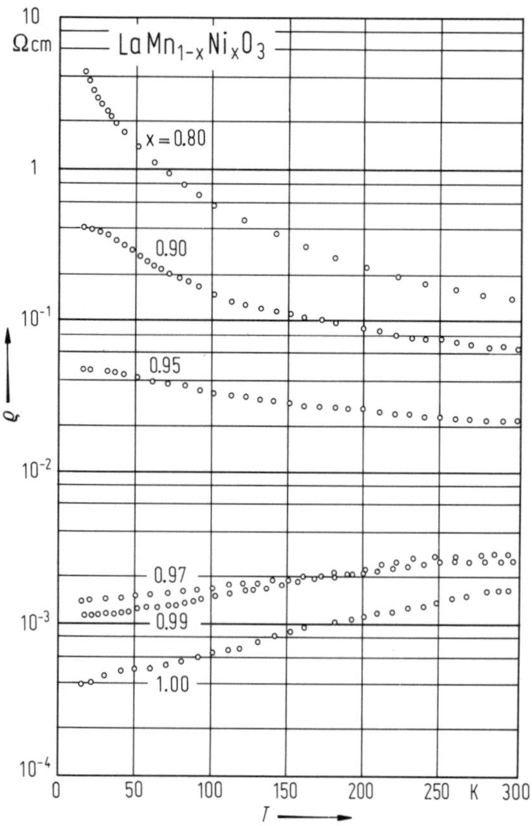

Fig. 177 $LaMn_{1-x}Ni_xO_3$. Electrical resistivity ρ vs. temperature T [84V3].

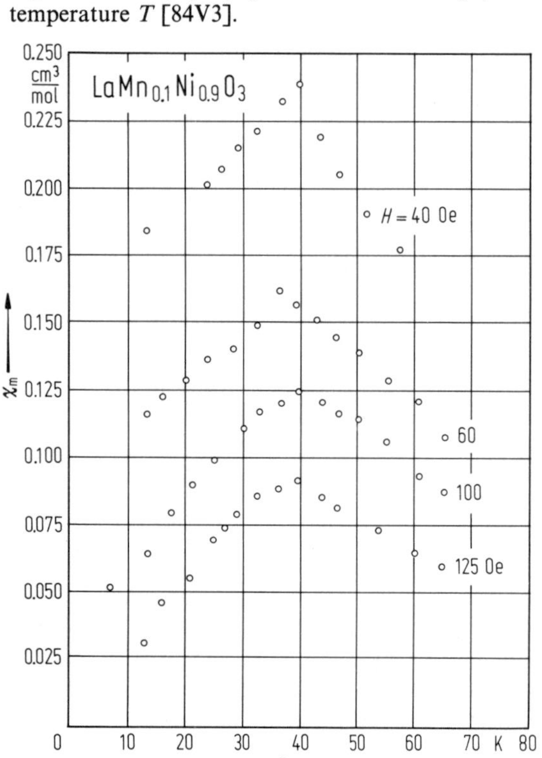

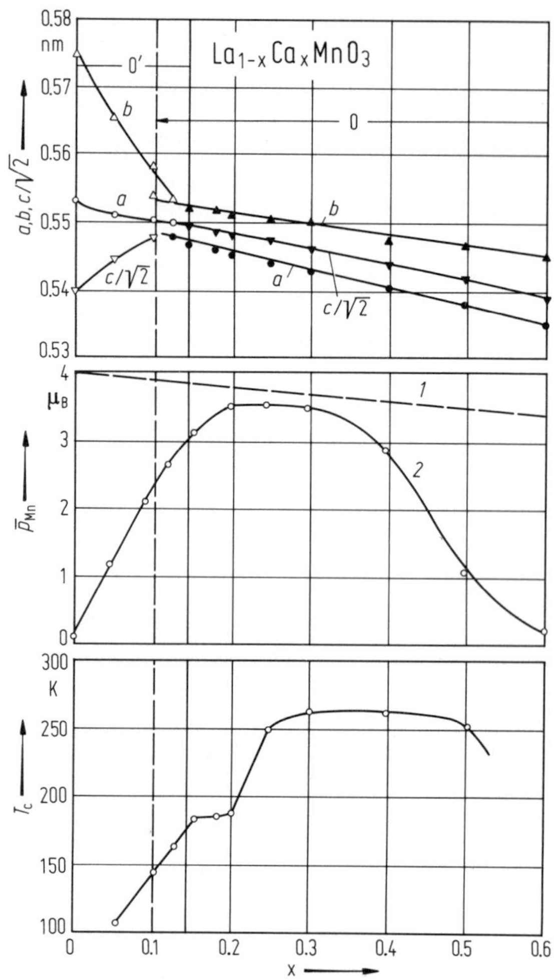

Fig. 180. $La_{1-x}Ca_xMnO_3$. Lattice parameters a, b, c at RT. Average magnetic moment per Mn ion $\bar{p}_{Mn^{3+}}$ at 4.2 K and magnetic ordering temperature T_c vs. composition parameter x. O′ and O denote the two kinds of orthorhombic structures. Curve *1* is calculated [86P1].

For Fig. 179, see next page.

Fig. 178 $LaMn_{0.1}Ni_{0.9}O_3$. Low-field dc molar magnetic susceptibility χ_m vs. temperature T for various measuring fields H [84V3].

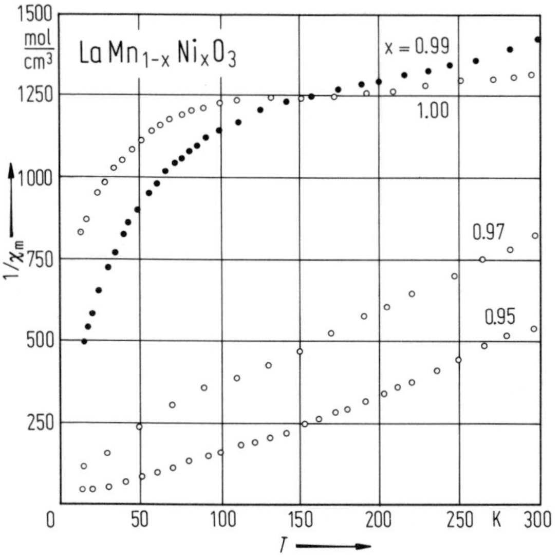

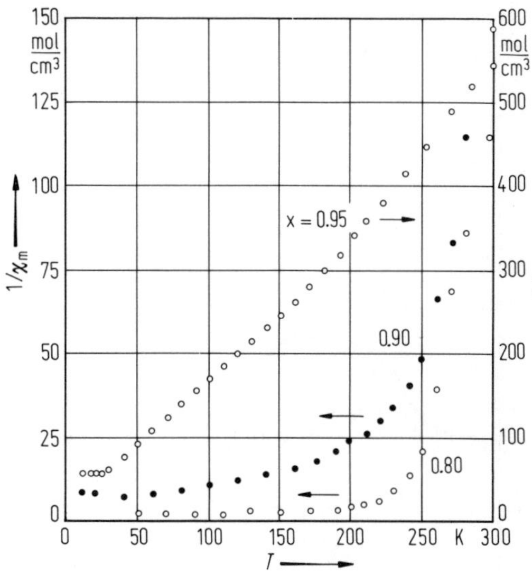

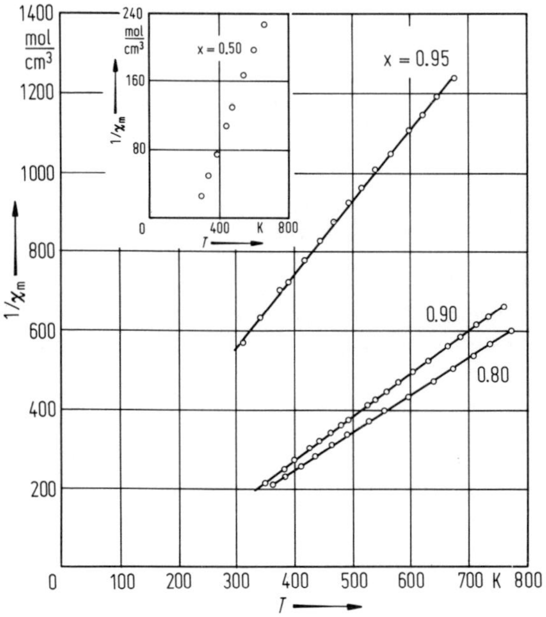

Fig. 179. $LaMn_{1-x}Ni_xO_3$. The inverse molar magnetic susceptibility χ_m^{-1} vs. temperature T for various values of the composition parameter x. For an analysis of the curves see Table 31 [84V3].

For Fig. 181, see next page.

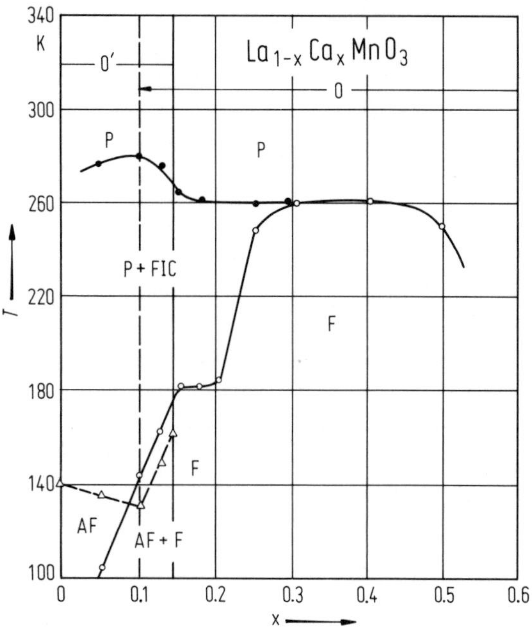

Fig. 182. $La_{1-x}Ca_xMnO_3$. Magnetic phase diagram. P: paramagnetism; F: ferromagnetism; AF: antiferromagnetism; FIC: ferromagnetic impurity clusters [86P1]. O, O′: see Fig. 180.

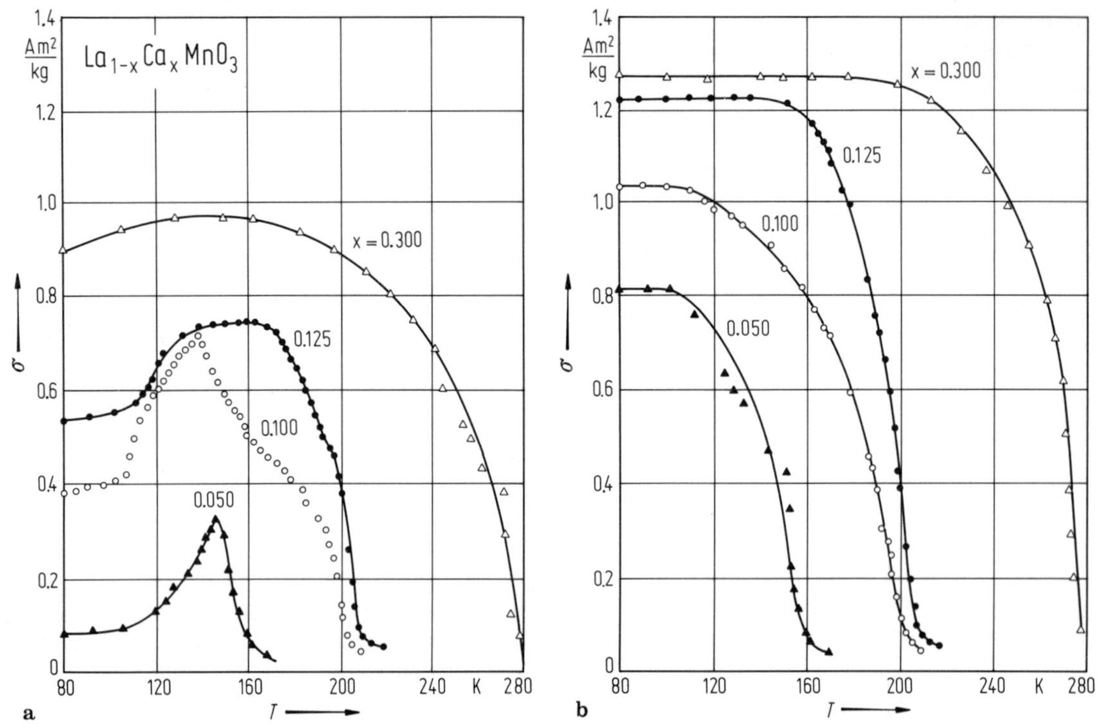

Fig. 181. $La_{1-x}Ca_xMnO_3$. Specific magnetization σ vs. temperature T, measured in an applied magnetic field of $H = 1.6 \cdot 10^3$ A/m. (a) After cooling the samples from high temperature in the absence of a magnetic field. (b) The same but now cooling in the presence of the magnetic measuring field [89T1].

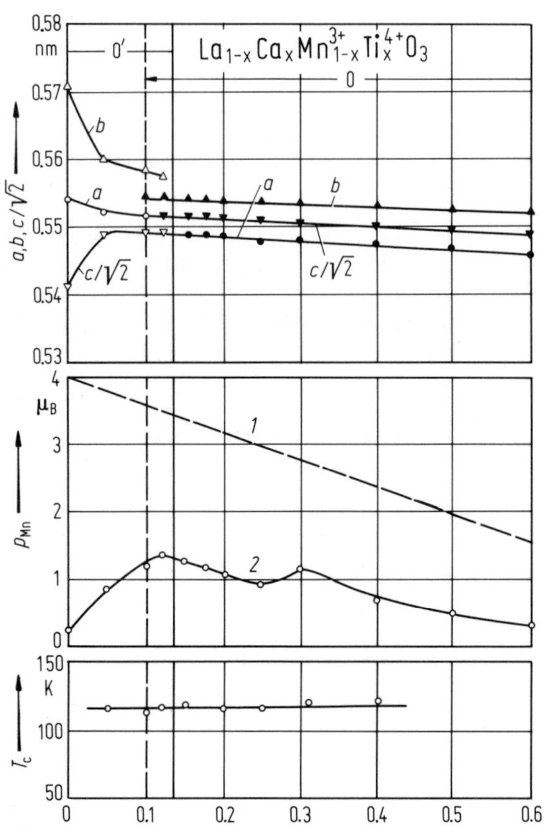

Fig. 183. $La_{1-x}Ca_xMn^{3+}_{1-x}Ti^{4+}_xO_3$. Lattice parameters a, b, c, at RT, magnetic moment per Mn ion $p_{Mn^{3+}}$ at 4.2 K and magnetic ordering temperatures T_c vs. composition x. O′ and O denote two kinds of orthorhombic structures [86P1].

Table 30. $LaMn_{1-x}Ni_xO_3$. O (orthorhombic) and Rh (rhombohedral) structures. Pseudo-cubic lattice parameter a and rhombohedral angle α at RT [84V3].

x	Structure	a [Å]	α [deg]
1.00	Rh	3.84	60.52
0.99	Rh	3.84	60.55
0.97	Rh	3.84	60.67
0.95	Rh	3.85	60.54
0.90	Rh	3.85	60.84
0.80	Rh	3.857	60.07
0.50	O	3.877	–

Table 31. $LaMn_{1-x}Ni_xO_3$. Curie-Weiss parameters Θ_p, p_{eff} and temperature independent part of the susceptibility, χ_{TIP}, derived from the high-temperature parts of the curves of Fig. 179 [84V3].

x	C_m [$cm^3\,K\,mol^{-1}$]	p_{eff} [μ_B/FU]	Θ_p [K]	χ_{TIP} [$cm^3\,mol^{-1}$]
1	0.0082		−5	$7.7 \cdot 10^{-4}$
0.99	0.0237		−4	$6.6 \cdot 10^{-4}$
0.97[a])	0.19		3	$6.96 \cdot 10^{-4}$
		1.85	−55	–
0.95		2.07	0	–
0.90		2.70	155	–
0.80		2.91	140	–
0.50		3.57	300	–

[a]) Analysis with and without χ_{TIP}.

$Pr_{1-x}Mn_{1+x}O_{3+y}$ (see also LB III/4a, p. 235 and III/12a, p. 449)

I RT lattice parameters Table 32. Magnetization Fig. 184. Antiferromagnetic below T_N, this ordering is accompanied by the onset of a weak ferromagnetic moment, increasing with x [80P1].

II Susceptibility data Fig. 185 and Table 32.

$Pr_{1-x}Ca_xMnO_3$, x = 0.1

I Lattice constants at RT, derived from neutron diffraction patterns: a = 5.442 Å, b = 5.617 Å and c = 7.635 Å. At 5K the Mn moments order is of the $F_{xy}A_{yx}$ type with the parallel components of the magnetic moment equal to 2.95(6)μ_B, and the antiparallel moment 1.58(3)μ_B. The magnitude of the moment is 3.35 μ_B. The Mn moments are arranged ferromagnetically in one xy layer, the magnetic axes of these layers are inclined alternatively about + 30° and − 30° to the [110] axis, lying close to the x and y axes, respectively. The Pr moments order is of the F_{xy} type with respective moments of 0.45(3) or 0.50(3) μ_B. The compound exhibits two magnetic transitions. T_C = 100 K is a ferromagnetic Curie point with easy y axis. Below T = 70 K a non-collinear structure with Mn moments in the xy plane. Below about 30 K the Pr moments order. At 5 K the resultant Mn and Pr moments are parallel to the [110] direction [79J1]. For similar detailed information for compounds with other x values, see [85J1] and Fig. 186.

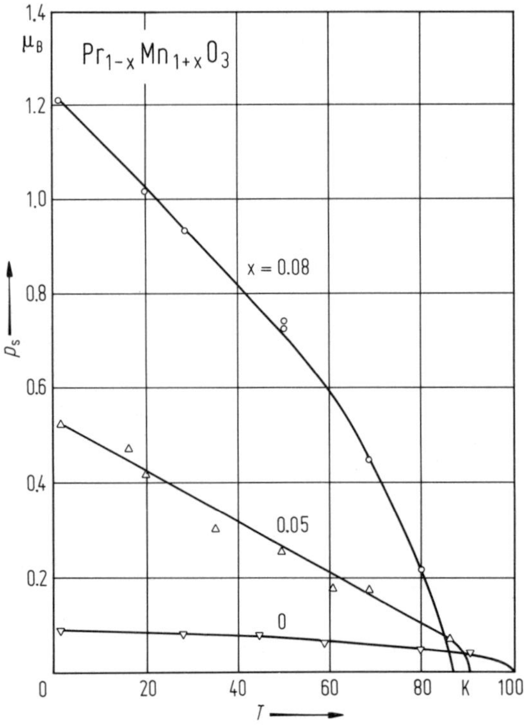

Fig. 184. $Pr_{1-x}Mn_{1+x}O_3$. Spontaneous magnetic moment per formula unit p_s vs. temperature T [80P1].

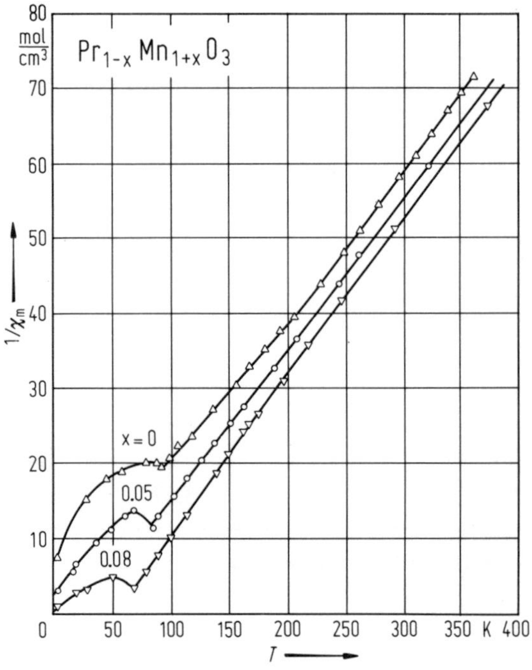

Fig. 185. $Pr_{1-x}Mn_{1+x}O_3$. Reciprocal molar magnetic susceptibility χ_m^{-1} vs. temperature T [80P1].

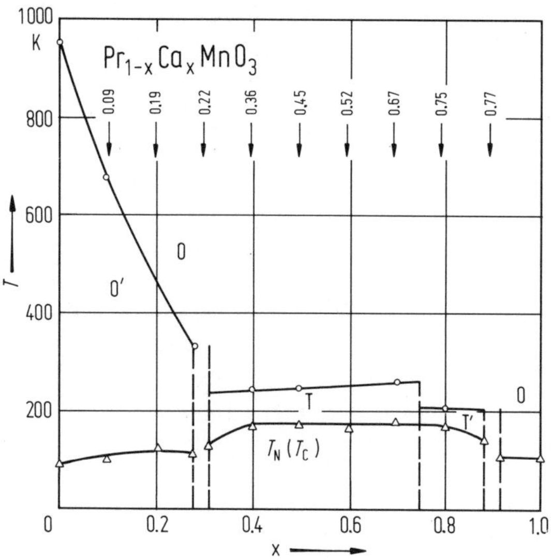

Fig. 186. $Pr_{1-x}Ca_xMnO_3$. Phase transition temperatures between the orthorhombic phases O, O', and the tetragonal phases T and T' indicated by open circles vs. composition. Magnetic transition temperatures T_N and T_C indicated by triangles [85J1]. The actual concentrations of Mn^{3+} ions given by chemical analysis are shown in the upper part of the figure.

Table 32. $Pr_{1-x}Mn_{1+x}O_{3+y}$. Lattice parameters *a*, *b*, *c* as derived from RT neutron diffraction patterns of powders. Measured and calculated Curie-Weiss constants C_m and $C_{m,calc}$, respectively, paramagnetic Curie point Θ_p as well as Néel temperature T_N for various compositions [80P1].

Composition x	y	*a* [a]) [Å]	*b* [a]) [Å]	*c* [a]) [Å]	C_m [$cm^3\,K\,mol^{-1}$]	$C_{m,calc}$ [$cm^3\,K\,mol^{-1}$]	T_N [K]	Θ_p [K]
0	0.017	5.447	5.809	7.585	4.56	4.55	95	− 12
0.03	0.03	5.445	5.766	7.592				
0.05	0.00	5.445	5.747	7.597	4.69	4.69	84	25
0.08	0.00	5.442	5.710	7.608	4.72	4.73	70	54

[a]) Standard deviation of the lattice parameters *a*, *b*, *c* is 0.002 Å.

$NdMnO_3$

I Phase transitions and non-stoichiometry [84K3]. Electrical conductivity and Jahn-Teller type phase transitions [83K4].

$Nd_{1-x}Ca_xMnO_3$

II Susceptibility vs. temperature curves Fig. 187. The slope of the curves χ_g^{-1} vs. T for x = 0.5, 0.6, 0.7, 0.8 and 0.9 changes at the temperatures of about 620, 490, 440, 280 and 230 K, respectively [88T1]. Metal-insulator transitions [88T1].

IV X-ray photoelectron spectroscopy [88T2].

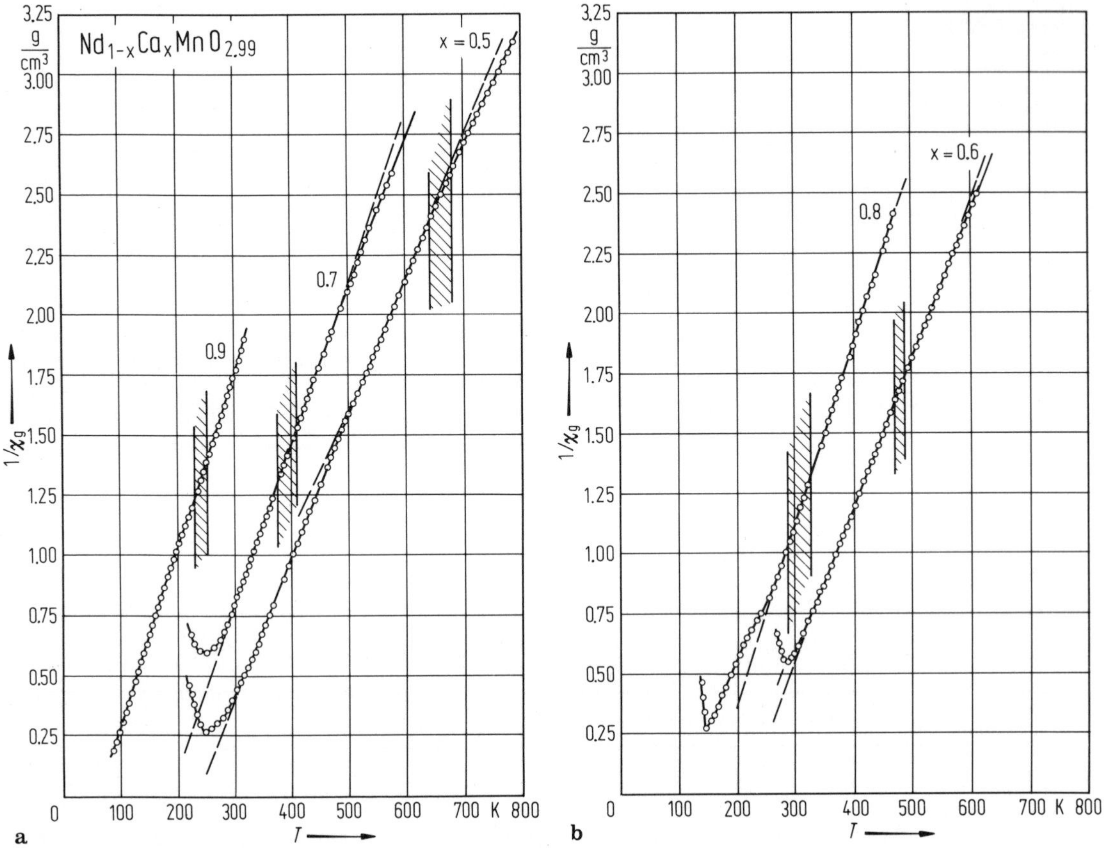

Fig. 187. $Nd_{1-x}Ca_xMnO_{2.99}$. Inverse magnetic mass susceptibility χ_g^{-1} vs. temperature T. The hatched areas indicate the metal-insulator transition temperature region [88T1].

$EuMnO_3$

III ^{151}Eu Mössbauer data [84S3; see also 81G6].

T [K]	δ*) [$mm\,s^{-1}$]	ΔE_Q [$mm\,s^{-1}$]	η
294.8	0.71(5)	− 5.59(27)	0.74(10)
77.9	0.77(4)	− 5.67(20)	0.95(20)

*) With respect to the source ^{151}Sm (SmF_3).

$DyMn_{1-x}Fe_xO_3$

I Spontaneous magnetization Fig. 188, magnetic phase diagram Fig. 189.

$ErMn_{1-x}Fe_xO_3$

I Spontaneous magnetization Fig. 190, magnetic phase diagram Fig. 191.

$TmMn_xFe_{1-x}O_3$

The compound $TmFeO_3$ shows a Γ_4–Γ_2 SRT in the *ac* plane at $T_R = 80 \cdots 90$ K. This temperature is increased to $93 \cdots 110$ and $110 \cdots 136$ K for x = 0.12 and 0.21, respectively. For neutron diffraction data see [83K3, see also 83K5]. For magnetization data Fig. 192.

$YMn_xFe_{1-x}O_3$

I Magnetic phase diagram Fig. 189.

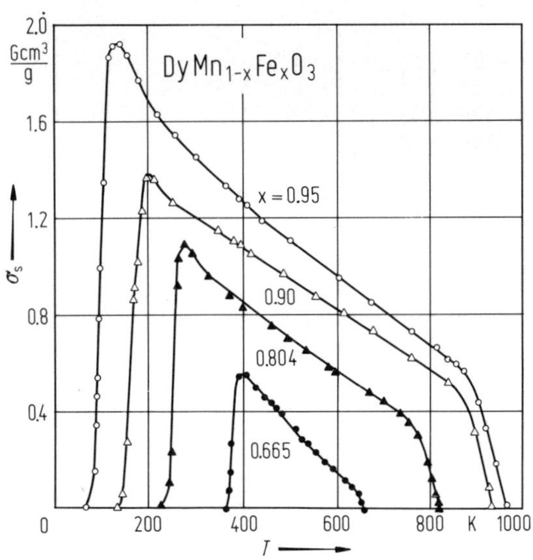

Fig. 188. $DyMn_{1-x}Fe_xO_3$, single crystals. Spontaneous mass magnetization σ_s vs. temperature T [82K1].

◀

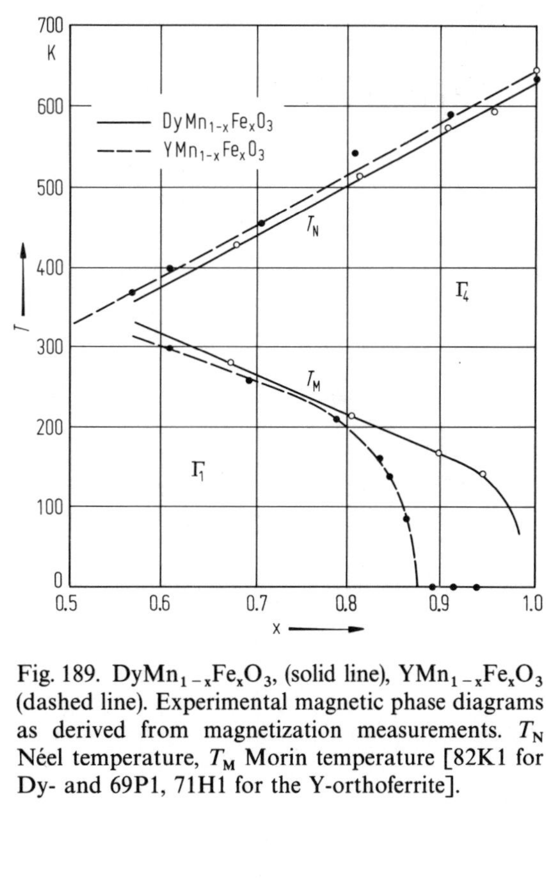

Fig. 189. $DyMn_{1-x}Fe_xO_3$, (solid line), $YMn_{1-x}Fe_xO_3$ (dashed line). Experimental magnetic phase diagrams as derived from magnetization measurements. T_N Néel temperature, T_M Morin temperature [82K1 for Dy- and 69P1, 71H1 for the Y-orthoferrite].

For Fig. 190, see next page.

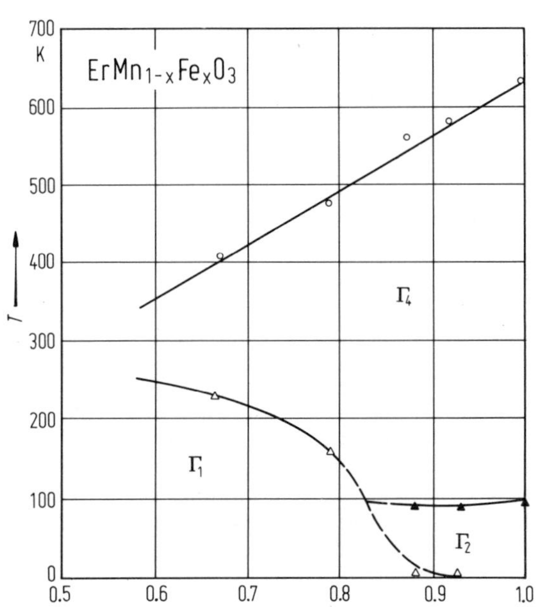

◀

Fig. 191. $ErMn_{1-x}Fe_xO_3$. Experimental magnetic phase diagram as derived from magnetization measurement [82K1].

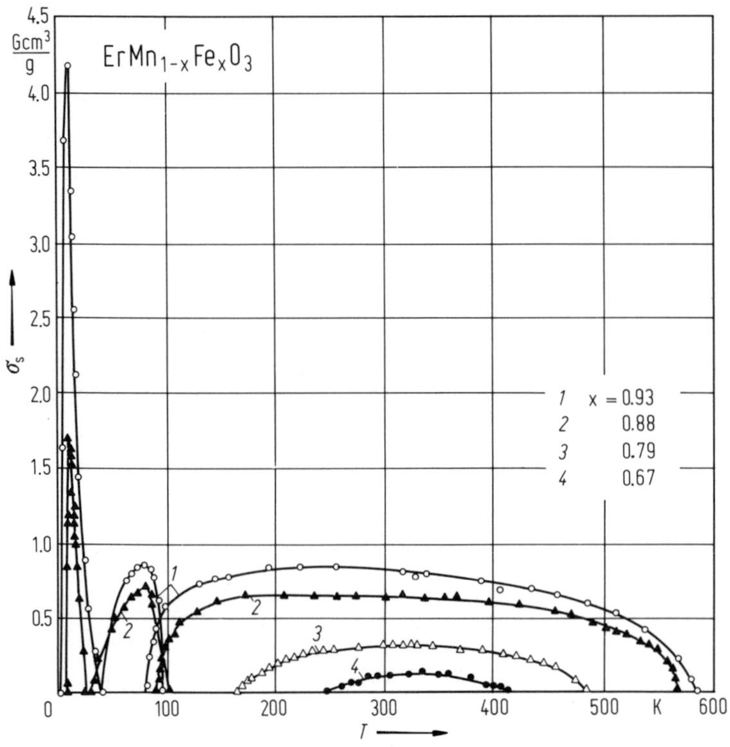

Fig. 190. $ErMn_{1-x}Fe_xO_3$, single crystals. Spontaneous mass magnetization σ_s vs. temperature T [82K1].

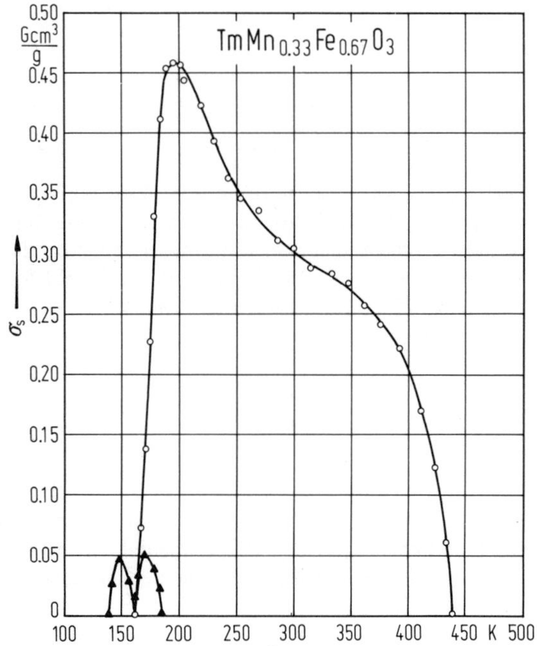

Fig. 192. $TmMn_{0.33}Fe_{0.67}O_3$. Spontaneous mass magnetization σ_s in crystallographic c direction (circles) and in the a direction (triangles). Complicated SRT takes place near $T = 160$ K. Below $T = 141$ K the compound is purely antiferromagnetic [83K3, see also 83K5].

5.5.8 $RFeO_3$ compounds

I For a survey of the spontaneous magnetic structures in the rare earth orthoferrites see Fig. 193. Mechanisms of the magnetic phase transitions of the order-order spin-reorientation type, in which a rotation of the magnetization vector or of the antiferromagnetism vector with respect to the crystallographic axes occurs upon a change of temperature, applied magnetic field or external pressure, are treated extensively in [76B1]. Monoclinic distortions of the lattice during the spin-flip transition have been observed and calculated [81K5]. For calculations of the magnetic dipole anisotropy energy and of the magnetostriction see [78M1, 79M2].
See also Table C and Fig. 2 in subsection 5.5.1.2.
For structural data see Fig. 90 and for the pressure-dependence of the hyperfine field [87D2].

III, IV Magnon frequencies Fig. 194 and Table 33 and [80W1]. Muon precession frequencies Table 34 and their temperature dependence Fig. 195. Birefringence and elastooptic properties [88M1], Fig. 196.

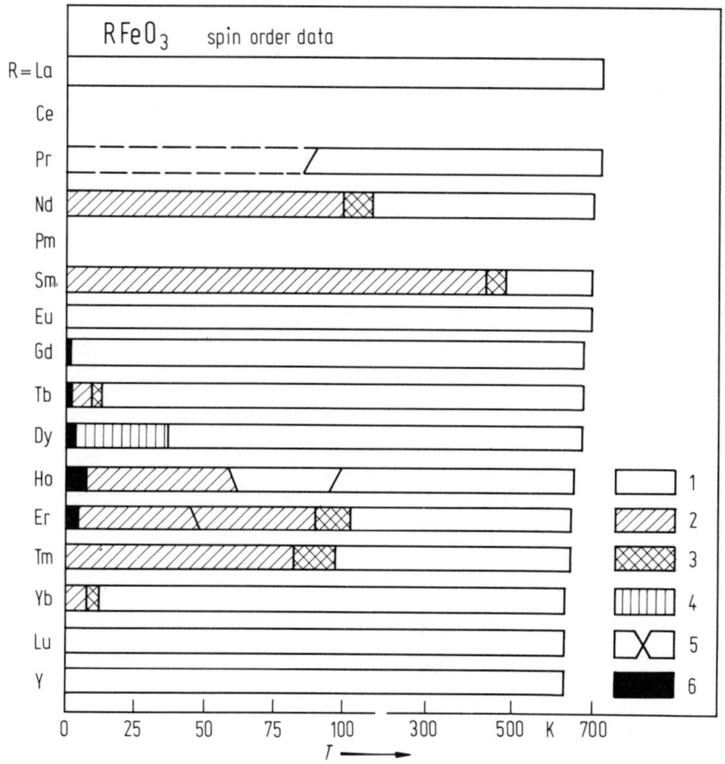

Fig. 193. $RFeO_3$. Diagram indicating the spontaneous magnetic orientation phases and phase transitions. 1) $\Gamma_4(G_x,F_z,A_y)$, 2) $\Gamma_2(G_z,F_x,C_y)$, 3) spin reorientation ($\Gamma_4 \rightarrow \Gamma_2$), 4) $\Gamma_1(G_y,A_x,C_z)$, 5) $\sigma^R + \sigma^{Fe} = 0$ antiferromagnetism, 6) R^{3+} re-ordering region [81G4].

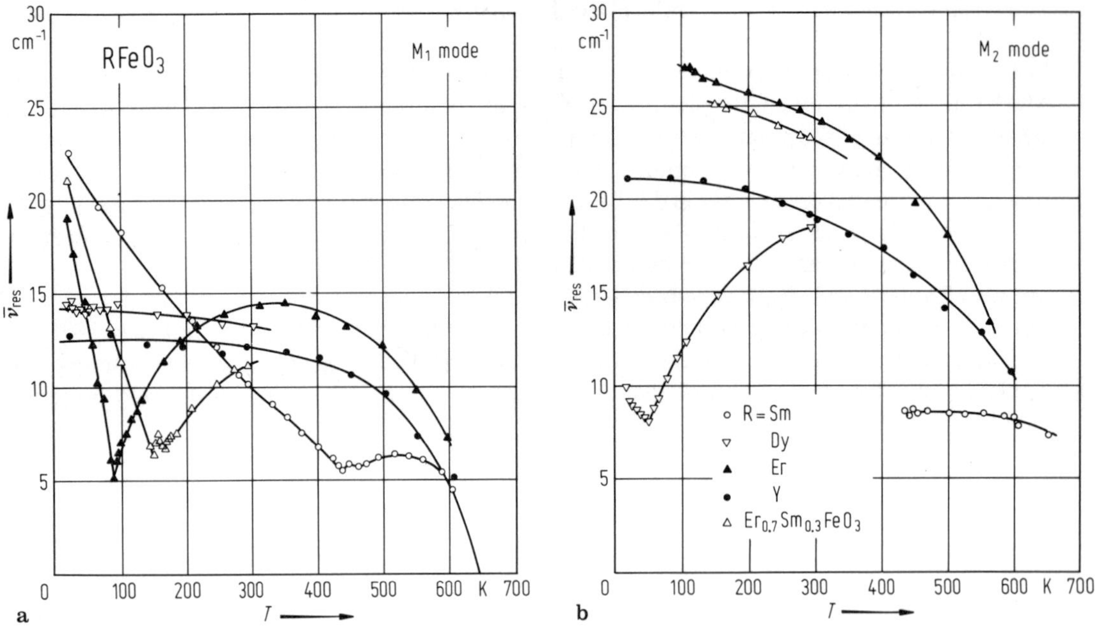

Fig. 194. $RFeO_3$, single crystals. Temperature dependences of the $k = 0$ magnon frequencies (given as wave number $\bar{\nu}_{res}$) of the modes M_1 and M_2 [83K6].

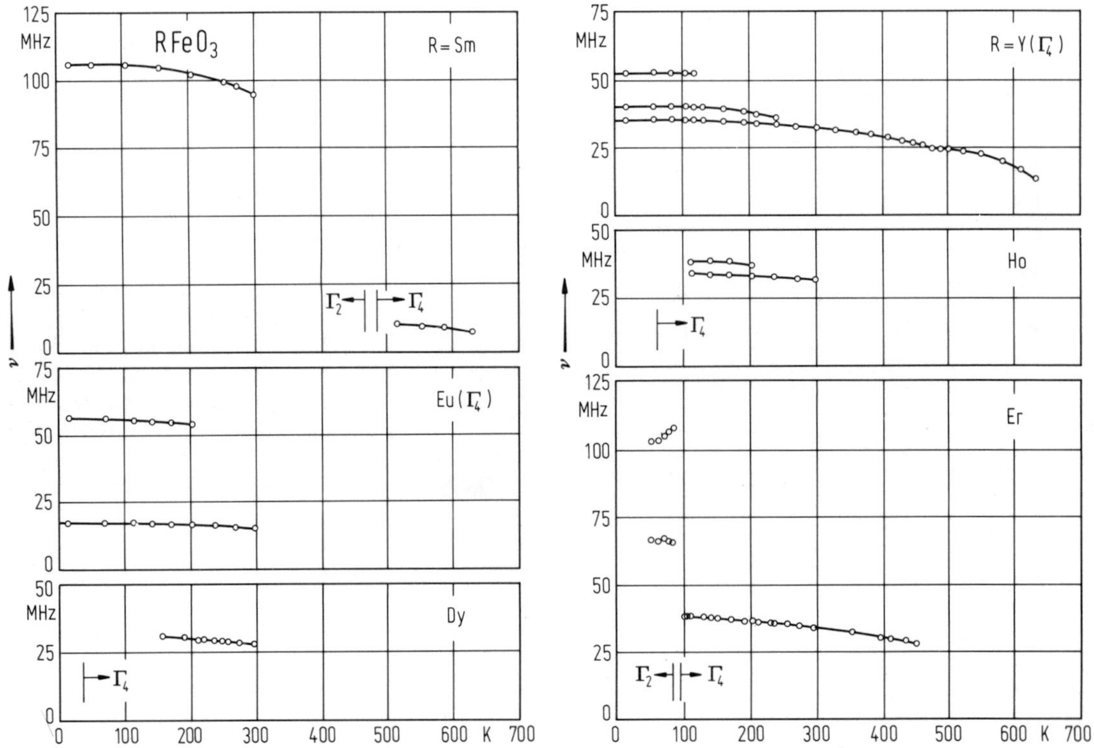

Fig. 195. $RFeO_3$. Muon precession frequencies ν in zero applied field vs temperature T. The iron-spin configurations are indicated [83H1, see also 84H2, 81H1].

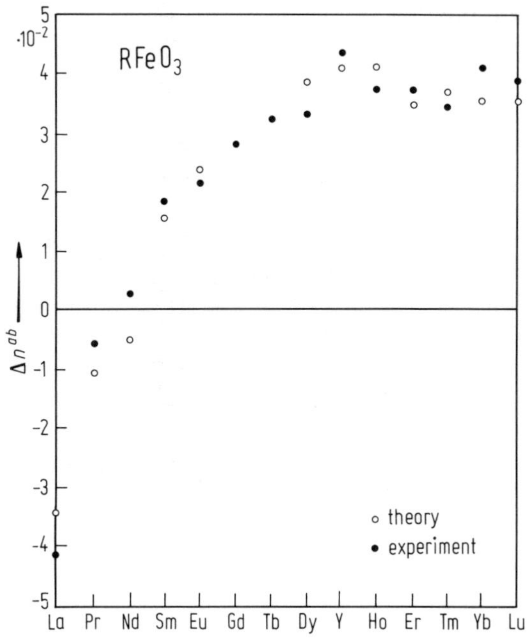

Fig. 196. $RFeO_3$. Birefringe $\Delta n^{a,b} = n^a - n^b$ in the *ab* plane at RT. The calculated values are based on a simple deformation model [88M1].

Table 34. $RFeO_3$. Muon precession frequency ν and corresponding local magnetic field B_μ for the Γ_4 iron-spin configuration at RT [83H1].

Compound	ν [MHz]	B_μ [kG]
$SmFeO_3$	13.5*)	1.00
$EuFeO_3$	15.8	1.17
$DyFeO_3$	27.9	2.06
$HoFeO_3$	31.0	2.29
$ErFeO_3$	34.0	2.51
$YFeO_3$	32.0	2.36

*) Extrapolated from high-temperature results using Mössbauer data of [67E1].

Table 33. $RFeO_3$. Magnon modes M_1 and M_2 (the $k = 0$ frequencies $\hbar\omega_1$ and $\hbar\omega_2$ of two spin-wave modes) at RT and the spin-flip transition temperature T_R derived from the temperature dependence of these frequencies.

Compound	$\hbar\omega_1$ [cm^{-1}]		$\hbar\omega_2$ [cm^{-1}]		T_R [K]
	[82W1]	[83K6]	[82W1]	[83K6]	[83K6]
$NdFeO_3$		8.3		18.0	90 ··· 120
$SmFeO_3$	8.9	6.2a)	18.4	8.4a)	436 ··· 456
$Sm_{0.3}Er_{0.7}FeO_3$		11.0		23.0	148 ··· 166
$EuFeO_3$		11.7		17.5	–
$DyFeO_3$	9.1	13.5	16.5	18.5	50
$HoFeO_3$	13.0		18.0		–
$ErFeO_3$	13.75	13.3	22.7	23.7	90 ··· 103
$YFeO_3$	11.1	12.0	17.4	18.7	–

a) At 500 K.

5.5.8.1 $LaFeO_3$ compounds

$LaFeO_3$ (see LB III/4a, p. 238 and III/12a, p. 451)
I Chemical stability at high temperature [79N1].
III Effective magnetic hyperfine field derived from ^{119}Sn Mössbauer spectra Table 35.

$La_{0.3}Dy_{0.7}FeO_3$
I Magnetization curves Figs. 197⋯199. The components of the saturation magnetization along the *a* and *b* axis at 4.2 K are $\sigma^a \approx 77\ \mathrm{G\,cm^3\,g^{-1}}$ and $\sigma^b \approx 114\ \mathrm{G\,cm^3\,g^{-1}}$, respectively. The dilution of the concentration of Dy^{3+} ions by non-magnetic La^{3+} ions changes only slightly the anisotropy of the *g* factor and of the R-Fe exchange interaction of the remaining Dy^{3+} ions. A replacement of Dy^{3+} ions (radius 1.22Å) by La^{3+} ions (radius 1.07 Å) reduces the orthorhombic distortion. This may result in a considerable change in the magnetic anisotropy constants of the iron ions reflected by the appearance of the $G_{xy}F_z \rightarrow G_zF_x$ spin reorientation and a change in the nature of the transition in the *ab* plane ($G_xF_z \rightarrow G_{xy}F_z$). Three spin reorientations of the iron spins are found: below $T \approx 60$ K the reorientation $G_xF_z \rightarrow G_{xy}F_z$ in the *ab* plane, resulting at ≤ 10K in a configuration close to G_y; below this temperature a mode G_zF_x appears (vector ***G*** moving out of the *ab* plane), disappearing again in the region of antiferromagnetic Dy^{3+} ordering at $T \leq 3.3$ K, where ***G*** returns to the *ab* plane [79D1].

$LaFe_{1-x}Co_xO_3$
I Resistivity Fig. 200.
II Magnetic susceptibility Fig. 201.

$LaFe_{1-x}Ni_xO_3$
II Magnetic susceptibility Fig. 202.

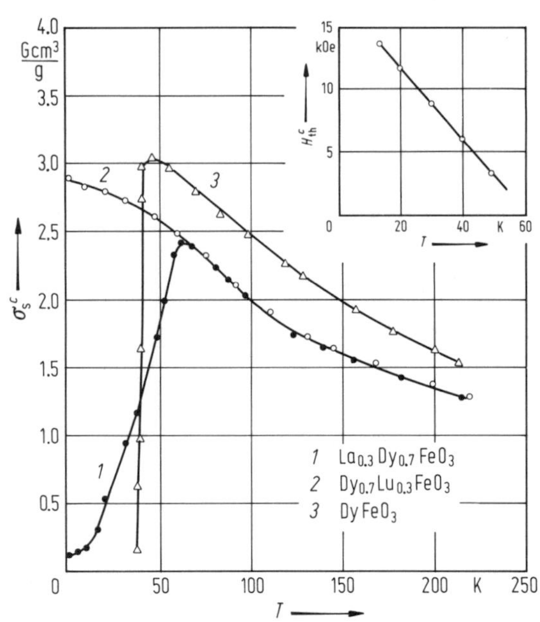

Fig. 197. $La_{0.3}Dy_{0.7}FeO_3$, $Dy_{0.7}Lu_{0.3}FeO_3$, $DyFeO_3$. Temperature dependence of the spontaneous weak ferromagnetic moment σ_s^c along the *c* axis. Inset: temperature dependence of the threshold field H_{th}^c inducing the spin reorientational phase transition $G_{xy}F_z \rightarrow G_xF_z$ in $La_{0.3}Dy_{0.7}FeO_3$ [79D1].

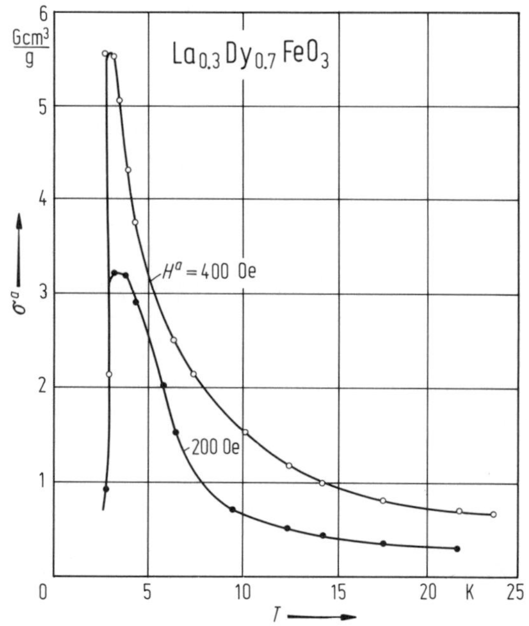

Fig. 198. $La_{0.3}Dy_{0.7}FeO_3$. Temperature dependence of the magnetization σ^a caused by a magnetic field H^a in the direction of the *a* axis [79D1].

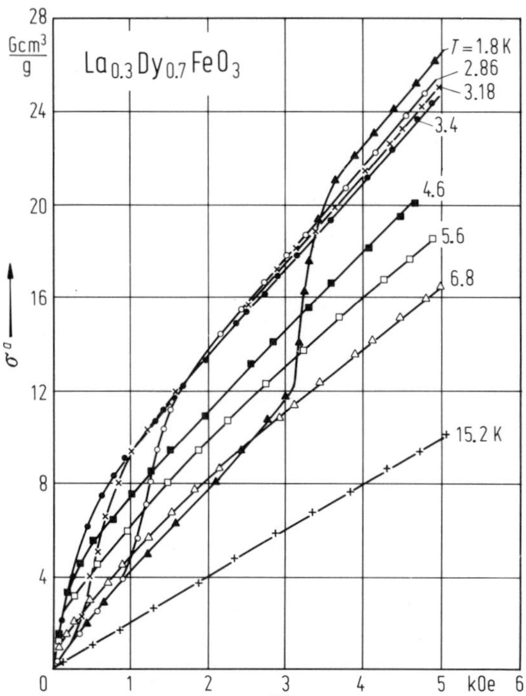

Fig. 199. $La_{0.3}Dy_{0.7}FeO_3$. Magnetization curves measured along the *a* axis at various temperatures *T* [79D1].

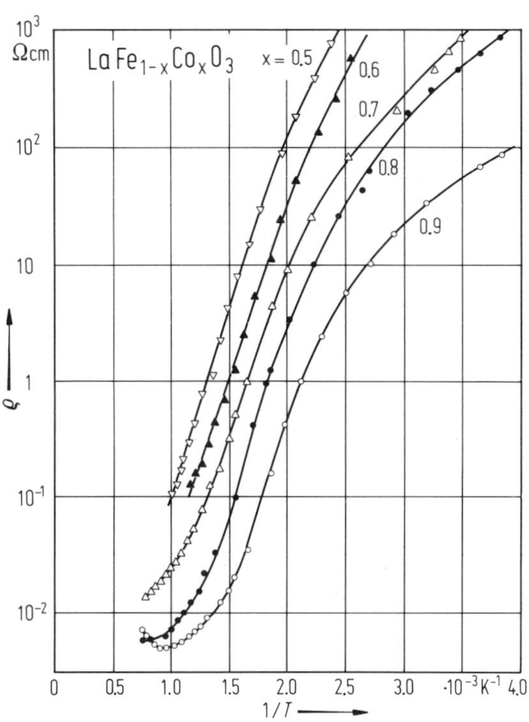

Fig. 200. $LaFe_{1-x}Co_xO_3$. The resistivity ρ vs. the reciprocal temperature *T* [75R1].

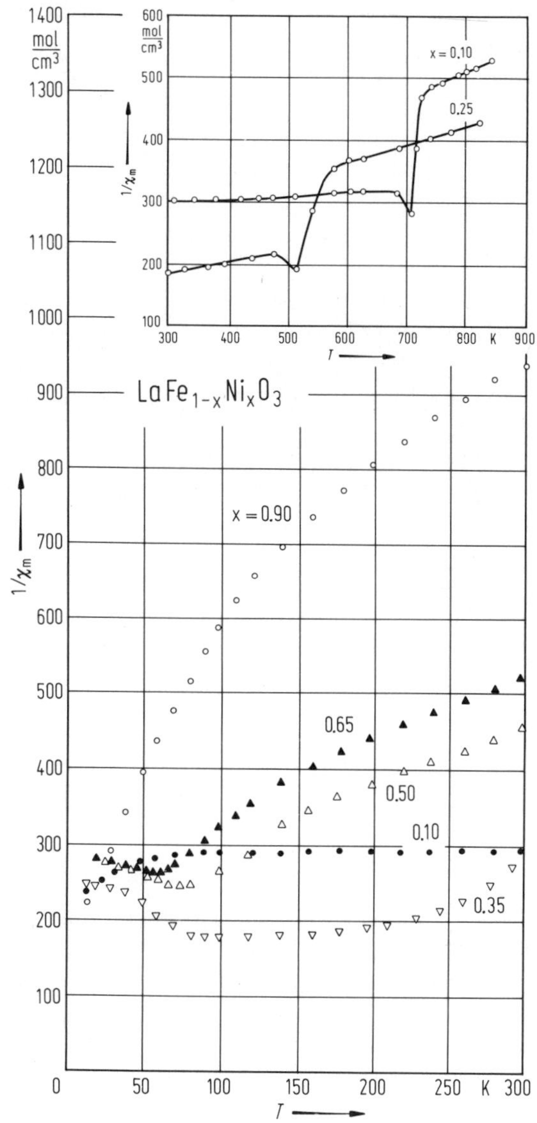

Fig. 202. $LaFe_{1-x}Ni_xO_3$. Inverse molar magnetic susceptibility χ_m^{-1} vs. temperature *T* [84V2, see also 78A1].

For Fig. 201, see next page.

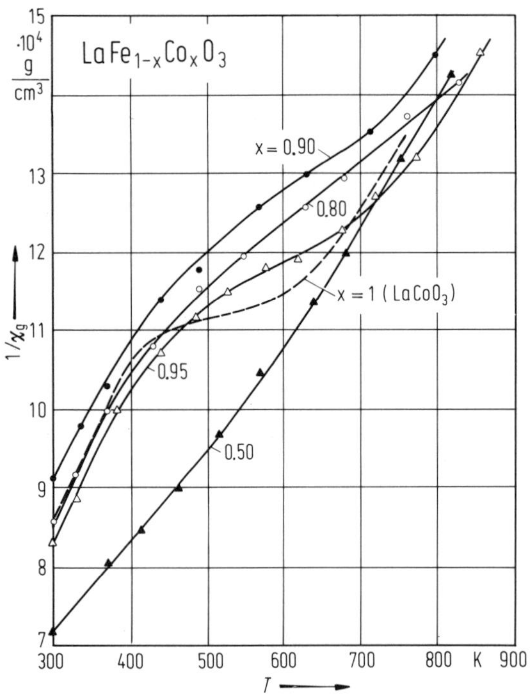

Fig. 201. $LaFe_{1-x}Co_xO_3$. Inverse magnetic mass susceptibility χ_g^{-1} vs. temperature T [75R1].

Table 35. $LaFeO_3$, $EuFeO_3$, $LuFeO_3$. Effective magnetic hyperfine fields (in kOe) derived from Mössbauer spectra of ^{119}Sn doped samples with M substitutions. H^6_{hyp} is the field corresponding to six Fe^{3+} nearest neighbors and H^5_{hyp} corresponds to five nearest neighbors of Fe^{3+} ions and an additional M^{3+} ion [83G4].

M	$LaFeO_3$ H^6_{hyp}	H^5_{hyp}	$EuFeO_3$ H^6_{hyp}	H^5_{hyp}	$LuFeO_3$ H^6_{hyp}	H^5_{hyp}
Sn	252	215	196	173	158	145
Al	258	222	198	184	–	–
Ga	253	212	192	160	153	124
Sc	251	203	?	?	–	–
Cr	254	221	195	181	155	142
Mn	258	247	197	187	–	–
Co	254	221	195	183	156	140
Ni	255	235	194	185	–	–

5.5.8.2 $CeFeO_3$

(See LB III/4a, p. 238 for magnetic properties and III/4a, p. 165 for crystallographic properties.)

5.5.8.3 $PrFeO_3$ (see also LB III/4a, p. 238 and III/12a, p. 452)

I Refined data on crystal and magnetic structure parameters, based on neutron diffraction at 8 and 293 K, are given in [85S1] and [82S1], respectively, see Table 36. The iron magnetic moment at RT is parallel to the *a* axis (G_x type) and $p_{Fe^{3+}} = p_x = 3.73\,(3)\,\mu_B$. The same holds at 8 K: $p_{Fe^{3+}} = p_x = 4.14\,(4)\,\mu_B$. Neutron diffraction at $T = 1.5 \cdots 80$ K does not reveal long-range order of the magnetic moments of the Pr^{3+} ions [87S2]. For susceptibility curves, see Fig. 203.

IV For the complex Kerr effect components for photon energies between 2 and 5.6 eV see [83V1].

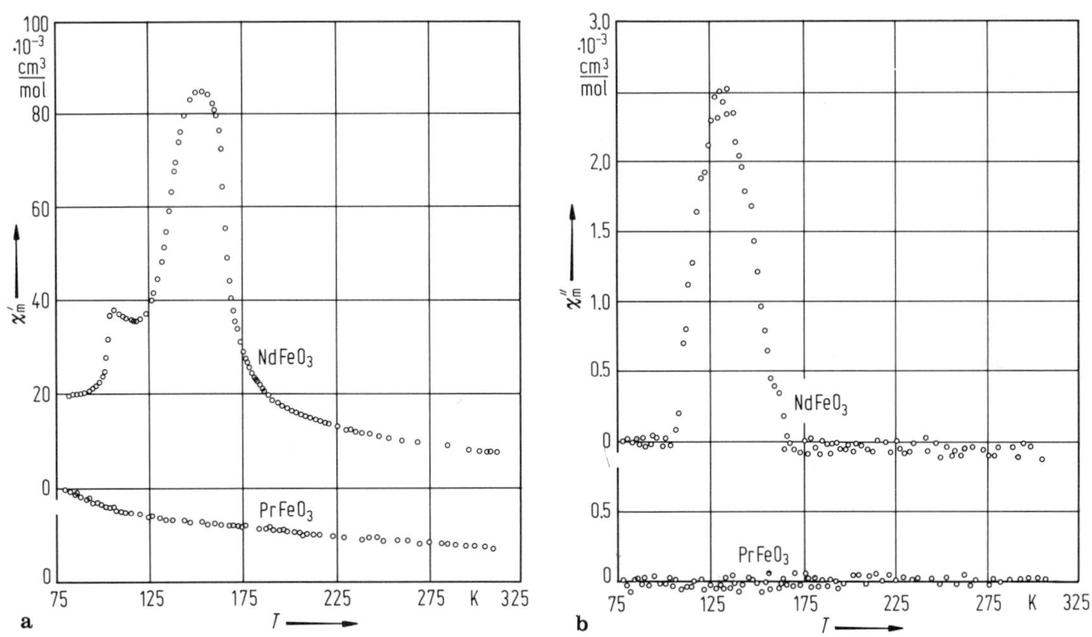

Fig. 203. $PrFeO_3$; $NdFeO_3$. Polycrystalline sample. Temperature dependence of real and imaginary part of the ac susceptibility χ' and χ'', respectively, at $h = 0.3$ Oe and a frequency of 76 Hz [83C1].

Table 36. $PrFeO_3$. Crystallographic and magnetic data at 8 K and at RT [85S1].

	Neutron diffraction		X-rays
Ref.	85S1	82S1	70M1
T [K]	8	293	293
a [Å]	5.486 (4)	5.483 (5)	5.482
b [Å]	5.591 (4)	5.578 (5)	5.578
c [Å]	7.783 (5)	7.788 (7)	7.786
$p_{Fe^{3+}}$ [μ_B]	4.14 (4) in x direction	3.73 (3) in x direction	–

5.5.8.4 $NdFeO_3$ compounds

$NdFeO_3$ (see also LB III/4a, p. 239 and III/12a, p. 452)

I Parameters of crystal and magnetic structure derived from neutron diffraction experiments are given in Table 37.

At RT the direction of the magnetic moment makes an angle of 15(5)° with the orthorhombic x axis [82S2, see also 76K1]. Neutron diffraction reveals that the reorientation from G_x via G_xG_z to G_z takes place in a large temperature range, 112 ··· 187K [77K4], 70 ··· 170K [82S1] or 105 ··· 200 K [86S1]. Results of susceptibility measurements suggest a spin reorientation range 103 ··· 165 K, Fig. 203 [83C1]. The extended range seems to be related to inhomogeneities in the crystal [91B1]. For spin reorientation range derived from magnetization measurements see Table 38 [90G1]. A possible onset of magnetic order of the Nd^{3+} moments may be the cause of two weak extra lines in the neutron diffraction pattern at 1.5 K. The ordered magnetic moment of Fe^{3+} is $p_{Fe^{3+}} = 4.24(2)\mu_B$ at 1.5 K [87S2]. For INS data see [88L1].

$Nd_{1-x}Yb_xFeO_3$

I Phase diagram Fig. 204 and Table 38. Spontaneous magnetization Fig. 205

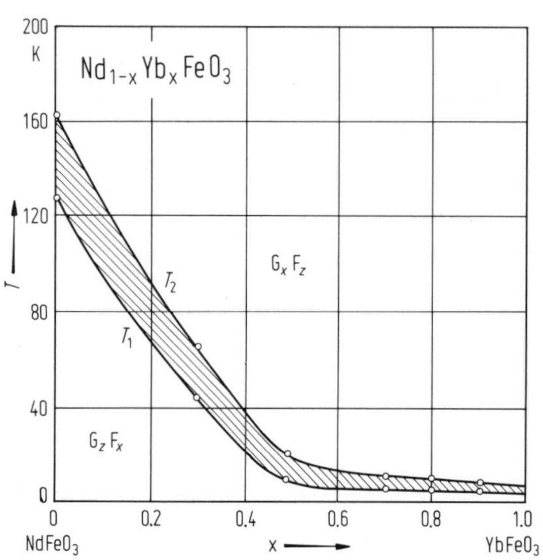

Fig. 204. $Nd_{1-x}Yb_xFeO_3$. Magnetic phase diagram. The hatched region indicates the width of the temperature interval ΔT of the SR process [90G1].

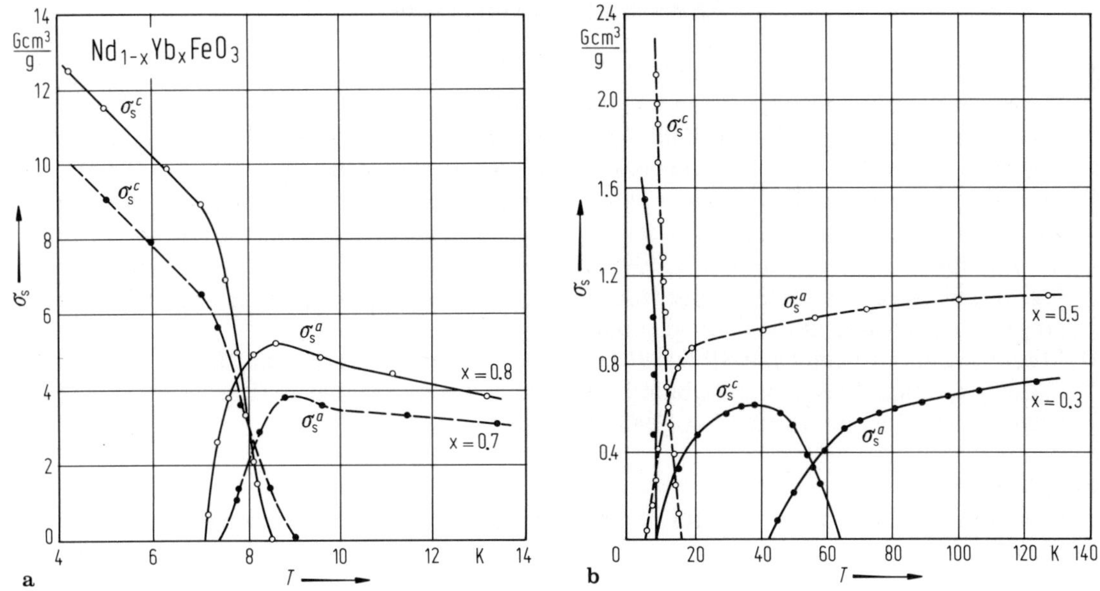

Fig. 205. $Nd_{1-x}Yb_xFeO_3$. Spontaneous mass magnetization σ_s vs. temperature T for the crystallographic a and c directions [90G1]. (a) x = 0.7 and 0.8, (b) x = 0.3 and 0.5.

Table 37. $NdFeO_3$. Crystallographic and magnetic parameters at various temperatures.

	Neutron diffraction			X-rays
Ref.	82S1	87S2	82S1	70M1
T [K]	8	1.5	293	293
a [Å]	5.442 (5)	5.443 (2)	5.452 (5)	5.453
b [Å]	5.587 (5)	5.588 (2)	5.591 (5)	5.584
c [Å]	7.746 (7)	7.749 (3)	7.767 (7)	7.768
$p_{Fe^{3+}}$ [μ_B]	4.13 (4) in z direction	4.24 (2) in z direction	3.82 (4) in x direction	

Table 38. $Nd_{1-x}Yb_xFeO_3$. Phase transition temperatures T_N and T_1, T_2 (see Fig. 204) and magnetic compensation point T_{comp} in K [90G1].

Composition	T_N	T_1	T_2	$\Delta T = T_2 - T_1$	T_{comp}
$YbFeO_3$	627	6.5	7.4	0.9	–
$Yb_{0.8}Nd_{0.2}FeO_3$	640	7.1	8.5	1.4	–
$Yb_{0.7}Nd_{0.3}FeO_3$	647	7.2	8.9	1.7	–
$Yb_{0.5}Nd_{0.5}FeO_3$	670	9.7	17.5	7.8	–
$Yb_{0.3}Nd_{0.7}FeO_3$	–	47	76	29	10
$NdFeO_3$	698	125	167	42	8

5.5.8.5 $SmFeO_3$ compounds

$SmFeO_3$ (see also LB III/4a, p. 239 and III/12a p. 453)

I The spin reorientation upon cooling is $\Gamma_4(G_xF_z) \rightarrow \Gamma_2(F_xG_z)$. The temperature dependence of the angle θ between the orientation of the weak ferromagnetism vector $\boldsymbol{F}$ and the direction of the c axis is shown in Fig. 206. The spin orientation during the transition is $G_{zx}F_{xz}$ [84B1]. The spontaneous magnetization decreases about 5% in the SR range, the boundaries of the SR range are 457.2 (5) and 465.7 (5) K [77A1], see Fig. 207. For the field dependence of the magnetization see Fig. 208. Magnetostriction Fig. 209. Anisotropy constants Fig. 210. Susceptibility curves Fig. 211. Between 7 and 50 MHz the susceptibility is almost independent on frequency in the SR range. For anomalies in the thermal conductivity see Fig. 212.

III Microwave spectroscopy and SRT [89B1]. For spin wave modes see Fig. 215.

IV Faraday rotation (Figs. 213, 214), birefringence, light absorption, linear dichroism and magnetic circular dichroism are discussed in [79G1 and 90L1].

$Sm_{1-x}Eu_xFeO_3$

I Monoclinic distortions at the SRT for x = 0.5 [87M1]. Magnetic susceptibility at the SRT [90K2].

$Sm_{1-x}Gd_xFeO_3$

I The spin rotation intervals for x = 0.6 and 0.4 are 177 ⋯ 233 and 267 ⋯ 321 K, respectively, see also Fig. 216. For sound propagation in the low-symmetry phase Γ_{24} see Fig. 217. At RT the absolute values of the velocities v_s^o of the shear waves at 10 MHz are 3.85 (15) and 4.6 (2) $\cdot 10^3$ ms^{-1} for the above-mentioned compositions, respectively [84B1, 87B1]. For thermal conductivity see Fig. 218.

$Sm_{1-x}Tb_xFeO_3$

I A planar class of an orthorhombic crystal system, space group C_{2v}^9–Pn2a, class mm2 is suggested [78S1], lattice parameters Fig. 219. For x = 0.55 the spin reorientation $\Gamma_2 \leftrightarrows \Gamma_4$ takes place in the temperature range 250 ⋯ 300 K [87K6], see Fig. 220

IV Optical birefringence [78S1, 79S2].

$Sm_{0.65}Tb_{0.2}Er_{0.15}FeO_3$

I Orientation of weak ferromagnetism Fig. 221. Thermal conductivity Fig. 222.

$Sm_{1-x}Er_xFeO_3$

I Spin orientation at 430 K for x = 0.2 [87R1].

III Spin wave modes for x = 0.7 Fig. 223.

$Sm_{1-x}Tm_xFeO_3$

III Microwave spectroscopy and SRT [89B1].

$Sm_{1-x}Yb_xFeO_3$

I For x = 0.3 the SR region is 159 ⋯ 295 K, see Fig. 217. At RT the absolute value of the velocity v_s^o of the 10 MHz shear wave is 3.24 (13) $\cdot 10^3$ ms^{-1} [87B1]. Microwave spectroscopy and SRT [89B1]. Magnetic phase diagram and composition dependence of the compensation temperature Fig. 224.

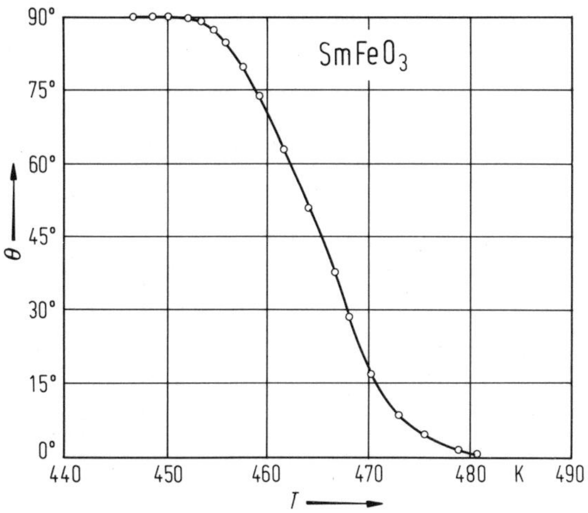

Fig. 206. $SmFeO_3$. Temperature dependence of the angle θ between the c axis and the direction of the weak ferromagnetism in the spin reorientation region as derived from torque measurements [84B1].

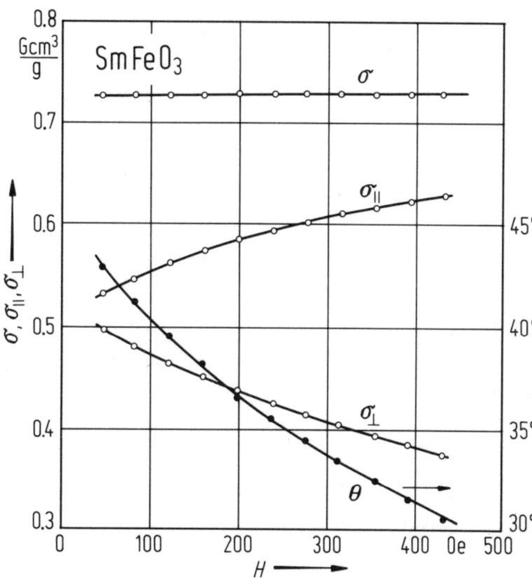

Fig. 208. $SmFeO_3$. Field dependence at $T = 466$ K of the mass magnetization σ and its components in the field direction, $\sigma_{\parallel}$ (i.e. the [101] direction), and perpendicular to this direction ($\sigma_{\perp}$), as well as the angle θ between the magnetization and the c axis [77A1].

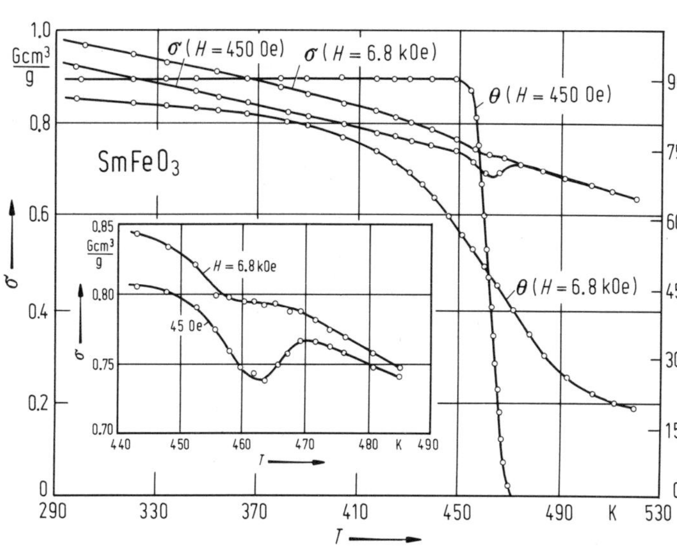

Fig. 207. $SmFeO_3$. Temperature dependence of the magnitude σ of the magnetization and its angle θ (with respect to the c axis) under the influence of an externally applied magnetic field H in the [101] direction [77A1].

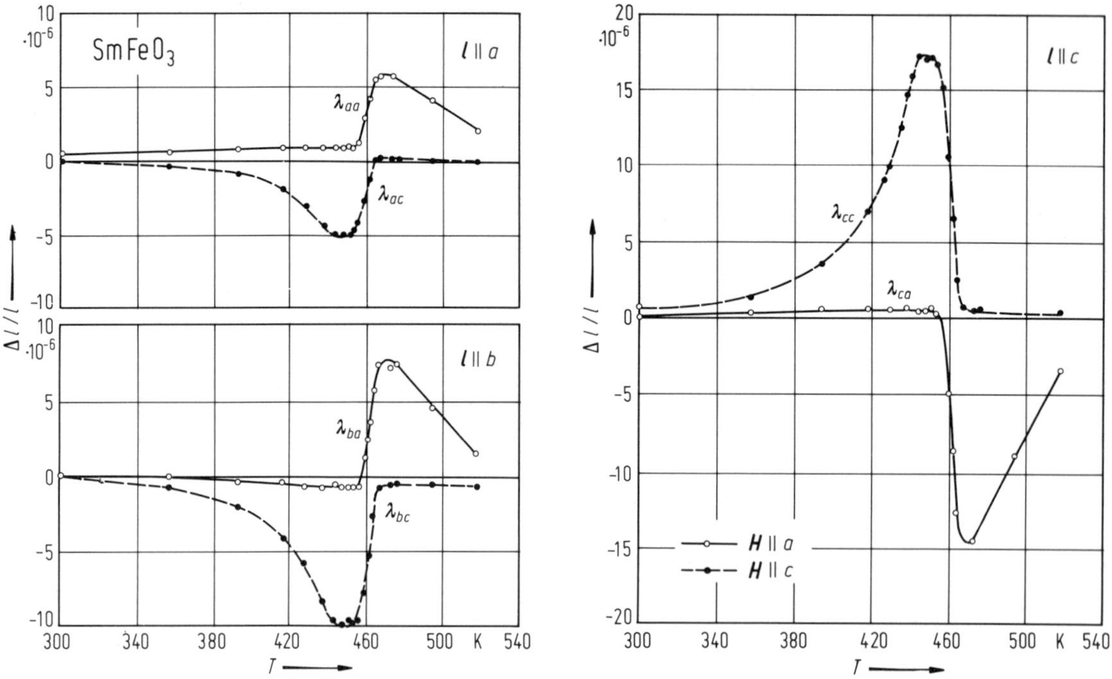

Fig. 209. $SmFeO_3$. Magnetostriction $\lambda_{ij} = \Delta l/l$ in an applied magnetic field $H = 6.8$ kOe as a function of temperature T. λ_{ij} is the magnetostriction along the i axis for an applied field along the j axis [77A2].

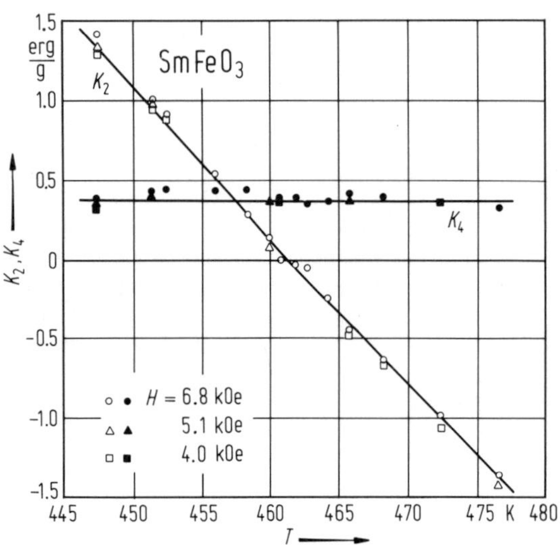

Fig. 210. $SmFeO_3$. Temperature dependence of the second order and fourth order magnetic anisotropy constants K_2 and K_4, respectively. Magnetic torque measurements for various applied fields H. [77A1].

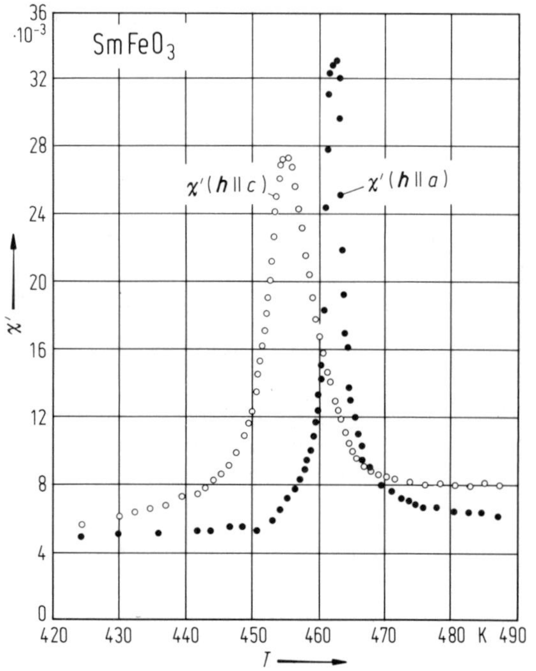

Fig. 211. $SmFeO_3$. The real part of the magnetic susceptibility, χ', vs. temperature for a frequency of 12 MHz. The susceptibility is measured for two cylinders with their long axis parallel to the ac field $\boldsymbol{h}$ and parallel to the crystallographic a or c axis [77H3].

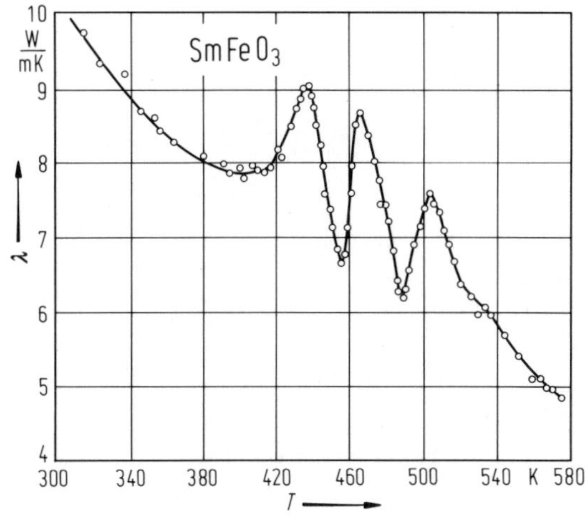

Fig. 212. $SmFeO_3$. Thermal conductivity λ vs. T in the spin reorientation region [84B1].

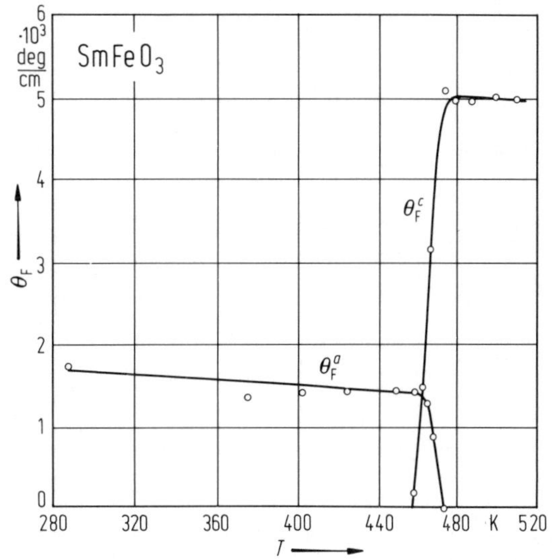

Fig. 213. $SmFeO_3$. Faraday rotation. θ_F along the a and the c axis as dependent on temperature. Wave length 0.65 μm (1.91 eV) [79G1].

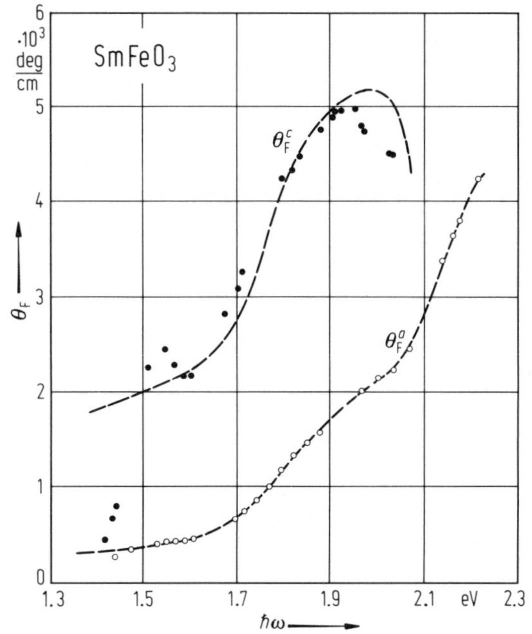

Fig. 214. $SmFeO_3$. Energy spectra of the Faraday rotation. θ_F measured along the a axis at $T = 290$ K and along the c axis at $T = 483$ K [79G1]. Dashed lines are calculated.

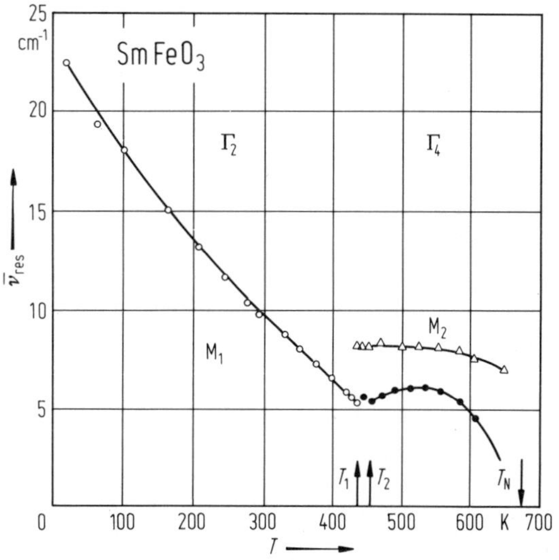

Fig. 215. $SmFeO_3$. Temperature dependence of the frequencies (given as wavenumber $\bar{\nu}_{res}$ in cm^{-1}) of the AFMR modes M_1 and M_2 as derived from Raman scattering from one-magnon excitations [88K3].

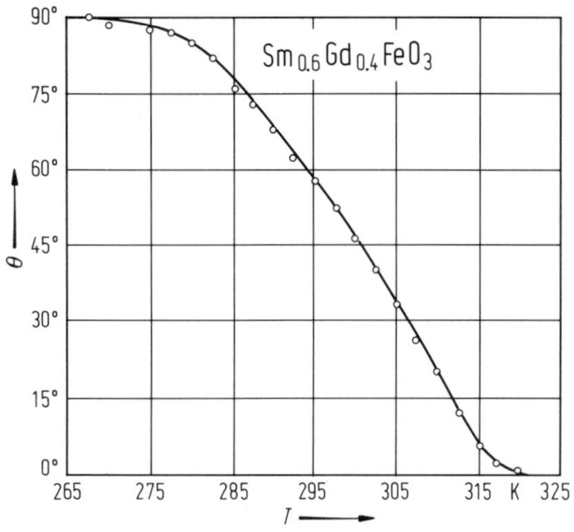

Fig. 216. $Sm_{0.6}Gd_{0.4}FeO_3$. Temperature dependence of the angle θ between the c axis and the direction of the weak ferromagnetism in the spin reorientation region, as derived from torque measurements [84B1].

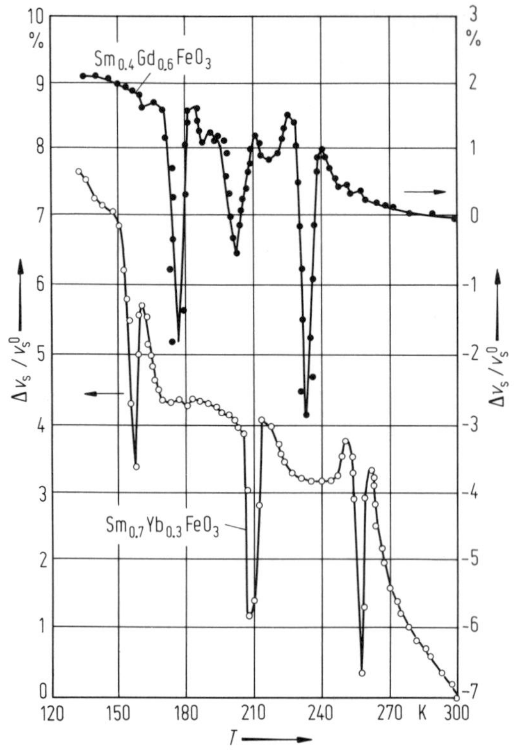

Fig. 217. $Sm_{0.4}Gd_{0.6}FeO_3$; $Sm_{0.7}Yb_{0.3}FeO_3$. Temperature dependence of the relative velocity changes $\Delta v_s/v_s^0$ of 10 MHz shear waves in the SR range, $\boldsymbol{k}\|c$ axis [87B1] $v_s^0 = v_s\,(300\ \text{K})$.

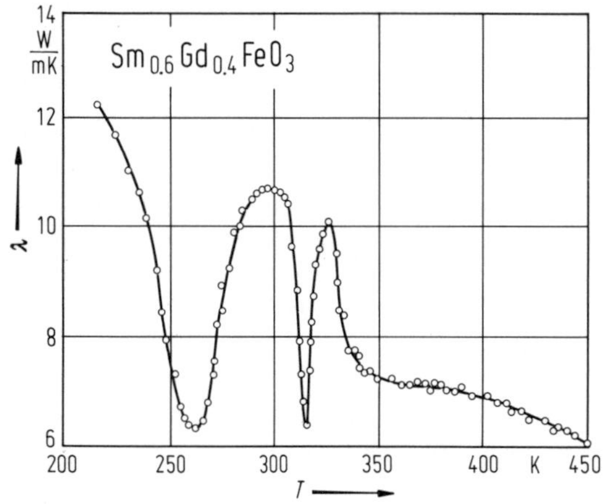

Fig. 218. $Sm_{0.6}Gd_{0.4}FeO_3$. Thermal conductivity λ vs. T in the spin reorientation region [84B1].

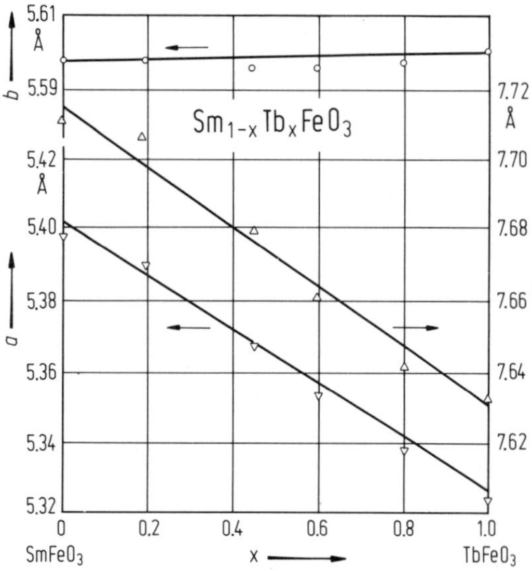

Fig. 219. $Sm_{1-x}Tb_xFeO_3$. Lattice parameters at RT [78S1].

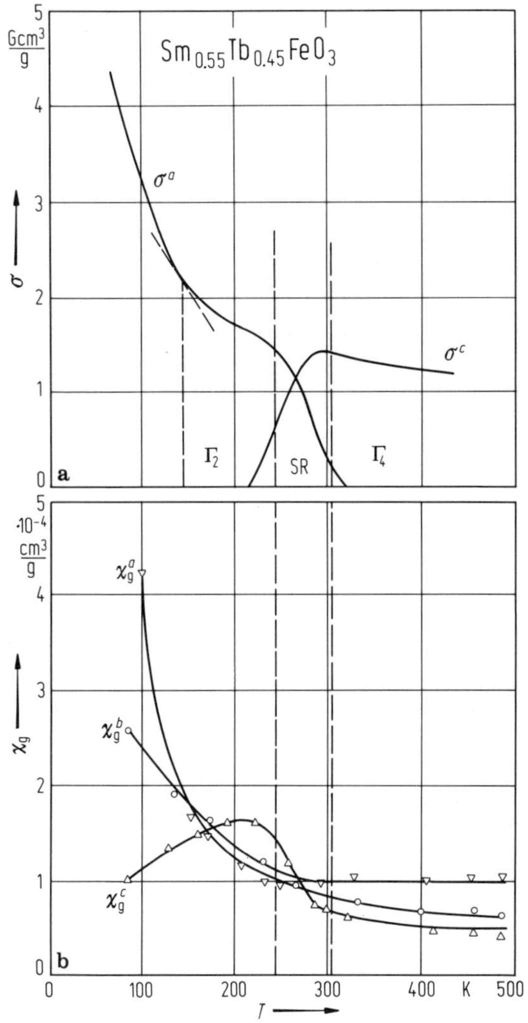

Fig. 220. $Sm_{0.55}Tb_{0.45}FeO_3$. (a) Temperature dependence of the mass magnetization σ in a magnetic field of 12 kOe measured along the directions of the a and c axes. (b) Temperature dependence of the high-field magnetic mass susceptibilities χ_g along the main crystallographic axes [87K6].

◀

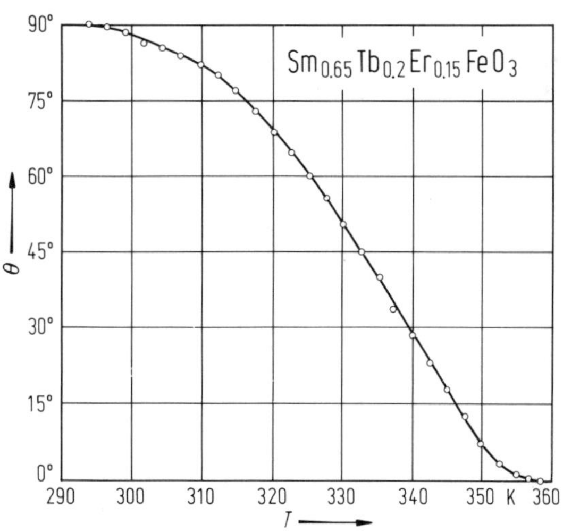

Fig. 221. $Sm_{0.65}Tb_{0.2}Er_{0.15}FeO_3$. Temperature dependence of the angle θ between the c axis and the direction of the weak ferromagnetism in the SR range as derived from torque measurements [84B1].

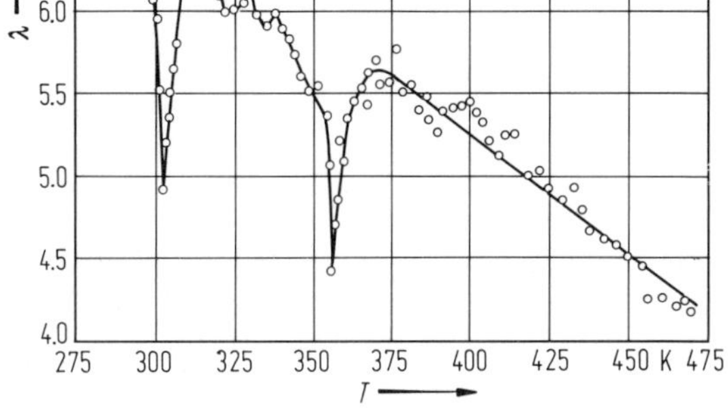

◀

Fig. 222. $Sm_{0.65}Tb_{0.2}Er_{0.15}FeO_3$. Thermal conductivity λ vs. T in the SR range [84B1].

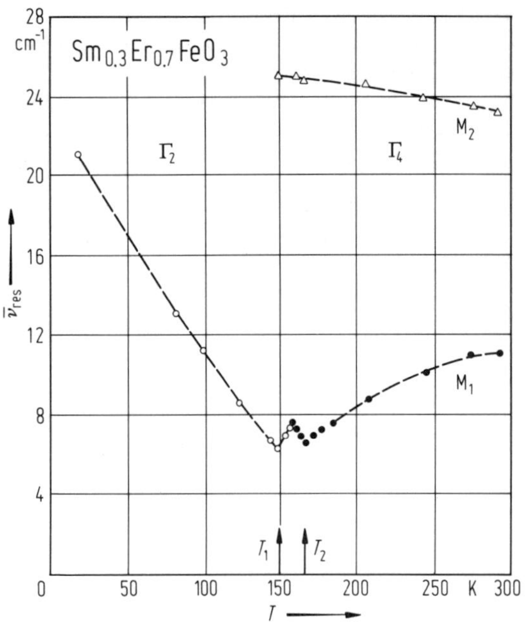

Fig. 223. $Sm_{0.3}Er_{0.7}FeO_3$. Temperature dependence of the frequencies (given as wavenumbers $\bar{\nu}_{res}$ in cm^{-1}) of the AFMR modes M_1 and M_2 as derived from Raman scattering from one-magnon excitations [88K3].

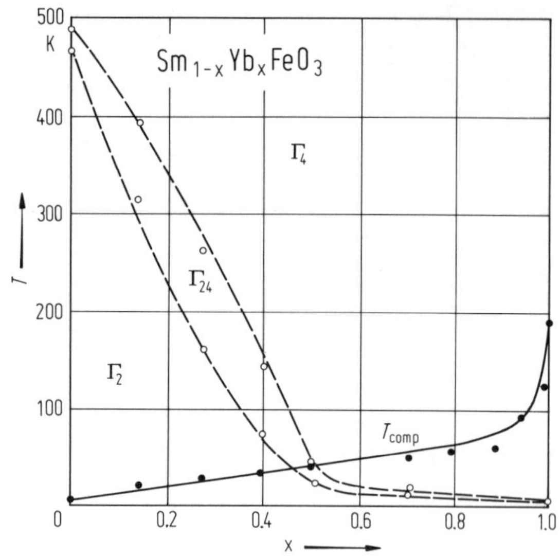

Fig. 224. $Sm_{1-x}Yb_xFeO_3$. Magnetic phase diagram, showing the limits of the spin reorientation range (broken curves) and the magnetization compensation temperature T_{comp} (continuous curve). Points experimental, curves calculated [90B3].

5.5.8.6 $EuFeO_3$ compounds

$EuFeO_3$ (see also LB III/4a, p. 240 and III/12a, p. 454)

I $T_N = 622$ K, from ^{57}Fe Mössbauer spectra [83G2]. Lattice parameters Table 39.

III ^{151}Eu Mössbauer parameters [84S3; see also 81G6] and Table 40.

T [K]	δ*) [$mm\,s^{-1}$]	ΔE_Q [$mm\,s^{-1}$]	η
297.5	0.72 (6)	− 6.09 (30)	0.89 (20)
78.1	0.90 (6)	− 7.07 (32)	0.80 (23)

*) With respect to the source ^{151}Sm (SmF_3).

H_{hyp} equals 555 and 508 kOe at 85 and 295 K, respectively [83G2]. Quadrupole interaction $\varepsilon \approx 0.01$ $mm\,s^{-1}$ for the Fe^{3+} sites at 295 K. Effective magnetic hyperfine field derived from ^{119}Sn Mössbauer spectra, Table 35.

IV Muon spin resonance Table 34 and Fig. 195. Magnon modes Table 33.

$EuFe_{1-x}Co_xO_3$

I Néel temperatures (± 5K) as derived from the collapse of the hyperfine structure of the ^{57}Fe Mössbauer spectra [83G3].

x	0.1	0.2	0.3	0.4	0.5	0.6
T_N [K]	605	524	439	405	234	< 85

III Mössbauer spectra. Replacement of a neighboring Fe^{3+} by Co^{3+} results in a reduction of H_{hyp} at ^{57}Fe nucleus of 10.2 (5) kOe extrapolated to 0 K. From the ^{151}Eu Mössbauer spectra for x = 0.5 at 295 K is derived $\Delta E_Q = -6.79(9)\,mm\,s^{-1}$, $\eta = 0.81(3)$ [83G3].

$Eu_{1-x}Gd_xFeO_3$

I Lattice parameters Table 39.

III Mössbauer parameters Table 40.

Table 39. $Eu_{1-x}Gd_xFeO_3$. RT lattice parameters in [Å] of the orthorhombic structure (± 0.005 Å) [91L1].

x	*a*	*b*	*c*
0	5.372	5.608	7.681
0.2	5.364	5.602	7.675
0.4	5.360	5.607	7.671
0.6	5.363	5.601	7.673
0.8	5.354	5.613	7.663

Table 40. $Eu_{1-x}Gd_xFeO_3$. Mössbauer parameters at RT [91L1].

Compound	^{151}Eu parameters			^{57}Fe parameters		
x	$\delta^{a)}$ [mm/s] (± 0.01)	ΔE_Q [mm/s] (± 0.15)	η (± 0.07)	$\delta^{b)}$ [mm/s] (± 0.01)	ΔE_Q [mm/s] (± 0.02)	H_{hyp} [kOe] (± 0.3)
0	− 0.32	− 6.14	0.91	0.35	− 0.02	503.3
0.2	− 0.30	− 6.20	0.91	0.31	0.06	503.9
0.4	− 0.29	− 5.92	0.90	0.34	− 0.02	502.7
0.6	− 0.27	− 5.83	0.68	0.35	− 0.03	502.4
0.8	− 0.32	− 6.54	0.87	0.35	− 0.03	501.2

a) With respect to Eu_2O_3.
b) With respect to α-Fe.

5.5.8.7 $GdFeO_3$ compounds

$GdFeO_3$ (see also LB III/4a, p. 240 and III/12a, p. 454).

I For temperature dependence of the spontaneous magnetization see Fig. 225, magnetostriction Fig. 226. From the kinks in the magnetostriction vs. field curves, which correspond to spin reorientations $\Gamma_4 \rightarrow \Gamma_2$ the phase diagram of Fig. 227 can be constructed. The observed properties can only be explained by taking into account an anisotropic Gd-Fe exchange additional to the isotropic exchange and dipole interaction [81B1].

II The exchange integral between Fe^{3+} and Gd^{3+} as derived from an analysis of paramagnetic Curie-Weiss law behavior of Gd^{3+} ions in the temperature range 77 ··· 650 K is $J/k = 0.23$ K [79S1].

III In the temperature range 1.1 ··· 7 K only one ^{57}Fe NMR line is found with a frequency of 74.358(12) MHz corresponding to a magnetic field of 540.0(1) kOe at the Fe^{3+} nuclei, independent of temperature. The configuration $\Gamma_4(G_x F_z)$ was retained down to 1.1 K [87G1]. See also [88V1]. Spin ordering of the Gd^{3+} moments at $T_{N_2} = 2.48(5)$ K, derived from ^{57}Fe NMR spectra [87G1, 88V1]. Field induced spin reorientation [77D2]. Canting angle versus temperature Fig. 228, critical fields Table 41.

Measurements of the NMR line width and the rf susceptibility along the *a* axis in the absence of a magnetic field seem to reveal a not well-defined correlation of electron spin fluctuations at $T = 2.44(5)$ K [87G1]. Submillimeter spectra reveal a quasiferromagnetic and a quasianttiferromagnetic mode of the AFMR, Fig. 229. The peculiar temperature dependences seem to be related to the fact that the splitting of the ground state of Gd^{3+} at $H = 0$ is as small as less than 0.1 cm^{-1}, whereas the AFMR frequencies in the Fe subsystem correspond to $\approx 10\ cm^{-1}$ [86B1]. Quadrupole splitting of Gd ground state $\Delta E_Q = 2.73(8)\ mm\,s^{-1}$ [79B1]. The Gd^{3+}–Fe^{3+} exchange interaction derived from Mössbauer spectra is $J/k = 0.39$ K [77D2].

$GdFe_{0.99}Co_{0.01}O_3$

I For magnetization vs. temperature see Fig. 225.

$Gd_{0.95}Y_{0.05}FeO_3$

IV Optical properties [83V1].

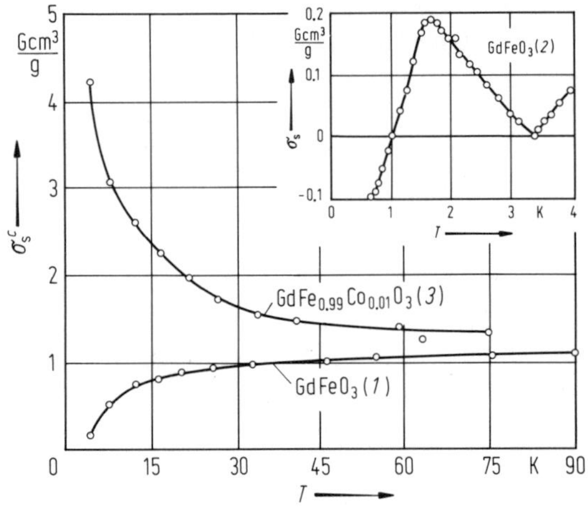

Fig. 225. $GdFeO_3$. Spontaneous mass magnetization σ_s^c along the *c* axis vs. temperature *T*. Curve *1*: [75D1]. *2*: [70C1]. *3*: For 1% Co-substituted $GdFeO_3$ [81B1].

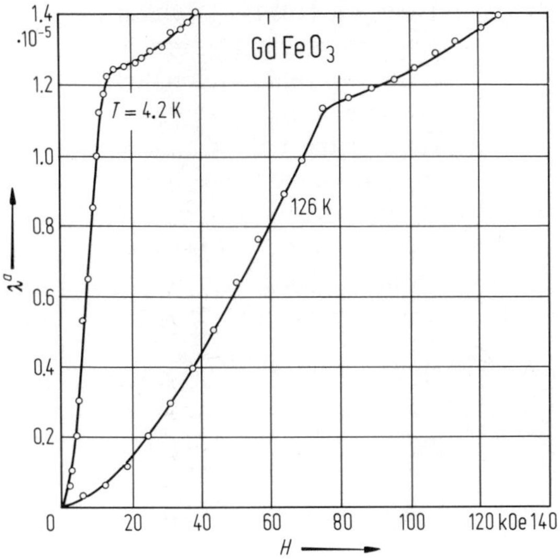

Fig. 226. $GdFeO_3$. Longitudinal magnetostriction λ^a measured at two temperatures in a magnetic field parallel to the a axis [81B1].

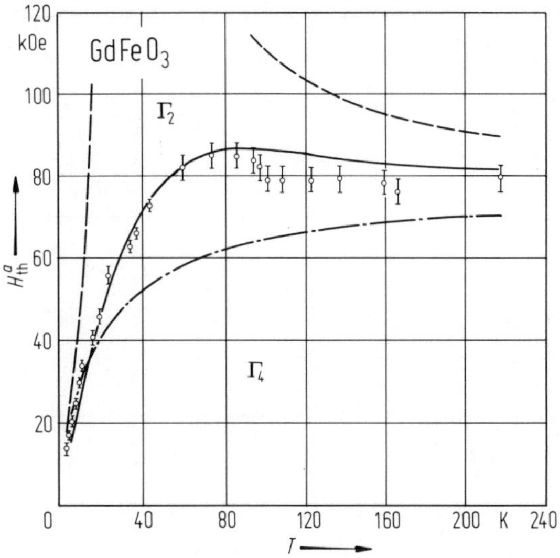

Fig. 227. $GdFeO_3$. Phase diagram (H, T) for an applied magnetic field $\boldsymbol{H} \parallel a$ axis. The experimental points indicate the phase transitions $\Gamma_4(G_xF_z) \rightarrow \Gamma_2(G_zF_x)$. The dashed and dot-dash curves are calculated taking into account only isotropic Gd–Fe exchange with positive or negative interaction, respectively. The continuous curve was obtained for an additional anisotropic Gd–Fe exchange [81B1].

For Fig. 228, see next page.

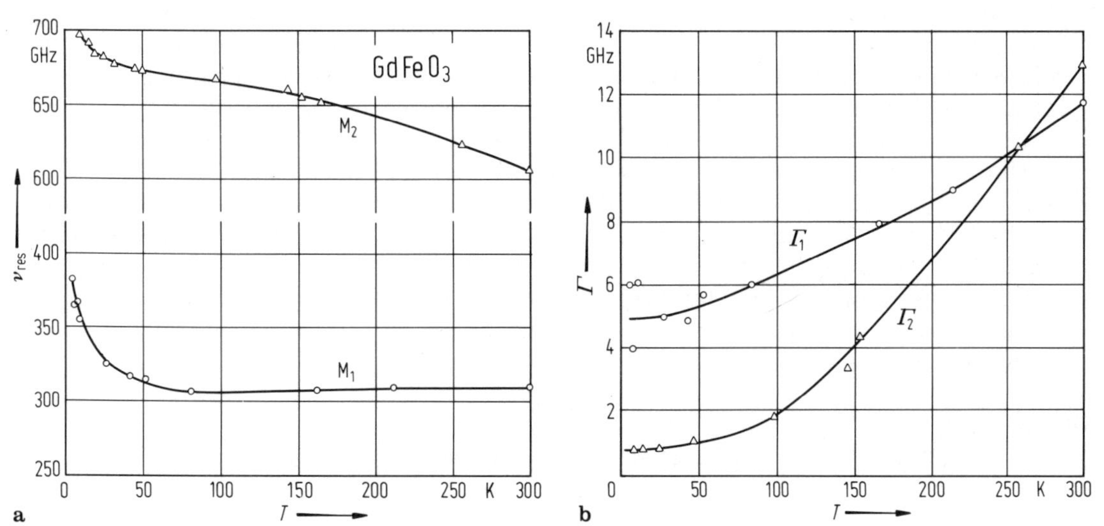

Fig. 229. $GdFeO_3$. Temperature dependence of the AFMR modes M_1 and M_2 (a) and the resonance linewidths Γ_1 and Γ_2 (b) for the quasiferromagnetic and quasiantiferromagnetic modes, respectively [86B1].

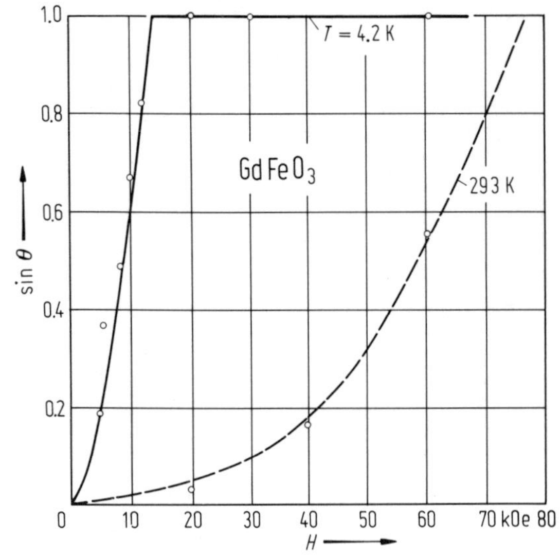

Fig. 228. $GdFeO_3$. Sinus of the angle θ between the direction of the magnetization and the crystallographic *a* axis for applied fields *H* in the *a* direction [77D2].

Table. 41. $GdFeO_3$. Critical magnetic fields H_{cr} in the crystallographic *a* direction to cause complete reorientation of the Fe^{3+} magnetic moments from the direction of the *a* axis to the *c* axis, as derived from Mössbauer spectra [77D2].

T [K]	H_{cr} [kOe]
293	76 (2)
77	81 (2)
4.2	12 (2)
2.0	7 (1)

5.5.8.8 $TbFeO_3$ compounds

$TbFeO_3$ (see also LB III/4a, p. 240 and III/12a, p. 454, 455)

I For spontaneous magnetization at $T < T_N = 651$ K see Fig. 37. For magnetization and magnetostriction isotherms at low temperature see Figs. 230 ··· 233. From these data the magnetic phase diagrams of Fig. 234 and Table 42 have been derived, see also [75B3]. A theoretical phase diagram at zero field strength based on dipolar interactions is described in [75B4], and based on the energy spectra and wave functions of the Tb^{3+} ion in [79B2].

High-field susceptibilities Fig. 235. Domain wall velocity [77C2].

IV For Raman spectra, see [85V1] Fig. 236 shows the temperature dependence of the magnon modes M_1 and M_2 with RT frequencies of 11.8(2) and 18.5(2) cm^{-1}, respectively. The equatorial Kerr effect, Fig. 237 gives information on surface magnetization states. It is concluded that at temperatures below 50 K the Morin-type transition $G_xF_z \rightarrow G_y$ takes place. The application of a magnetic field of 3 kOe parallel to the [100] direction "erases" the surface magnetic state for the region $T = 35 \cdots 50$ K [88B1].

$Tb_{1-x}Dy_xFeO_3$

For x = 0.75 a $G_xF_z \rightarrow G_y$ SRT occurs, which is not the case for x = 0.50 or 0.25, Figs. 238, 239.

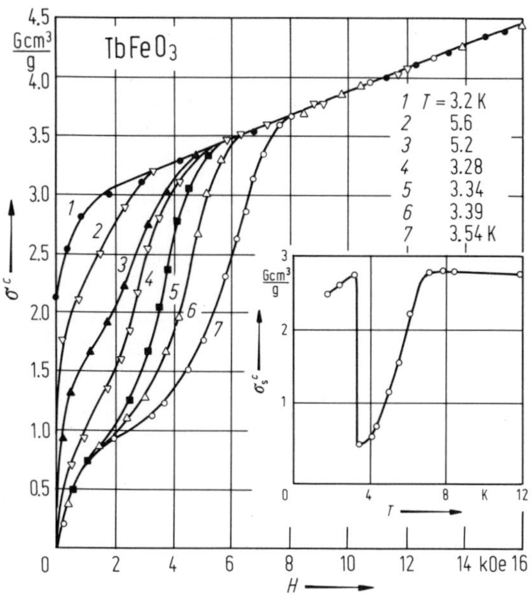

Fig. 230. $TbFeO_3$. Mass magnetization isotherms along the c axis of the crystal, σ^c, for various temperatures T. Inset: Temperature dependence of the spontaneous weak mass magnetization σ_s^c along the c axis in the spin-reorientation region [80D1].

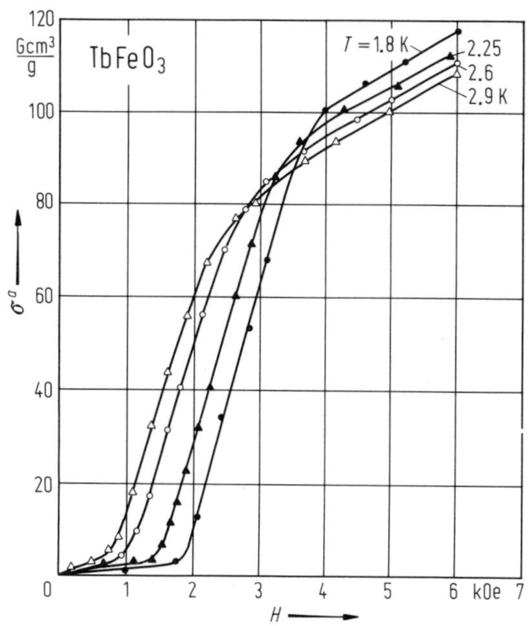

Fig. 231. $TbFeO_3$. Magnetization isotherms along the a axis, showing the metamagnetic transitions at various temperatures [80D1]; see also Figs. 18, 20.

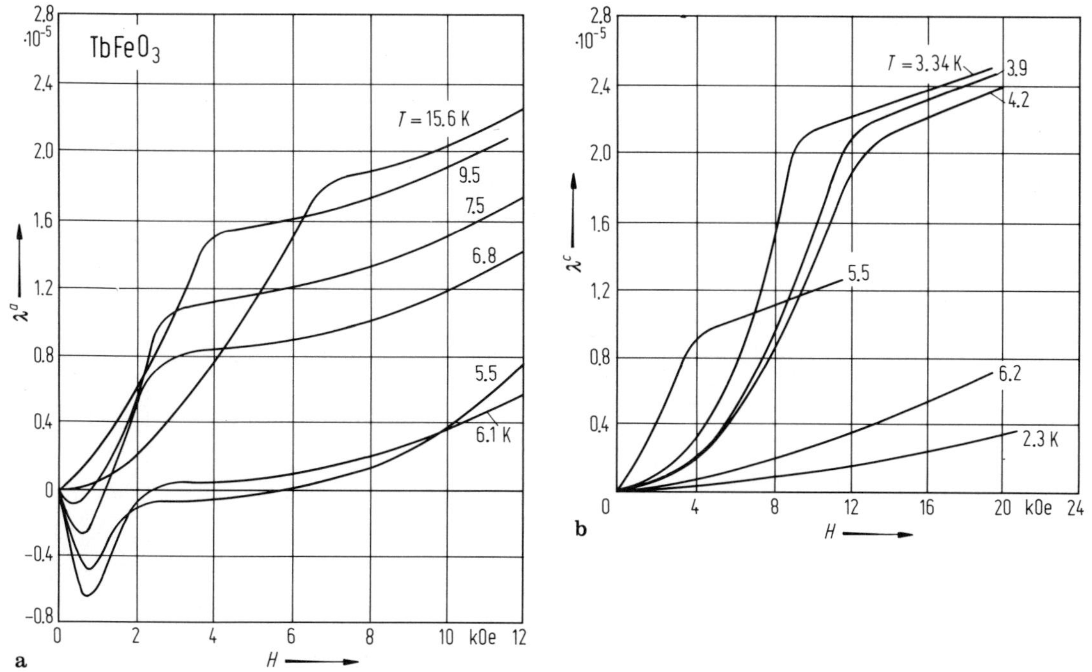

Fig. 232. $TbFeO_3$. Longitudinal magnetostriction isotherms for various temperatures T along (a) the a axis and (b) the c axis of the crystal [80D1].

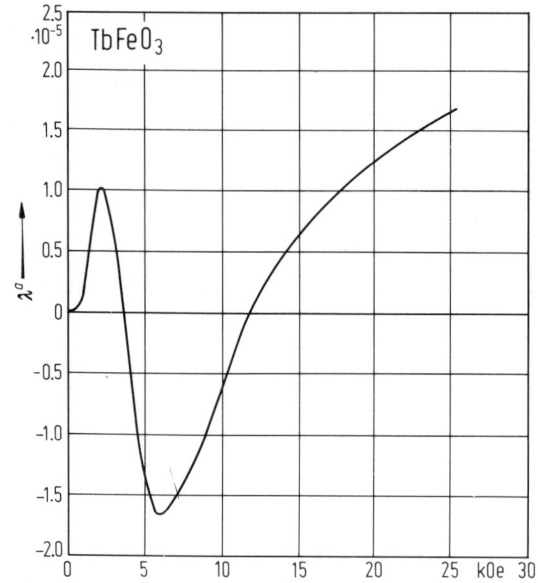

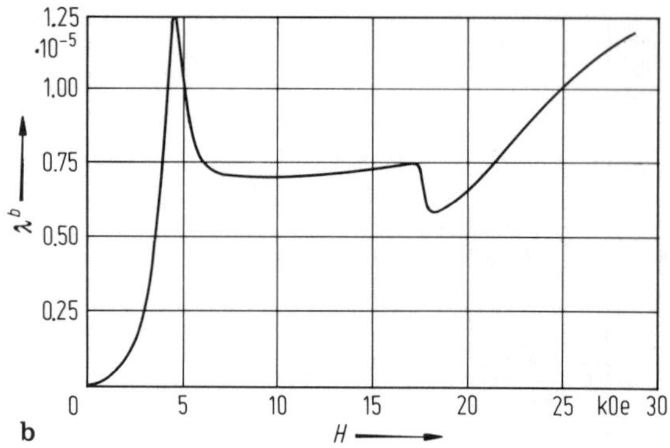

Fig. 233. $TbFeO_3$. Field dependence of the longitudinal magnetostriction for (a) $\boldsymbol{H} \parallel a$ axis at 2.7 K and (b) $\boldsymbol{H} \parallel b$ axis at 2.88 K [79B3].

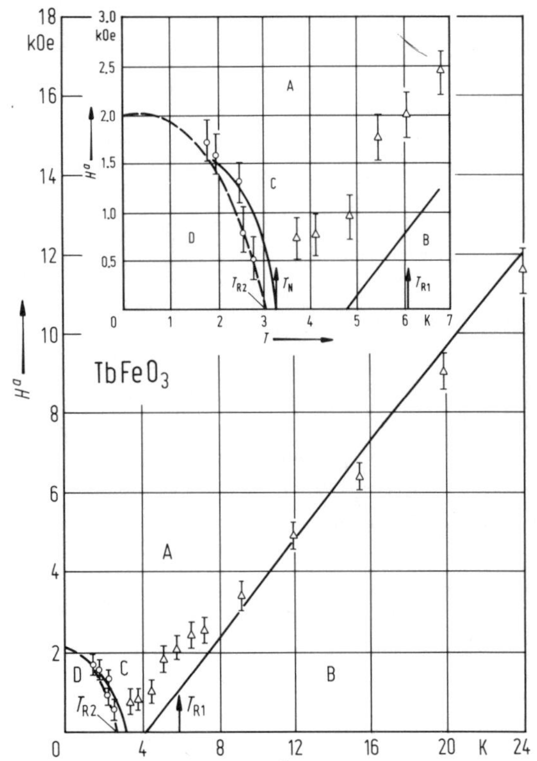

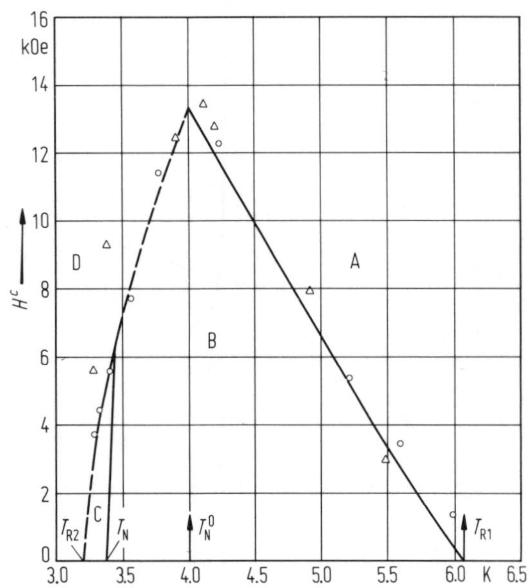

Fig. 234. $TbFeO_3$. H–T phase diagrams derived from magnetization (circles) and magnetostriction (triangles) data. Fig. (a) for $\boldsymbol{H} \parallel a$ axis, Fig. (b) for $\boldsymbol{H} \parallel c$ axis. The curves have a theoretical basis with $T_{R1} = 6.1$, $T_{R2} = 3.2$ and $T_N^0 = 4.2$ K. Fully drawn curves apply to second order and dashed curves to first order phase transitions, respectively. The various phases A, B, C and D are explained in Table 42 [80D1].

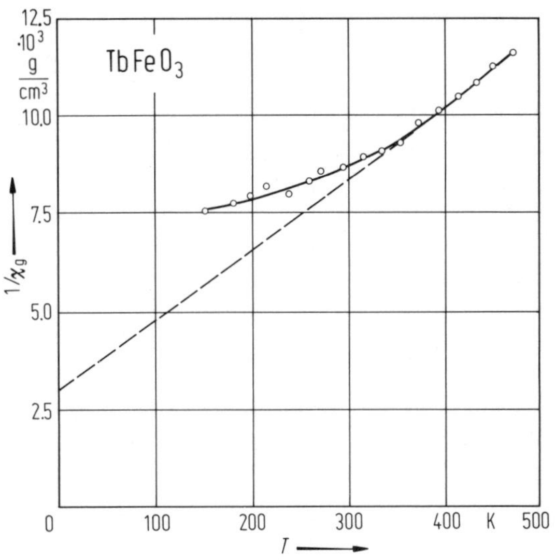

Fig. 235. $TbFeO_3$. Reciprocal magnetic mass susceptibility χ_g^{-1} vs. temperature T. The Curie-Weiss approximation gives $p_{eff} = 10.7\ \mu_B$/FU and $\Theta_p = -138.6$ K [80E1].

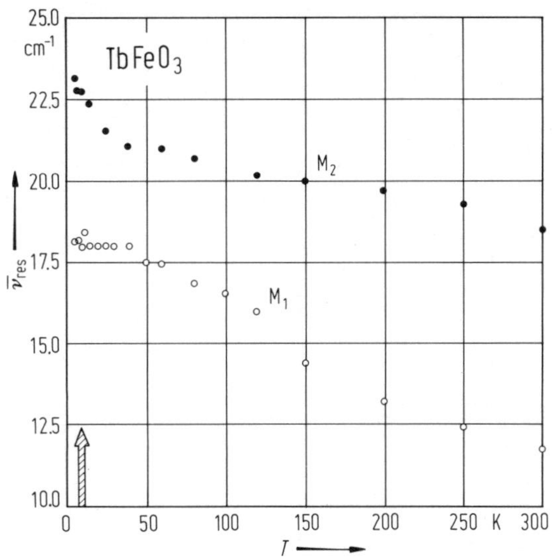

Fig. 236. $TbFeO_3$. Temperature dependence of the M_1 and M_2 magnon frequencies (given as wavenumbers $\bar{\nu}_{res}$ in cm^{-1}) as derived from Raman scattering. Shaded arrow: SR transition region. [85V1].

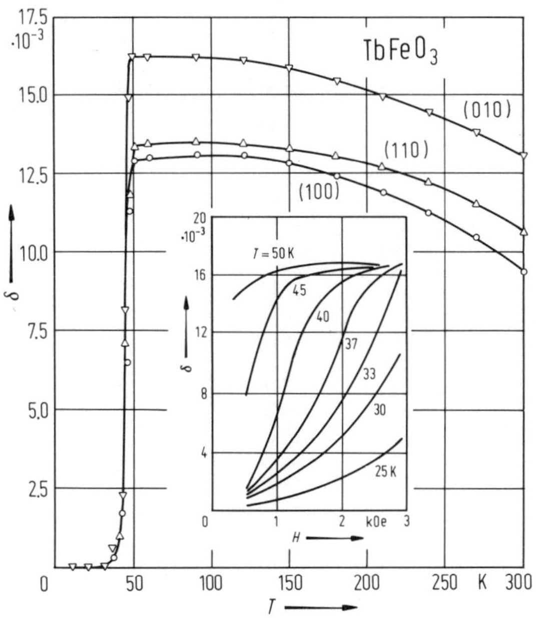

Fig. 237. $TbFeO_3$. Temperature dependence of the equatorial Kerr effect δ in the region of its spectral maximum 0.3 eV. Angle of incidence light 65°. Applied field $H = 0.5$ kOe and parallel [001] axis. Sample orientations are given. Inset: Field dependence of δ for sample (010) at different temperatures [88B1].

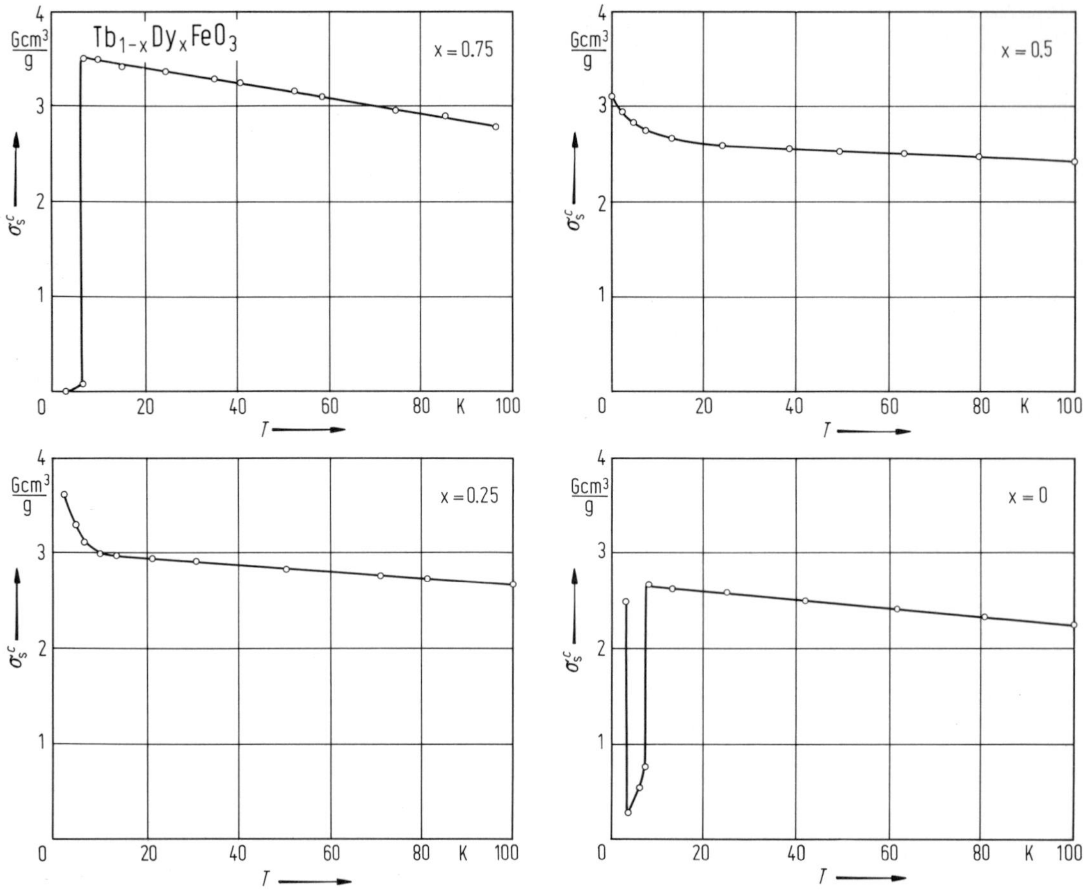

Fig. 238. $Tb_{1-x}Dy_xFeO_3$. Spontaneous magnetization in the crystallographic *c* direction, σ_s^c, vs. temperature *T* [77D1].

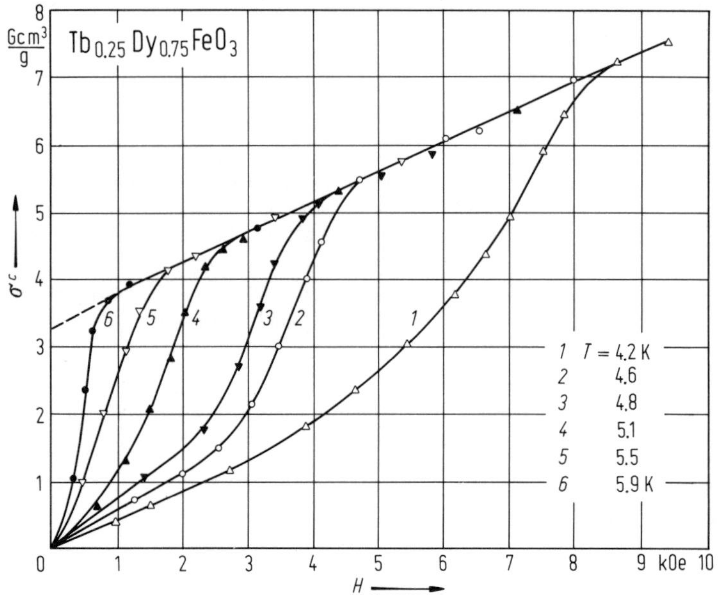

Fig. 239. $Tb_{0.25}Dy_{0.75}FeO_3$. Magnetization curves at various temperatures for an applied field parallel to the *c* axis [77D1].

Table 42. $TbFeO_3$. Magnetic symmetry and phase designations for the phase diagrams of Fig. 234 [80D1].
a) Applied magnetic field $\boldsymbol{H} \| a$ axis.

Phase	Fe^{3+} ions		R^{3+} ions	
	Representation of D_{2h}^{16} group	Magnetic configuration	Representation of D_{2h}^{16} group	Magnetic configuration
A	Γ_2	G_xF_z	Γ_2	F_xC_y
B	Γ_{42}	$G_{xz}F_{zx}$	Γ_{42}	F_xC_y
C	Γ_2	G_zF_x	Γ_{82}	A_xG_y, F_xC_y
D	Γ_{42}	$G_{xz}F_{zx}$	Γ_{842}	A_xG_y, F_xC_y

Note: In the magnetic configurations of the R subsystem are omitted the small components F_z of the magnetic moment of the Tb^{3+} Ising ions along the c axis, which is of Van Vleck origin.

b) Applied magnetic field $\boldsymbol{H} \| c$ axis.

Phase	Fe^{3+} ions		R^{3+} ions	
	Representation of D_{2h}^{16} group	Magnetic configuration	Representation of D_{2h}^{16} group	Magnetic configuration
A	Γ_4	G_xF_z	Γ_4	
B	Γ_{42}	$G_{xz}F_{zx}$	Γ_{42}	F_xC_y
C	Γ_{42}	$G_{xz}F_{zx}$	Γ_{842}	A_xG_y, F_xC_y
D	Γ_4	G_xF_z	Γ_{84}	A_xG_y

5.5.8.9 $DyFeO_3$ compounds

$DyFeO_3$ (see also LB III/4a, p. 241 and III/12a, p. 456).

I At the Morin temperature T_M the high-temperature weak ferromagnetism phase $\Gamma_4(F_zG_x)$ changes into the antiferromagnetic Γ_1 (G_y) phase. For the rotation angle θ as derived from Mössbauer spectra see Fig. 240. From magnetization curves, Belov [74B3] concluded $T_M \approx 41$ K. For a theoretical analysis of the spin reorientations see [79Z1]. For influence of Ga^{3+} substitution on T_M, Fig. 241 and [87S1]. At about 4 K an antiferromagnetic ordering of the Dy ions occurs into the mode A_yG_x, the spin configuration of the Fe ions remaining unchanged [74B3]. Applied magnetic fields induce various phase transitions:

For $\boldsymbol{H} \| c$ and $T < T_M$ the AF $\Gamma_1(G_y)$ phase changes via the phase $\Gamma_{14}(F_zG_{xy})$ into $\Gamma_4(F_zG_x)$, Fig. 242; see also Figs. 25, 39. For $\boldsymbol{H} \| b$ and $T < T_M$ the transition is $\Gamma_1(G_y) \rightarrow \Gamma_4(F_zG_x)$, Fig. 243.

For $\boldsymbol{H}\|a$ the situation is complicated, see Figs. 243, 244 (see also Fig. 257). For a magnetic field applied in the (*ab*) plane under an angle α with the *a* axis the situation is still more complicated, Fig. 245. The curve for $\alpha = 0$ corresponds to the $\Gamma_{12} \leftrightarrow \Gamma_{24}$ transition, for $\alpha = 90°$ it corresponds to the $\Gamma_1 \leftrightarrow \Gamma_4$ transition. For the general case $\boldsymbol{H} = (H_x, H_y, 0)$ the curves correspond to first order transitions between the twin phases of the canted configuration $\Gamma_{124}(G_{x,y,z})$ [88E1, 89G2].

Visual investigation of the spin reorientation shows a phase transition of the first order, practically free of hysteresis and accompanied by the coexistence of two AF phases with their axes parallel to the crystallographic *a* or *b* axis, being the phases above and below the Morin temperature $T_M = 51(1)$ K, respectively [77B4]. For similar data at 7 K see [85P1].

For the speed of formation of a transition front from the AF to the weak ferromagnetic phase under the influence of an applied magnetic field see [89K2, 90G2]. Temperature dependence of the spontaneous magnetization Figs. 38, 197, 241 (see also Fig. 23) and magnetization curves Fig. 246.

Thermal effects during magnetization [90B1]. Spontaneous magnetostriction at RT: $\lambda_s^a \approx \lambda_s^b \approx -3 \cdot 10^{-3}$ and $\lambda_s^c \approx +6 \cdot 10^{-3}$ [78G2] and Fig. 247. Linear magnetostriction [74B3, 85Z1].

Exchange fields and anisotropy fields Table 43. Domain wall velocity Fig. 248, domain wall mobility at RT: $\mu_w \cong 1600\ \mathrm{cm\,s^{-1}Oe^{-1}}$.

III For high-frequency magnetic properties see [85B2]. The frequencies of the AFMR modes M_1 and M_2 at RT are 12.67(4) and 17.00(4) $\mathrm{cm^{-1}}$, see Fig. 249.

For AFMR in strong magnetic fields see [86B2, 88B2, 88B3, 88B4]. Hyperfine parameters derived from Mössbauer spectra at 4.2 K: $H_{hyp} = 554(2)$ kOe, $\Delta E_Q = 0.07(4)\ \mathrm{mm\,s^{-1}}$. Isomer shift at RT relative to α iron: $\delta = 0.39(3)\ \mathrm{mm\,s^{-1}}$ [80P2], see also Fig. 250. See Figs. 251, 252 for the influence of dilution with diamagnetic Ga^{3+}. The muon spin rotation (µSR) technique reveals an internal magnetic field H_μ at the muon increasing from 2.059(2) kOe at RT to 2.28(2) at 155 K, see also Fig. 253.

Domain wall NMR spectra [75Z2].

IV Faraday effect in a magnetic field [77C3]. Optical pumping [89K3].

$DyFe_{0.998}Co_{0.002}O_3$

I ^{57}Fe NMR spectra establish that the iron ions have a spin configuration of the G_y type below 138.9 K and of the G_xF_x type above 140.6 K. These are the AF and WF phases, respectively [77K4]. The region of phase coexistence is influenced by an external field, Fig. 254. For a theory of these orientational magnetic phase transitions see [90G3].

$Dy_{1-x}Ho_xFeO_3$ (see also LB III/12a, p. 456)

I, III Spontaneous magnetization and magnetic phases in the absence of a magnetic field, see Figs. 255, 256. For x = 0.5 the compound is weakly ferromagnetic above $T = 45$ K with the magnetization parallel to the *c* axis (G_xF_z). Between 25 and 45 K the material is antiferromagnetic (G_y). In the temperature range from 25 ⋯ 18 K a transition occurs which leads to a low-temperature weak ferromagnet with the magnetic moment parallel to the *a* axis. For x = 0.25 the crystal is a pure antiferromagnet between 13 ⋯ 20 K. For x = 0.75 the mixed Γ_{124} configuration occurs between 63 ⋯ 20 K as a consequence of a three dimensional rotation. The phase diagram is confirmed by neutron diffraction [77V2]. Magnetic phase diagrams in the presence of a magnetic field Fig. 257.

$Dy_{1-x}Er_xFeO_3$

I In these compounds two transitions occur in the magnetic state: the Morin transition at temperature T_M, phase transition $G_xF_z \leftrightarrow G_y$, and the spin reorientation at T_R, for the transition $G_xF_z \leftrightarrow G_zF_x$, see Fig. 258. Threshold fields inducing these transitions are shown in Fig. 259 for x = 0.2.

II NMR frequencies Fig. 268.

$Dy_{1-x}Tm_xFeO_3$

Similar to $Dy_{1-x}Er_xFeO_3$, see Fig. 258.

$Dy_{0.7}Lu_{0.3}FeO_3$

For the magnitude of the weak magnetism, see Fig. 197.

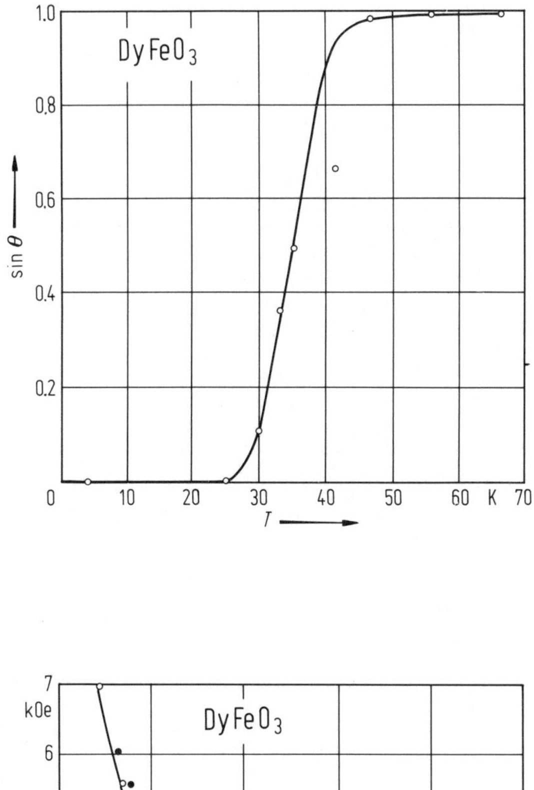

Fig. 240. $DyFeO_3$. Temperature dependence of sin θ, where θ is the angle between the Fe^{3+} antiferromagnetic vector and the crystal b axis. The Morin temperature $T_M \doteq 35(3)$K [80P2].

◀

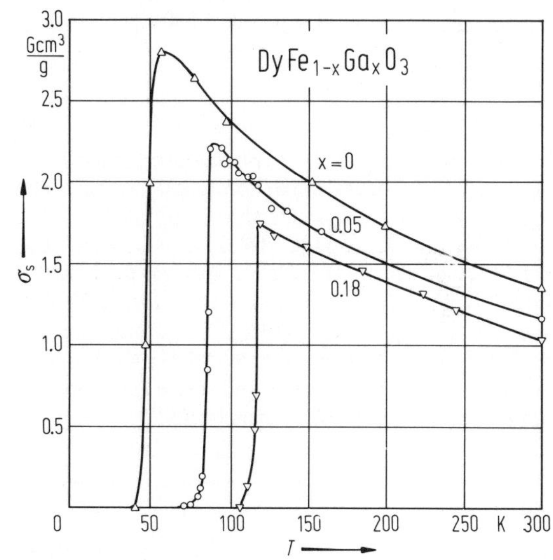

Fig. 241. $DyFe_{1-x}Ga_xO_3$, single crystals. Spontaneous mass magnetization σ_s vs. temperature T [84N1].

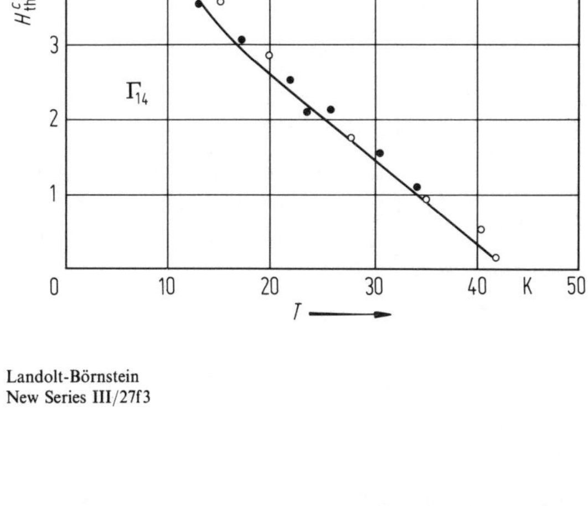

◀

Fig. 242. $DyFeO_3$. Threshold field H^c_{th} for the phase transition $\Gamma_1(G_y) \rightarrow \Gamma_4$ (F_zG_x) as derived from magnetization measurements [74B3]. From Mössbauer spectra at 4.2 K a value $H^c_{th} = 7(1)$ kOe is found [80P2].

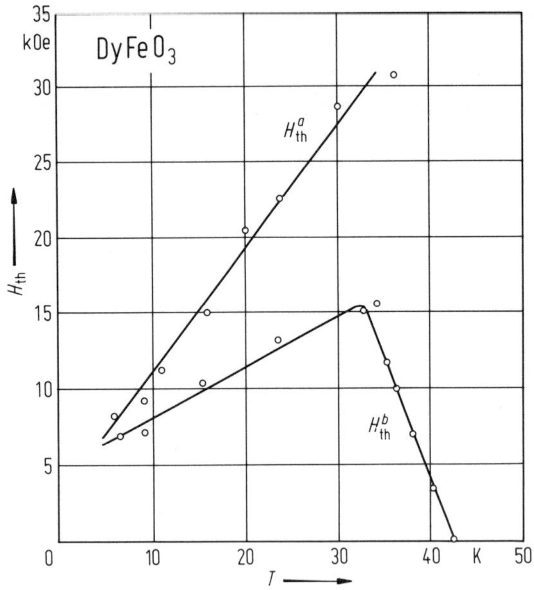

Fig. 243. $DyFeO_3$. Similar to Fig. 242 but now for the same phase transitions in the crystallographic *a* and *b* directions [74B3]. From Mössbauer spectra at 4.2 K a value $H^b_{th} = 7.5(10)$ kOe is found [80P2].

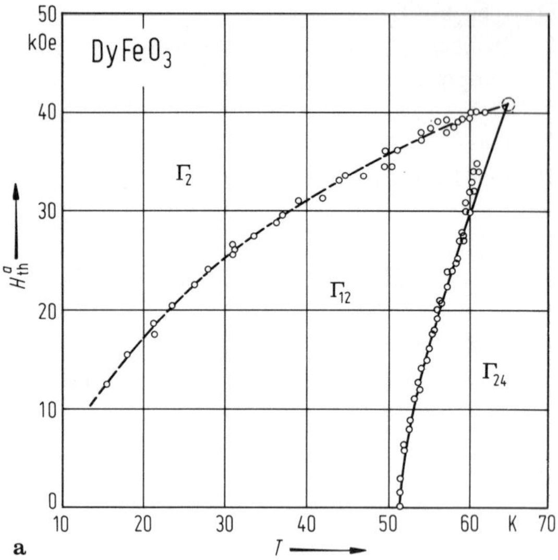

a

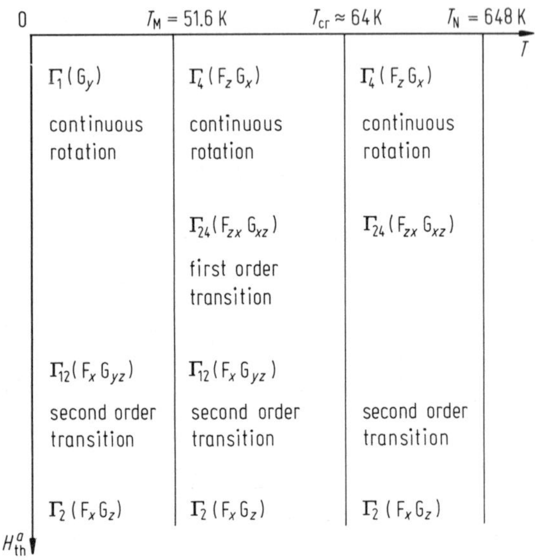

b

Fig. 244. $DyFeO_3$. (a) $(H-T)$ phase diagram for $\boldsymbol{H} \parallel a$ axis as derived from magneto-optical studies. The coordinates of the multicritical point are $H_{cr} = 41$ kOe and $T_{cr} \approx 64$ K [89G2, see also 83A2 and 87E1]. Analysis of Mössbauer spectra lead to the values $H^a_{th1} = 45(5)$, 65(5) and 93(10) kOe at 77, 130 and 270 K, respectively [80P2]. For similar results derived from AFMR spectra see [88B3]. (b) Schematic survey of magnetic phases and phase transitions in the H, T field, where $\boldsymbol{H} \parallel a$ axis [89G2].

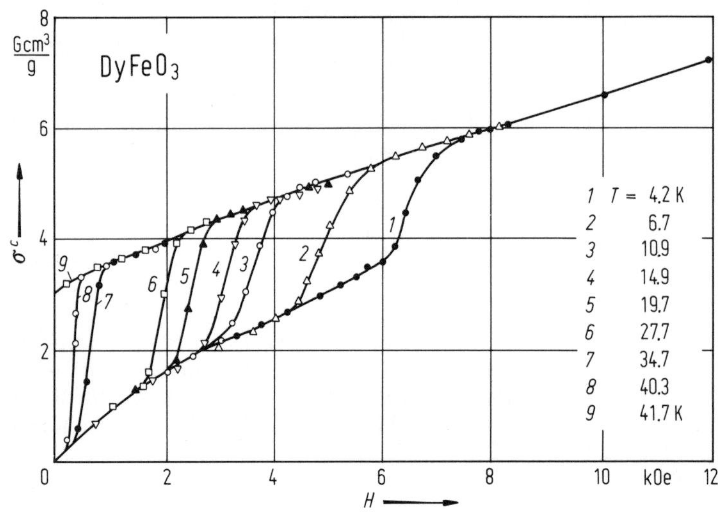

For Fig. 245, see next page.

Fig. 246. $DyFeO_3$. Mass magnetization σ^c at low temperatures for applied fields along the *c* axis [74B3].

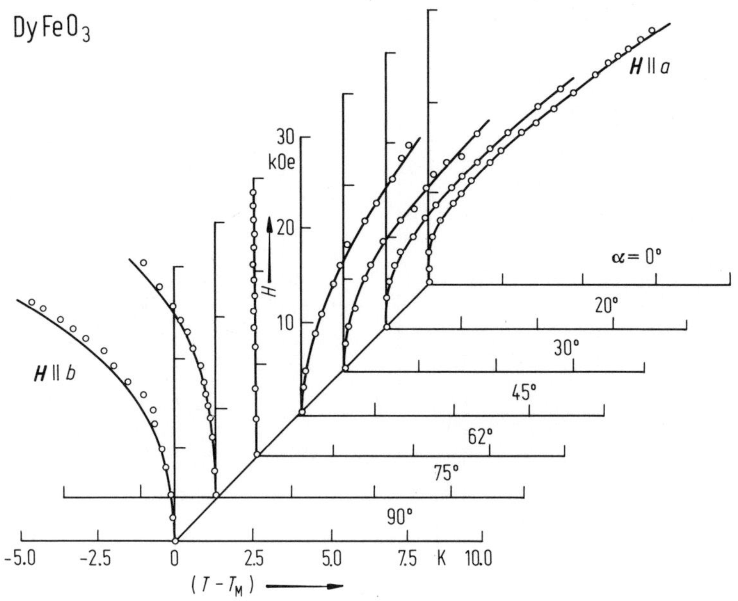

Fig. 245. DyFeO$_3$. H–T phase diagram for different orientations of the applied field in the ab plane. α is the angle of H with the a axis. $T_M = 51.6$ K [89G3].

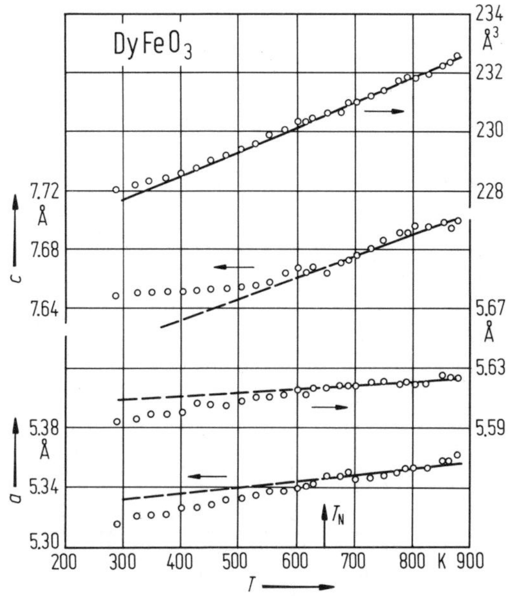

Fig. 247. DyFeO$_3$. Lattice parameters a, b and c and unit cell volume V vs. T just above and below the Néel temperature $T_N = 646$ K, showing the anomaly of the thermal expansion [78G2].

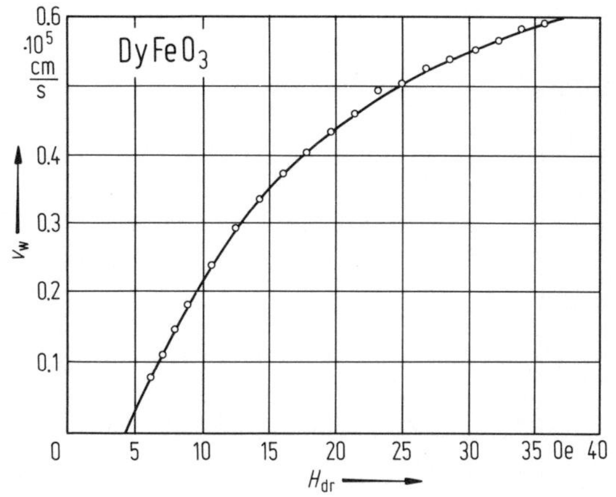

Fig. 248. DyFeO$_3$. RT domain wall velocity v_w vs. applied magnetic driving field H_{dr} [78T1].

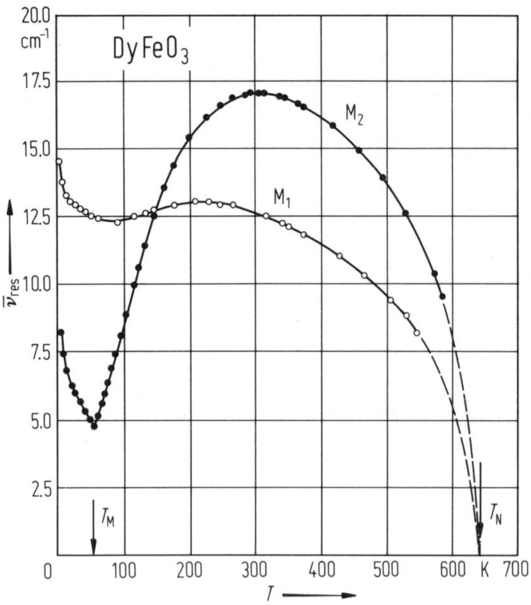

Fig. 249. $DyFeO_3$. Temperature dependence of the AFMR frequencies (given as wavenumbers $\bar{\nu}_{res}$ in cm^{-1}) of the quasi-ferromagnetic mode M_1, $\boldsymbol{h} \perp c$ and the quasi-antiferromagnetic mode M_2, $\boldsymbol{h} \parallel c$. h: rf magnetic field. Data derived from optical transmission spectra [85B2, see also 82K4]. For data from Raman scattering see [88K3].

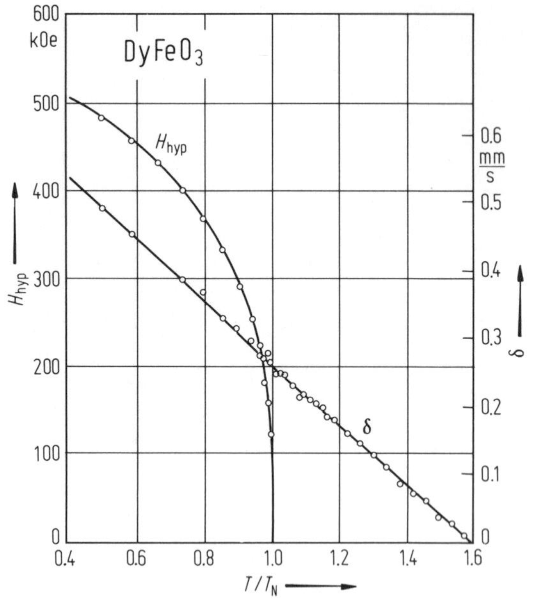

Fig. 250. $DyFeO_3$. Mössbauer data. Isomer shift δ and effective magnetic hyperfine field H_{hyp} vs. temperature T. The isomer shift of the ^{57}Fe nuclei is measured relative to the line in Cr at RT. $T_N = 645$ K [75N1].

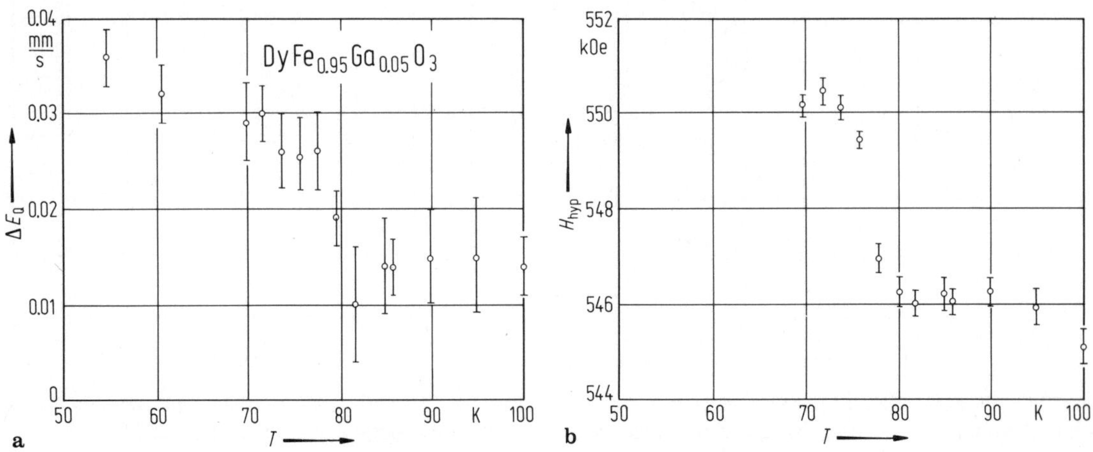

Fig. 251. $DyFe_{0.95}Ga_{0.05}O_3$. Hyperfine interaction parameters ΔE_Q and H_{hyp} vs. temperature T [84N1].

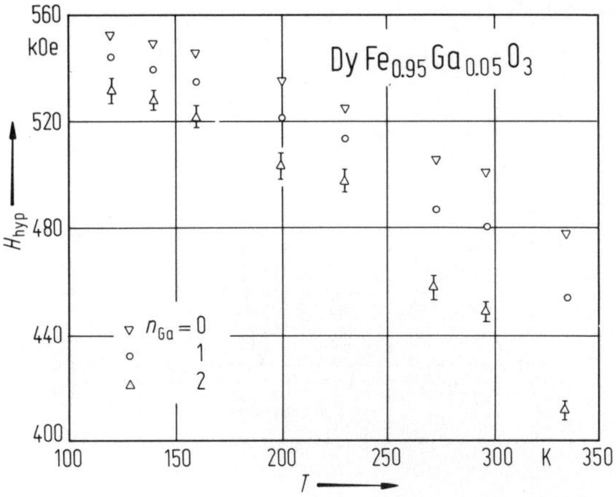

Fig. 252. $DyFe_{0.95}Ga_{0.05}O_3$. Temperature dependence of the magnetic hyperfine field H_{hyp} at an Fe^{3+} ion when the nearest neighborhood contains n_{Ga} gallium ions [84N1].

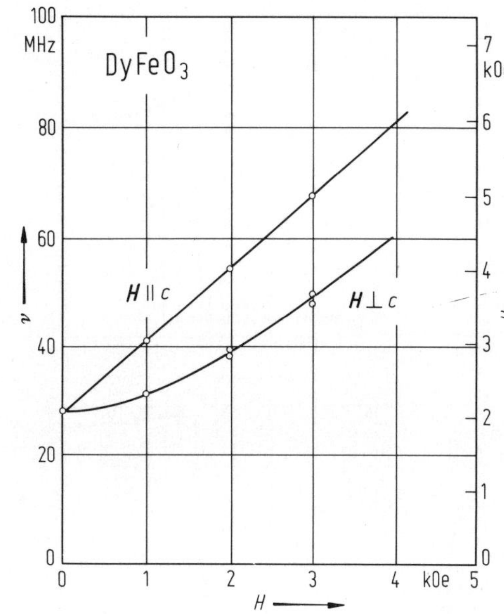

Fig. 253. $DyFeO_3$. Field dependence of the muon spin rotation frequency ν and the corresponding local fields H_μ for two directions of the applied field H with respect to the c axis of the crystal. RT [80H2].

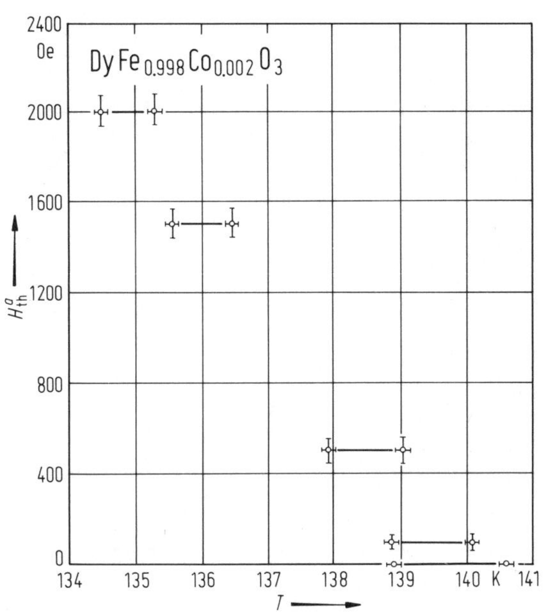

Fig. 254. $DyFe_{0.998}Co_{0.002}O_3$. Coexistence region of the low-temperature AF iron phase and the high-temperature weak ferromagnetic iron phase, as derived from ^{57}Fe NMR spectra [77K4].

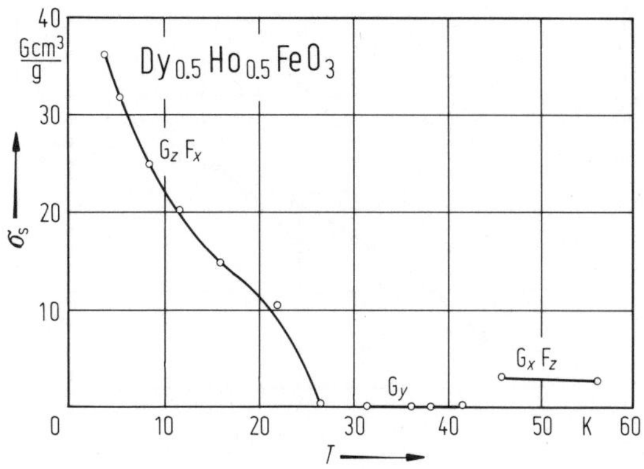

Fig. 255. $Dy_{0.5}Ho_{0.5}FeO_3$. Temperature dependence of the spontaneous mass magnetization, σ_s. The magnetic structures in the various temperature regions are indicated [85C1, 77B2].

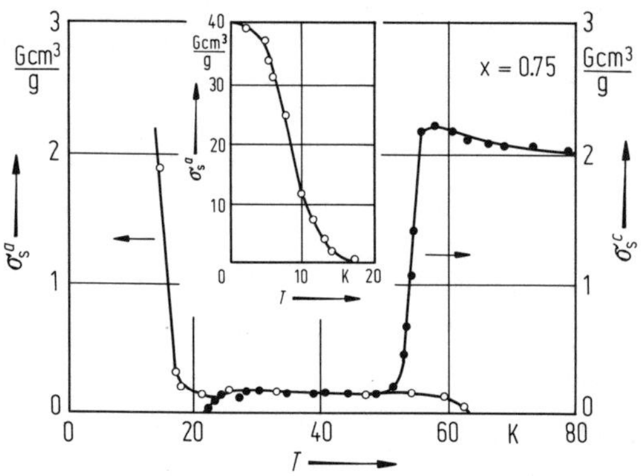

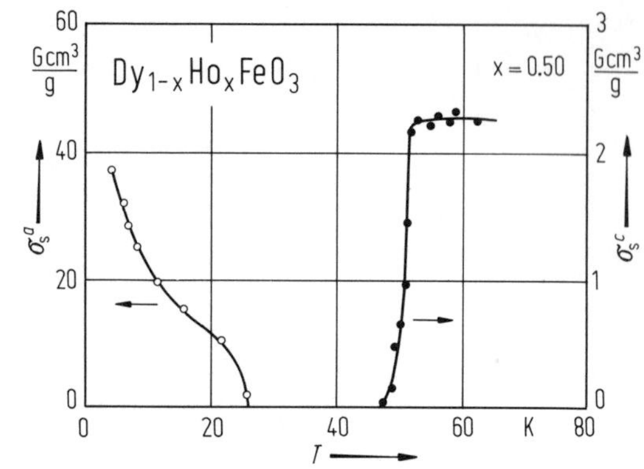

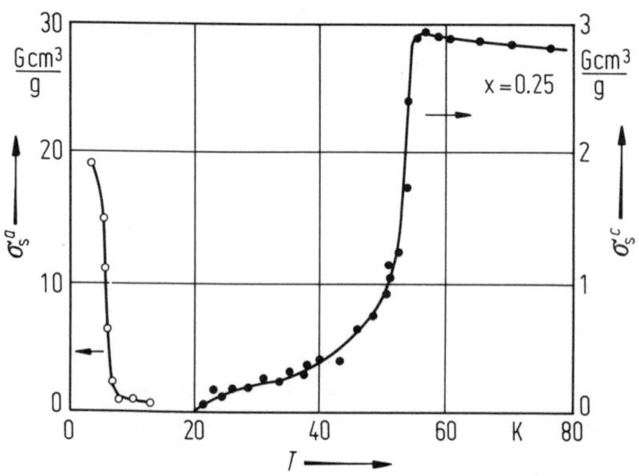

Fig. 256. $Dy_{1-x}Ho_xFeO_3$. Spontaneous specific magnetizations σ_s^a and σ_s^c along the crystallographic *a* and *c* axes, respectively, as a function of temperature measured on single crystals [85C1].

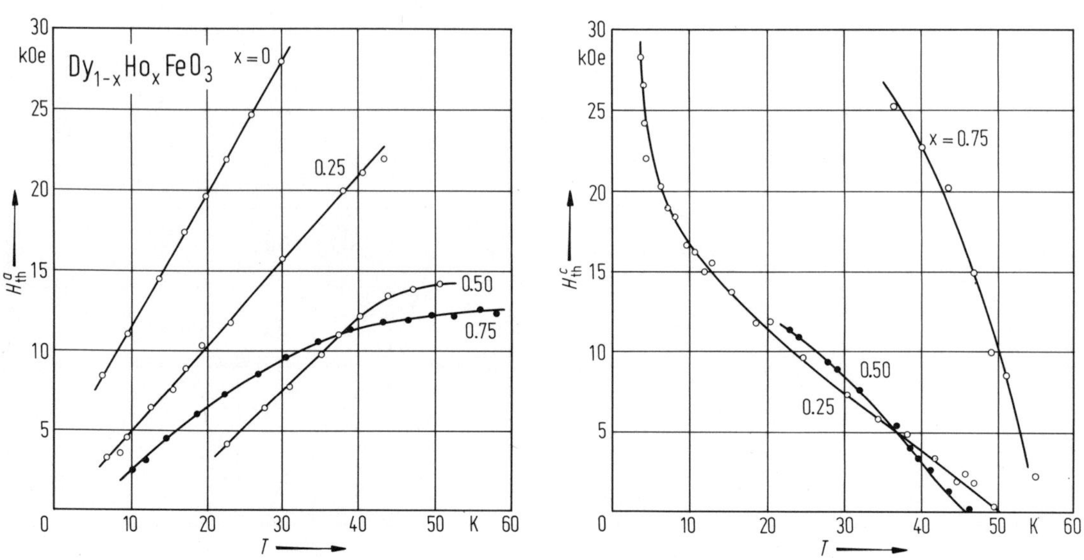

Fig. 257. $Dy_{1-x}Ho_xFeO_3$. Magnetic threshold fields in the crystallographic *a* and *c* directions, H_{th}^a and H_{th}^c, respectively, as a function of temperature [85C1].

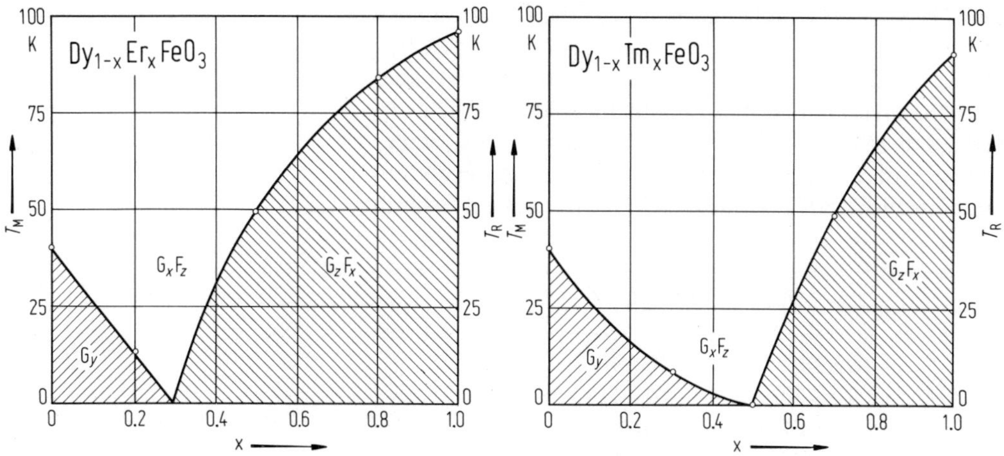

Fig. 258. $Dy_{1-x}Er_xFeO_3$; $Dy_{1-x}Tm_xFeO_3$. The Morin transition temperature T_M and the spin reorientation temperature T_R for the magnetic phase transitions $G_xF_z \leftrightarrow G_y$ and $G_xF_z \leftrightarrow G_zF_x$, respectively, as dependent on the composition parameter x [76D1].

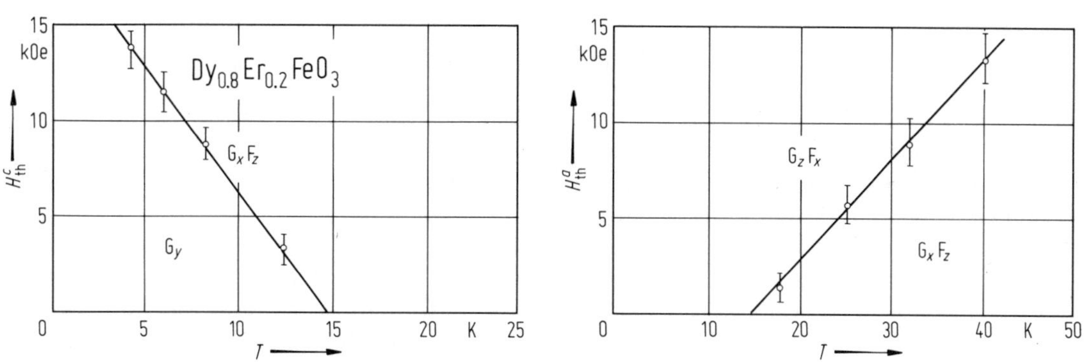

Fig. 259. $Dy_{0.8}Er_{0.2}FeO_3$. Temperature dependence of the threshold fields H_{th} inducing the magnetic order transitions $G_y \leftrightarrow G_xF_z$ and $G_xF_z \leftrightarrow G_zF_x$ for applied magnetic fields parallel to the *a* and *c* axis of the crystal, respectively [76D1].

Table 43. $DyFeO_3$. The exchange field, H_{ex}, the Dzyaloshinsky field H_D and the second and fourth order anisotropy fields H_{K2} and H_{K4} at various temperatures derived from Mössbauer spectra in the presence of applied magnetic fields. The field H^a_{th} is the critical field for the phase transformation $\Gamma_4 \rightarrow \Gamma_2$ under the influence of an applied magnetic field in the crystallographic *a* direction [80P2, see also 74B3]. All magnetic fields in Tesla.

Parameter	$T = 270$ K	130 K	77 K
H_{ex}	672(34)	670(30)	–
H_D	10.8(20)	10.4(20)	–
H_{K2}	0.14(4)	0.084(50)	–
H_{K4}	0.051(46)	0.049(20)	–
H^a_{th}	9.3(10)	6.5(6)	4.5(5)

5.5.8.10 $HoFeO_3$ compounds

$HoFeO_3$ (see also LB III/4a, p. 241 and III/12a p. 456)
Temperature dependence of lattice constants Fig. 260.

I An analysis of high-frequency AFMR spectra has revealed the existence of four phases of $HoFeO_3$ with different types of spin ordering, Fig. 261.

The spin flip from the high-temperature ordered $\Gamma_4(F_zG_x)$ phase to the low temperature ordered $\Gamma_2(F_xG_z)$ phase takes place in three consecutive steps: at $T_1 = 58(2)$ K the ferromagnetism ***F*** vector remains in the *c* direction, but the vector ***G*** rotates in the (*ac*) plane from the *a* axis into the direction of the *c* axis, resulting in phase $\Gamma_{24}(F_zG_{xz})$, which exists in the temperature range $T_1 \cdots T_2$, where $T_2 = 51(1)$ K. At the latter temperature the vector ***G*** jumps to the (*bc*) plane and the ferromagnetism vector ***F*** jumps to the *a* axis, resulting in phase $\Gamma_{12}(F_xG_{yz})$. At $T_3 = 39(2)$ K the vector ***G*** is parallel to the *c* axis, phase $\Gamma_2(F_xG_z)$ [89B2]. These phases have also been the basis for the explanation of magnetization and magnetostriction curves [89V1] and of acoustic properties [89D1].

The phase transition temperatures depend on the presence of a magnetic field, see Fig. 262 for $\boldsymbol{H} \| a$ or *c* axis. In the case of $\boldsymbol{H} \| b$ axis the phase relations are much more complicated, see [90V2]. Saturation magnetization Figs. 263, 264. Magnetostriction Fig. 265, Young's modulus, internal friction and thermal expansion Fig. 266. Domain-wall mobility at RT: $\mu_w \cong 450\ \mathrm{cm\,s^{-1}Oe^{-1}}$ [78T1].

For the muon spin rotation frequencies see Fig. 195. For phonon spectra [85V1].

III AFM resonance modes Fig. 261. The RT frequencies of the spin wave modes M_1 and M_2 are 390 and 540 GHz, respectively [82W1].

IV Kerr effect Fig. 267, for the influence of an applied magnetic field [87B2]. Equatorial Kerr effect spectra near SRT [91G1].

$Ho_{0.7}Er_{0.3}FeO_3$
NMR frequencies Fig. 268.

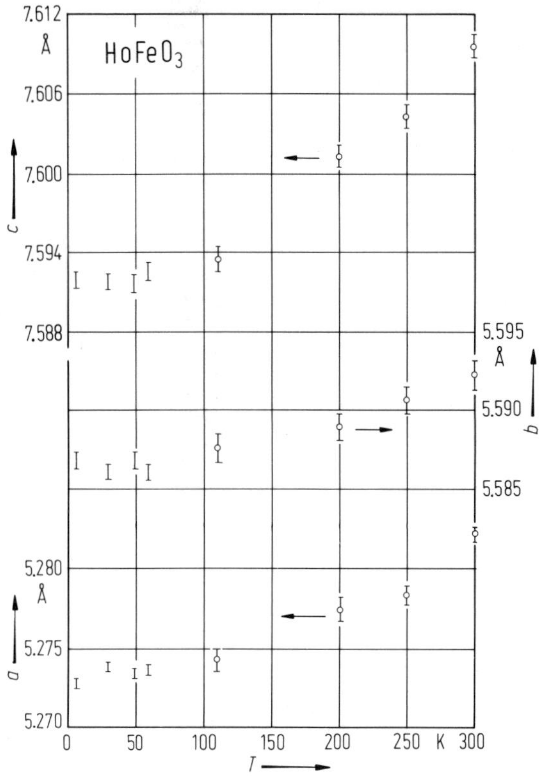

Fig. 260. $HoFeO_3$. Temperature dependence of lattice constants *a*, *b* and *c* as derived from X-ray spectra. [R. Kleinberger in 74B4].

◀

For Fig. 261, see next page.

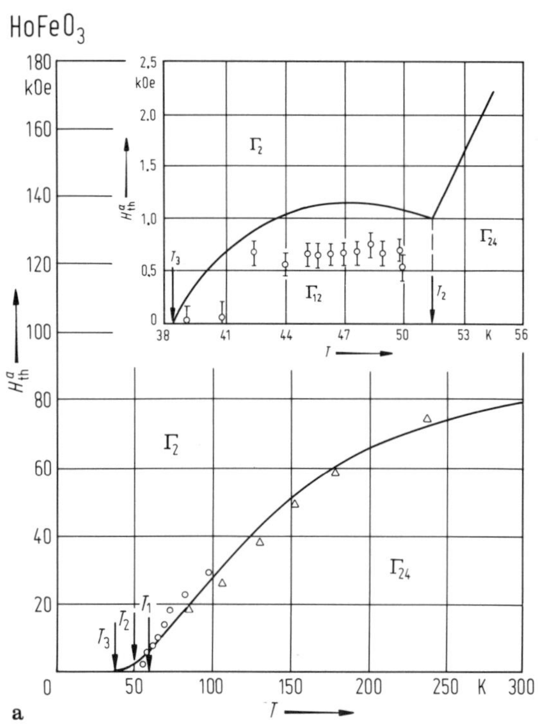

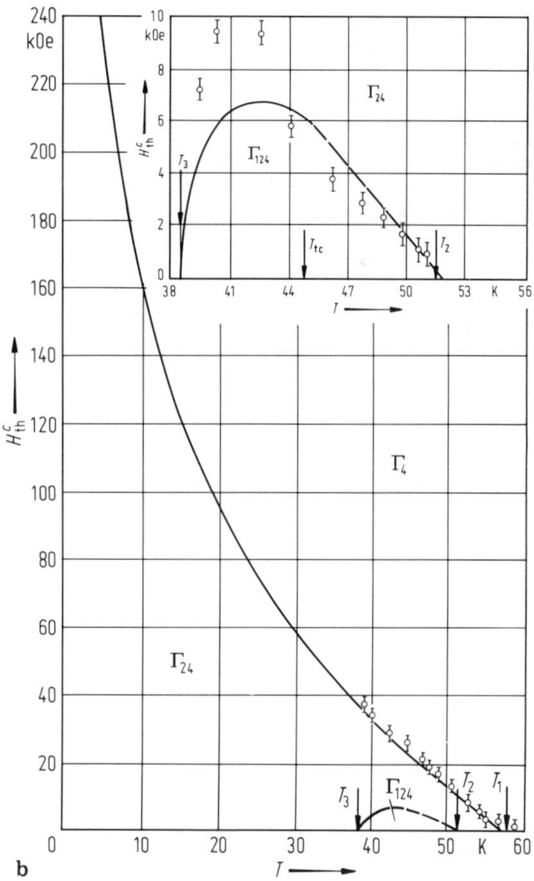

Fig. 262. $HoFeO_3$. Magnetic phase diagrams for an applied field parallel to the *a* axis (a) and parallel to the *c* axis (b) [89V1]. Triangles are from other literature. Solid lines: second order phase transitions; dashed curves: first order phase transition. T_{tc}: temperature corresponding to a tricritical point. All curves theoretical with adapted parameters.

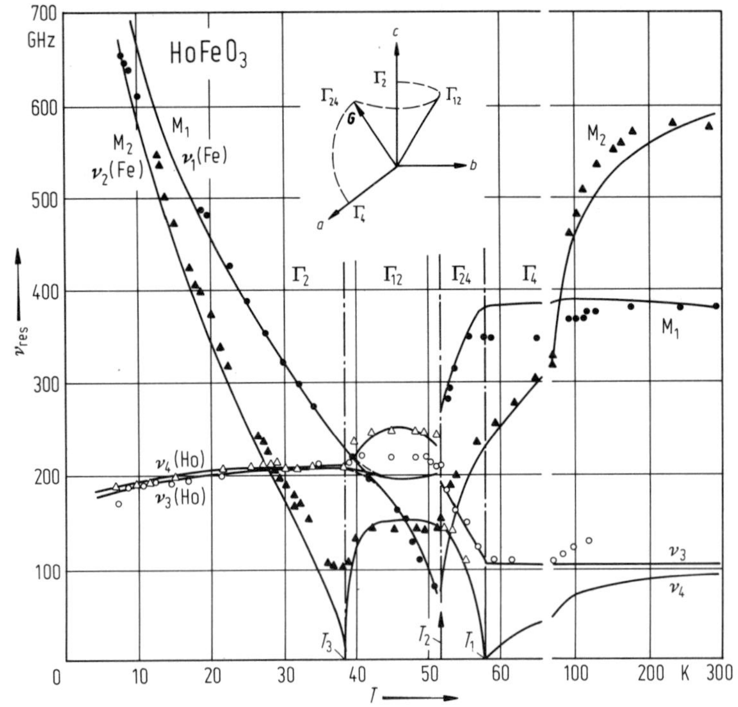

Fig. 261. $HoFeO_3$. Temperature dependence of the frequencies ν_1 and ν_2 of the Fe^{3+}-AFMR modes M_1 and M_2, respectively. The resonance frequencies ν_3 and ν_4 correspond to two modes due to electronic transitions inside the ground quasidoublet of the Ho^{3+} ions, the so called R modes. The curves have been calculated for adapted magnetic interaction parameters. Inset: scheme of phase transitions showing the character of rotation of the G-vector [89B2]. RT values for M_1 and M_2 of 390 and 540 GHz, respectively, have been reported by [82W1]. See also [86B3].

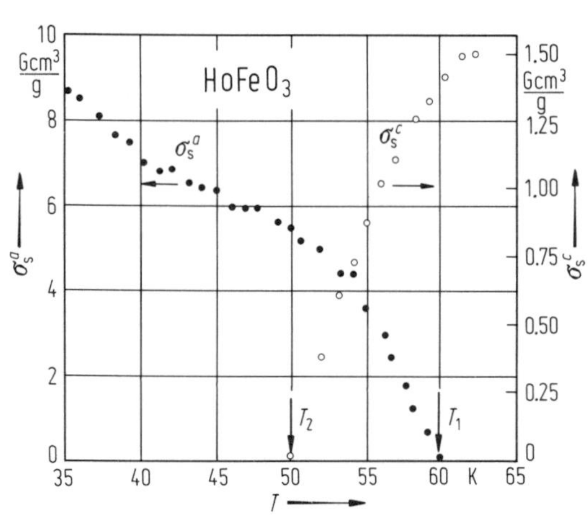

Fig. 263. $HoFeO_3$. Temperature dependence of the spontaneous magnetic mass magnetizations σ_s^a and σ_s^c, along the crystallographic a and c direction, respectively [89V1].

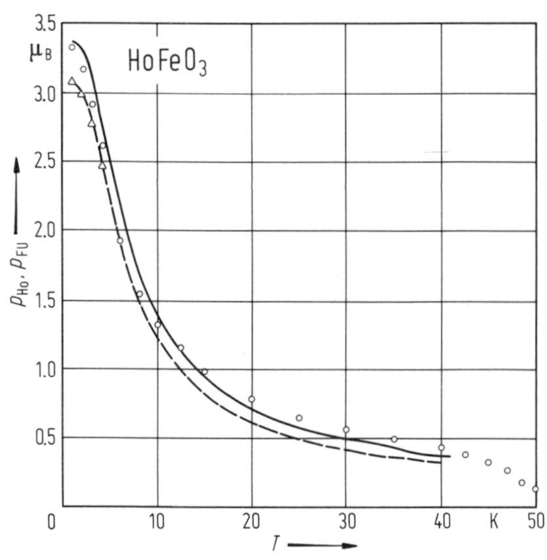

Fig. 264. $HoFeO_3$. Temperature dependence of the spontaneous magnetic moment along the a axis. Circles: bulk magnetization measurement (p_{FU}). Triangles: neutron diffraction data, Ho^{3+} moment only ($p_{Ho^{3+}}$). Curves calculated on the basis of dipolar interactions; fully drawn curve: total moments, dashed curve: Ho^{3+} moments only. [74B4].

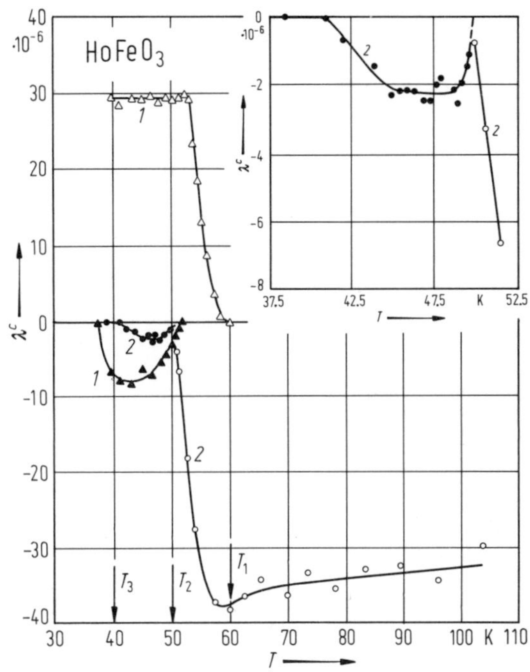

Fig. 265. $HoFeO_3$. Temperature dependence of the linear magnetostriction along the c axis λ^c. The values are obtained from λ^c vs. H curves by extrapolation to zero field strength. Curves *1*: $\boldsymbol{H} \parallel c$, closed triangles correspond to the phase transition $\Gamma_{124} \rightarrow \Gamma_{24}$, open triangles to $\Gamma_{24} \rightarrow \Gamma_4$. Curves *2*: $\boldsymbol{H} \parallel a$, closed circles correspond to $\Gamma_{12} \rightarrow \Gamma_2$, open circles to $\Gamma_{24} \rightarrow \Gamma_2$ [89V1]. For $\boldsymbol{H} \parallel b$ see [90V2].

For Fig. 267, see next page.

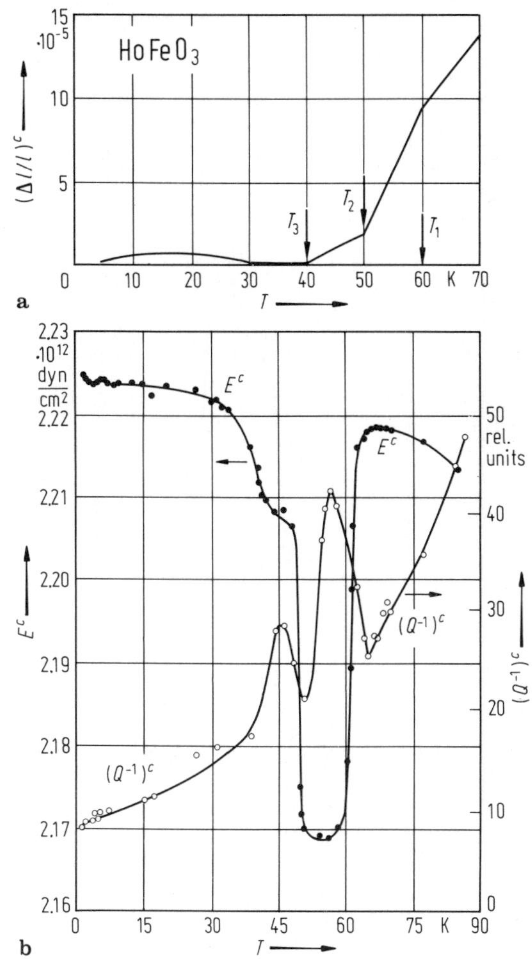

Fig. 266. $HoFeO_3$. Temperature dependence of (a) thermal expansion coefficient $(\Delta l/l)^c$, (b) Young's modulus E^c and internal friction coefficient $(Q^{-1})^c$, all measured along the c direction of the crystal [89V1]. $1\ \mathrm{dyn/cm^2} = 10^{-1}$ Pa.

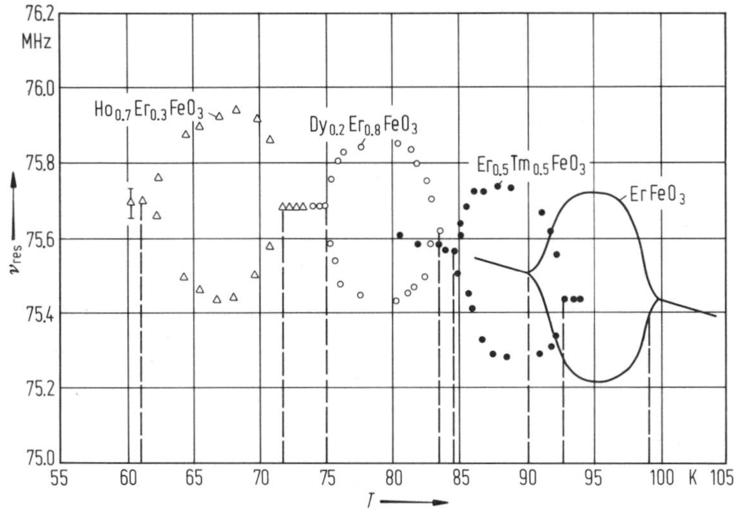

Fig. 268. $ErFeO_3$, $Er_{0.5}Tm_{0.5}FeO_3$, $Dy_{0.2}Er_{0.8}FeO_3$, $Ho_{0.7}Er_{0.3}FeO_3$. Temperature dependence of the ^{57}Fe NMR frequencies in the spin reorientation range [77V2], the curve for $ErFeO_3$ is from [72K1].

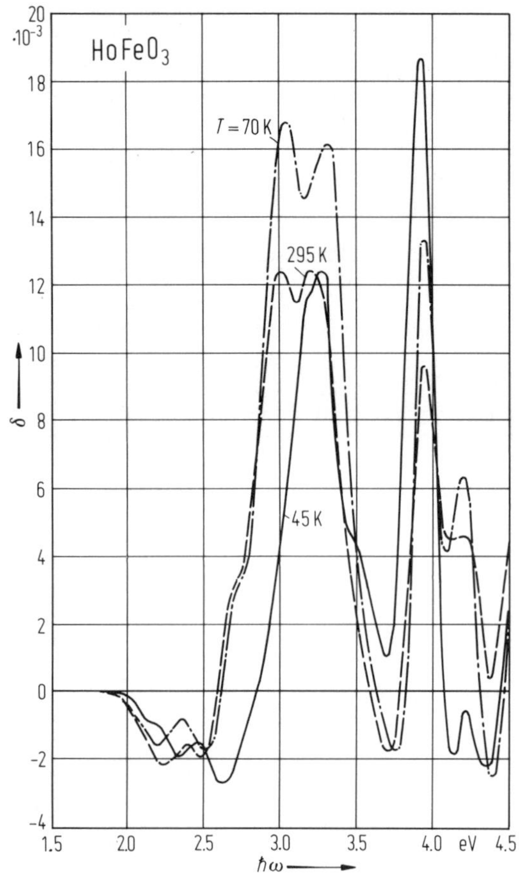

Fig. 267. $HoFeO_3$. Frequency dependence of the equatorial Kerr effect δ at various temperatures. Crystal plates parallel to (010) plane. Angle of incidence of light $\varphi = 65°$. The magnetization was parallel to the surface of the sample and perpendicular to the plane of incidence of the light [87B2].

5.5.8.11 $ErFeO_3$ compounds

$ErFeO_3$ (see also LB III/4a, p. 242 and III/12a, p. 457)

I The magnetic structure below $T_{N1} = 643$ K is antiferromagnetic $\Gamma_4(F_zG_x)$ for the iron spins with a weak ferromagnetism along the c axis. The erbium spin lattice is paramagnetic down to liquid helium temperatures. The magnetization of the erbium sublattice induced by the iron sublattice is directed opposite to the weak ferromagnetism. In the reorientation region $T_R \approx 90 \cdots 100$ K (in [76V1] neutron diffraction shows that this transition takes place in the temperature range between 96 and 79 K) a phase transition $\Gamma_4(F_zG_x) \rightarrow \Gamma_{42}(F_{zx}G_{xz}) \rightarrow \Gamma_2(F_xG_z)$ takes place, [77V2] and Fig 268; see also Fig. 191. The weak antiferromagnetism component A_y changes to C_y. The temperature dependence of the A_y component weak antiferromagnetism (hidden canting) is measured by NMR, Fig. 269. Neutron diffraction shows that below $T_{N2} = 4.4(1)$ K [76V1] the erbium magnetic sublattice orders antiferromagnetically parallel to the c axis (configuration C_z) whereas the iron moments turn gradually away from the c axis into the direction of the b axis [73G1], see Fig. 270. The latter has been confirmed by [76V1]. The low temperature phase is $\Gamma_{12}(G_{yz}F_x; C_z)$. At 1.5 K the angle ϕ in Fig. 270 equals 51 ° [73G1], see also Fig. 271. For a theory of low-temperature spin reorientation see [78V1], and for the same in the presence of a magnetic field [80K6]. Applied magnetic fields have a great influence on the phase transitions. Below the compensation point for the weak ferromagnetic moment at $T_{comp} = 45$ K, new orientational transitions take place in a magnetic field, causing a change in orientation of the antiferromagnetic vector. These are first-order transitions between two different canted phases, as was revealed by magnetostriction measurements [88M2].

Threshold fields for phase transitions near T_R, see Fig. 272. Magnetic phase diagrams below T_{N2}, Figs. 273, 274.

Temperature dependence of the magnetization Fig. 275 and Young's modulus Figs. 276, 277. Sound velocity Fig. 278. For a study of domain walls with NMR see [75Z2].

III Magnon excitations Fig. 279. Room temperature one-magnon frequencies for M_1 and M_2 modes are 13.75 and 22.7 cm^{-1}, respectively [82W1]. Magnetic resonances in the intermediate state [79D2, 85D1]. NMR measurements see Fig. 268. Muon spin rotation frequency Table 34, Figs. 195, 280. Muon site determination [84H2].

IV Kerr effect Fig. 281. Sound velocity under pressure near T_{N2} [78G3].

$Er_{0.5}Tm_{0.5}FeO_3$

III ^{57}Fe NMR frequencies Fig. 268.

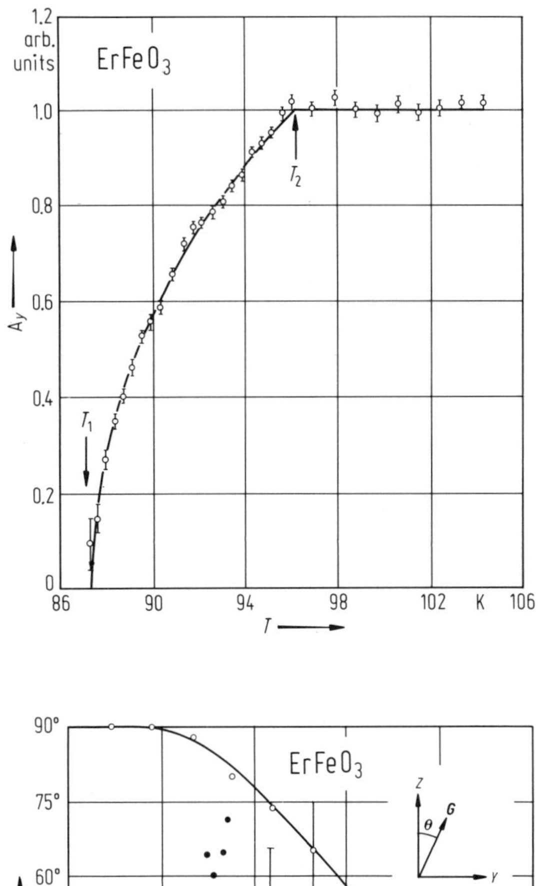

Fig. 269. $ErFeO_3$. Temperature dependence at the spin reorientation of the weak antiferromagnetism (hidden canting) A_y in arbitrary units, as derived from neutron diffraction [82P1].

◀

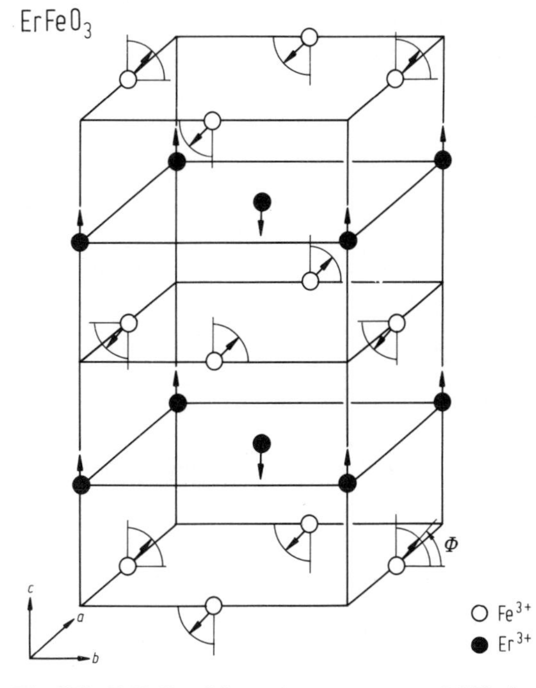

Fig. 270. $ErFeO_3$. Magnetic structure at 1.5 K derived from neutron diffraction patterns [73G1].

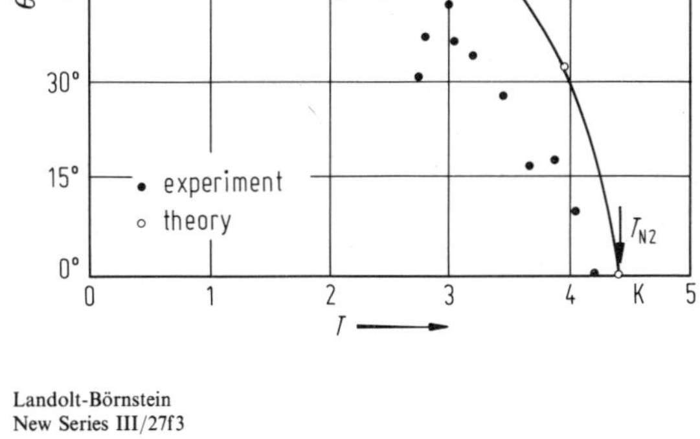

◀

Fig. 271. $ErFeO_3$. Angle θ between direction of iron magnetic moment and z axis as derived from neutron diffraction spectra at various temperatures. The curve is the result of a calculation based on interaction parameters. The vertical bars indicate allowed deviations. $T_{N_2} = 4.4(1)$ K [76V1].

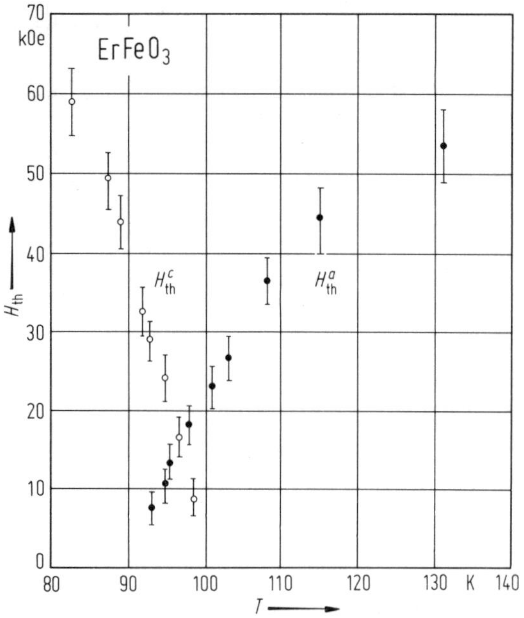

Fig. 272. $ErFeO_3$. Threshold fields H^c_{th} and H^a_{th} for the phase transitions $\Gamma_2 \rightarrow \Gamma_4$ and $\Gamma_4 \rightarrow \Gamma_2$, respectively [76B1].

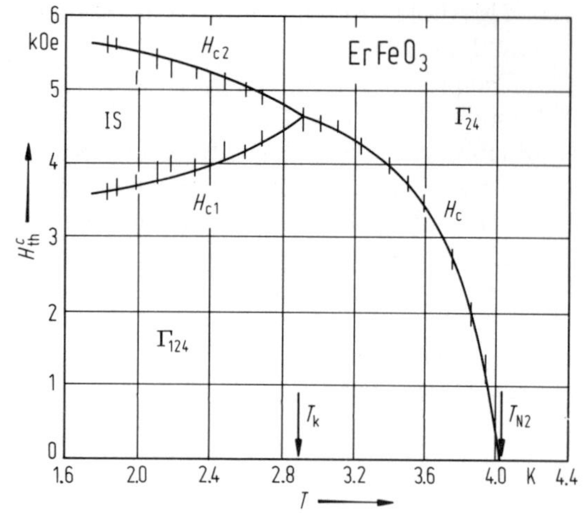

Fig. 273. $ErFeO_3$. Magnetic phase diagram for applied magnetic fields parallel to the c axis. Spin configurations for Er subsystem. IS: so-called meta magnetic intermediate states or Γ_{1234} phase. $T_K = 2.9(1)$K is the tricritical temperature. H_{c1} and H_{c2} are the boundaries of the first order transitions, H_c is the temperature dependence of the second-order phase transition. $T_{N2} = 4.05(5)$K. Results derived from magnetization measurements [86D1], for similar earlier and later results derived from NMR see [81K4, 89V2], from dielectric resonance experiments [84D2, 86D2] and from high-frequency magnetic susceptibility [85D2].

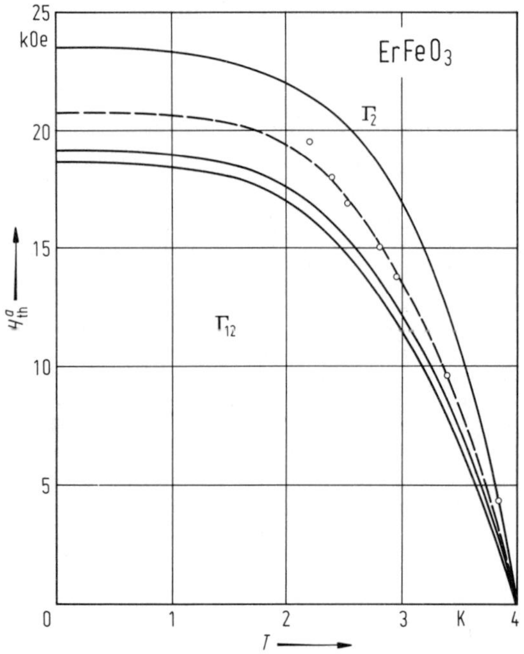

Fig. 274. $ErFeO_3$. Phase diagram for $\boldsymbol{H} \| a$ axis, derived from magnetostriction isotherms. Circles experimental, dashed and solid curves calculated on the basis of various estimated interaction properties [80K6].

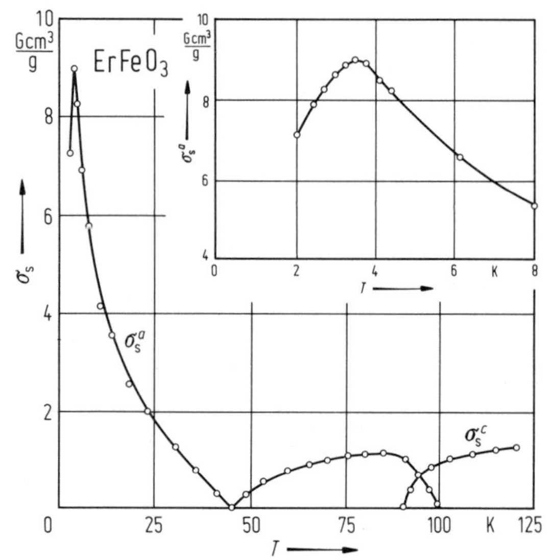

Fig. 275. $ErFeO_3$. Temperature dependence of the spontaneous mass magnetization σ_s along the crystallographic a and c axes [76B1].

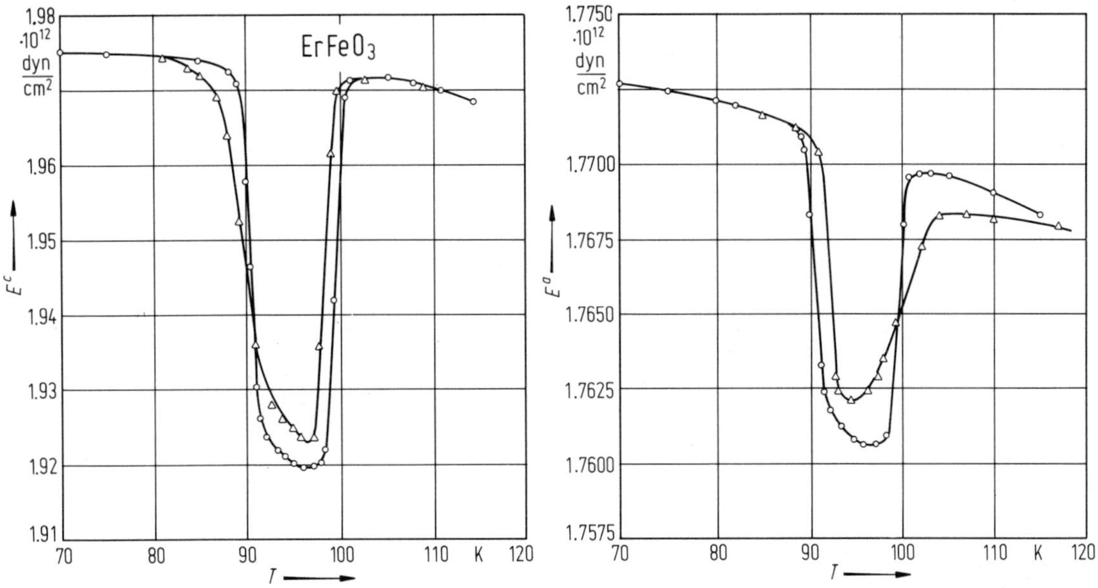

Fig. 276. $ErFeO_3$. Temperature dependence of the Young's modulus E along two crystallographic axes, with (Δ) and without (○) an applied magnetic field [76B1].

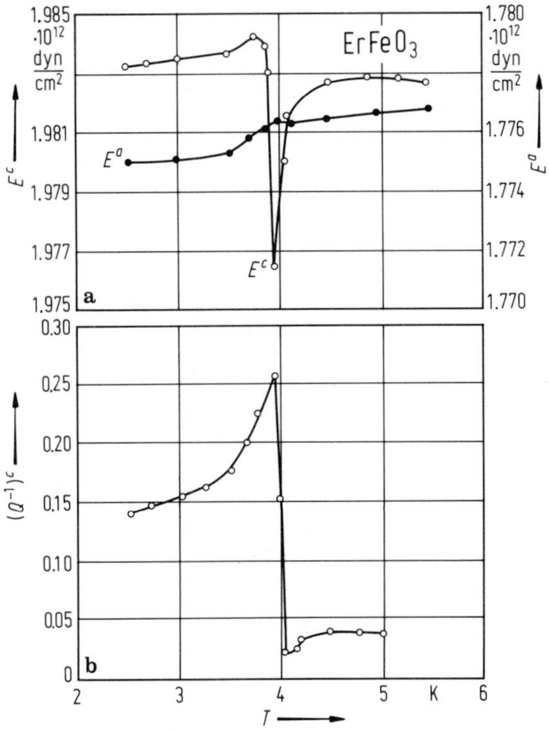

Fig. 277. $ErFeO_3$. Temperature dependence of (a) Young's modulus E and (b) internal friction Q^{-1} measured along the crystallographic a axis (●) and c axis (○) [76B1].

For Fig. 278, see next page.

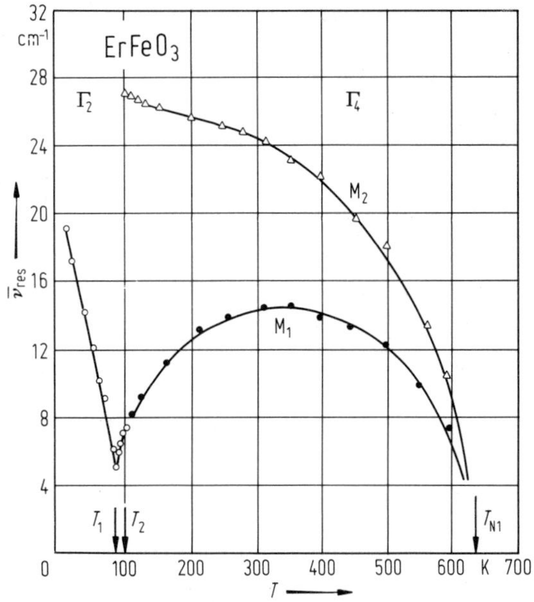

Fig. 279. $ErFeO_3$. Temperature dependence of the frequency (given as wavenumber $\bar{\nu}_{res}$ in cm^{-1}) of the one-magnon spin wave modes M_1 and M_2. Line M_2 was not observed in the low-temperature Γ_2 (G_zF_x) spin configuration below T_1 [88K3, see also 82K4].

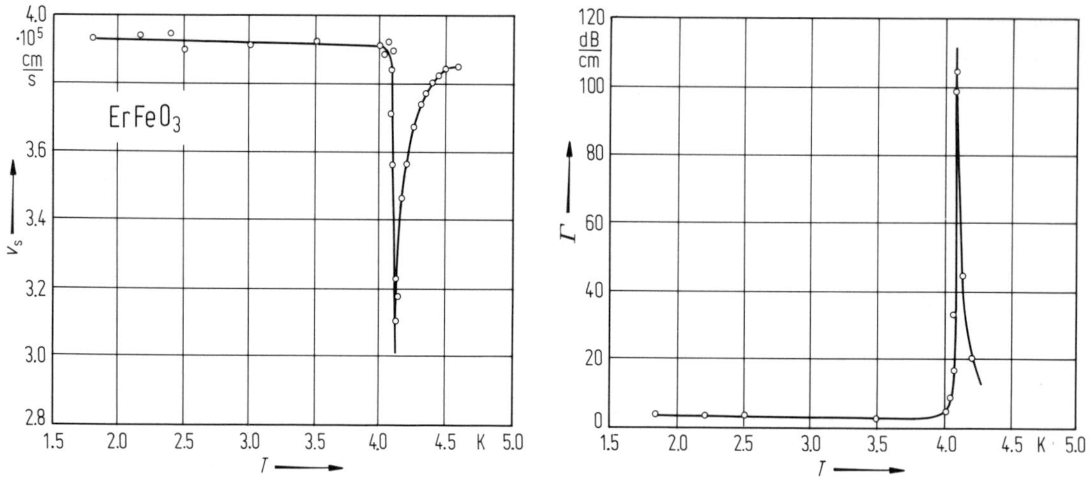

Fig. 278. $ErFeO_3$. Temperature dependence of the velocity v_s and the attenuation coefficient Γ of a transverse ultrasonic wave, $f = 25$ MHz, characterized by the wave vector $\boldsymbol{k} \parallel c$ and the polarization $\boldsymbol{E} \parallel b$ [89B3].

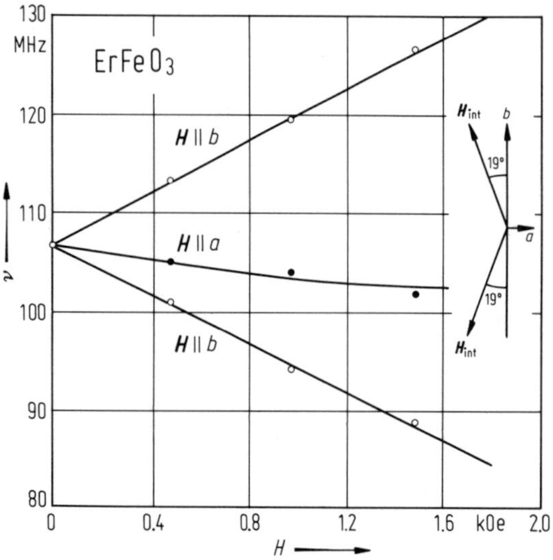

Fig. 280. $ErFeO_3$. Muon-spin-rotation frequency ν vs. applied magnetic field H in the a and b crystallographic directions at 80 K. The internal field is in the ab plane with components $H_{x,\,int} = -2.53(7)$ kOe and $H_{y,\,int} = \pm 7.44(4)$ kOe [83H1].

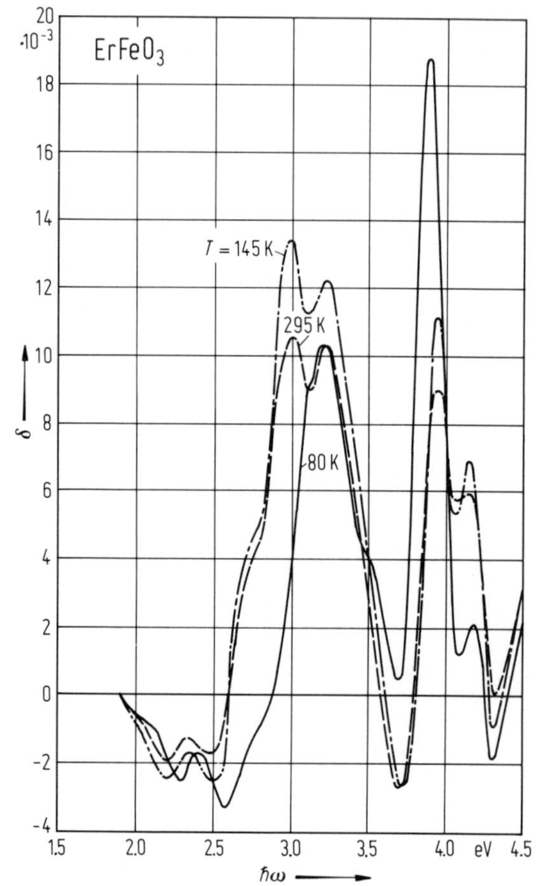

Fig. 281. $ErFeO_3$. Equatorial Kerr effect δ vs. photon energy $\hbar\omega$ at various temperatures. Angle of incidence of light $\varphi = 65°$. Magnetic field parallel to c axis for $T = 295$ and 145 K, and parallel to a axis for $T = 80$ K [87B2].

5.5.8.12 $TmFeO_3$

$TmFeO_3$ (see also LB III/4a, p. 242 and III/12a, p. 458)

I Domain-wall energy density Fig. 282. Calculated temperature dependence of magnetocrystalline anisotropy constants Fig. 283, The spin reorientation $\Gamma_2 \rightarrow \Gamma_4$ occurs in the temperature range 94 ··· 80 K, corresponding to $K_2 = 0$ at 94 K and $K_2 + 2\ K_4 = 0$ at 80 K. At RT is found $K_2 = 1.3(1) \cdot 10^6$ erg cm^{-3} [84S2].

The change of sound velocity in the SR range is shown in Fig. 284. For non-linear acoustic phenomena see [83L1]. Surface magnetoelastic waves on the *ab* surface along the *a* axis show velocity dips of 2.5% in the SR range, where $v_o = 3.02(5) \cdot 10^5$ cm s^{-1} [81G5]. Domain wall velocity depends linearly on the magnitude of the applied magnetic field up to a velocity of $2 \cdot 10^6$ cm s^{-1} (applied field about 2 Oe) [82C1, 81C1].

III ^{57}Fe NMR at 96.5 K reveals the "overt" canting angles $\phi = 1.10(5)°$ leading to weak ferromagnetism and the "hidden" canting angle $\chi = 0.54(5)°$ [82L1], see Fig. 2.

Mössbauer data at various temperatures Figs. 31, 285, 286. The angle θ between the principal EFG axis *a* and H_{hyp} is close to 90° at temperatures above 95 K. Below 80 K this axis lies in the *ab* plane, making an angle $|\beta| = 65(15)°$ with the *a* axis [84H1].

The submillimeter spectra show five magnetic resonance lines, see Fig. 287. The M_1 and M_2 modes reveal SR transition temperatures T_{R1} and T_{R2} of 84 and 93.6 K, respectively; the RT magnon frequencies are 13.8(5) and 24.4(2) cm^{-1} [83V2, 85V1].

Domain ^{57}Fe NMR in the SR region and in domain walls [86N1, 91G1]. AFMR frequencies [84B2].

IV Kerr effect Fig. 288.

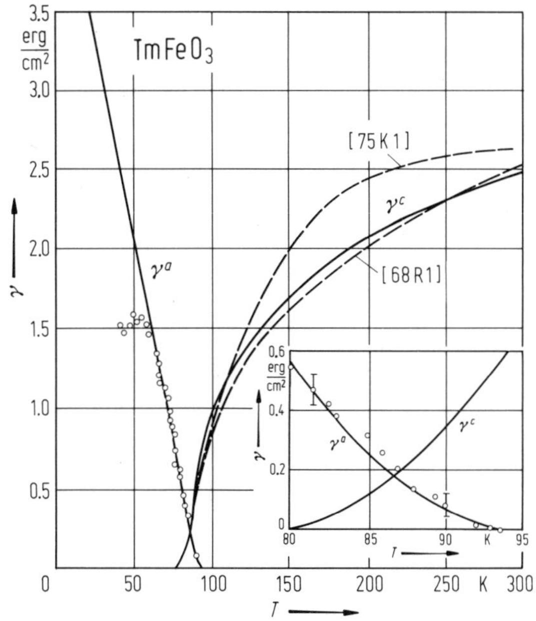

Fig. 282. $TmFeO_3$. Temperature dependence of the domain wall energy density γ measured for domain walls parallel to the *ab* plane. Curves γ^a and γ^c represent calculated values for platelets cut perpendicular to the *a* and *c* axis, respectively [84S2]. Broken curves according to [75K1] and [68R1].

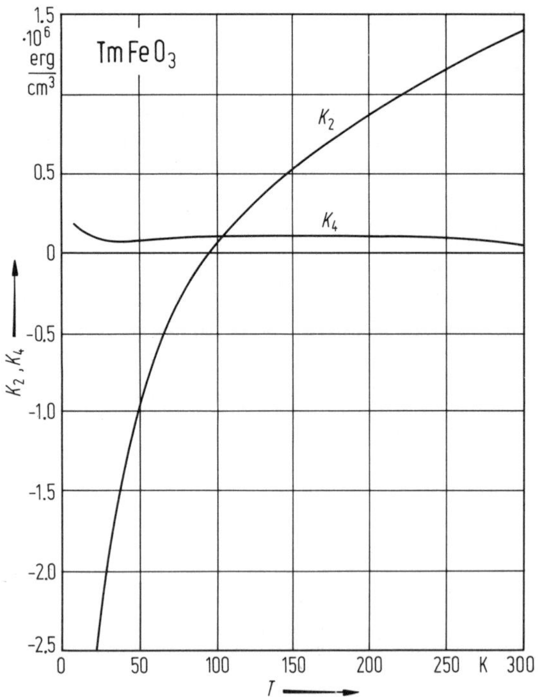

Fig. 283. $TmFeO_3$. Magnetocrystalline anisotropy constants K_2 and K_4 vs. T as derived from magneto-optical investigations of stripe domain wall configurations [84S2].

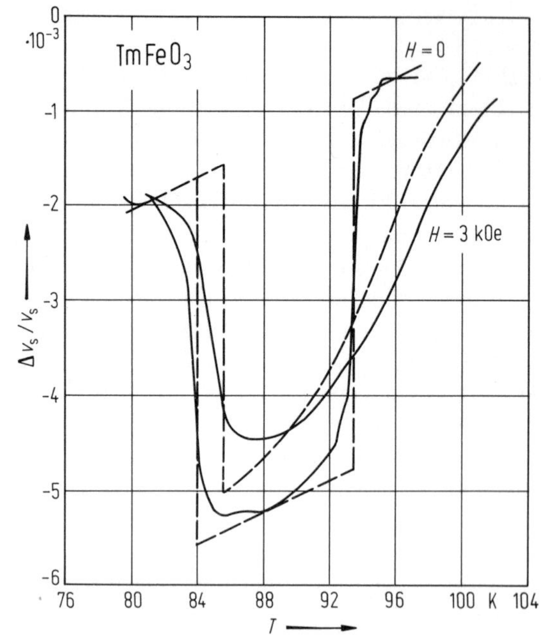

Fig. 284. $TmFeO_3$. Relative shift of the velocity $\Delta v_s/v_s$ of the longitudinal 90MHz sound waves propagating along the c axis in the spin-reorientation temperature region with and without an applied magnetic field H. Dashed curves according to model calculations [80G3, see also 76G3].

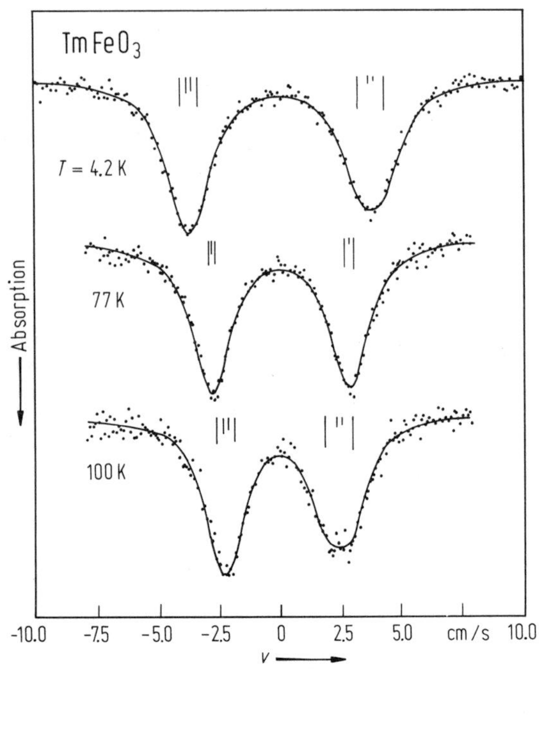

Fig. 285. $TmFeO_3$. Mössbauer $^{169}Tm^{3+}$ absorption line shapes (T_N = 630 K) at various temperatures T. The induced magnetic moment of the Tm ions is parallel to the c axis at 100 K and parallel to the a axis below 80 K [84H1].

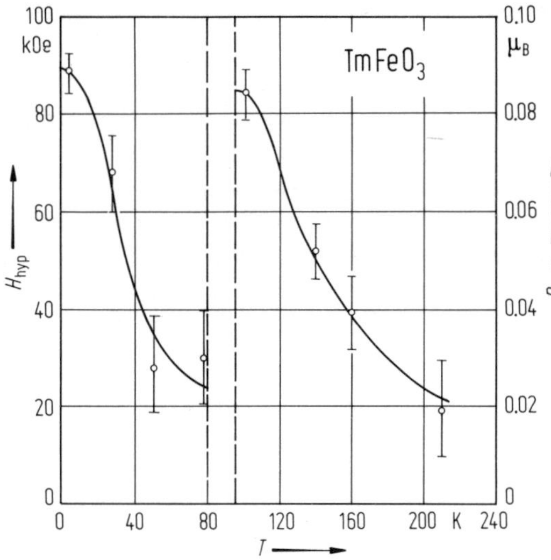

Fig. 286. $TmFeO_3$. Temperature dependence of the $^{169}Tm^{3+}$ effective magnetic hyperfine field H_{hyp} and the associated magnetic moment $p_{Tm^{3+}}$ of the Tm^{3+} ions. Above 95 K is $\boldsymbol{p}_{Tm^{3+}} \| c$ and below 80 K holds $\boldsymbol{p}_{Tm^{3+}} \| a$ [84H1].

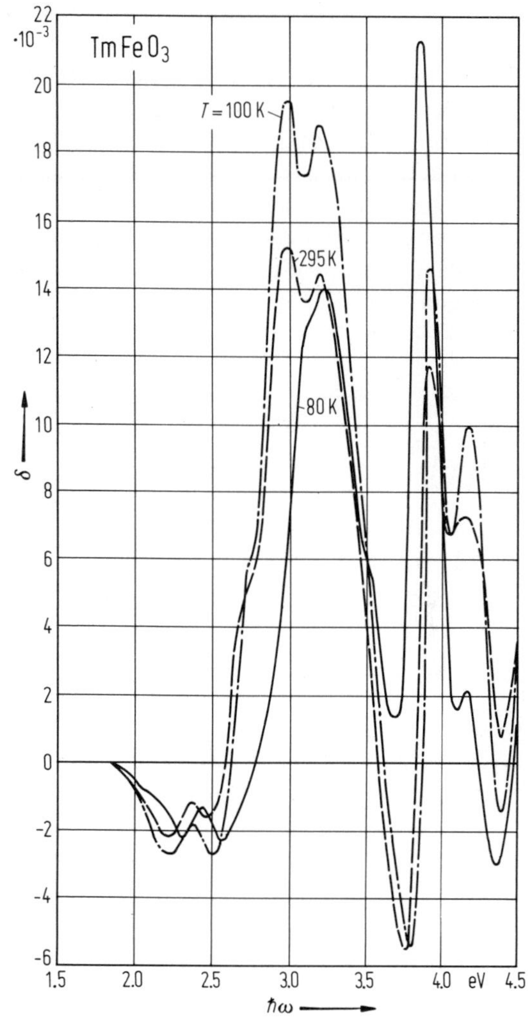

Fig. 288. $TmFeO_3$. Equatorial Kerr effect δ vs. photon energy $\hbar\omega$ at various temperatures T. Angle of incidence of light 65°. Magnetic field parallel to c axis for $T = 295$ and 100 K, and parallel to a axis for $T = 80$ K [87B2].

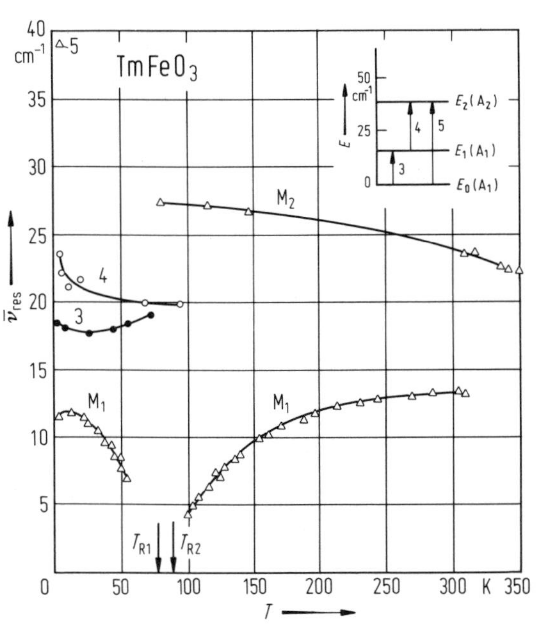

Fig. 287. $TmFeO_3$. Temperature dependence of the AFMR frequencies (given as wavenumbers $\bar{\nu}_{res}$ in cm^{-1}) of the iron ion modes M_1 and M_2 and of the rare earth ion modes 3, 4 and 5. The inset shows the transitions between the lower singlets of the principal multiplet of the Tm^{3+} ion [85B3].

5.5.8.13 $YbFeO_3$

$YbFeO_3$ (see also LB. III/4a, p. 244 and III/12a, p. 459)

I NMR. The relative value of the weak antiferromagnetic component A_y/G_x (i.e. the hidden canting of the iron spins) is $1.59(7) \cdot 10^{-2}$ at RT [83P1].
Phase transition temperatures from magnetization measurements Table 38.

III The temperature dependence of the soft magnetic resonance mode reveals the transition temperatures $T_1 = 6.80$ and $T_2 = 7.95$ K between the phases $\Gamma_2 \leftrightarrow \Gamma_{24}$ and $\Gamma_{24} \leftrightarrow \Gamma_4$ of the Yb^{3+}-spin system, respectively. See Fig. 289 and [90B2]. For magnetoacoustic effects in the transition regions see [87D1].

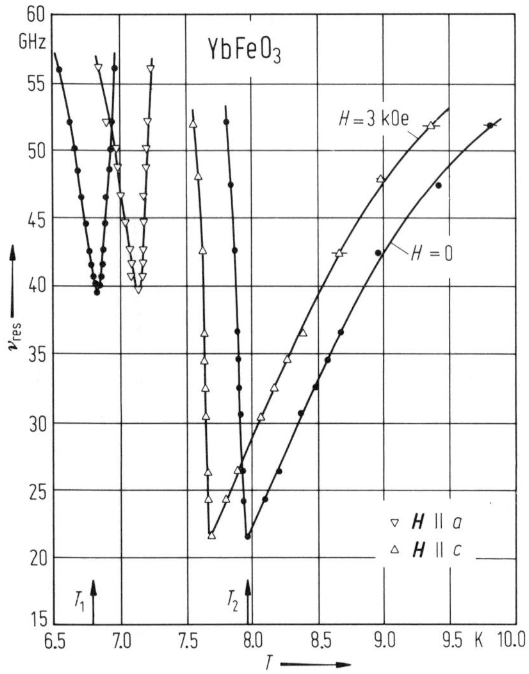

Fig. 289. $YbFeO_3$. Temperature dependence of the soft magnetic resonance mode frequency ν_{res} in the $\Gamma_2 \rightarrow \Gamma_4$ SR range of the Yb ions with and without the presence of a magnetic field H [87D1].

5.5.8.14 $LuFeO_3$ compounds

$LuFeO_3$ (see also LB III/4a, p. 244 and III/12a, p. 459)

I For calculated spontaneous magnetization see Fig. 165.

III Magnetic hyperfine field Table 35. For NMR data of domain walls, see [76Z3].

$Lu_{0.5}Y_{0.5}FeO_3$

I Magnetostriction Fig. 290.
Threshold fields Fig. 166.

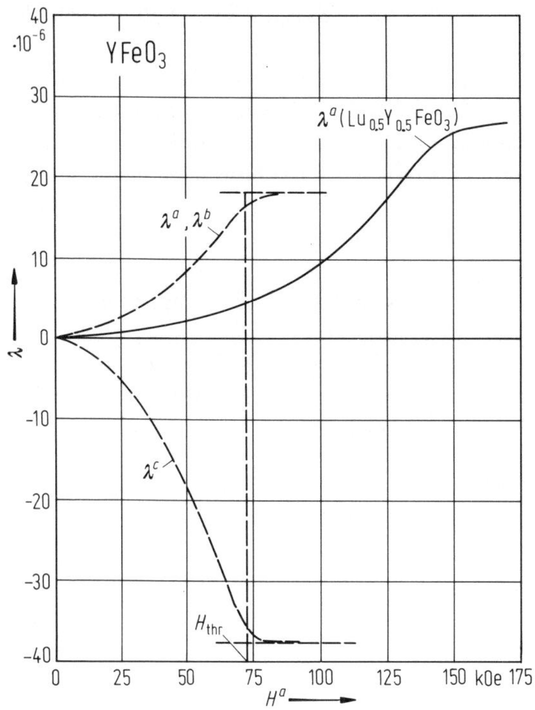

Fig. 290. $Lu_{0.5}Y_{0.5}FeO_3$ (solid curve), $YFeO_3$ (dashed curves). Field dependence of magnetostriction constants in the three crystallographic directions as a consequence of a magnetic field H^a applied parallel to the *a* axis [81K3].

5.5.8.15 $YFeO_3$ compounds

$YFeO_3$ (see also LB III/4a, p. 244 and III/12a, p. 460)

I Noncollinear antiferromagnet, weak ferromagnetism along the *c* axis. G_xF_z configuration between 4.2 K and $T_N = 645$ K. Temperature dependence of magnetization Fig. 75; see also Fig. 170. Spin rotation $G_xF_z \rightarrow G_zF_x$ in a magnetic field along the AF axis (*a* axis), threshold field $H^a_{th} = 75(4)$ kOe at 80 K [81K3], see also Fig. 166. From neutron diffraction analyses follows $A_y/G_x = 1.93(18) \cdot 10^{-2}$ at RT [81P1, 82P1]. Magnetostriction Fig. 290 and [81K5]. The magnetocrystalline anisotropy constant $K_2 = 0.12\ \mathrm{cm}^{-1}$ per Fe^{3+} ion at 80 K [81K3]. For a discussion of dipolar interactions [74B4]. The domain walls can move at very high speed under the influence of an applied magnetic field. Up to the transverse sound velocity, i.e. 4000 m s^{-1} a moving wall remains straight [82C1, 82C3]. For data on the velocity Fig. 291. For details on the low speed part [78T1, 78T2]. NMR in domain wall [80N2].

III Muon spin rotation frequency and corresponding local magnetic field at RT Table 34, Figs. 292, 195. Muon site determination [84H2].

^{57}Fe NMR at 4.2 K reveals the "overt" canting angle $\phi = 0.74(2)^\circ$ leading to weak ferromagnetism and the "hidden" canting angle $\psi = 1.25(5)^\circ$ [80L1 and 82L1] defined in Fig. 2.

Mössbauer data Fig. 293 and Table 44.

Reorientation of the antiferromagnetic vector from the *a* axis towards the *c* axis occurs continuously with the increasing applied field, see Fig. 294.

IV Raman scattering shows spin wave frequencies at 11.0 cm^{-1} and 17.3 cm^{-1}, implying an in-plane anisotropy field of 1047 Oe [80W1].

AFMR frequencies derived from Raman spectra Fig. 295 and [84K2, 84V1, 85B4]. See also [85V1, 82W1]. Microcrystalline particles [86A1].

Principal values of the RT refractive index in the range 220 ··· 350 GHz: $n^a = 4.807(5)$, $n^b = 4.578(5)$ and $n^c = 5.050(5)$. The absorption index $k = 0.001(2)$, independent on direction and frequency [84K2]. See also Fig. 296. Linear birefringe and its field dependence [82A2, 80K1, 86K1]. Faraday effect [84K2, 77C3]. Kerr effect Fig. 297. Light-induced changes of magnetization [78N2, 87B4].

$YFe_{1-x}Co_xO_3$

I Magnetic properties derived from neutron diffraction of powders Fig. 298, and from torque measurements on crystals Fig. 299. Magnetocrystalline anisotropy Fig. 300.

The compound x = 0.0024 has the spin configuration G_xF_z above 240 K. Between 80 ··· 4.2 K the configuration is $G_{xz}F_{zx}$, the magnetic moment at low temperature is stably oriented in a direction of 53° with respect to the c axis. Apparently only one second-order phase transition occurs at the moment of departure of the spins from the c axis [77B3], in contrast to what is found for simultaneous substitutions of Ti and Co [77A3]. For a detailed investigation of the magnetostriction for the latter compounds in the SR range see [77A2]. For the influence of an applied field Fig. 301.

III NMR spin echo data in domains and domain boundaries [77V1].

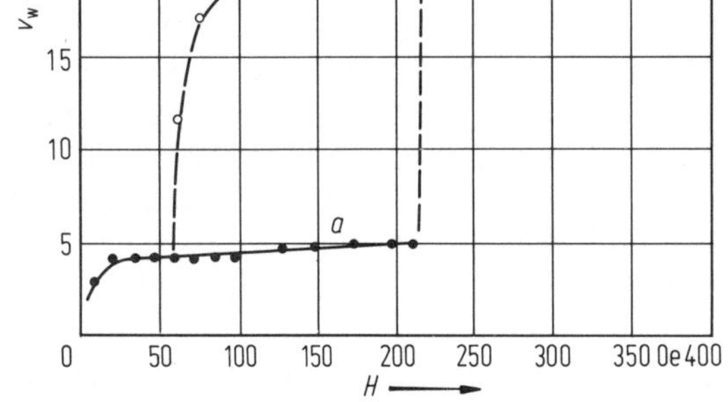

Fig. 291. $YFeO_3$. A 100 μm thick platelet cut perpendicular to the optical axis. Velocity of a domain wall under the influence of a pulsed field H. Curve a: straight wall; b, c: maximum speed of leading parts of deformed wall; d: straightened-out wall [82C1].

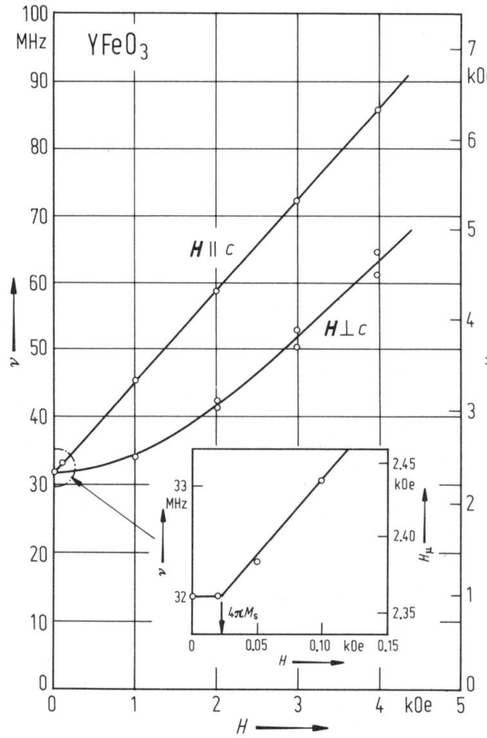

Fig. 292. $YFeO_3$, single crystal. Muon spin rotation frequency ν and local field H_μ at the muon at RT as a function of an external field H parallel or perpendicular to the c axis. Inset: low field region. $4\pi M_s \approx 20$ G is the spontaneous magnetization of the compound [81H1].

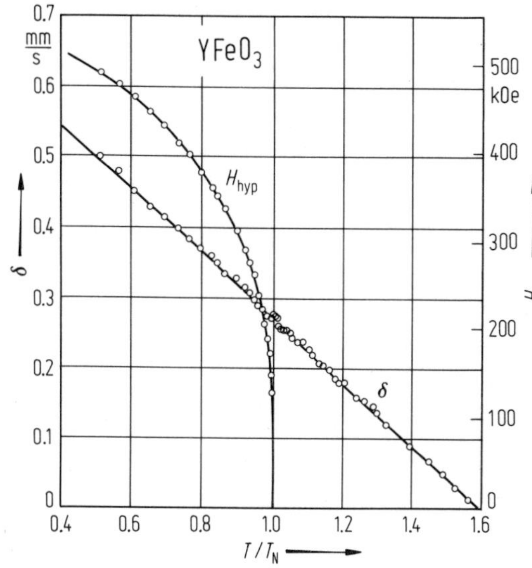

Fig. 293. $YFeO_3$. Mössbauer data. Isomer shift δ and effective magnetic hyperfine field H_{hyp} at ^{57}Fe nucleus vs. temperature. The isomer shift of the ^{57}Fe nuclei is measured relative to the line in Cr at RT. $T_N = 643$ K [75N1].

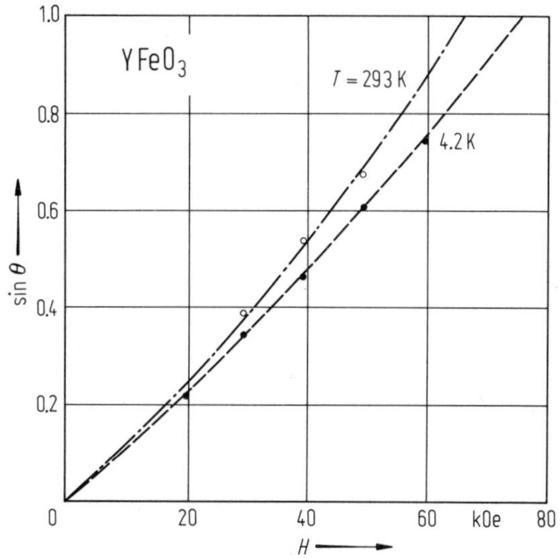

Fig. 294. $YFeO_3$. Orientation angle θ of Fe^{3+} spins with respect to the a axis vs. applied magnetic field in the c direction at two temperatures, as derived from Mössbauer spectra [75D2].

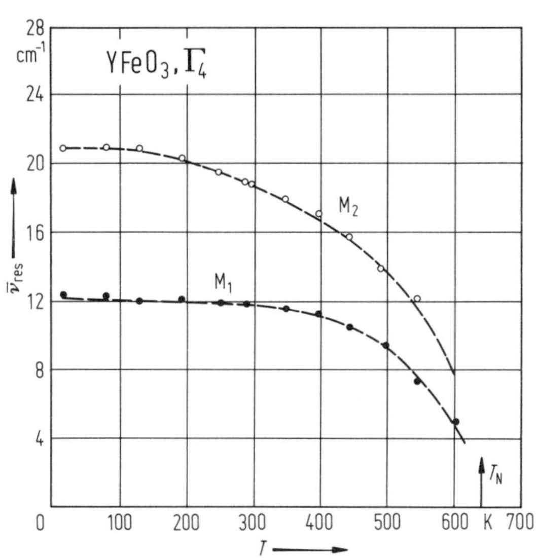

Fig. 295. $YFeO_3$. Temperature dependence of the AFMR frequencies (given as wavenumber $\bar{\nu}_{res}$ in cm^{-1}) as obtained from Raman spectra [88K3].

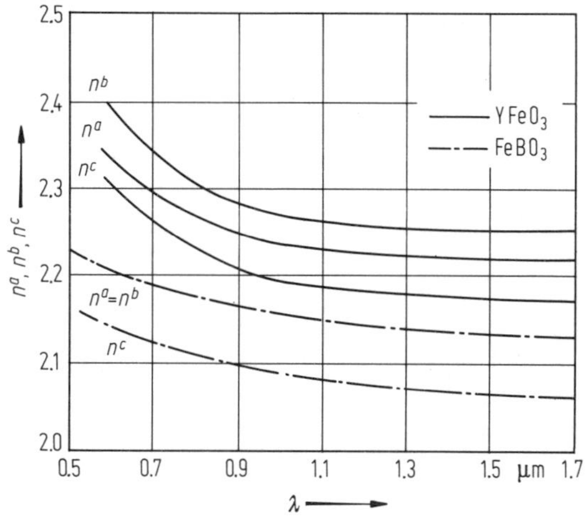

Fig. 296. $YFeO_3$; $FeBO_3$. Refractive indices n^a, n^b, n^c along the crystallographic axes as dependent on the wavelength in the IR region [82A2].

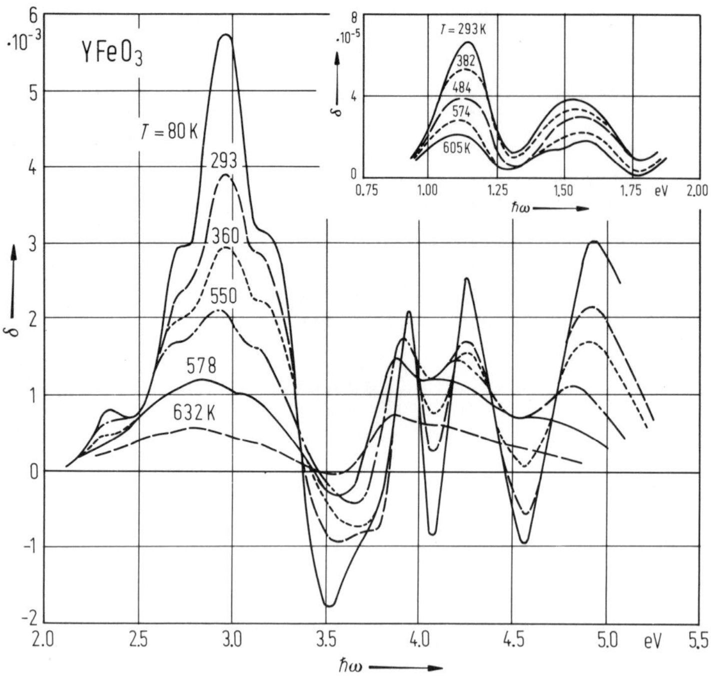

Fig. 297. $YFeO_3$. Equatorial Kerr effect δ vs. photon energy $\hbar\omega$ at various temperatures T [84Z1].

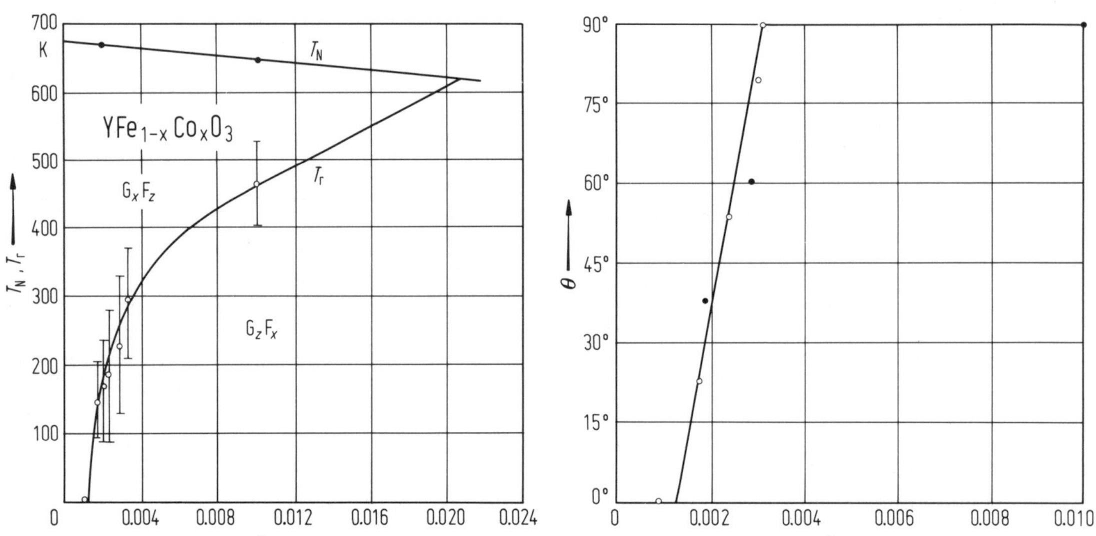

Fig. 298. $YFe_{1-x}Co_xO_3$. Composition dependence of the Néel temperature T_N, the magnetic reorientation temperature T_r and the angle θ at which the magnetization orients itself at low temperature with respect to the c axis. ● from neutron diffraction analysis of powder sample; ○ from magnetostriction data of single crystals [77B3].

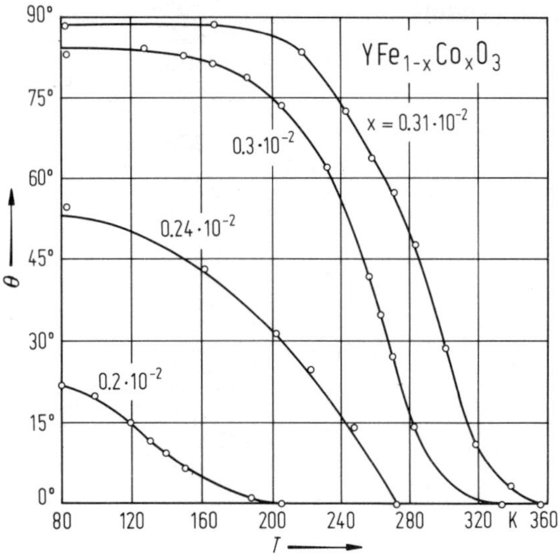

Fig. 299. $YFe_{1-x}Co_xO_3$. Temperature dependence of the angle θ between the direction of the spontaneous magnetic moment and the c axis for various compositions [76B2].

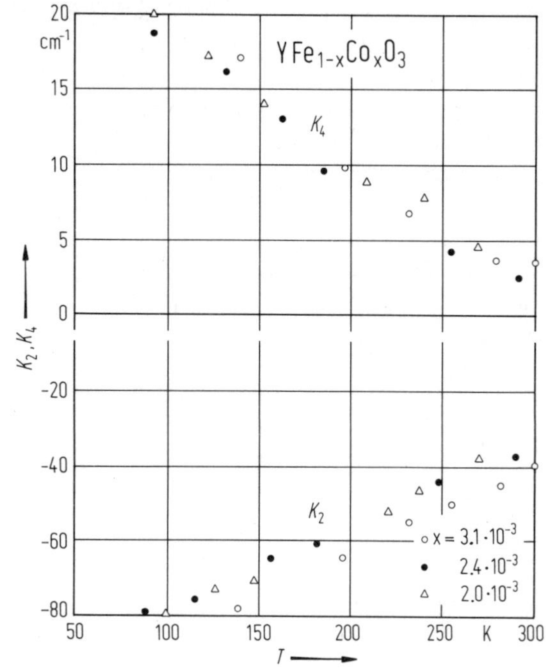

Fig. 300. $YFe_{1-x}Co_xO_3$. Temperature dependence of the second order (K_2) and fourth order (K_4) anisotropy constants per Co^{2+} ion for various compositions derived from torque measurements [77B3].

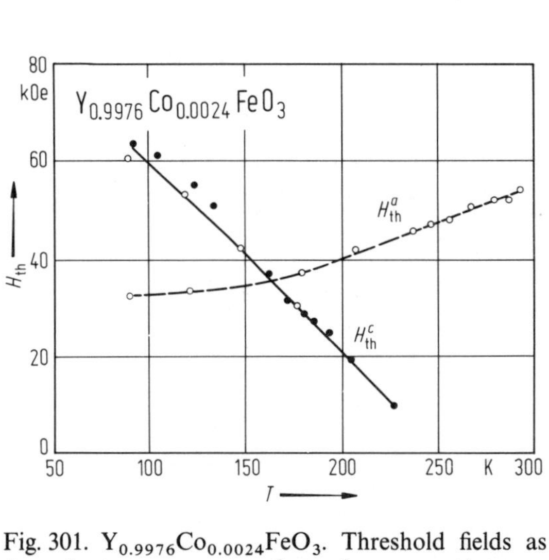

Fig. 301. $Y_{0.9976}Co_{0.0024}FeO_3$. Threshold fields as dependent on temperature for various magnetic transitions. $\boldsymbol{H}\|a$ axis: $G_{xz}F_{zx} \rightarrow G_zF_x$ at $T < 240$ K; $G_xF_z \rightarrow G_zF_x$ at $T > 240$ K. $\boldsymbol{H} \parallel c$ axis: $G_xF_z \rightarrow G_zF_x$ [77B3, see also 76B2]. Open circles: from isotherms of the rotating moment; full circles: from isotherms of the magnetostriction.

Table 44. $YFeO_3$. Mössbauer data. Effective hyperfine field H_{hyp}, isomer shift δ and quadrupole splitting ΔE_Q [75D2].

Parameter	$T = 4.2$ K	293 K
H_{hyp} [kOe]	552(5)	512(5)
δ [a]) [mm s^{-1}]	0.50(1)	0.30(1)
$\Delta E_Q = \frac{1}{2}e^2qQ$ [mm s^{-1}]	+ 0.08(2)	− 0.04(2)

[a]) Relative to α iron at RT.

5.5.9 $RCoO_3$ compounds

$RCoO_3$

I, Curie-Weiss constants Table 45.

II Resistivities Fig. 302.

Oxygen non-stoichiometry [87P2].

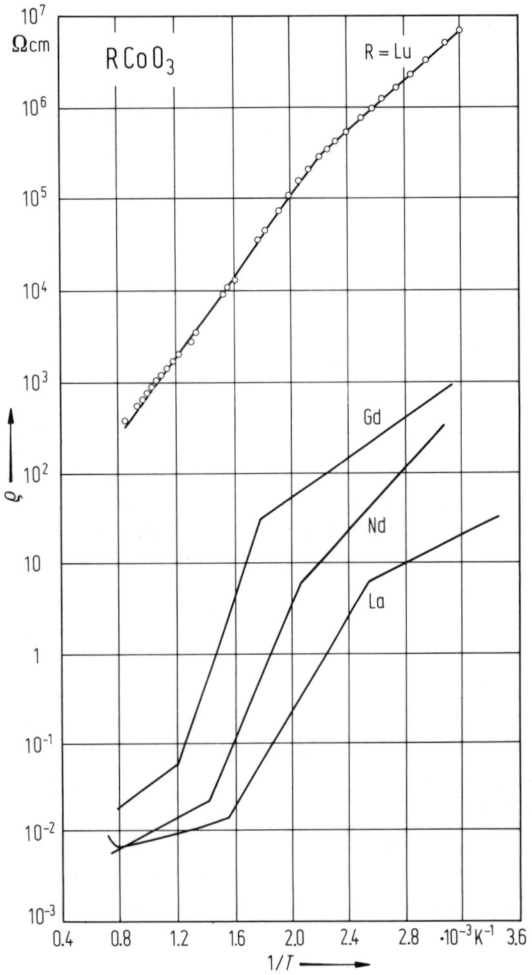

Fig. 302. $RCoO_3$. Electrical resistivity ρ vs. reciprocal temperature $1/T$ [76B3].

Table 45. $RCoO_3$. Curie-Weiss constants. Paramagnetic Curie temperature Θ_p and effective paramagnetic moment p_{eff} per FU expressed in Bohr magnetons μ_B for powder samples [80M1].

Compound	Temperature range [K]	p_{eff} [μ_B/FU]	Θ_p [K]
$PrCoO_3$	< 190	3.73	35
	> 700	6.36	524
$NdCoO_3$	< 250	3.71	51
$GdCoO_3$	< 370	7.80	3
	370 ··· 500	8.56	70
$TbCoO_3$	< 350	10.38	2
	350 ··· 675	11.03	48
	> 675	12.52	245
$DyCoO_3$	< 365	10.41	3
	365 ··· 625	11.02	47
	> 645	12.53	245
$HoCoO_3$	< 350	10.58	4
	350 ··· 650	11.25	45
	> 650	12.33	176
$YbCoO_3$	< 375	5.38	77
	375 ··· 850	5.67	107

$LaCoO_3$ (see also LB III/4a, p. 246–248 and III/12a, p. 463–465).

I X-ray analysis of low-temperature phases [90E1]: microstructure is tetragonal with $a = 3.84$Å and $c = 11.81$Å. Neutron diffraction of $R\bar{3}c$ phase between 4 ··· 1248 K [86T1]. At RT $a = 5.3778(1)$Å, $\alpha = 60.798(1)°$.

Electrical conductivity Fig. 303.

Indications for temperature-induced magnetic moments as proposed in the thermally populated high-spin model are found in low temperature neutron spectra [89A1]. The effect of near-neighbor impurity effects on the low-temperature spin state of the cobalt ions is shown in [80M2] and [83V3].

II Magnetic susceptibility Figs. 304, 305. See also Fig. 201. $p_{eff} = 4.01\ \mu_B$/FU, $\Theta_p = -310$ K [79R1, see also 83V3].

$LaCo_{1-x}Mg_xO_3$

II Magnetic susceptibility Fig. 305, Table 46.

$LaCo_{1-x}Ni_xO_3$

I For transition temperature see Fig. 95.

II Magnetic susceptibility Fig. 306.

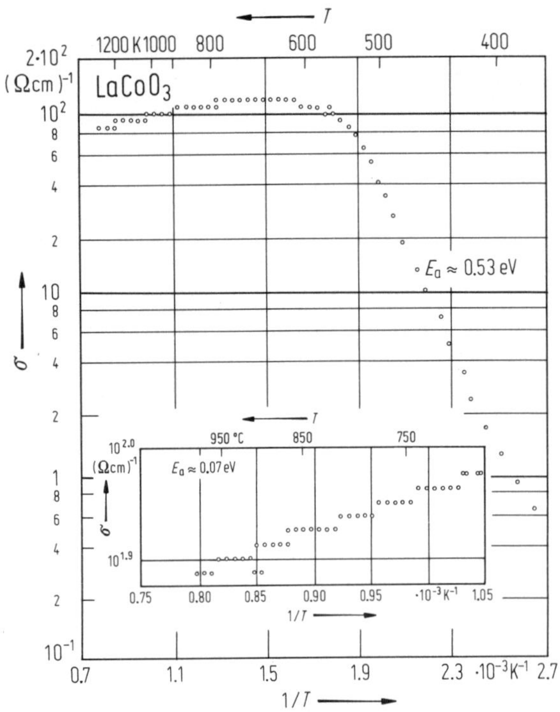

Fig. 303. $LaCoO_3$. Electrical conductivity σ vs. reciprocal temperature T^{-1}. Activation energy approximately 0.53 eV in the semiconducting branch. The data in the metallic regime ($T > 950$ K) are shown in detail in the insert [82T2].

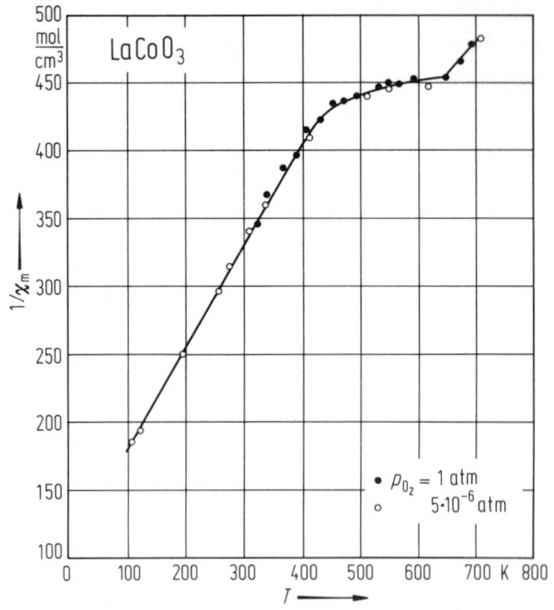

Fig. 304. $LaCoO_3$. Inverse molar susceptibility χ_m^{-1} vs. temperature T, measured for a compound $La^{3+}(Co^{2+}_{0.14}\ Co^{3+}_{0.23}\ Co^{III}_{0.63})\ O^{2-}_{2.93}$ in two different atmospheres. Co^{III} stands for the low spin state of Co^{3+}. The molar Curie constant $C_m = 1.32$ $cm^3\,K\,mol^{-1}$ for $T < 425$ K [82C2]. 1 atm = 101325 Pa. p_{O_2}: oxygen partial pressure.

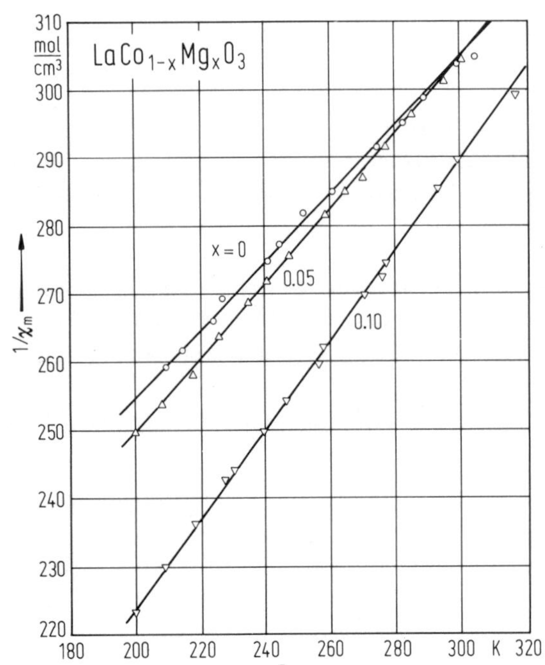

Fig. 305. $LaCo_{1-x}Mg_xO_3$. Inverse molar magnetic susceptibility χ_m^{-1} vs. temperature T [79R1].

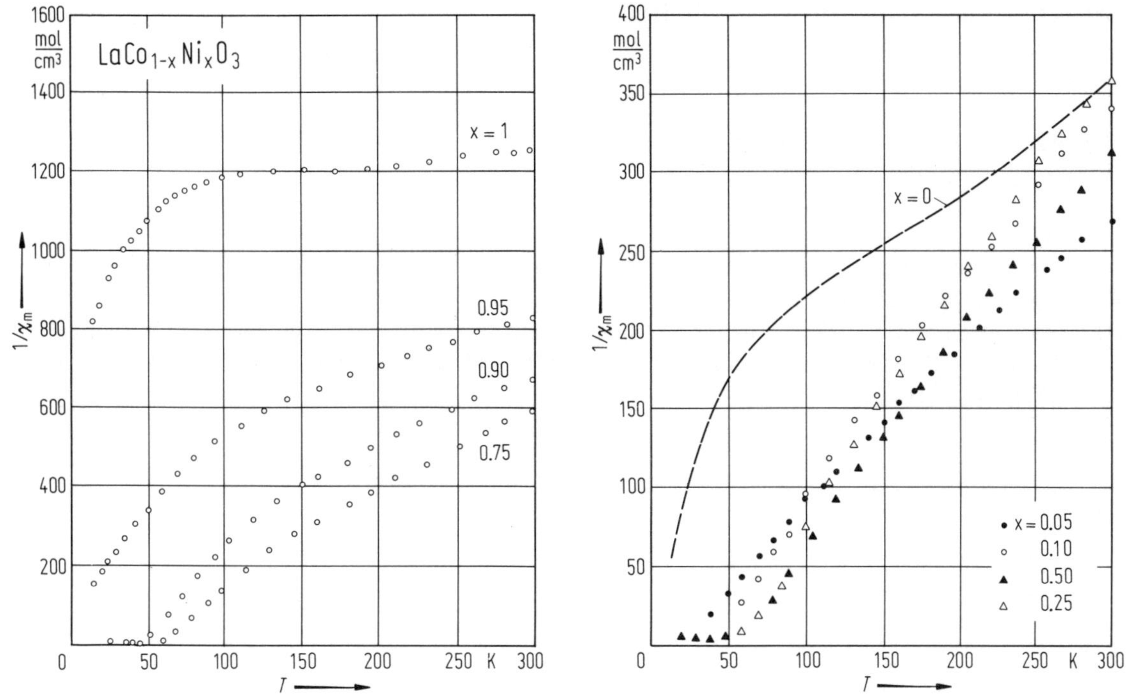

Fig. 306. $LaCo_{1-x}Ni_xO_3$. Inverse magnetic molar susceptibility χ_m^{-1} vs. temperature T [84V2, see also 77A4].

Table 46. $LaCo_{1-x}Mg_xO_3$. Curie-Weiss constants derived from data in Fig. 305 [79R1].

x	p_{eff} [μ_B/FU]	Θ_p [K]
0	4.01	− 310
0.05	3.82	− 252
0.10	3.47	− 134

$EuCoO_3$

III ^{151}Eu Mössbauer data [84S3].

T [K]	δ*) [$mm\,s^{-1}$]	$\Delta E_Q = eV_{zz}Q$ [$mm\,s^{-1}$]	η
297.1	0.71(5)	− 5.76(19)	0.71(18)
77.4	0.81(4)	− 6.06(26)	0.53(28)

*) With respect to the source $^{151}Sm(SmF_3)$.

$GdCoO_3$ (see also LB III/4a, p. 248 and III/12a, p. 465)
I Resistivity Fig. 302.
II Susceptibility Fig. 12, Table 45.

$HoCoO_3$ (see also LB III/12a, p. 465)
II Susceptibility Fig. 12, Table 45.

$ErCoO_3$
I Neutron diffraction: no magnetic ordering down to 0.3 K [70K1]. Lattice parameters Fig. 307.
II Magnetic susceptibility Fig. 308.
III Mössbauer spectra [75J1]. The signals due to high-spin Co^{3+} and low spin Co^{III} are of equal intensity in the temperature range RT to 600 K. RT isomer shift of ^{57}Fe with respect to $K_4Fe(CN)_6 \cdot 3H_2O$ is 0.73 mm s^{-1}.

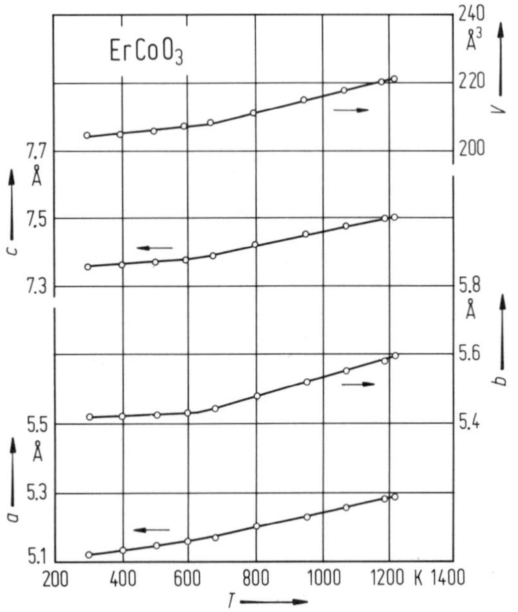

Fig. 307. $ErCoO_3$. Lattice parameters a, b and c and unit cell volume V as dependent on temperature T [75J1].

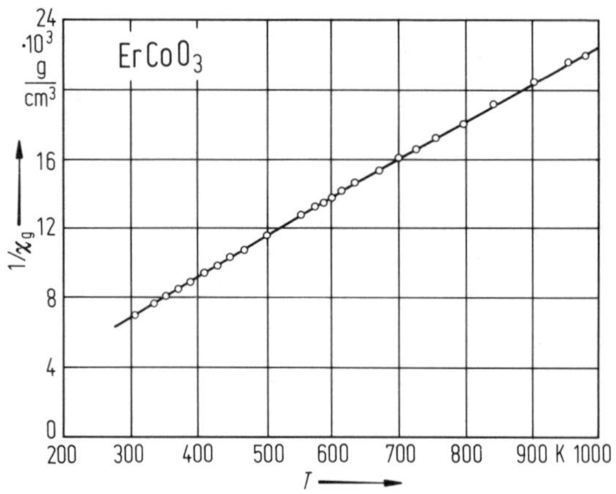

Fig. 308. $ErCoO_3$. Inverse magnetic mass susceptibility χ_g^{-1} vs. temperature T [75J1].

$YbCoO_3$
I Lattice parameters at RT: $a = 5.080$Å, $b = 4.422$Å and $c = 7.318$Å [82B1].
II Magnetic susceptibility Fig. 309. See also Table 45. $p_{eff} = 3.85$ and 3.96 μ_B below and above about 550 K, respectively. The ratio of Co^{3+}/Co^{III} is less than 1 below 550 K and about equal to 1 above 750 K [82B1].

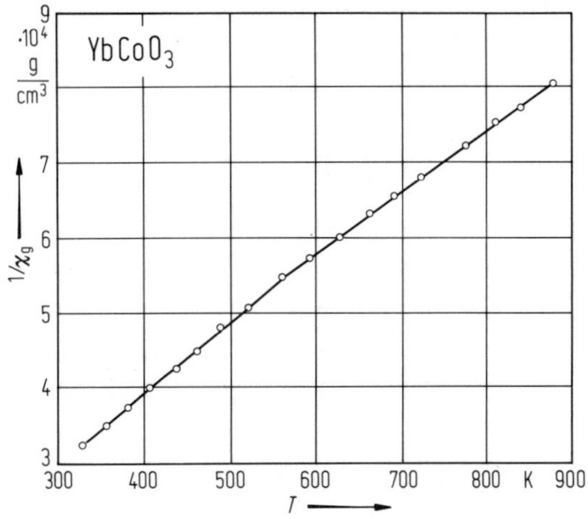

Fig. 309. $YbCoO_3$. Inverse magnetic mass susceptibility χ_g^{-1} vs. temperature T [82B1].

$LuCoO_3$

I Lattice parameters at RT: $a = 5.017$ Å, $b = 5.341$ Å and $c = 7.295$ Å [76B3]. Resistivity Fig. 302.

II Magnetic mass susceptibility Fig. 310. The low spin Co^{III} ions partially transform to the high spin Co^{3+} ions up to about 450 K, above this temperature the ratio of the two spin states remains constant [76B3].

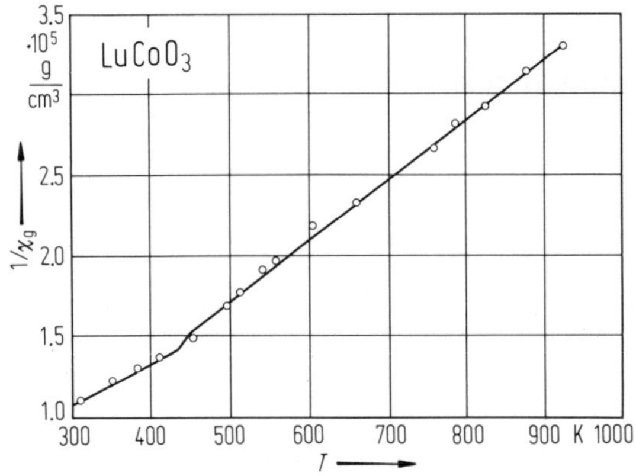

Fig. 310. $LuCoO_3$. Inverse magnetic mass susceptibility χ_g^{-1} vs. temperature [76B3].

$YCoO_3$

I Lattice parameters Fig. 311.

II Magnetic susceptibility Fig. 312.

III Mössbauer spectra [75J1]. At low temperature the presence of high spin of Co^{3+} and low spin Co^{III} is found. Above 500 K three distinct Mössbauer resonances are found. Isomer shift with respect to $K_4Fe(CN)_6 \cdot 3H_2O$ is 0.70 mm s^{-1} at 78 K [75J1].

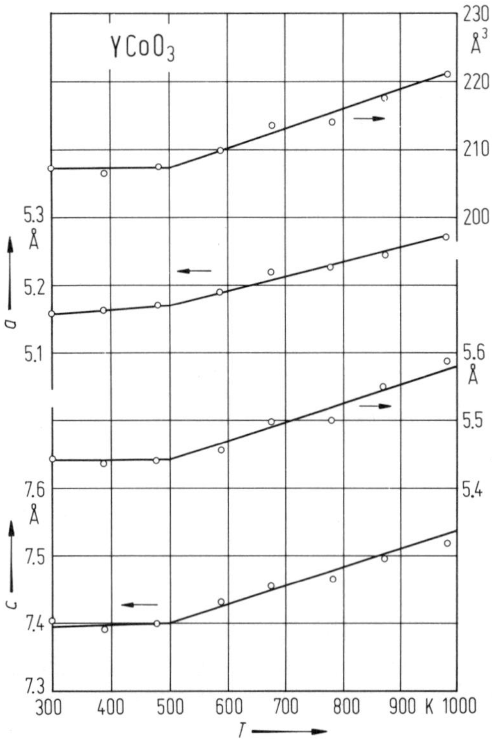

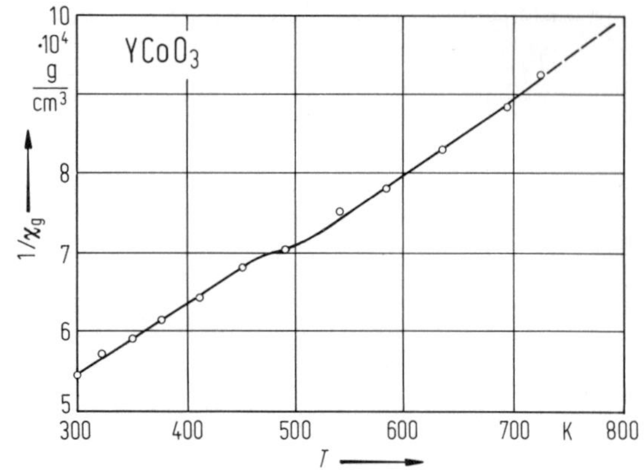

Fig. 312. $YCoO_3$. Inverse magnetic mass susceptibility χ_g^{-1} vs. temperature T [75J1].

◀

Fig. 311. $YCoO_3$. Lattice parameters a, b and c and unit cell volume as dependent on temperature [75J1].

5.5.10 $RNiO_3$ compounds

$LaNiO_3$

I Lattice parameters Table 30.
Resistivity vs. temperature Fig. 177.

II Paramagnetic properties Table 31 and Figs. 179, 306.

$PrNiO_3$

I Lattice parameters Table 47.

Table 47. $PrNiO_3$. Crystal structures and lattice parameters at various temperatures [90H1].

T [°C]	RT	200	400	500	600
Space group	Pbnm orthorhombic			R$\bar{3}$c rhombohedral	
a [Å]	5.4146(1)	5.4294(1)	5.4456(1)	5.4577(1)	5.4614(1)
b [Å]	5.3757(1)	5.3849(1)	5.3952(1)	–	–
c [Å]	7.6199(2)	7.6362(2)	7.6520(2)	13.1058(3)	13.1319(2)

Acknowledgement

Many thanks are due to Professor E. Burzo and to Professor R. Szymczak for their help in finding the relevant literature.

5.5.11 References for 5.5

56 B1 Bozorth, R.M., Williams, H.J., Walsh, D.E.: Phys. Rev. **103** (1956) 572.

63 B1 Bertaut, E.F.: Magnetism, Vol. 3, Rado, G.T., Suhl, H., (eds.) New York: Academic Press (1963), p. 149.

66 B1 Bertaut, E.F., Mareschal, J., de Vries, G., Aléonard, R., Pauthenet, R., Rebouillat, J.P., Zarubicka, V.: IEEE Trans. Magn. **2** (1966) 453.

67 E1 Eibschütz, M., Shtrikman, S., Treves, D.: Phys. Rev. **156** (1967) 562.

68 C1 Cashion, J.D., Cooke, A.H., Leask, M.J.M., Thorp, T.L., Wells, M.R.: J. Mater. Sci. **3** (1968) 402.

68 C2 Combarieu, A.de, Mareschal, J., Michel, J.C., Peyrard, J., Sivardière, J.: C.R. Acad. Sci. (Paris) Ser. B **267** (1968) 1169.

68 D1 De Combarieu, A., Mareschal, J., Michel, J.C., Sivardière, J.: Solid State Commun. **6** (1968) 257.

68 R1 Rossol, F.C.: J. Appl. Phys. **39** (1968) 5263.

69 P1 Polyakov, V.P., Bystrov, M.V., Seliber, G.G.: Fiz. Tverd. Tela **11** (1969) 1409; Sov. Phys. Solid State (English Transl.) **11** (1969) 1147.

70 C1 Cashion, J.D., Cooke, A.H., Martin, D.M., Wells, M.R.: J. Phys. **C3** (1970) 1612.

70 G1 Goodenough, J.B., Longo, J.M.: Landolt-Börnstein, New Series, III/Vol. **4a**, Berlin, Heidelberg, New York: Springer-Verlag, 1970, p. 126.

70 H1 Holmes, L., Eibschütz, M., van Uitert, J.: Appl. Phys. **41** (1970) 1184.

70 K1 Kappatsch, A., Quezel-Ambrunaz, S., Sivardière, J.: J. Phys. (Paris) **31** (1970) 369.

70 M1 Marezio, M., Remeika, J.P., Dernier, P.D.: Acta Crystallogr. **B26** (1970) 2008.

70 P1 Pataud, P., Sivardière, J.: J. Phys. (Paris) **31** (1970) 803.

70 S1 Starovoitov, A.T., Ozhogin, V.I., Bokhov, V.A., Zonn, Z.N., Loginov, G.M.: Fiz. Tverd. Tela **12** (1970) 301; Sov. Phys. Solid State (English Transl.) **12** (1970) 245.

71 H1 Holmes, L.M., van Uitert, L.G., Hecker, R.R.: Am. Inst. Phys. Conf. Proc. **5** (1971) 690.

71 S1 Sivardière, J., Quezel-Ambrunaz, S.: C. R. Acad. Sci. (Paris) Ser. **B273** (1971) 619.

71 S2 Shtrikman, S., Wanklyn, B.M., Yaeger, I.: Int. J. Magn. **1** (1971) 327.

72 C1 Courths, R., Hüfner, S., Pelzl, J., van Uitert, L.G.: Z. Phys. **249** (1972) 445.

72 H1 Hornreich, R.M., Wanklyn, B.M., Yaeger, I.: Int. J. Magn. **2** (1972) 77.

72 K1 Kovtun, N.M., Karnachev, A.S., Solovev, E.E., Chervonenkis, A.Ya., Shemyakov, A.A.: Fiz. Tverd. Tela **14** (1972) 2150; Sov. Phys. Solid State (English Transl.) **14** (1972) 1856.

73 F1 Finkman, E., Cohen, E., van Uitert, L.G.: Phys. Rev. **B7** (1973) 2899.

73 G1 Gorodetsky, G., Hornreich, R.M., Yaeger, I., Pinto, H., Shachar, G., Shaked, H.: Phys. Rev. **B8** (1973) 3398.

74 B1 Bazuev, G.V., Matveenko, I.I., Shveikin, G.P.: Fiz. Tverd. Tela **16** (1974) 240; Sov. Phys. Solid State (English Transl.) **16** (1974) 155.

74 B2 Bonville, P., Gonzalez-Jimenez, F., Imbert, P., Varret, F.: J. Phys. (Paris) **12** (1974) C6-575.
74 B3 Belov, K.P., Zvezdin, A.K., Kadomtseva, A.M., Krynetskii, I.B.: Zh. Eksp. Teor. Fiz. **67** (1974) 1983; Sov. Phys. JETP (English Transl.); **40** (1975) 980.
74 B4 Bidaux, R., Bouree, J.E., Hammann, J.: J. Phys. Chem. Solids **35** (1974) 1645.
74 T1 Tsushima, K., Tamaki, T., Yamaura, R.: Proc. Intern. Conf. MKM-73 on Magnetism Nauka, Vol. **5** (1974) 270.
74 Y1 Yamaguchi, T.: J. Phys. Chem. Solids **35** (1974) 479.
75 B1 Bazuev, G.V., Iutin, N.N., Matveenko, I.I., Shveikin, G.P.: Fiz. Tverd. Tela **17** (1975) 1167; Sov. Phys. Solid State (English Transl.) **17** (1975) 747.
75 B2 Bazuev, G.V., Shveikin, G.P.: Fiz. Tverd. Tela **17** (1975) 3453; Sov. Phys. Solid State (English Transl.) **17** (1976) 2261.
75 B3 Bidaux, R., Bouree, J.E., Hammann, J.: J. Phys. (Paris) **36** (1975) 803.
75 B4 Bidaux, R., Bouree, J.E., Hammann, J.: J. Phys. Chem. Solids **36** (1975) 655.
75 C1 Courths, R., Hüfner, S.: Z. Physik B **22** (1975) 245.
75 D1 Derkachenko, V.N., Kadomtseva, A.M., Ovchinnikova, T.L., Timofeeva, V.A., Khokhlov, V.A.: Fiz. Nizk. Temp. **1** (1975) 1007; Sov. J. Low. Temp. Phys. (English Transl.) **1** (1975) 483.
75 D2 Durbin, G.W., Johnson, C.E., Thomas, M.F.: J. Phys. C: Solid State Phys. **8** (1975) 3051.
75 H1 Hasson, A., Hornreich, R.M., Komet, Y.: Phys. Rev. B **12** (1975) 5051.
75 J1 Jadhao, V.G., Singru, R.M., Rama Rao, G., Bahadur, D., Rao, C.N.R.: J. Chem. Soc. Faraday Trans. II **71** (1975) 1885.
75 J2 Johnston, D.C.: Ph.D. thesis, University of California, San Diego, 1975.
75 K1 Kandaurova, G.S., Babushkin, O.A., Vaskowskii, V.O., Raev, V.K.: Fiz. Tverd. Tela **17** (1975) 1631; Sov. Phys. Solid State (English Transl.) **17** (1975) 1066.
75 N1 Nikolaev, V.I., Rusakov, V.S.: Kristallografiya **20** (1975) 845; Sov. Phys. Crystallogr. (English Transl.) **20** (1976) 519.
75 R1 Rao, C.N.R., Parkash, O.M., Ganguly, P.: J. Solid State Chem. **15** (1975) 186.
75 W1 Washimiya, S., Yamaguchi, T.: J. Phys. Soc. Jpn. **38** (1975) 1302.
75 Z1 Zubkov, V.G., Bazuev, G.V., Belousov, E.I., Shveikin, G.P.: Fiz. Tverd. Tela **17** (1975) 318; Sov. Phys. Solid State (English Transl.) **17** (1975) 193.
75 Z2 Zalesskii, A.V., Savvinov, A.M., Zheludev, I.S., Ivashchenko, A.N.: Zh. Eksp. Teor. Fiz. **68** (1975) 1449; Sov. Phys. JETP (English Transl.) **41** (1976) 723.
76 B1 Belov, K.P., Kadomtseva, A.M., Kovtun, N.M., Derkachenko, V.N., Melov, V.N., Khokhlov, V.A.: Phys. Status Solidi (a) **36** (1976) 415.
76 B2 Belov, K.P., Kadomtseva, A.M., Milov, V.N., Lukina, M.M., Ovchinnikova, T.L., Khafizova, N.A.: Fiz. Tverd. Tela **18** (1976) 3129; Sov. Phys. Solid State (English Transl.) **18** (1976) 1825.
76 B3 Bahadur, D.: Indian J. Chem. **14** A (1976) 204.
76 B4 Belov, K.P., Zvezdin, A.K., Kadomtseva, A.M., Levitin, R.Z.: Usp. Fiz. Nauk **119** (1976) 447; Sov. Phys. Usp. (English Transl.) **19** (1976) 574.
76 C1 Cook, D.C., Cashion, J.D.: J. Phys. C: Solid State Phys. **9** (1976) L 97.
76 C2 Cooke, A.H., England, N.J., Preston, N.F., Swithenby, S.J., Wells, M.R.: Solid State Commun. **18** (1976) 545.
76 C3 Courths, R., Hüfner, S.: Z. Physik B **24** (1976) 193.
76 D1 Derkachenko, V.N., Zvezdin, A.K., Kadomtseva, A.M., Kovtun, N.M., Matveev, V.M., Timofeeva, V.A., Khokhlov, V.A.: Fiz. Tverd. Tela **18** (1976) 3399; Sov. Phys. Solid State (English Transl.) **18** (1976) 1978.

76G1 Ganguly, P., Parkash, O.M., Rao, C.N.R.: Phys. Status Solidi (a) **36** (1976) 669.
76G2 Gordon, J.D., Hornreich, R.M., Shtrikman, S.: Phys. Rev. B **13** (1976) 3012.
76G3 Gorodetsky, G., Shaft, S.: Phys. Rev. B **14** (1976) 2051.
76H1 Hornreich, R.M., Shtrikman, S., Wanklyn, B.M., Yaeger, I.: Phys. Rev. B **13** (1976) 4046.
76K1 Kaun, L.P., Lippold, B., Lukina, M.M., Matz, W., Savenko, B.N., Henning, K.: Kristallografiya **21** (1976) 387; Sov. Phys. Crystallogr. (English Transl.) **21** (1976) 212.
76S1 Sirota, N.N., Karavay, A.P., Pavlov, V.I.: Krist. Tech. **11** (1976) 861.
76V1 Vigneron, F.: J. Phys. (Paris) **37** (1976) 103.
76Z1 Zubkov, V.G., Bazuev, G.V., Shveikin, G.P.: Fiz. Tverd. Tela **18** (1976) 2002; Sov. Phys. Solid State (English Transl.) **18** (1976) 1165.
76Z2 Zalesskii, A.V., Zheludev, I.S.:At. Energy Rev. **14** (1976) 133.
76Z3 Zalesskii, A.V., Zvezdin, A.K., Zheludev, I.S., Savvinov, A.M., Lebedev, A.F.: Phys. Status Solidi B **73** (1976) 317.
77A1 Abe, M., Gomi, M., Shono, K., Mori, Y., Nomura, S.: Jpn. J. Appl. Phys. **16** (1977) 279.
77A2 Abe, M., Kaneta, K., Gomi, M., Mori, Y., Nomura, S.: Jpn. J. Appl. Phys. **16** (1977) 1799.
77A3 Abe, M., Gomi, M., Shono, K., Nomura, S.: Jpn. J. Appl. Phys. **16** (1977) 523.
77A4 Asai, K., Sekizawa, H., Mizushima, K., Iida, S.: J. Phys. Soc. Jpn. **43** (1977) 1093.
77B1 Bazuev, G.V., Shveikin, G.P.: Zh. Neorg. Khim. **22** (1977) 1239; Russ. J. Inorg. Chem. (English Transl.) **22** (1977) 675.
77B2 Belov, K.P., Zvezdin, A.K., Kadomtseva, A.M., Krynetskii, I.B., Matveev, V.M.: Fiz. Tverd. Tela **19** (1977) 259; Sov. Phys. Solid State (English Transl.) **19** (1977) 149.
77B3 Belov, K.P., Kadomtseva, A.M., Krén, E., Lukina, M.M., Milov, V.N., Shvab, E.: Zh. Eksp. Teor. Fiz. **72** (1977) 363; Sov. Phys. JETP (English Transl.) **45** (1977) 192.
77B4 Belyaeva, A.I., Stel'makhov, Yu.N., Potakova, V.A.: Fiz. Tverd. Tela **19** (1977) 3124; Sov. Phys. Solid State (English Transl.) **19** (1977) 1826.
77C1 Cooke, A.H., England, N.J., Swithenby, S.J., Wells, M.R.: Physica **86-88**B (1977) 562.
77C2 Chetkin, M.V., Shalygin, A.N., Kampa, A.de la: Fiz. Tverd. Tela **19** (1977) 3470; Sov. Phys. Solid State (English Transl.) **19** (1977) 2029.
77C3 Chetkin, M.V., Kiryushin, A.V.: Fiz. Tverd. Tela **19** (1977) 1895; Sov. Phys. Solid State (English Transl.) **19** (1977) 1111.
77D1 Derkachenko, V.N., Kadomtseva, A.M., Kovtun, N.M., Timofeyeva, V.A., Khokhlov, V.A., Kazakevich, O.V.: Fiz. Nizk. Temp. **3** (1977) 895; Sov. J. Low Temp. Phys. (English Transl.) **3** (1977) 435.
77D2 Durbin, G.W., Johnson, C.E., Thomas, M.F.: J. Phys. C: Solid State Phys. **10** (1977) 1975.
77G1 Gorodetsky, G., Hornreich, R.M., Shaft, S., Sharon, B., Shaulov, A., Wanklyn, B.M.: Phys. Rev. B **16** (1977) 515.
77H1 Hornreich, R.M., Scheel, H.J., Sharon, B.: Phys. Rev. B **16** (1977) 1112.
77H2 Hammann, J., Ocio, M.: Acta Crystallogr. A **33** (1977) 975.
77H3 Hamasaki, T., Hashimoto, T., Abe, M., Mori, Y.: J. Phys. Soc. Jpn. **42** (1977) 1190.
77K1 Kadomtseva, A.M., Zvezdin, A.K., Lukina, M.M., Milov, V.N, Mukhin, A.A., Ovchinnikova, T.L.: Zh. Eksp. Teor. Fiz. **73** (1977) 2324; Sov. Phys. JETP (English Transl.) **46** (1977) 1216.
77K2 Kaneko, M., Kurita, S., Tsushima, K.: J. Phys. C: Solid State Phys. **10** (1977) 1979.

77 K3 Kadomtseva, A.M., Moskvin, A.S., Bostrem, I.G., Wanklyn, B.M., Khafizova, N.A.: Zh. Eksp. Teor. Fiz. **72** (1977) 2286; Sov. Phys. JETP (English Transl.) **45** (1977) 1202.
77 K4 Karnachev, A.S., Kovtun, N.M., Lukina, M.M., Solovev, E.E.: Fiz. Tverd. Tela **19** (1977) 1627; Sov. Phys. Solid State (English Transl.) **19** (1977) 950.
77 R1 Rohrer, H., Gerber, Ch.: Phys. Rev. Lett. **38** (1977) 909.
77 S1 Sakai, T., Adachi, G., Shiokawa, J.: J. Appl. Phys. **48** (1977) 379.
77 S2 Shamir, N., Melamud, M., Shaked, H., Shtrikman, S.: Physica **90** B (1977) 217.
77 S3 Shamir, N., Shaked, H., Shtrikman, S.: Physica **90** B (1977) 211.
77 S4 Shamir, N., Melamud, M., Shaked, H., Shtrikman, S.: Physica **86–88** B (1977) 1037.
77 S5 Sirota, N.N., Pavlov, V.I., Lashkov, E.S., Karavai, A.P.: Fiz. Tverd. Tela **19** (1977) 1748; Sov. Phys. Solid State (English Transl.) **19** (1977) 1020.
77 S6 Sirota, N.N., Karavay, A.P., Kofman, N.A., Pavlov, V.I.: Krist. Tech. **12** (1977) 945.
77 T1 Tamaki, T., Tsushima, K., Yamaguchi, Y.: Physica **86–88** B (1977) 923.
77 V1 Vanchikov, V.V., Zalesskii, A.V., Karnachev, A.S., Krivenko, V.G., Solov'ev, E.E.: Fiz. Tverd. Tela **19** (1977) 3640; Sov. Phys. Solid State (English Transl.) **19** (1977) 2126.
77 V2 Vanchikov, V.V., Zalesskii, A.V., Kazakevich, O.V., Karnachev, A.S., Solov'ev, E.E., Timofeeva, V.A.: Fiz. Tverd. Tela **19** (1977) 1508; Sov. Phys. Solid State (English Transl.) **19** (1977) 882.
78 A1 Asai, K., Sekizawa, H., Mizushima, K., Iida, S.: J. Phys. Soc. Jpn. **45** (1978) 1417.
78 B1 Bazuev, G.V., Shveikin, G.P.: Izv. Akad. Nauk SSSR, Neorg. Mater. **14** (1978) 267; Inorg. Mater. (USA) (English Transl.) **14** (1978) 201.
78 B2 Borukhovich, A.S., Zubkov, V.G., Bazuev, G.V.: Fiz. Tverd. Tela **20** (1978) 1816; Sov. Phys. Solid State (English Transl.) **20** (1978) 1049.
78 B3 Bombik, A., Leśniewska, B., Oleś, A.: Phys. Status Solidi (a) **50** (1978) K 17.
78 B4 Bonville, P., Hodges, J.A., Imbert, P.: Phys. Rev. B **18** (1978) 2196.
78 C1 Cook, D.C., Cashion, J.D.: Hyperfine Interact. **5** (1978) 479.
78 C2 Clark, J.B., Richter, P.W., du Toit, L.: J. Solid State Chem. **23** (1978) 129.
78 G1 Greedan, J.E., MacLean, D.A.: Inst. Phys. Conf. Ser. No. **37** (1978) 249.
78 G2 Gordienko, V.A., Guseinov, M.M., Korchazhkin, V.V., Nikolaev, V.I., Rusakov, V.S., Tereshonok, V.A.: Kristallografiya **23** (1978) 868; Sov. Phys. Crystallogr. (English Transl.) **23** (1978) 490.
78 G3 Gorodetsky, G., Missell, F.P., Garland, C.W.: J. Phys. Chem. Solids **39** (1978) 723.
78 H1 Hornreich, R.M.: J. Magn. Magn. Mater. **7** (1978) 280.
78 K1 Kadomtseva, A.M., Milov, V.N., Moskvin, A.S., Paradavi-Khorvat M.: Fiz. Tverd. Tela **20** (1978) 817; Sov. Phys. Solid State (English Transl.) **20** (1978) 474.
78 K2 Kadomtseva, A.M., Lukina, M.M., Moskvin, A.S., Khafizova, N.A.: Fiz. Tverd. Tela **20** (1978) 2139; Sov. Phys. Solid State (English Transl.) **20** (1978) 1235.
78 K3 Karnachev, A.S., Klochan, V.A., Kovtun, N.M., Sanina, V.A., Solov'ev, E.E.: Fiz. Nizk. Temp. **4** (1978) 90; Sov. J. Low Temp. Phys. (English Transl.) **4** (1978) 45.
78 K4 Kamata, K., Nakajima, T., Hayashi, T. Nakamura, T.: Mater. Res. Bull. **13** (1978) 49.
78 M1 Moskvin, A.S., Sinitsyn, E.V., Smirnov, A.Y.: Fiz. Tverd. Tela **20** (1978) 3465; Sov. Phys. Solid State (English Transl.) **20** (1978) 2002.
78 N1 Nomura, S.: Landolt-Börnstein, New Series, III/Vol. **12a**, Berlin, Heidelberg, New York: Springer-Verlag, 1978, p. 368.
78 N2 Nadolski, S.: IEEE Trans. Magn MAG-**14** (1978) 912.
78 S1 Sirota, N.N., Lomako, I.D.: Krist. Tech. **13** (1978) 873.
78 T1 Tsang, C.H., White, R.L.: J. Appl. Phys. **49** (1978) 6052.
78 T2 Tsang, C.H., White, R.L., White, R.M.: J. Appl. Phys. **49** (1978) 1838.

78 V1 Vitebskii, I.M., Yablonskii, D.A.: Fiz. Tverd. Tela **20** (1978) 2300; Sov. Phys. Solid State (English Transl.) **20** (1978) 1327.

78 W1 Washimiya, S., Satoko, C.: J. Phys. Soc. Jpn. **45** (1978) 1204.

78 Z1 Zuk, J., Piotrowski, K.: Solid. State. Commun **28** (1978) 381.

79 A1 Asai, K., Sekizawa, H., Iida, S.: J. Phys. Soc. Jpn. **47** (1979) 1054.

79 B1 Barton, W.A., Cashion, J.D.: J. Phys. C: Solid State Phys. **12** (1979) 2897.

79 B2 Belov, K.P., Zvezdin, A.K., Mukhin, A.A.: Zh. Eksp. Teor. Fiz. **76** (1979) 1100; Sov. Phys. JETP (English Transl.) **49** (1979) 557.

79 B3 Belov, K.P., Zvezdin, A.K., Kadomtseva, A.M., Krynetskii, I.B. Mukhin, A.A.: Zh. Eksp. Teor. Fiz. **76** (1979) 1421; Sov. Phys. JETP (English Transl.) **49** (1979) 723.

79 D1 Derkachenko, V.N., Zvezdin, A.K., Kadomtseva, A.M., Kazakevich, O.V., Kovtun, N.M., Mukhin, A.A., Timofeeva, V.A., Khokhlov, V.A.: Fiz. Tverd. Tela **21** (1979) 669; Sov. Phys. Solid State (English Transl.) **21** (1979) 394.

79 D2 Dasnshin, N.K., Kovtun, N.M., Polyanenko, A.P.: Zh. Eksp. Teor. Fiz. **77** (1979) 1058; Sov. Phys. JETP (English Transl.) **50** (1979) 532.

79 G1 Gomi, M., Abe, M., Nomura, S.: Jpn. J. Appl. Phys. **18** (1979) 739.

79 J1 Jirák, Z., Vratislav, S., Zajícek, J.: Phys. Status Solidi (a) **52** (1979) K 39.

79 K1 Karnachev, A.S., Soboleva, T.K., Solov'ev, E.E., Stefanovskii, E.P.: Fiz. Tverd. Tela **21** (1979) 1451; Sov. Phys. Solid State (English Transl.) **21** (1979) 836.

79 M1 MacLean, D.A., Hok-Nam, Ng., Greedan, J.E.: J. Solid State Chem. **30** (1979) 35.

79 M2 Moskvin, A.S., Bostrem, I.G.: Fiz. Tverd. Tela **21** (1979) 1080; Sov. Phys. Solid State (English Transl.) **21** (1979) 628.

79 N1 Nakamura, T., Petzow, G., Gauckler, L.J.: Mat. Res. Bull. **14** (1979) 649.

79 R1 Ramadass, N., Gopalakrishnan, J., Sastri, M.V.C.: J. Less-Common Met. **65** (1979) 129.

79 S1 Sakata, T., Enomura, A.: Phys. Status Solidi (a) **52** (1979) 311.

79 S2 Sirota, N.N., Lomako, I.D.: Krist. Tech. **14** (1979) 243.

79 T1 Toyokawa, K., Kurita, S., Tsushima, K.: Phys. Rev. B **19** (1979) 274.

79 W1 Wong, C.P., Yaeger, I., Morrish, A.H., Wanklyn, B.M.: Phys. Rev. B **20** (1979) 3879.

79 Z1 Zvezdin, A.K., Matveev, V.M.: Zh. Eksp. Teor. Fiz. **77** (1979) 1076; Sov. Phys. JETP (English Transl.) **50** (1979) 543.

80 B1 Bonville, P., Hodges, J.A., Imbert, P.: J. Phys. (Paris) **41** (1980) 1213.

80 C1 Cook, D.C., Cashion, J.D.: J. Phys. C: Solid State Phys. **13** (1980) 4199.

80 D1 Derkachenko, V.N., Zvezdin, A.K., Krynetskii, I.B., Kadomtseva, A.M., Mukhin, A.A., Khokhlov, V.A.: Fiz. Tverd. Tela **22** (1980) 1753; Sov. Phys. Solid State (English Transl.) **22** (1980) 1021.

80 E1 Enomura, A., Sakata, T.: Phys. Stat. Sol. (a) **58** (1980) 539.

80 G1 Golovenchits, E.I., Laikhtman, B.D., Sanina, V.A.: Pis'ma Zh. Eksp. Teor. Fiz. **31** (1980) 243; JETP Lett. (English Transl.) **31** (1980) 223.

80 G2 Golovenchits, E.I., Sanina, V.A.: Fiz. Tverd. Tela **23** (1980) 1677; Sov. Phys. Solid State (English Transl.) **23** (1981) 977.

80 G3 Gorodetsky, G., Shtrikman, S.: J. Appl. Phys. **51** (1980) 1127.

80 H1 Hammann, J., Ocio, M.: J. Magn. Magn. Mater. **15–18** (1980) 39.

80 H2 Holzschuh, E., Boekema, C., Denison, A.B., Kündig, W., Meier, P.F., Rüegg, K.: Hyperfine Interact. **8** (1980) 77.

80 H3 Howning, W.-Y., Thorn, R.J.: J. Phys. Chem. Solids **41** (1980) 75.

80 K1 Krichevtsov, B.B., Pisarev, R.V., Rubinshtein, M.M.: Fiz. Tverd. Tela **22** (1980) 2128; Sov. Phys. Solid State (English Transl.) **22** (1980) 1240.
80 K2 Kojima, N., Tsushima, K., Tsujikawa, I.: J. Phys. Soc. Jpn. **49** (1980) 1449.
80 K3 Kojima, N., Tsushima, K., Kurita, S., Tsujikawa, I.: J. Phys. Soc. Jpn. **49** (1980) 1456.
80 K4 Kojima, N., Aoyagi, K., Tsushima, K., Tsujikawa, I., Sugano, S.: J. Phys. Soc. Jpn. **49** (1980) 1463.
80 K5 Karavay, A.P.: Krist. Tech. **15** (1980) 11.
80 K6 Kadomtseva, A.M., Krynetskii, I.B., Matveev, V.M.: Zh. Eksp. Teor. Fiz. **79** (1980) 1451; Sov. Phys. JETP (English Transl.) **52** (1980) 732.
80 L1 Lütgemeier, H., Bohn, H.G., Brajiczewska, M.: J. Magn. Magn. Mater. **21** (1980) 289.
80 M1 Madhusudan, W.H., Jagannathan, K., Ganguly, P., Rao, C.N.R.: J. Chem. Soc. Dalton Trans. **8** (1980) 1397.
80 M2 Madhusudan, W.H., Vasanthacharya, N.Y., Ganguly, P.: Indian J. Chem. **19A** (1980) 1037.
80 N1 Nadolski, S., Lütgemeier, H., Bohn, H.G.: Bull. Magn. Reson. **2** (1980) 248.
80 N2 Nadolski, S., Szymczak, H.: J. Magn. Magn. Mater. **21** (1980) 167.
80 P1 Pollert, E., Jirák, Z.: J. Solid State Chem. **35** (1980) 262.
80 P2 Prelorendjo, L.A., Johnson, C.E., Thomas, M.F., Wanklyn, B.M.: J. Phys. C: Solid State Phys. **13** (1980) 2567.
80 S1 Szymczak, R., Tsushima, K.: J. Magn. Magn. Mater. **21** (1980) 275.
80 T1 Turner, C.W., Greedan, J.E.: J. Solid State Chem. **34** (1980) 207.
80 T2 Turner, C.W., Greedan, J.E., Collins, M.F.: J. Magn. Magn. Mater **20** (1980) 165.
80 W1 White, R.M., Nemanich, R.J., Tsang, C.: J. Magn. Magn. Mater. **15–18** (1980) 773.
80 Z1 Zvezdin, A.K., Kadomtseva, A.M., Mukhin, A.A.: Izv. Akad. Nauk SSSR, Ser. Fiz. **44** (1980) 1348; Bull. Acad. Sci USSR, Phys. Ser (English Transl.) **44** (1980) 18.
80 Z2 Zubkov, V.G., Bazuev, G.V., Shveikin, G.P.: Kristallografiya **25** (1980) 180; Sov. Phys. Crystallogr. (English Transl.) **25** (1980) 103.
81 A1 Artem'ev, G.G., Kadomtseva, A.M., Krynetskii, I.B., Ledneva, T.M., Lukina, M.M., Ovchinnikova, T.L.: Fiz. Tverd. Tela **23** (1981) 863; Sov. Phys. Solid State (English Transl.) **23** (1981) 490.
81 B1 Belov, D.V., Zvezdin, A.K., Kadomtseva, A.M., Krynetskii, I.B., Moskvin, A.S., Mukhin, A.A.: Fiz. Tverd. Tela **23** (1981) 2831; Sov. Phys. Solid State (English Transl.) **23** (1981) 1654.
81 B2 Balkevich, V.L., Satanovobei, A.V., Morin, Y.M., Salnikov, V.E., Andronov, M.A.: Steklo i Keram. **38** (1981) 86.
81 C1 Chetkin, M.V., Akhutina, A.I., Ermilova, N.N., Kuzmenko, A.P., Didosyan, Yu.S.: Zh. Eksp. Teor. Fiz. **81** (1981) 2206; Sov. Phys. JETP (English Transl.) **54** (1981) 1172.
81 G1 Greedan, J.E., Seto, K.: Mater. Res. Bull **16** (1981) 1479.
81 G2 Garrett, J.D., Greedan, J.E., MacLean, D.A.: Mater. Res. Bull. **16** (1981) 145.
81 G3 Golovenchits, E.I., Sanina, V.A., Shaplygina, T.A.: Zh. Eksp. Teor. Fiz. **80** (1981) 1911; Sov. Phys. JETP (English Transl.) **53** (1981) 992.
81 G4 Golovenchits, E.I., Sanina, V.A.: Fiz. Tverd. Tela **23** (1981) 1677; Sov. Phys. Solid State (English Transl.) **23** (1981) 977.
81 G5 Gorodetsky, G., Shaft, S., Remeika, J.P.: J. Appl. Phys. **52** (1981) 7353.
81 G6 Gibb, T.C.: J. Chem. Soc. Dalton Trans. (1981) 2245.

81 H1 Holzschuh, E., Boekema, C., Kündig, W., Rüegg, K., Patterson, B.D.: Hyperfine Interact. **8** (1981) 615.

81 K1 Kurita, S., Toyokawa, K., Tsushima, K., Sugano, S.: Solid State Commun. **38** (1981) 235.

81 K2 Kadomtseva, A.M., Agafonov, A.P., Lukina, M.M., Milov, V.N., Moskvin, A.S., Semenov, V.A.: Fiz. Tverd. Tela **23** (1981) 3554; Sov. Phys. Solid State **23** (English Transl.) (1981) 2065.

81 K3 Kadomtseva, A.M., Agafonov, A.P., Lukina, M.M., Milov, V.N., Moskvin, A.S., Semenov, V.A., Sinitsyn, E.V.: Zh. Eksp, Teor. Fiz. **81** (1981) 700; Sov. Phys. JETP (English Transl.) **54** (1981) 374.

81 K4 Klochan, V.A., Kovtun, N.M., Troitskii, G.A., Khmara, V.M.: Zh. Eksp. Teor. Fiz. **81** (1981) 627; Sov. Phys. JETP (English Transl.) **54** (1981) 335.

81 K5 Kadomtseva, A.M., Agafonov, A.P., Milov, V.N., Moskvin, A.S., Semenov, V.A.: Pisma Zh. Eksp. Teor. Fiz. **33** (1981) 400; JETP Lett. (English Transl.) **33** (1981) 383.

81 M1 MacLean, D.A., Seto, K., Greedan, J.E.: J. Solid State Chem. **40** (1981) 241.

81 M2 MacLean, D.A., Greedan, J.E.: Inorg. Chem. **20** (1981) 1025.

81 M3 Morishita, T., Tsushima, K.: Phys. Rev. B **24** (1981) 341.

81 P1 Plakhty, V.P., Chernenkov, Yu. P., Shveitser, Zh., Bedrisova, M.N.: Zh. Eksp. Teor. Fiz. **80** (1981) 2465; Sov. Phys. JETP (English Transl.) **53** (1981) 1291.

81 R1 Radhakrishna, P., Hammann, J., Ocio, M., Pari, P., Allain, Y.: Solid State Commun. **37** (1981) 813.

81 S1 Shamir, N., Shaked, H., Shtrikman, S.: Phys. Rev. B **24** (1981) 6642.

81 S2 Sirota, N.N., Pavlov, V.I., Bogush, A.K., Karavay, A.P.: Cryst. Res. Technol. **16** (1981) 1151.

81 T1 Turner, C.W., Collins, M.F., Greedan, J.E.: J. Magn. Magn. Mater. **23** (1981) 265.

82 A1 Andronenko, S.I., Koroleva, L.N., Bondar, I.A., Ioffe, V.A.: Fiz. Tverd. Tela **24** (1982) 1538; Sov. Phys. Solid State (English Transl.) **24** (1982) 881.

82 A2 Abe, M., Gomi, M., Nomura, S.: Proc. 3rd Int. Conf. Ferrites, Kyoto, Japan, Watanabe, H. et al. (eds.) (1982) 783.

82 B1 Bahadur, D.: Bull. Mater. Sci. **4** (1982) 7.

82 C1 Chetkin, M.V., Gadetskii, S.N., Akhutkina, A.I.: Pis'ma Zh. Eksp. Teor. Fiz. **35** (1982) 373; JETP Lett. (English Transl.) **35** (1982) 459.

82 C2 Coustumer, Le, L.R., Barbaux, Y., Bonnelle, J.P.: Nouv. J. Chim. **6** (1982) 7.

82 C3 Chetkin, M.V., Akhutkina, A.I., Kuzmenko, A.P., Gadetsky, S.N.: J. Appl. Phys. **53** (1982) 7864.

82 G1 Goral, J.P., Greedan, J.E.: J. Solid State Chem. **43** (1982) 204.

82 G2 Goral, J.P., Greedan, J.E., MacLean, D.A.: J. Solid State Chem. **43** (1982) 244.

82 G3 Golovenchits, E.I., Sanina, V.A.: Fiz. Tverd. Tela **24** (1982) 375; Sov. Phys. Solid State (English Transl.) (1982) **24** 213.

82 K1 Kadomtseva, A.M., Bostrem, I.G., Krynetskii, I.B., Moskvin, A.S., Ovchinnikova, T.L., Terziev, V.G.: Fiz. Tverd. Tela **24** (1982) 2344; Sov. Phys. Solid State (English Transl.) **24** (1982) 1331.

82 K2 Kamigaki, K., Kaneko, T., Ohashi, K., Yoshida, H., Abe, S.: Proc. 3rd Int. Conf. Ferrites, Kyoto, Japan, Watanabe, H. et al. (eds.) (1982) 406.

82 K3 Kojima, N., Tsujikawa, I.: Proc. 3rd Int. Conf. Ferrites, Kyoto, Japan, Watanabe, H. et al. (eds.) (1982) 769.

82 K4 Koshizuka, N., Hayashi, K., Tsushima, T.: Proc. 3rd Int. Conf. Ferrites, Kyoto, Japan, Watanabe, H. et al. (eds.) (1982) 760.

82 L1 Lütgemeier, H., Bohn, H.G., Nadolski, S.: Proc 3rd Int. Conf. Ferrites, Kyoto, Japan, Watanabe, H. et al. (eds.) (1982) 774.

82 P1 Plakhty, V.P., Chernenkov, Y.P., Bedrizova, M.N.: AIP. Conf. Proc. **89** (1982) 330.

82 Q1 Quézel, S., Rossat-Mignod, J., Tchéou, F.: Solid State Commun. **42** (1982) 103.

82 S1 Sosnowska, I., Fischer, P.: AIP Conf. Proc. **89** (1982) 346.

82 S2 Sosnowska, I., Steichele, E.: AIP Conf. Proc. **89** (1982) 309.

82 T1 Turner, C.W., Collins, M.F., Greedan, J.E.: The rare earths in Mod. Sc. Techn. Vol. **3** (1982) 443, Mc Carthy, Silber, Rhyne (eds.).

82 T2 Thornton, G., Tofield, B.C., Williams, D.E.: Solid State Commun. **44** (1982) 1213.

82 T3 Tripathi, A.K., Lal, H.B.: Indian J. Pure Appl. Phys. **20** (1982) 271.

82 T4 Tamaki, T., Tushima, K., Adachi, T.: Proc. 3rd Int. Conf. Ferrites, Kyoto, Japan, Watanabe, H. et al. (eds.) (1982) 765.

82 W1 White, R.M., Nemanich, R.J., Herring, C.: Phys. Rev. B **25** (1982) 1822.

82 Z1 Zubkov, V.G., Berger, I.F., Bazuev, G.V., Artamonova, A.M.: Fiz. Tverd. Tela **24** (1982) 1176; Sov. Phys. Solid State (English Transl.) **24** (1982) 663.

83 A1 Aegerter, M.A., Matsuoka, M., Panepucci, H.: Radiation Effects **73** (1983) 115.

83 A2 Aubert, G., Michelutti, B.: J. Magn. Magn. Mater. **31–34** (1983) 859.

83 B1 Bogush, A.K., Pavlov, V.I., Balyko, L.V.: Cryst. Res. Technol. **18** (1983) 589.

83 C1 Chmielowski, M., Sosnowska, I.: Solid State Commun. **48** (1983) 1007.

83 D1 Dräger, K.: Z. Naturforsch. **38** A (1983) 848.

83 F1 Fierro, J.L.G., Gonzalez-Tejuca, L.: J. Colloid Surface Sci. **96** (1983) 107.

83 G1 Goral, J.P., Greedan, J.E.: J. Magn. Magn. Mater. **37** (1983) 315.

83 G2 Gibb, T.C.: J. Chem. Soc. Dalton Trans. (1983) 2031.

83 G3 Gibb, T.C.: J. Chem. Soc. Dalton Trans. (1983) 873.

83 G4 Gibb, T.C.: J. Chem. Soc. Dalton Trans. (1983) 2035.

83 H1 Holzschuh, E., Denison, A.B., Kündig, W., Meier, P.F., Patterson, B.D.: Phys. Rev. B **27** (1983) 5294.

83 K1 Karnachev, A.S., Klechin, Y.I., Kovtun, N.M., Moskvin, A.S., Solov'ev, E.E., Tkachenko, A.A.: Zh. Eksp. Teor. Fiz. **85** (1983) 670; Sov. Phys. JETP (English Transl.) **58** (1983) 390.

83 K2 Kadomtseva, A.M., Artem'ev, G.G., Lukina, M.M., Milov, V.N., Nedel'ko, V.I., Semenova, M.V., Sinitsyn, E.V.: Pis'ma Zh. Eksp. Teor. Fiz. **38** (1983) 383; JETP Lett. (English Transl.) **38** (1983) 463.

83 K3 Kadomtseva, A.M., Krynetskii, I.B., Krezhov, K., Moskvin, A.S., Ovchinnikova, T.L., Lukina, M.M.: Fiz. Tverd. Tela **25** (1983) 877; Sov. Phys. Solid State (English Transl.) **25** (1983) 502.

83 K4 Kamegashira, N., Miyazaki, Y.: Phys. Status Solidi **76** A (1983) K39.

83 K5 Krezhov, K.A., Mikhov, M.T., Kadomtseva, A.M., Krinetskii, I.B., Lukina, M.M.: Phys. Status Solidi B **115** (1983) 47.

83 K6 Koshizuka, N., Hayashi, K.: J. Magn. Magn. Mater. **31–34** (1983) 569.

83 L1 Lebedev, A.Yu., Ozhogin, V.I., Safonov, V.L., Yakubovskii, A. Y: Zh. Eksp. Teor. Fiz. **85** (1983) 1059; Sov. Phys. JETP (English Transl.) **58** (1983) 616.

83 M1 Matsuoka, M., Aegerter, M.A., Panepucci, H., Terrile, M.C., Helman, J.S., Scheel, H.J.: Phys. Rev. Lett. **50** (1983) 204.

83 P1 Plakhty, V.P., Chernenkov, Yu. P., Bedrizova, M.N.: Solid State Commun. **47** (1983) 309.

83 S1 Sinitsyn, E.V., Kadomtseva, A.M., Moskvin, A.S., Bostrem, I.G., Timofeeva, V.A., Kazakevich, O.V.: Fiz. Tverd. Tela **25** (1983) 285; Sov. Phys. Solid State (English Transl.) **25** (1983) 161.

83 T1 Tamaki, T., Tsushima, K.: J. Magn. Magn. Mater. **31** (1983) 571.
83 T2 Turner, C.W., Greedan, J.E.: J. Magn. Magn. Mater. **36** (1983) 242.
83 V1 Visnovsky, S., Wanklyn, B.M.: J. Phys. C: Solid State Phys. **16** (1983) L 655.
83 V2 Venugopalan, S., Dutta, M., Ramdas, A.K., Remeika, J.P.: Phys. Rev. B **27** (1983) 3115.
83 V3 Vasanthacharya, N.Y., Ganguly, P.: Bull. Mater. Sci. **5** (1983) 307.
83 V4 Vorotilova, L.S., Dmitrieva, L.V., Ioffe, V.A., Rakhmankulov, R.M.: Fiz. Tverd. Tela **25** (1983) 1726; Sov. Phys. Solid State (English Transl.) **25** (1983) 994.
84 A1 Agafonov, A.P., Zorin, I.A., Kadomtseva, A.M., Moskvin, A.S., Timofeeva, V.A.: Fiz. Tverd. Tela **26** (1984) 2131; Sov. Phys. Solid State (English Transl.) **26** (1984) 1292.
84 B1 Barilo, S.N., Ges, A.P., Soshnikov, L.E., Smirnova, T.V.: Phys. Status Solidi (b) **122** (1984) 47.
84 B2 Balbashov, A.M., Kozlov, G.V., Lebedev, S.P., Prokhorov, A.M., Prokhorov, A.S.: Pisma Zh. Eksp. Teor. Fiz. **39** (1984) 461; JETP Lett. **39** (1984) 560.
84 C1 Coutures, J., Coutures, J.P.: J. Solid State Chem. **52** (1984) 95.
84 D1 Derkachenko, V.N., Zvezdin, A.K., Kadomtseva, A.M., Kovtun, N.M., Lukina, M.M., Mukhin, A.A.: Phys. Status Solidi (a) **84** (1984) 215.
84 D2 Dan'shin, N.K., Kovtun, N.M., Sdvizhkov, M.A.: Fiz. Tverd. Tela **26** (1984) 3635; Sov. Phys. Solid State (English Transl.) **26** (1984) 2185.
84 G1 Ges', A.P., Derkachenko, V.M., Fedotova, V.V.: Phys. Status Solidi (a) **84** (1984) K63.
84 G2 Greedan, J.E.: J. Magn. Magn. Mater. **44** (1984) 299.
84 G3 Greedan, J.E., Turner, C.W., Goodings, D.A.: J. Magn. Magn. Mater. **42** (1984) 255.
84 H1 Hodges, J.A., Imbert, P., Schuhl, A.: J. Magn. Magn. Mater. **43** (1984) 101.
84 H2 Holzschuh, E., Kündig, W., Denison, A.B.: Hyperfine Interact. **17–19** (1984) 345.
84 J1 Jedryka, E., Nadolski, S., Wojcik, M.: J. Magn. Magn. Mater. **40** (1984) 303.
84 K1 Kojima, N., Tsujikawa, I., Hori, H., Nishimura, H., Date, M., Tsushima, K.: J. Phys. Soc. Jpn. **53** (1984) 2875.
84 K2 Kocharyan, K.N., Martirosyan, R.M., Prpryan, V.G., Sarkisyan, E.L.: Zh. Eksp. Teor. Fiz. **86** (1984) 639; Sov. Phys. JETP (English Transl.) **59** (1984) 373.
84 K3 Kasvegashira, N., Miyazki, Y.: Mat. Res. Bull. **19** (1984) 1201.
84 M1 Menyhard, N.: Solid State Commun. **52** (1984) 31.
84 N1 Nikolov, O., Ruskov, T., Tomov, T., Kadomtseva, A.M., Krinetskii, I.B., Lukina, M.M.: J. Magn. Magn. Mater. **44** (1984) 181.
84 S1 Sanina, V.A., Golovenchits, E.I., Smolenskii, G.A.: Pis'ma Zh. Eksp. Teor. Fiz. **40** (1984) 110; JETP Lett. (English Transl.) **40** (1984) 857.
84 S2 Szymczak, R., Balbashov, A.M.: Physica **125** B (1984) 40.
84 S3 Stadnik, Z.M., Boer, E.de: Solid State Commun. **50** (1984) 335.
84 V1 Volkov, A.A., Goncharov, Yu.G., Kozlov, G.V., Kocharyan, K.N., Lebedev, S.P., Prokhorov, A.S., Prokhorov, A.M.: Pis'ma Zh. Eksp. Teor. Fiz. **39** (1984) 140; JETP Lett. (English Transl.) **39** (1984) 166.
84 V2 Vasanthacharya, N.Y., Ganguly, P., Rao, C.N.R.: J. Solid State Chem. **53** (1984) 140.
84 V3 Vasanthacharya, N.Y., Ganguly, P., Goodenough, J.B., Rao, C.N.R.: J. Phys. C: Solid State Phys. **17** (1984) 2745.
84 Z1 Zubov, V.E., Lyskov, V.A., Krinchik, G.S.: Fiz. Tverd. Tela **26** (1984) 2335; Sov. Phys. Solid State (English Transl.) **26** (1984) 1416.
85 B1 Belov, D.V., Kolmakova, N.P., Krynetskii, I.B., Milov, V.N., Mukhin, A.A., Semenov, V.A.: Zh. Eksp. Teor. Fiz. **88** (1985) 1063; Sov. Phys. JETP (English Transl.) **61** (1985) 624.

85 B2 Balbashov, A.M., Volkov, A.A., Lebedev, S.P., Mukhin, A.A., Prokhorov, A.S.: Zh. Eksp. Teor. Fiz. **88** (1985) 974; Sov. Phys. JETP (English Transl.) **61** (1985) 573.

85 B3 Balbashov, A.M., Volkov, A.A., Kozlov, G.V., Lebedev, S.P., Mukhin, A.A., Pronin, A.Yu., Prokhorov, A.S., Prokhorov, A.M.: Pis'ma Zh. Eksp. Teor. Fiz. **42** (1985) 456; JETP Lett. (English Transl.) **42** (1985) 564.

85 B4 Balbashov, A.M., Volkov, A.A., Kozlov, G.V., Lebedev, S.P., Prokhorov, A.S.: Fiz. Tverd. Tela **27** (1985) 270; Sov. Phys. Solid State (English Transl.) **27** (1985) 165.

85 C1 Chernenkov, Yu.P., Plakhtii, V.P., Kadomtseva, A.M., Krynetskii, I.B., Matveev, V.M., Derkachenko, V.M.: Fiz. Tverd. Tela **27** (1985) 349; Sov. Phys. Solid State (English Transl.) **27** (1985) 214.

85 D1 Dan'shin, N.K., Kovtun, N.M., Sdvizhkov, M.A.: Zh. Eksp. Teor. Fiz. **89** (1985) 203; Sov. Phys. JETP (English Transl.) **62** (1985) 115.

85 D2 Dan'shin, N.K., Kovtun, N.M., Sdvizhkov, M.A.: Fiz. Tverd. Tela **27** (1985) 3635; Sov. Phys. Solid State (English Transl.) **27** (1985) 2189.

85 G1 Greedan, J.E.: J. Less Common Met. **111** (1985) 335.

85 J1 Jirák, Z., Krupicka, S., Simsa, Z. Dlouha, M., Vratislav, S.: J. Magn. Magn. Mater. **53** (1985) 153.

85 K1 Kojima, N., Tsujikawa, I., Aoyagi, K., Tsushima, K.: J. Phys. Soc. Jpn. **54** (1985) 4804.

85 K2 Kojima, N., Tsujikawa, I., Tsushima, K.: J. Phys. Soc. Jpn. **54** (1985) 4794.

85 M1 Milov, V.N., Artem'ev, G.G., Nedel'ko, V.I., Prun, A.F., Semenova, M.V., Sinitsyn, E.V.: Zh. Eksp. Teor. Fiz. **88** (1985) 272; Sov. Phys. JETP (English Transl.) **61** (1985) 159.

85 P1 Piotrowski, K., Szewczyk, A., Szymczak, R., Eremenko, V.V., Gnatchenko, S.L., Kharchenko, N.F.: Acta Phys. Pol. A **68** (1985) 139.

85 S1 Sosnowska, I., Fischer, P.: J. Less Common Met. **111** (1985) 109.

85 V1 Venugopalan, S., Dutta, M., Ramdas, A.K., Remeika, J.P.: Phys. Rev. B **31** (1985) 1490.

85 Z1 Zvezdin, A.K., Zorin, I.A., Kadomtseva, A.M., Krynetskii, I.B., Moskvin, A.S.: Zh. Eksp. Teor. Fiz. **88** (1985) 1098; Sov. Phys. JETP (English Transl.) **61** (1985) 645.

86 A1 Ayyub, P., Multani, M., Vijayaraghavan, R.: Phys. Lett. A **119** (1986) 95.

86 B1 Balbashov, A.M., Fuik, V.H., Lebedev, S.P., Kozlov, G.V., Mukhin, A.A., Pronin, A.Yu., Prokhorov, A.S.: Zh. Eksp. Teor. Fiz. **90** (1986) 299; Sov. Phys. JETP (English Transl.) **63** (1986) 172.

86 B2 Balbashov, A.M., Berezin, A.G., Gufan, Yu. M., Kolyadko, G.S., Marchukov, P.Yu., Nikolaev, I.V., Rudashevskii, E.G.: Pis'ma Zh. Eksp. Teor. Fiz. **43** (1986) 205; JETP Lett. (English Transl.) **43** (1986) 259.

86 B3 Balbashov, A.M., Kozlov, G.V., Lebedev, S.P., Mukhin, A.A., Pronin, A.Yu., Prokhorov, A.S., Prokhorov, A.M.: Pisma Zh. Eksp. Teor. Fiz. **43** (1986) 33; JETP Lett. (English Transl.) **43** (1986) 41.

86 D1 Dan'shin, N.K., Derkachenko, V.N., Kovtun, N.M., Sdvizhkov, M.A.: Fiz. Tverd. Tela **28** (1986) 2609; Sov. Phys. Solid State (English Transl.) **28** (1986) 1461.

86 D2 Dan'shin, N.K., Kovtun, N.M., Sdvizhkov, M.A.: Fiz. Tverd. Tela **28** (1986) 1200; Sov. Phys. Solid State (English Transl.) **28** (1986) 672.

86 G1 Golovenchits, E.I., Sanina, V.A.: Fiz. Tverd. Tela **28** (1986) 713; Sov. Phys. Solid State (English Transl.) **28** (1986) 397.

86 H1 Hori, H., Kojima, N., Nishimura, H., Tsujikawa, I., Tsushima, K., Date, M.: J. Phys. Soc. Jpn. **55** (1986) 1380.

86 K1 Kocharyan, K.N., Martirosyan, R.M., Prpryan, V.G., Sarkisyan, E.L.: Fiz. Tverd. Tela **28** (1986) 2872; Sov. Phys. Solid State (English Transl.) **28** (1986) 1611.
86 M1 Motokawa, M., Hori, H. Nishimura, H., Tsushima, K., Date, M.: J. Magn. Magn. Mater. **59** (1986) 243.
86 N1 Nadolski, S., Lütgemeier, H., Bohn, H.G., Wojcik, M.: Bull. Magn. Res. **8** (1986) 206.
86 P1 Pavlov, V.I., Bychkov, G.L., Bogush, A.K.: Cryst. Res. Technol. **21** (1986) 487.
86 S1 Sosnowska, I., Steichele, E., Hewat, A.: Physica **136** B (1986) 394.
86 T1 Thornton, G., Tofield, B.C., Hewat, A.W.: J. Solid State Chem. **61** (1986) 301.
87 B1 Barilo, S.N., Ges', A.P., Luginets, A.M., Soshnikov, L.E.: Fiz. Tverd. Tela **29** (1987) 1541; Sov. Phys. Solid State (English Transl.) **29** (1987) 883.
87 B2 Balykina, E.A., Gan'shina, E.A., Krinchik, G.S.: Zh. Eksp. Teor. Fiz. **93** (1987) 1879; Sov. Phys. JETP (English Transl.) **66** (1987) 1073.
87 B3 Belov, D.V., Zorin, I.A., Kadomtseva, A.M., Lukina, M.M., Matveev, V.M., Artemev, G.G.: Fiz. Tverd. Tela **29** (1987) 1592; Sov. Phys. Solid State (English Transl.) **29** (1987) 916.
87 B4 Balbashov, A.M., Zon, B.A., Kupershmidt, V.Ya., Pakhomov, G.V., Urazbaev, T.T.: Fiz. Tverd. Tela **29** (1987) 1297; Sov. Phys. Solid State (English Transl.) **29** (1987) 743.
87 D1 Dan'shin, N.K., Zherlitsyn, S.V., Zvada, S.S., Kramarchuk, G.G., Sdvizhkov, M.A.: Zh. Eksp. Teor. Fiz. **93** (1987) 2151; Sov. Phys. JETP (English Transl.) **66** (1987) 1227.
87 D2 Doroshev, V.D., Ivanov, S.F., Molchanov, A.N., Moskvin, A.S.: Pis'ma Zh. Eksp. Teor. Fiz. **45** (1987) 583; JETP Lett. (English Transl.) **45** (1987) 743.
87 E1 Eremenko, V.V., Gnatchenko, S.L., Kharchenko, N.F., Lebedev, P.P., Piotrowski, K., Szymczak, H., Szymczak, R.: Europhys. Lett. **11** (1987) 1327.
87 G1 Ges', A.P., Doroshev, V.D., Kovtun, N.M., Troitskii, G.A., Fedotova, V.V., Khmara, V.M.: Fiz. Tverd. Tela **29** (1987) 3436; Sov. Phys. Solid State (English Transl.) **29** (1987) 1968.
87 K1 Kadomtseva, A.M., Artem'ev, G.G., Ledneva, T.M., Lukina, M.M., Ovchinnikova, T.L., Milov, V.N., Moskvin, A.S., Semenova, M.V.: Fiz. Tverd. Tela **29** (1987) 679; Sov. Phys. Solid State (English Transl.) **29** (1987) 389.
87 K2 Kadomtseva, A.M., Artem'ev, G.G., Ledneva, T.M., Lukina, M.M., Moskvin, A.S., Ovchinnikova, T.L.: Fiz. Tverd. Tela **29** (1987) 325; Sov. Phys. Solid State (English Transl.) **29** (1987) 185.
87 K3 Kadomtseva, A.M., Zvezdin, A.K., Mukhin, A.A., Zorin, I.A., Krynetskii, I.B., Kuz'min, M.D., Lukina, M.M.: Zh. Eksp. Teor. Fiz. **92** (1987) 179; Sov. Phys. JETP (English Transl.) **65** (1987) 101.
87 K4 Kurita, S., Tazaka, Y., Tsushima, K.: J. Phys. Soc. Jpn. **56** (1987) 612.
87 K5 Karnachev, A.S., Klechin, Yu.I., Kovtun, N.M., Moskvin, A.S., Solov'ev, E.E., Tkachenko, A.A.: Zh. Eksp. Teor. Fiz. **92** (1987) 279; Sov. Phys. JETP (English Transl.) **65** (1987) 157.
87 K6 Kocharyan, K.N., Martirosyan, R.M., Karneeva, S.S., Kalantaryan, V.P., Ges', A.P.: Fiz. Tverd. Tela **29** (1987) 924; Sov. Phys. Solid State (English Transl.) **29** (1987) 529.
87 M1 Milov, V.N., Semenova, M.V., Nedolko, V.I., Prun, A.F.: Fiz. Tverd. Tela **29** (1987) 1569; Sov. Phys. Solid State (English Transl.) **29** (1987) 902.
87 P1 Piotrowski, K., Szymczak, R., Eremenko, V.V., Gnatchenko, S.L., Kharchenko, N.F.: Acta Phys. Polon. A **72** (1987) 343.
87 P2 Petrov, A.N., Cherepanov, V.A., Zuev, A.Y.: Russ. J. Phys. Chem. **61** (1987) 326.
87 R1 Rao, K.V., Roig, A., Szymczak, H.: Acta Phys. Pol. **A72** (1987) 351.
87 S1 Sinitsyn, E.V., Nikolov, O., Tomov, T., Ruskov, T., Ivanov, S.I.: Fiz. Tverd. Tela **29** (1987) 1001; Sov. Phys. Solid State (English Transl.) **29** (1987) 573.

87 S2 Sosnowska, I., Fischer, P.: Phase Transitions **8** (1987) 319.

87 T1 Telechko, S.P., Kononyuk, I.F., Lyutsko, V.A., Zonov, Y.G.: Izv. Akad. Nauk SSSR, Neorg. Mater. **23** (1987) 1520; Inorg. Mater. (English Transl.) **23** (1987) 1342.

87 V1 Valiev, U.V., Klochkov, A.A., Lukina, M.M., Turganov, M.M.: Opt. Spektrosk. **63** (1987) 543; Opt. Spectrosc. (USSR) (English Transl.) **63** (1987) 319.

87 Z1 Zorin, I.A., Kadomtseva, A.M., Lukina, M.M., Mukhin, A.A.: Zh. Eksp. Teor. Fiz. **93** (1987) 266; Sov. Phys. JETP (English Transl.) **66** (1987) 153.

88 B1 Balykina, E.A., Gan'shina, E.A., Krinchik, G.S.: Fiz. Tverd. Tela **30** (1988) 570; Sov. Phys. Solid State (English Transl.) **30** (1988) 326.

88 B2 Balbashov, A.M., Marchukov, P.Yu., Nikolaev, I.V., Rudashevskii, E.G.: Fiz. Tverd. Tela **30** (1988) 675; Sov. Phys. Solid State (English Transl.) **30** (1988) 386.

88 B3 Balbashov, A.M., Marchukov, P.Yu., Nikolaev, I.V., Rudashevskii, E.G.: Zh. Eksp. Teor. Fiz. **94** (1988) 309; Sov. Phys. JETP (English Transl.) **67** (1988) 1910.

88 B4 Balbashov, A.M., Marchukov, P.Y., Rudashevskii, E.G.: Zh. Eksp. Teor. Fiz. **94** (1988) 358; Sov. Phys. JETP (English Transl.) **67** (1988) 1291.

88 E1 Eremenko, V.V., Gnatchenko, S.L., Kharchenko, N.F., Lebedev, P.P., Piotrowski, K., Szymczak, H., Szymczak, R.: J. Phys. (Paris) C 8, Suppl. 12, **49** (1988) 919.

88 H1 Hori, H., Date, M.: J. Phys. Soc. Jpn. **57** (1988) 2565.

88 K1 Kadomtseva, A.M., Kuz'min, M.D., Lukina, M.M., Mukhin, A.A.: Zh. Eksp, Teor. Fiz. **94** (1988) 251; Sov. Phys. JETP (English Transl.) **67** (1988) 1876.

88 K2 Kovtun, N.M., Moskvin, A.S., Troitskii, G.A., Khmara, V.M.: Fiz. Tverd. Tela **30** (1988) 3629; Sov. Phys. Solid State (English Transl.) **30** (1988) 2085.

88 K3 Koshizuka, N., Hayashi, K.: J. Phys. Soc. Jpn. **57** (1988) 4418.

88 L1 Loewenhaupt, M., Sosnowska, I., Frick, B.: J. Phys. (Paris). C8, Suppl. 12, **49** (1988) 921.

88 M1 Moskvin, A.S., Latypov, D.G., Gudkov, V.G.: Fiz. Tverd. Tela **30** (1988) 413; Sov. Phys. Solid State (English Transl.) **30** (1988) 235.

88 M2 Milov, V.N., Kolmakova, N.P., Mukhin, A.A., Nedelko, V.I., Semenova, M.V., Belov, D.V.: Fiz. Tverd. Tela **30** (1988) 3057; Sov. Phys. Solid State (English Transl.) **30** (1988) 1758.

88 P1 Patil, D.S., Venkatramani, N., Rohatgi, V.K.: J. Mater. Sci. Lett. **7** (1988) 413.

88 S1 Salem, Y., Joubert, M.-F., Linares, C., Jacquier, B.: J. Lumin. **40,41** (1988) 694.

88 T1 Taguchi, H., Nagao, M., Shimada, M.: J. Solid State Chem. **76** (1988) 284.

88 T2 Taguchi, H., Nagao, M., Shimada, M., Takeda, Y., Yamamoto, O.: J. Solid State Chem. **77** (1988) 336.

88 V1 Vitebskii, I.M., Kovtun, N.M., Troitskii, G.A., Khmara, V.M.: Izv. Akad. Nauk SSSR, Ser. Fiz. **52** (1988) 1739; Bull. Acad. Sci. USSR, Phys. Ser. (English Transl.) **52** (1988) 72.

88 Z1 Zorin, I.A., Kadomtseva, A.M., Krynetskii, I.B., Lukina, M.M., Mukhin, A.A.: Fiz. Tverd. Tela **30** (1988) 76; Sov. Phys. Solid State (English Transl.) **30** (1988) 41.

89 A1 Asai, K., Gehring, P., Chou, H., Shirane, G.: Phys. Rev. B **40** (1989) 10982.

89 B1 Barilo, S.N., Ges, A.P., Guretskii, S.A., Danshin, N.K., Kramarchuk, G.G., Louguinets, A.M., Sdvizhkov, M.A., Fedotova, V.V.: Acta Phys. Polon. A **76** (1989) 69.

89 B2 Balbashov, A.M., Kozlov, G.V., Lebedev, S.P., Mukhin, A.A., Pronin, A.Yu., Prokhorov, A.S.: Zh. Eksp. Teor. Fiz. **95** (1989) 1092; Sov. Phys. JETP (English Transl.) **68** (1989) 629.

89 B3 Balbashov, A.M., Dan'shin, N.K., Izotov, A.I., Sdvizhkov, M.A., Tsymbal, L.T.: Fiz. Tverd. Tela **31** (1989) 279; Sov. Phys. Solid State (English Transl.) **31** (1989) 1259.

89 D1 Dan'shin, N.K., Zherlitsyn, S.V., Zvada, S.S., Mukhin, A.A., Sdvizhkov, M.A., Fil', V.D.: Fiz. Tverd. Tela **31** (1989) 198; Sov. Phys. Solid State (English Transl.) **31** (1989) 832.

89 G1 Ges', A.P., Fedotova, V.V., Derkachenko, V.N., Buzhynsky, S.A.: S.A.: Acta Phys. Pol. A **76** (1989) 75.
89 G2 Gnatchenko, S.L., Kharchenko, N.F., Lebedev, P.P., Piotrowski, K., Szymczak, H., Szymczak, R.: J. Magn. Magn. Mater. **81** (1989) 125.
89 G3 Gnatchenko, S.L., Lebedev, P.P., Kharchenko, N.F., Piotrowski, K., Szymczak, R.: Fiz. Nizk, Temp. **15** (1989) 868; Sov. J. Low Temp. Phys. (English Transl.) **15** (1989) 480.
89 K1 Kadomtseva, A.M., Krynetskii, I.B., Kuz'min, M.D., Zvezdin, A.K.: J. Magn. Magn. Mater. **81** (1989) 196.
89 K2 Kharchenko, N.F., Gnatchenko, S.L., Chizhik, A.B.: J. Phys. (Paris) **50** (1989) 1153.
89 K3 Krynetskii, I.B., Sinitsyn, E.V.: Fiz. Tverd. Tela **31** (1989) 309; Sov. Phys. Solid State (English Transl.) **31** (1989) 899.
89 T1 Troyanchuk, I.O., Pastushonok, S.N.: Phys. Status Solidi (a) **115** (1989) K 225.
89 V1 Vorob'ev, G.P., Kadomtseva, A.M., Krynetskii, I.B., Mukhin, A.A.: Zh. Eksp. Teor. Fiz. **95** (1989) 1049; Sov. Phys. JETP (English Transl.) **68** (1989) 604.
89 V2 Vitebskii, I.M., Kovtun, N.M., Troitskii, G.A., Khmara, V.M.: Fiz. Nizk. Temp. **15** (1989) 1066; Sov. J. Low Temp. Phys. (English Transl.) **15** (1989) 591.
89 Z1 Zorin, I.A., Kadomtseva, A.M., Kozlov, G.V., Lebedev, S.P., Lukina, M.M., Mukhin, A.A., Prokhorov, A.S.: Fiz. Tverd. Tela **31** (1989) 184; Sov. Phys. Solid State (English Transl.) **31** (1989) 1201.
90 A1 Andronenko, S.I., Bazhan, A.N., Mezentseva, L.P.: Fiz. Tverd. Tela **32** (1990) 773; Sov. Phys. Solid State (English Transl.) **32** (1990) 455.
90 B1 Belov, K.P., Krynetskii, I.B., Sinitsin, E.V.: J. Magn. Magn. Mater. **84** (1990) 76.
90 B2 Barilo, S.N., Ges', A.P., Dan'shin, N.K., Kramarchuk, G.G., Luginets, A.M., Sdvizhkov, M.A., Fedotova, V.V.: Zh. Eksp. Teor. Fiz. **97** (1990) 1921; Sov. Phys. JETP (English Transl.) **70** (1990) 1083.
90 B3 Barilo, S.N., Ges', A.P., Kovalev, A.A., Luginets, A.M., Milov, V.N., Mukhin, A.A., Prun, A.F.: Fiz. Tverd. Tela **32** (1990) 2643; Sov. Phys. Solid State (English Transl.) **32** (1990) 1532.
90 E1 Echigoya, J., Hiratsuka, S., Suto, H.: Phys. Status Solidi (a) **118** (1990) 371.
90 G1 Ges', A.P., Derkachenko, V.N., Fedotova, V.V., Ignatenko, A.A., Solovykh, T.K.: Fiz. Tverd. Tela **32** (1990) 2212; Sov. Phys. Solid State (English Transl.) **32** (1990) 1286.
90 G2 Gnatchenko, S.L., Chizhik, A.B., Kharchenko, N.F.: Pisma Zh. Eksp. Teor. Fiz. **51** (1990) 282; JETP Lett. (English Transl.) **51** (1990) 324.
90 G3 Gufan, Y.M., Rafalovich, V.I.: Sov. Phys. Solid State **32** (1990) 1122.
90 H1 Huang, T.C., Parrish, W., Toraya, H., Lacorre, P., Torrance, J.B.: Mater. Res. Bull. **25** (1990) 1091.
90 K1 Kolmakova, N.P., Krynetskii, I.B., Lukina, M.M., Mukhin, A.A.: Phys. Status Solidi (b) **159** (1990) 845.
90 K2 Karneeva, S.S., Kalantaryan, V.P., Kacharyan, K.N., Martirossyan, R.M., Milov, V.N., Sinitsyn, E.V., Solovykh, T.K.: Fiz. Tverd. Tela **32** (1990) 2193; Sov. Phys. Solid State (English Transl.) **32** (1990) 1275.
90 L1 Lomako, I.D., Konstantinova, A.F., Ges', A.P.: Kristallografiya **35** (1990) 418: Sov. Phys. Crystallogr. (English Transl.) **35** (1990) 240.
90 V1 Vorob'ev, G.P., Kadomtseva, A.M., Mukhin, A.A., Lukina, M.M.: Fiz. Tverd. Tela **32** (1990) 413; Sov. Phys. Solid State (English Transl.) **36** (1990) 238.

90 V2 Vorob'ev, G.P., Kadomtseva, A.M., Krynetskii, I.B., Mukhin, A.A.: Zh. Eksp. Teor. Fiz. **98** (1990) 1726; Sov. Phys. JETP (English Transl.) **71** (1990) 969.

91 A1 Artemev, G.G., Kadomtseva, A.M., Kuzmin, M.P., Lukina, M.M., Moskvin, A.S., Churakaev, V.V.: Fiz. Tverd. Tela **33** (1991) 3387; Sov. Phys. Solid State (English Transl.) **33** (1991) 1910.

91 B1 Barilo, S.N., Ges', A.P., Guretskii, A.M., Dan'shin, N.K., Kramarchuk, G.G., Luginets, A.M., Sdvizhkov, M.A.: Fiz. Tverd. Tela **33** (1991) 621; Sov. Phys. Solid State (English Transl.) **33** (1991) 354.

91 G1 Ganshina, E.A., Zenkov, A.V., Krinchik, G.S., Moskvin, A.S., Trifonov, A.Yu.: Fiz. Tverd. Tela **33** (1991) 1122; Sov. Phys. Solid State (English Transl.) **33** (1991) 637.

91 L1 Liu, X., Jin, M., Liu, M.: Hyperfine Interact. **68** (1991) 241.

91 M1 Moskvin, A.S., Artemev, G.G., Kadomtseva, A.M., Ledneva, T.M., Lukina, M.M., Ovchinnikova, T.L.: Fiz. Tverd. Tela **33** (1991) 638; Sov. Phys. Solid State (English Transl.) **33** (1991) 366.

91 S1 Shirakawa, N., Ishikawa, M.: Jpn. J. Appl. Phys. **30** (1991) L 755.

92 D1 Devi, P.S., Rao, M.S.: J. Mater. Sci. Lett. **11** (1992) 226.

5.6 Oxides with corundum and ilmenite structures

5.6.1 Introduction

The most prominent representatives of magnetic compounds with corundum and ilmenite structures are α-Fe_2O_3 and $FeTiO_3$, respectively. Their physical properties have already been described in Landolt-Börnstein, New Series, subvolumes III/4a and III/12b, together with the properties of those related compounds in which a part of the constituting iron ions is substituted by another 3d element. The aim of the present supplement to the preceeding surveys is both an updating of this former work as well as an extension in the sense that compounds for which the iron ions are completely substituted by another transition element are included.

5.6.2 List of frequently used symbols and abbreviations

Symbols

Symbol	Unit	Property
a, b, c	Å	lattice parameters
a_r	Å	rhombohedral lattice parameter
B	G	magnetic induction
B_μ	G	local muon field
C_g	$cm^3\,K\,g^{-1}$	Curie-Weiss constant per unit mass
C_m	$cm^3\,K\,mol^{-1}$	molar Curie-Weiss constant
$C_{p,g}$	$J\,g^{-1}\,K^{-1}$	mass heat capacity at constant pressure
$C_{p,m}$	$J\,mol^{-1}\,K^{-1}$	molar heat capacity at constant pressure
$C_{v,m}$	$J\,mol^{-1}\,K^{-1}$	molar heat capacity at constant volume
E	$V\,m^{-1}$	electric field strength
E	meV, THz	spin wave energy (1 meV = 0.241799 THz)
ΔE_Q	$mm\,s^{-1}$	quadrupole splitting
f	Hz	frequency
g	–	spectroscopic splitting factor
h, k, l	–	reciprocal lattice indices
H	Oe	magnetic field
H_A	Oe	anisotropy field
H_c	Oe	critical field; coercive field
H_{hyp}	Oe	magnetic hyperfine field
H_{int}	Oe	internal magnetic field
I		intensity
$J_i, J_{\alpha...\varepsilon}$	eV	exchange energy (J/k in K)
k	$J\,K^{-1}$	Boltzmann constant
l	nm	length
M	G	magnetization
p	kbar	hydrostatic pressure
p_c	kbar	critical pressure for phase transition
p_{eff}	μ_B	effective magnetic moment
$p_{M^{3+}}$	μ_B	magnetic moment per M^{3+}
q	$Å^{-1}$	inverse spin wave length
S	–	spin quantum number

T	K, °C	temperature
T_C	K	Curie point, magnetic ordering temperature
T_c	K	critical temperature for phase transition
T_L	K	phase transition temperature (for definition see Fig. 93)
T_M	K	Morin temperature
T_N	K	Néel temperature
T_{SG}	K	spin glass freezing temperature
v	$mm\,s^{-1}$	velocity of absorber in Mössbauer effect
x	–	concentration
α, β, γ	deg	angles, triclinic symmetry
α_r	deg	rhombohedral angle
$\alpha_{\|\|}, \alpha_\perp$	–	magnetoelectric parameters (dimensionless in CGS)
β	deg	cone half angle
δ	$mm\,s^{-1}$	isomer shift
ε	$mm\,s^{-1}$	quadrupole shift*)
η	–	asymmetry parameter
Θ_D	K	Debye temperature
Θ_p	K	paramagnetic Curie temperature
μ_B	$J\,T^{-1}$, erg G^{-1}	Bohr magneton
σ	emu g^{-1} = G $cm^3\,g^{-1}$	magnetic moment per gram (specific magnetization)
σ_r	G $cm^3\,g^{-1}$	remanent specific magnetization
σ_s	G $cm^3\,g^{-1}$	spontaneous specific magnetization
χ_g	emu g^{-1} = $cm^3\,g^{-1}$	magnetic susceptibility per gram
χ_m	emu mol^{-1} = $cm^3\,mol^{-1}$	magnetic susceptibility per mole
$\chi_{\|\|}, \chi_\perp$	–	magnetic susceptibility parallel and perpendicular to a main crystallographic axis
χ', χ''	–	real, imaginary part of magnetic susceptibility

*) The displacements ε of the energy levels of the Mössbauer spectrum by the quadrupole interaction (the so-called quadrupole shift) is in the case of magnetic order given by

$$\varepsilon = (e^2qQ/4)\,(3\cos^2\theta - 1)/2,$$

where $-e$ is the electron charge, $eq = V_{zz}$ the magnetic component of the electric field gradient tensor along the symmetry axis, Q the quadrupole moment of the ion and θ the angle between V_{zz} and the direction of the electron spin quantisation.

For α-Fe_2O_3 with $\theta = 90°$ holds:

$$\varepsilon = -(1/8)e^2qQ.$$

For the case of the spin axis parallel to the EFG axis as e.g. in Cr_2O_3:

$$\varepsilon = +(1/4)\,e^2qQ.$$

The quadrupole shift above the magnetic ordering temperature is related to the EFG by the relation

$$\varepsilon = (1/2)e^2qQ.$$

Abbreviations

ac	alternating current
AF	antiferromagnetism, antiferromagnetic
AFR	antiferromagnetic resonance
dc	direct current
eff	effective
EFG	electric field gradient
ESR	electron spin resonance
FC	field cooled
FU	formula unit
INS	inelastic neutron scattering
IRM	isothermal remanent magnetism
IS	isomer shift
M, M′	metal
NMR	nuclear magnetic resonance
P	paramagnetism, paramagnetic
R	rhombohedral
RT	room temperature
SG	spin glass
SF	spin flop
T	triclinic
TRM	thermoremanent magnetism
WF	weak ferromagnetic
ZFC	zero-field cooled
$\perp$, $\parallel$	perpendicular, parallel to a crystallographic axis

5.6.3 Oxides with corundum structure

Corundums, general

A comprehensive survey of structure and transport (electric) properties of the sesquioxides with corundum structure, Ti_2O_3, V_2O_3, Cr_2O_3 and α-Fe_2O_3 is given by Goodenough and Hamnett in Landolt–Börnstein, New Series, subvolume III/17g, p. 129. Pictures of the rhombohedral corundum structure (space group D_{3d}^6 – $R\bar{3}c$) are given in the Figs. 12 and 16. Lattice parameters are listed in Table 1.

Table 1. Corundum oxides, space group $R\bar{3}c$. RT crystallographic parameters for the hexagonal symmetry (a, c) and the rhombohedral symmetry (a_r, α_r).

Compound	Hexagonal system		Rhombohedral system		Ref.
	a [Å]	c [Å]	a_r [Å]	α_r	
Ti_2O_3	5.149	13.642			62N1
V_2O_3	4.952	14.002			62N1
	4.9517	14.005	5.474	53°48′	78B1, 2
Cr_2O_3	4.9607	13.599			62N1
	4.95762(8)	13.5874(10)			67S2
α-Fe_2O_3	5.0345	13.749	5.427	55°16′	62N1
	5.03490(9)	13.7524(18)			67S2
$CrTiO_3$	5.050	13.630(5)			72R2
$CrVO_3$	4.986(1)	13.762(3)			72R2
α-$CuVO_3$	12.859(2)	7.186(2)	7.800(6)	111.02(1)°	70C1
$NiCrO_3$	4.925	13.504			69C1
$MnTiO_3$II	5.2051(5)	13.699(3)	5.4666(11)	56°52′(2)	69S3

5.6.3.1 Binary oxides M_2O_3; M = 3d transition element

Ti_2O_3

Early neutron diffraction work [63A1] found evidence for antiferromagnetic ordering, but later studies [68K1] showed that the apparent superlattice peaks were due to multiple Bragg scattering. There is essentially no localized moment on the Ti^{3+} ions, a result confirmed by neutron polarization studies [69M1]. For the magnetic susceptibility see Figs. 1, 2a; for specific heat see Fig. 2b.

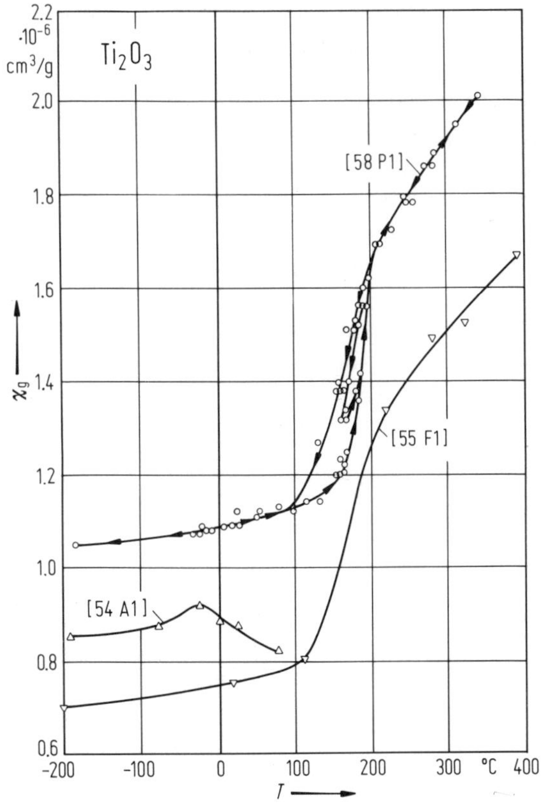

Fig. 1. Ti_2O_3, powder. Temperature dependence of the magnetic mass susceptibility χ_g according to three sources [58P1].

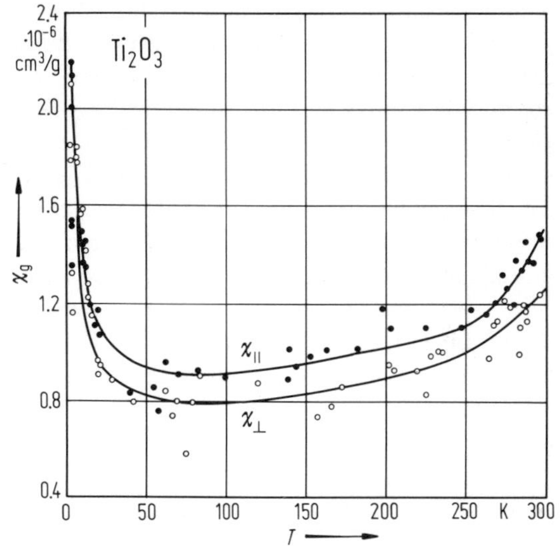

Fig. 2a. Ti_2O_3, single crystal. Temperature dependence of the principal magnetic mass susceptibilities. The increase at low T is attributed to magnetic impurities or lack of stoichiometry [60C1].

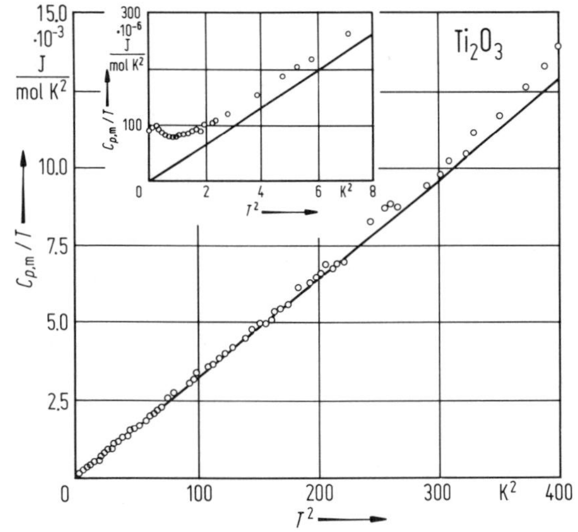

Fig. 2b. Ti_2O_3 (purity in parts per million by weight: C 50, Al 75, Cl 10, K 10 and Fe 13). Specific heat $C_{p,m}$ plotted as $C_{p,m}/T$ vs. T^2. Debye temperature $\Theta_D = 674$ K. The inset shows a small anomaly, probably of Schottky type and will be the result of magnetic impurities [73S2].

V_2O_3 and V_2O_{3+x}

Vanadium sesquioxide has the corundum structure at RT. At about 160 K a first order phase transition occurs, the rhombohedral unit cell transforms into a monoclinic unit cell, accompanied by an increase of resistivity of about 7 orders of magnitude from a metallic to an insulating state. At temperatures around about 500 K indications are found of a continuous anomalous variation of the lattice parameters of the corundum structure. Consequences for the magnetic phase diagram are shown in Fig. 3. This figure also shows the influence of an excess of oxygen on the magnetic phase transitions. For magnetic susceptibility, see Figs. 4–6, 42 and Table 2. Conventional neutron diffraction on powders and single crystals reveals an antiferromagnetic order in the monoclinic phase. The V moments are ferromagnetically coupled in monoclinic (010) layers or hexagonal (110) layers, with a reversal between adjacent layers, Fig. 7. The ordered moment is $1.2(1)\mu_B$ per V^{3+} ion. The magnetic and crystallographic transformations occur at the same temperature ($\approx$ 170 K). An extrapolation of the nuclear spectra yields an apparent Néel temperature $T_N = 285(10)$ K [79M1, see also 81Y1]. The anomaly of the magnetic susceptibility in the temperature range 500 ··· 570 K is accompanied by an anomalous behavior in the cold neutron small-angle scattering spectra. It is suggested to be caused by the formation of giant "correlation" clusters of size ≈ 600 Å and a magnetic moment density $\simeq 10^{-7}\mu_B$ Å^{-3} as the system suffers a broad metal-insulator transition [77P1]. From NMR data in magnetic fields at 1.8 K it is concluded that the magnetic moments are parallel to the hexagonal (110) layers, making an angle of 71(3)° with the c axis [75M1]. For hyperfine fields see Table 3. From the results of INS measurements at 142 K the spin-wave dispersion curves of Fig. 8 are derived. For the underlying exchange parameters and anisotropy constant see Table 4. For Mössbauer data see Table 5.

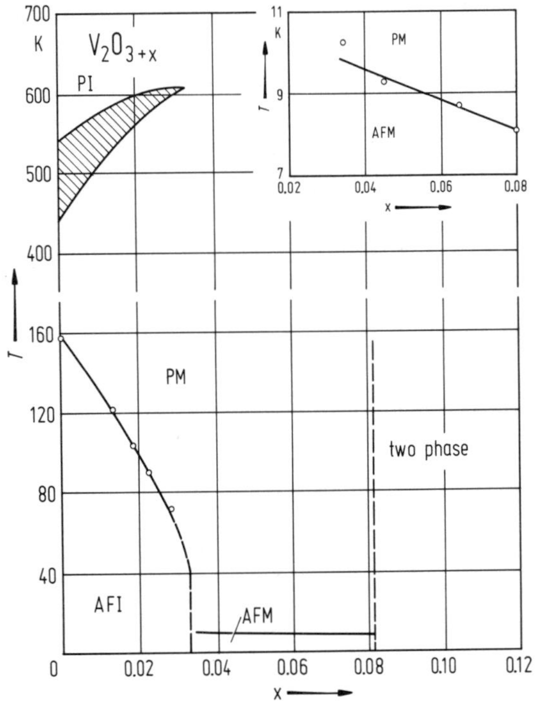

Fig. 3. V_2O_{3+x}. Magnetic phase diagram obtained from experimental results. PI: paramagnetic insulator, PM: paramagnetic metal, AFI: antiferromagnetic insulator, AFM: antiferromagnetic metal. Hatched area: continuous anomalous variation of the lattice parameters of the corundum structure [80U1].

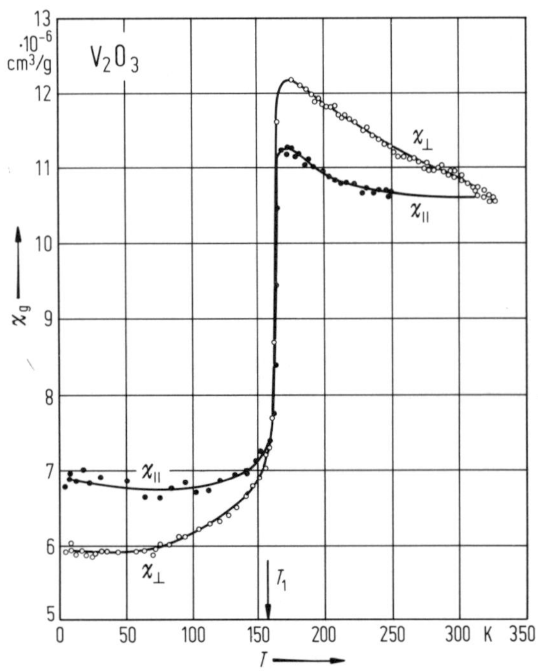

Fig. 4. V_2O_3, single crystal. Temperature dependence of the principal magnetic mass susceptibilities below RT [60C1]. T_1: phase transition temperature.

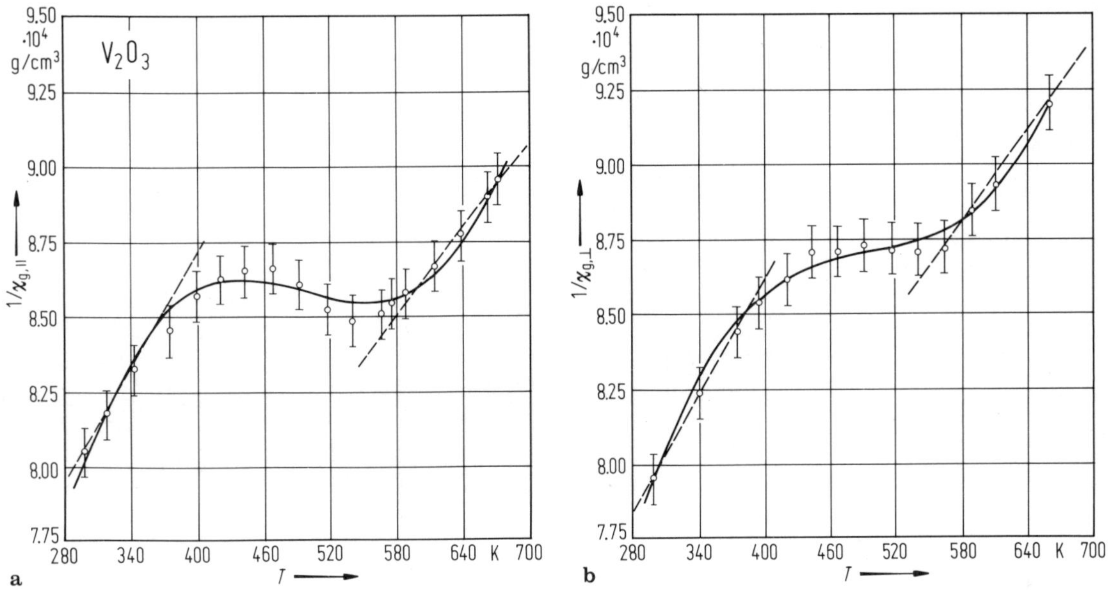

Fig. 5. V_2O_3, single crystal. Temperature dependence above RT of the reciprocal mass susceptibility along the trigonal axis (a) and perpendicular to the trigonal axis (b). Dashed lines are Curie-Weiss approximations (see Table 2) and solid lines are least-squares fits by a cubic equation [68A1].

For Fig. 6, see next page.

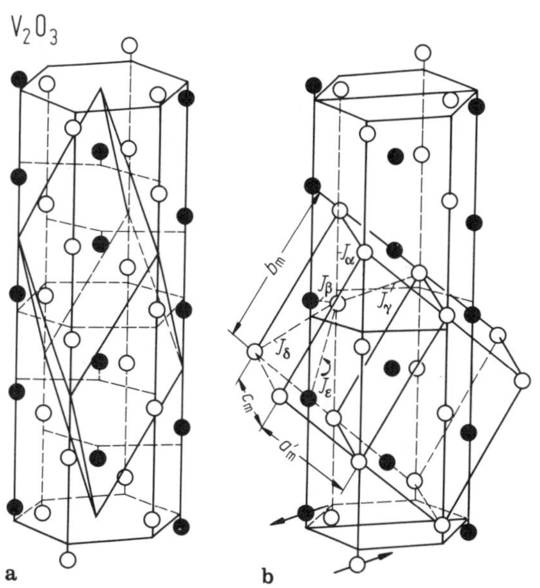

Fig. 7. V_2O_3. Relationships between V-atoms in hexagonal, rhombohedral and monoclinic unit cells, the oxygen atoms are omitted. The primitive rhombohedral axes are shown as bold lines in (a). The magnetic and crystallographic unit cells are shown in (b). Spins on the filled circles are opposite to those in the open circles. Assuming equivalent metal atoms, the magnetic monoclinic cell is half ($a'_m = \frac{1}{2} a_m$) of the crystallographic cell. $J_\alpha \ldots J_\varepsilon$ are magnetic coupling constants, see Table 4 [81W1].

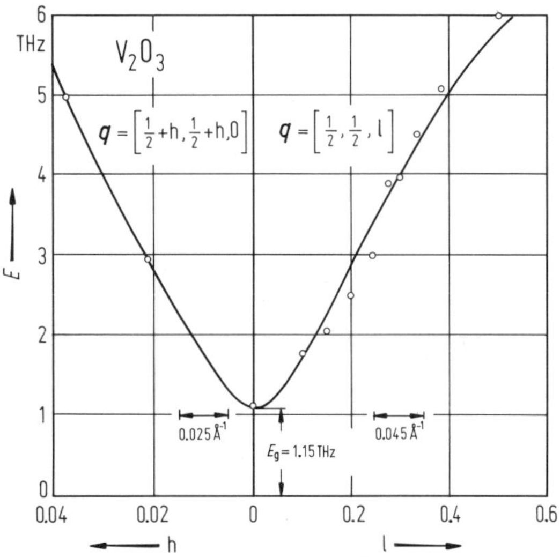

Fig. 8. V_2O_3, monoclinic antiferromagnetic phase. $T = 142$ K. Spin-wave dispersion relation in the direction of the [001] trigonal axis and along [hh0] axis. The curve represents the best fit with the parameters of Table 4. The energy gap E_g at the zone center is 4.75 meV [81W1].

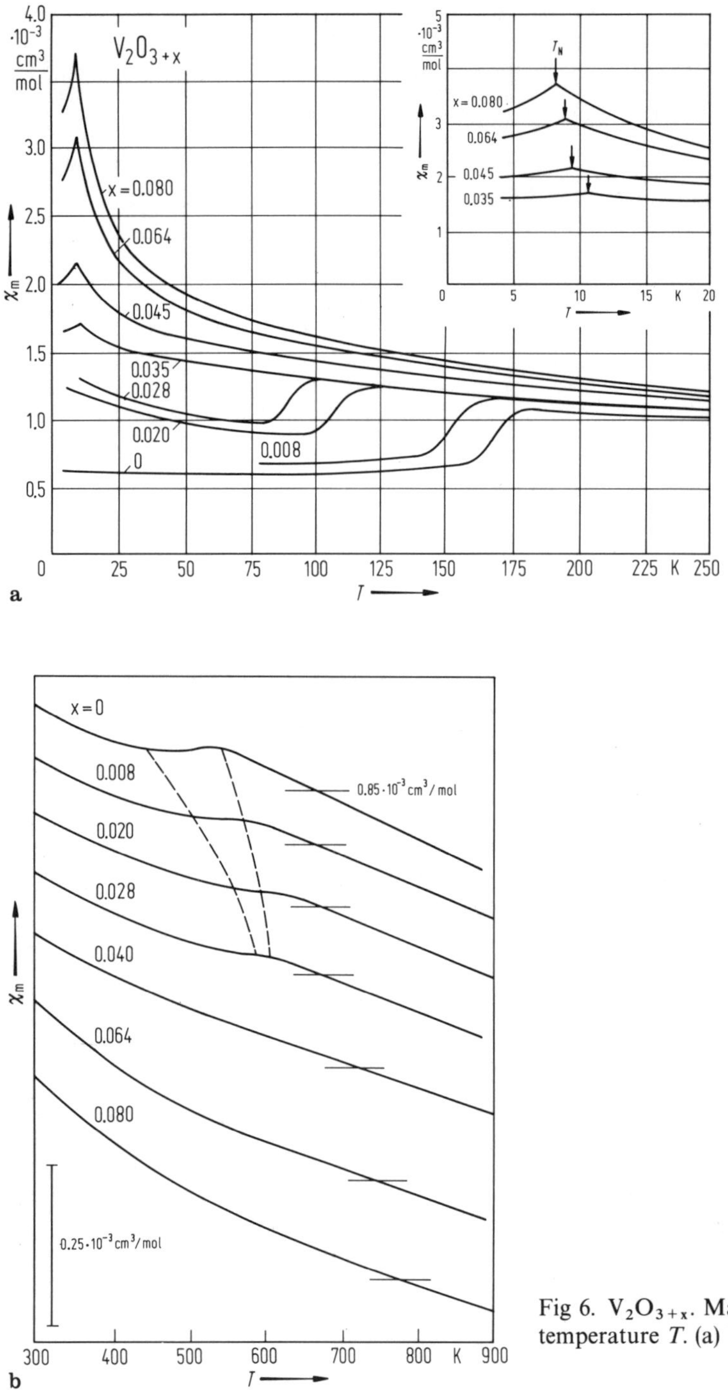

Fig 6. V_2O_{3+x}. Magnetic molar susceptibility χ_m vs. temperature T. (a) Below RT. (b) above RT [80U1].

Table 2. V_2O_3, single crystals. Magnetic parameters. For the metallic state the susceptibility curves in the Figs. 5a, b can be described by $\chi_g \approx \chi_{g,o} + C_g/(T - \Theta_p)$ with $\chi_{g,o} = 2.8 \cdot 10^{-16}\ cm^3\,g^{-1}$ and the following Curie-Weiss parameters. The free ion magnetic moment value of vanadium is $p_{V^{3+}} = 2.83\ \mu_B$ [68A1, see also 70M2].

	C_g [cm^3 K/g]	Θ_p [K]	p_{eff} [μ_B/V]
(a) Below 400 K			
$\chi_\perp$	0.00926	− 649	2.352
$\chi_\parallel$	0.00922	− 660	2.347
[65J1]*	0.00877	− 600	2.37
(b) Above 550 K			
$\chi_\perp$	0.01134	− 743	2.603
$\chi_\parallel$	0.01183	− 740	2.659

*) Powder sample.

Table 3. V_2O_3, V_2O_{3+x}. Internal magnetic fields. H_{int} as derived from nuclear resonance and scattering experiments.

Compound	T [K]	H_{int} [kOe]	Measuring method	Ref.
V_2O_3	0	185.9(1)	NMR	71Y1
V_2O_3		168	[d])	70A2
V_2O_3	105	175(15)	INS	70H1
V_2O_{3+x}, x = 0.04; 006	1.8	58(2)[a,c])	NMR	78U1
		9[b,c])	NMR	78U1
x = 0.01; 0.02	1.8	184.9(5)	NMR	78U1
		71(1)[a])	NMR	78U1

[a]) Assigned as being due to V^{4+}-like sites.
[b]) Assigned as being due to matrix V sites.
[c]) This would mean that also in the metallic V_2O_{3+x} antiferromagnetic order occurs.
[d]) Nuclear specific heat.

Table 4. V_2O_3, monoclinic antiferromagnetic phase, T = 142 K. Exchange constants (defined in Fig. 7) and uniaxial magnetic anisotropy energy constant A_u in meV [81W1].

J_α	J_β	J_γ	J_δ	J_ε	A_u
−27.2	45.9	−23.2	2.81	−4.76	9.51

Table 5. V_2O_3; V_2O_{3+x} Mössbauer data of polycrystalline ^{57}Fe-doped samples. "Para" denotes that the spectrum shows a single line, line width 0.4 mm s^{-1}. Hyperfine field H_{hyp} ± 2.5 kOe; isomer shift IS ± 0.03 mm s^{-1}, measured relative to pure iron metal [80U1; see also 65S1].

	T [K]	H_{hyp} [kOe]	IS [mm/s]
V_2O_3	300	para	0.67
	77	387	0.64
	4.2	458	0.62
$V_2O_{3.02}$	4.2	452	0.63
$V_2O_{3.04}$	4.2	273	0.73
$V_2O_{3.06}$	300	para	0.64
	77	para	0.78
	4.2	263	0.73
	2	283	0.76

Cr_2O_3

Magnetic susceptibility: Figs. 9, 10; see also Fig. 43. The compound is AF below the Néel temperature T_N, which is derived from the temperature dependences of various properties: magnetic powder susceptibility $T_N = 308$ K [65C1] and 307(2) K [78N1], magnetostriction $T_N = 307.3$ K [77Y1] and specific heat $T_N = 306.99$ K [77B1]. High-frequency ultrasonic wave experiments give detailed information on this phase transition [74B1].The influence of pressure p on T_N has been derived from the results of measurements of the magnetoelectric susceptibility, Fig. 11. For uniaxial pressure in the direction of the crystallographic c and a axes is found $dT_N/dp = 0.50(5)$ and 0.30(5) K/kbar, respectively [73G1]. This leads then for hydrostatic pressure to a value 1.10(15) K/kbar, whereas experimentally is found 1.50(5) K/kbar, in sharp disagreement with the results of neutron diffraction experiments: − 1.6(3) K/kbar [68W1]. In the AF phase the spins are directed along the [111] axis of the rhombohedral unit cell. Spins are paired antiparallel across a shared octahedral face, Fig. 12b. The chromium moment at 0 K is extrapolated to be $p_{Cr^{3+}} = 2.76(3)\,\mu_B$, [65C1], which is 8% lower than the spin-only value. At 77 K according to AFR measurements on a single crystal the g value is $g = 1.97(3)$ [63F1]; for a polycrystalline sample in the paramagnetic state is recorded $g \approx 1.97$ [53M1]. A magnetic field parallel to the c axis can cause a spin-flop phase SF as well as a paramagnetic phase P, see Fig. 13. The bicritical field is 121.0(15)kOe. At 4.2 K the SF transition field is 59.3(10)kOe. The accompanying length changes are shown in Fig. 14. Similar results have been found by studying the attenuation of ultrasonic waves, which exhibits a peak at the SF transition [69S2]. From the spin wave spectrum, Fig. 15, the exchange constants J_i of Table 6 have been derived. The temperature dependence of the sublattice magnetization is reproduced in Fig. 17. Antiferromagnetic resonance experiments [63F1] reveal a magnetic anisotropy field of about 370 Oe, the result of about equal contributions from magnetic dipole and crystalline or single-ion anisotropies [65A1].

For magnetoelectric parameters parallel and perpendicular to the crystallographic c axis, $\alpha_{||} = B_z/E_z$, $\alpha_\perp = B_x/E_x$, respectively, where $\boldsymbol{B}$ is the magnetic induction as a consequence of an applied electric field $\boldsymbol{E}$, see Fig. 18. For further experiments and theoretical considerations see [62R1]. For specific heat see Fig. 19.

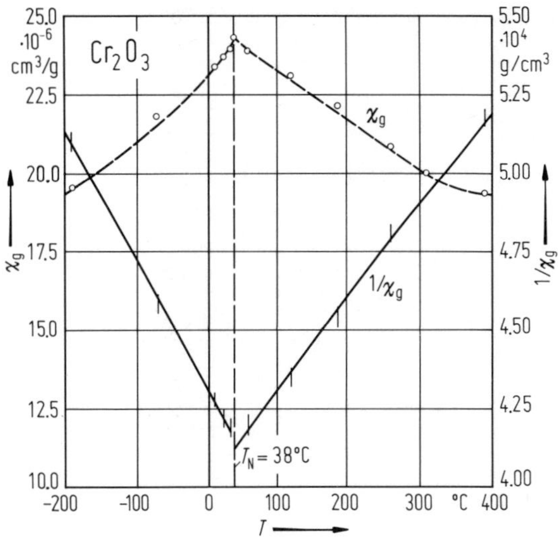

Fig. 9. Cr_2O_3, sintered polycrystalline sample. Magnetic mass susceptibility χ_g and its inverse as a function of temperature T. Measurements of T. R. McGuire in [50T1], see also [39F1] for temperatures up to 1300 K.

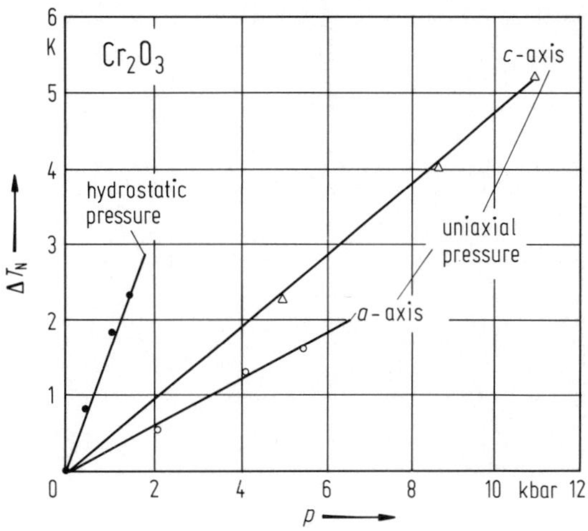

Fig. 11. Cr_2O_3, single crystal. Shift in the Néel temperature T_N as a function of uniaxial or hydrostatic pressure, p, as derived from magneto-electric data [73G1].

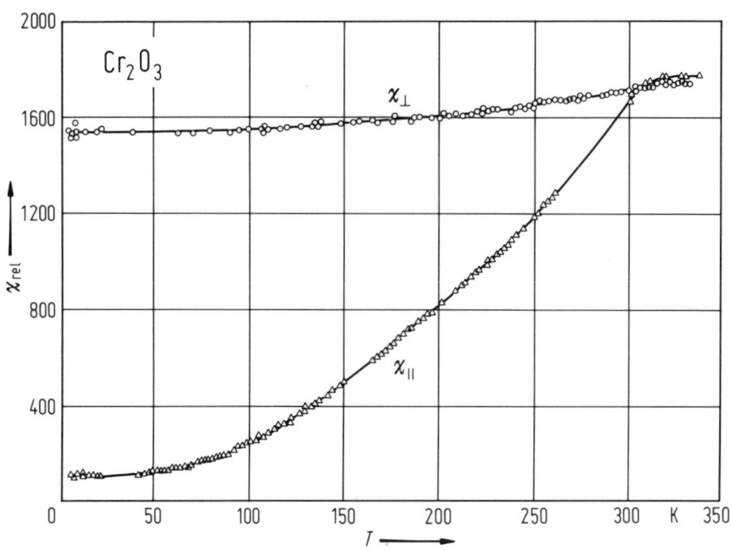

Fig. 10. Cr_2O_3, single crystal. Relative magnetic mass susceptibility χ_{rel} parallel and perpendicular to the crystallographic c axis vs. temperature T. $\chi_\perp = 22.4(4) \cdot 10^{-6}$ cm³/g at $T = 4.2$ K [63F1].

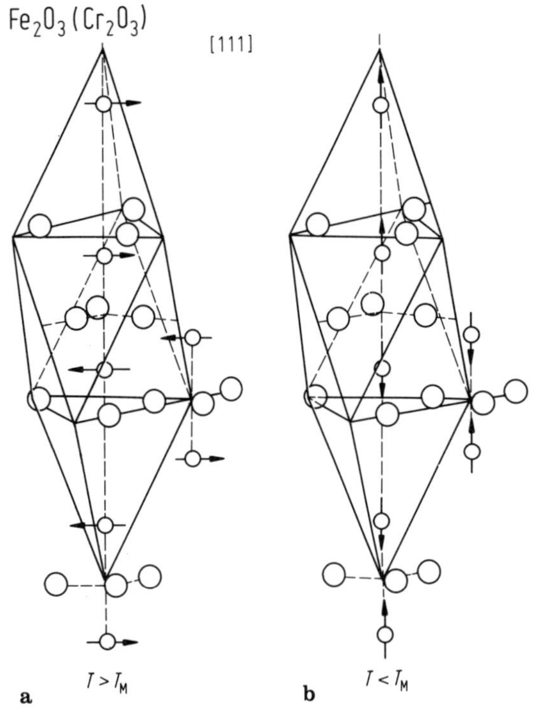

Fig. 12. Cr_2O_3. Magnetic structure as determined from neutron diffraction experiments for α-Fe_2O_3 for (a) $T > T_M$ and (b) $T < T_M$ [64C1].

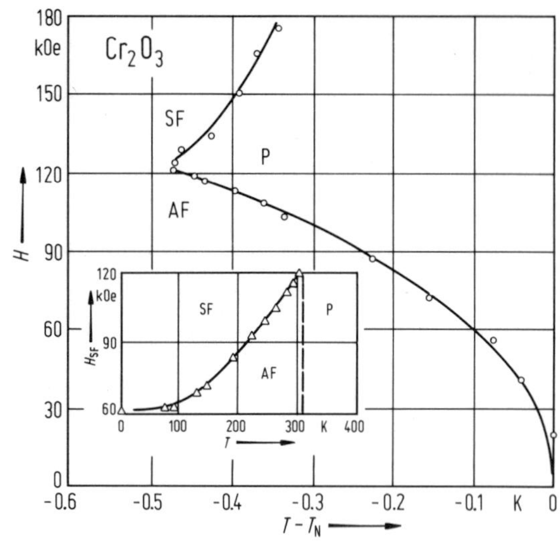

Fig. 13. Cr_2O_3, single crystal. Magnetic phase diagram for an applied magnetic field $H \parallel c$ axis, as derived from magnetostriction measurements in magnetic fields up to 175 kOe. $T_N = 307.3$ K. The inset shows the field $H_{SF}(T)$ for the first order spin-flop transition [77Y1]. The latter dependency has also been established by ultrasonic measurements [71S1].

For Fig. 15, see next page.

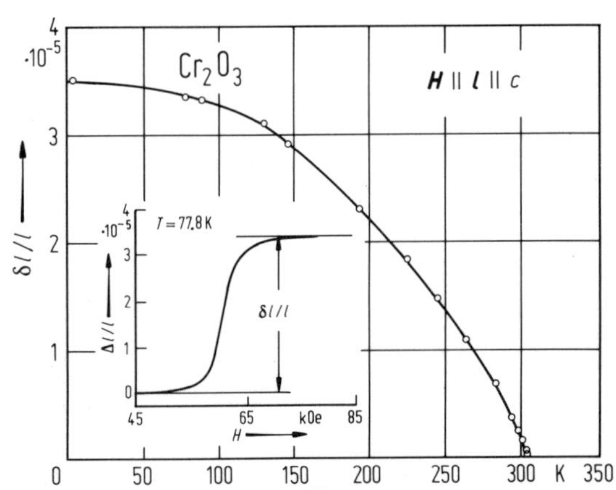

Fig. 14. Cr_2O_3, single crystal. Relative jump in length $\delta l/l$ at the spin-flop transition as a function of temperature. The inset shows the field dependence of the relative length $\Delta l/l$ at 77.8 K and the definition of $\delta l/l$ [77Y1], see also [71D1], which reports approximately 20% lower values. $\Delta l/l = (l(H) - l(0))/l(0)$; l is the length of the sample along the c axis.

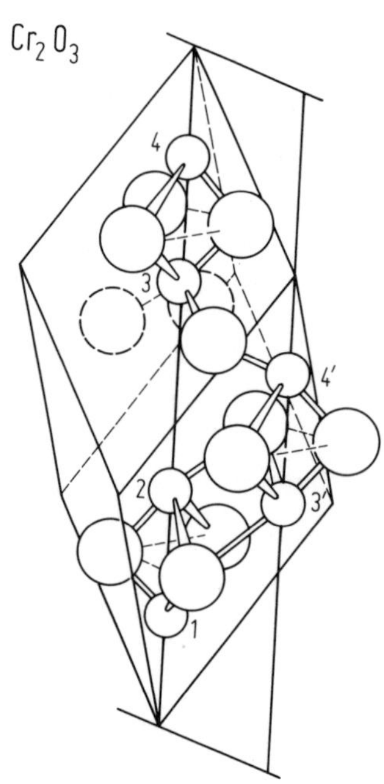

Fig. 16. Corundum structure. Rhombohedral unit cell. Small spheres: metal ions. Large spheres: oxygen ions. The correspondence between metal ion pairs and some of the exchange interaction constants is indicated in Table 6 [70S2].

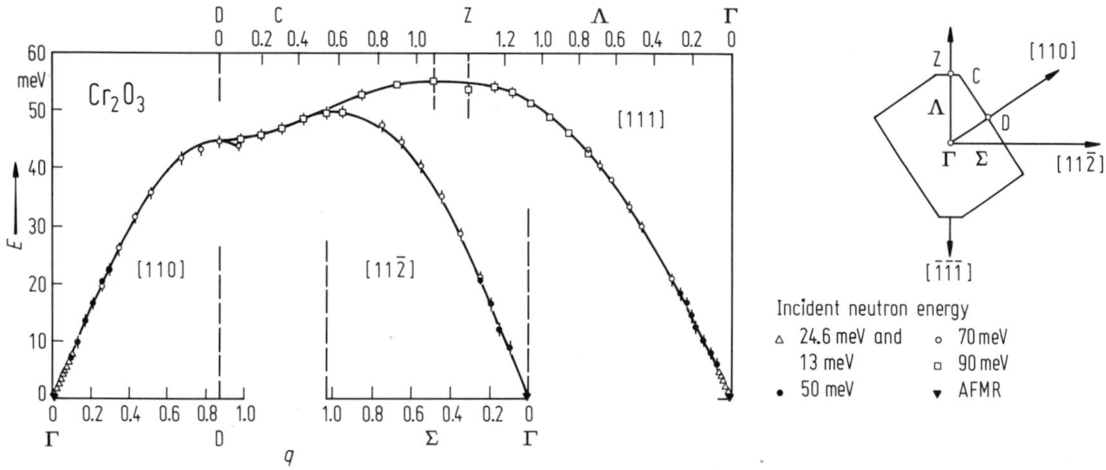

Fig. 15. Cr_2O_3. Spin-wave energy dispersion relation $E(q)$ in several crystallographic directions as derived from inelastic neutron scattering experiments at 78 K. The curves are calculated with the set of exchange parameters of Table 6. A sketch of the Brillouin zone is shown in the upper right. The upper curve relates to the upper scale and the lower curve to the lower scale, where appropriate [70S2].

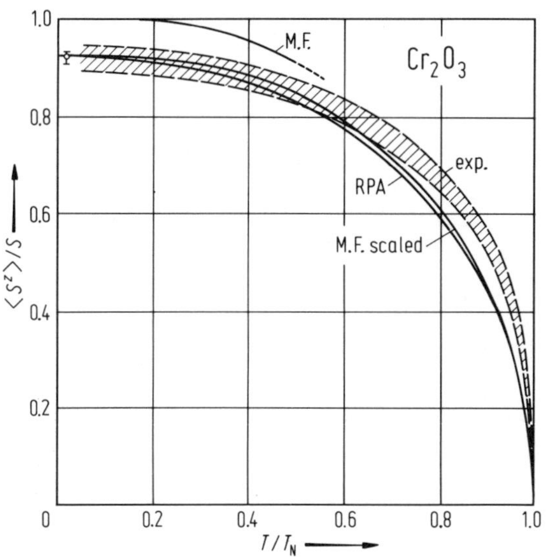

Fig. 17. Cr_2O_3. Temperature dependence of the average value of the z component of the atomic spins, $\langle S^z \rangle$ as measured (hatched region and the point $\langle S^Z \rangle = 1.400(15)$ at $T = 4.2$ K) by means of neutron diffraction by several authors [64C1, 65M1, 68S1]. Solid curves calculated using the random phase approximation (RPA) and the molecular-field theory (M.F.) [70S2].

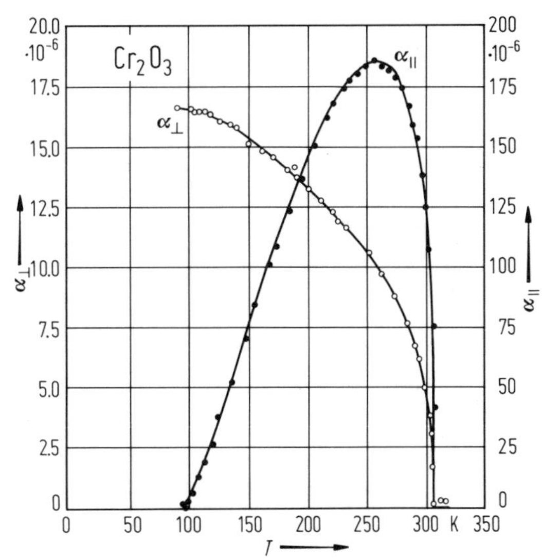

Fig. 18. Cr_2O_3, single crystal discs. Temperature dependence of the dimensionless magnetoelectric parameters $\alpha_\perp$ and $\alpha_\parallel$ in the Gaussian system [61F1].

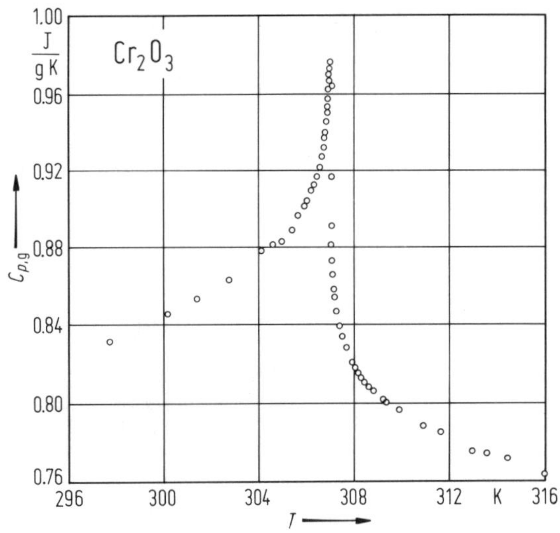

Fig. 19. Cr_2O_3, single crystal. Specific heat $C_{p,g}$ vs. temperature T near the antiferromagnetic transition temperature [77B1, also for further details].

Table 6. Cr_2O_3; α-Fe_2O_3. Identification and values of the nearest-neighbour exchange interaction constants J_i for various spin pairs i, j (given as J/k in K).

J_i	Spin (ion) pair see Fig. 16	Magnitude[a]) of J_i		
		Cr_2O_3		α-Fe_2O_3
		$T = 78$ K [70S2]	$T = 85$ K [69A1]	$T = 240$ K [70S1]
J_1	1–2	−87.3	−86	6.0(16)
J_2	2–3′	−39.6	−38	1.6(6)
J_3	3–4′	−0.9	−5.8	−29.7(20)
J_4	2–4′	0.2	−2.3	−23.2(10)
J_5	2–3	−2.2	–	−1.0(10)

[a]) If a temperature dependence of the exchange interactions is taken into account, somewhat different values are obtained [78K1].

α-Fe_2O_3

See Landolt-Börnstein, New Series, subvolumes III/4a, p. 8–16 and III/12b, p. 7–19. Magnetic structure: Fig. 12. Some additional data are given on Mössbauer parameters, Table 9 and Fig. 20a. The influence of radiation damage on the quadrupole interaction is shown in Fig. 20b. From a comparison of both figures might be concluded that the Morin temperature is reduced to about 190 K, and a new reverse Morin kind transition occurs at 80 K. The influence of applied external pressure on Mössbauer parameters is shown in the Figs. 20c–e. For the magnetic phases with and without an externally applied magnetic field see Figs. 21, 22, for muon behavior Fig. 23 and for the heat capacity Figs. 24, 25.

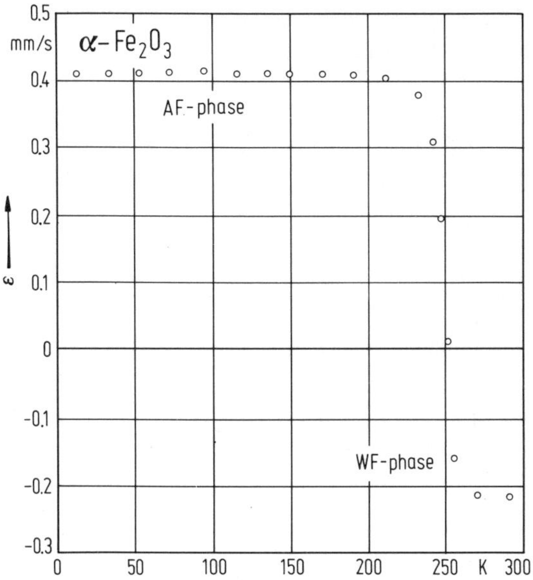

a

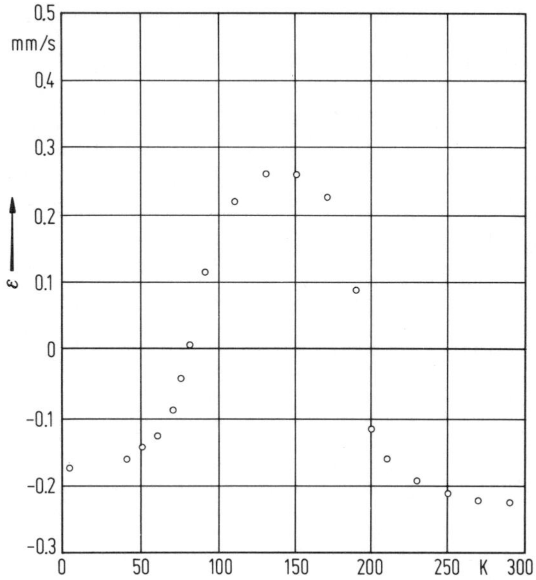

b

Fig. 20a, b. α-Fe_2O_3. Quadrupole shift ε vs. temperature T for (a) non-irradiated sample and (b) sample irradiated by the continuous spectrum of reactor neutrons giving a displacement of 0.0020 A for the atoms [92B1].

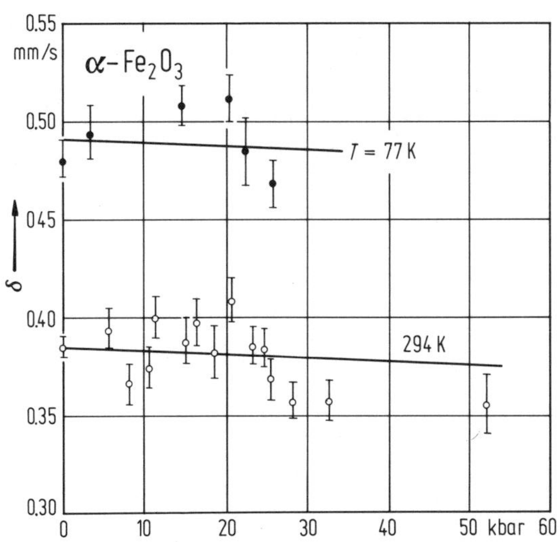

Fig. 20c. α-Fe_2O_3. Isomer shift δ relative to α-iron vs. externally applied pressure p at two temperatures. The straight lines indicate Debye-model results with $\Theta_D = 600$ K [83B1, see also 67V1].

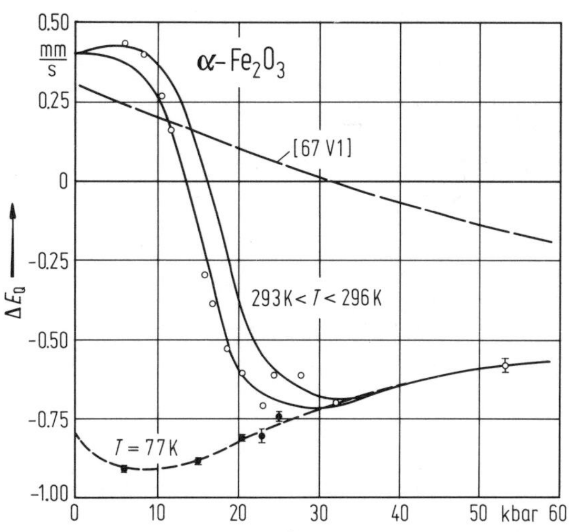

Fig. 20d. α-Fe_2O_3. Quadrupole interaction ΔE_Q vs. externally applied pressure p at two temperatures. The solid lines are theoretical fits taking various relaxation phenomena into account [83B1]. Dashed curve without measuring points is from [67V1].

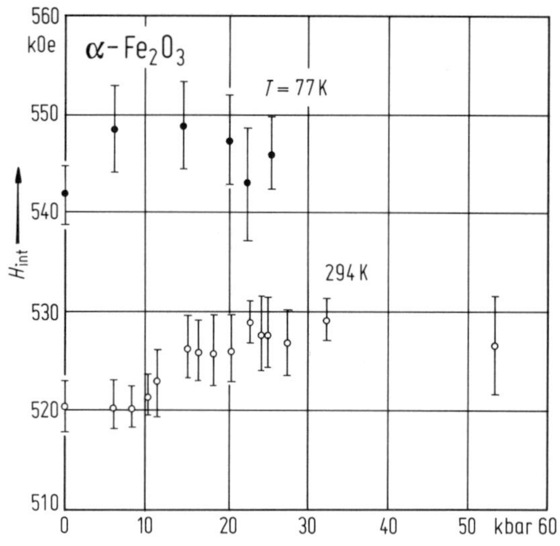

Fig. 20e. α-Fe_2O_3 powder, particle size about 90 nm. Internal magnetic field H_{int} as derived from Mössbauer spectra at various magnitudes of an externally applied pressure p[83B1].

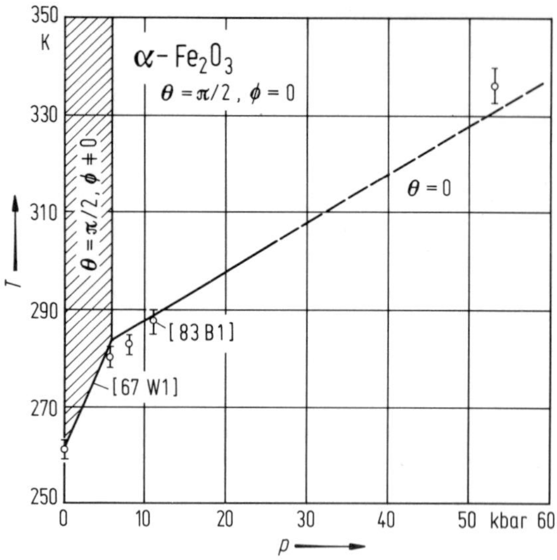

Fig. 21a. α-Fe_2O_3. Magnetic phase diagram based on Mössbauer spectra. θ and ϕ are the axial and azimuthal angles between c axis and spin direction. In the hatched region, the magnetization is non-zero, with a second order transition to the $\theta = \pi/2$, $\phi = 0$ region, and a first order transition between the $\theta = \pi/2$ and $\theta = 0$ regions. The Morin transition line is the boundary between the $\theta = \pi/2$ and $\theta = 0$ regions [83B1].

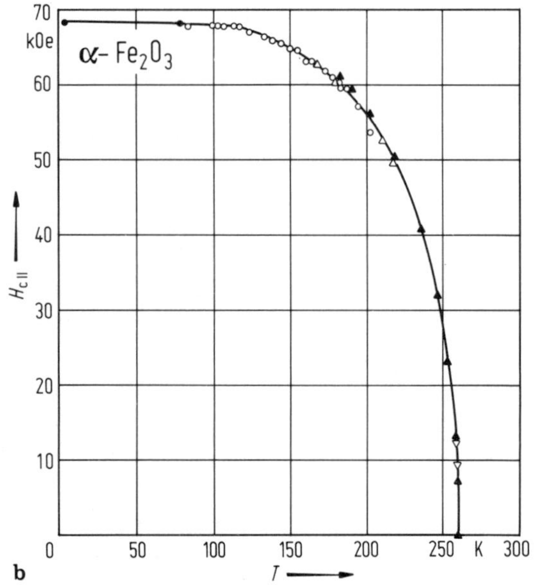

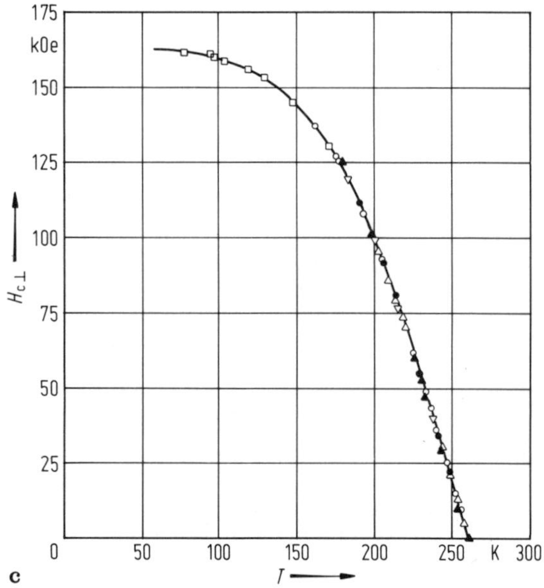

Fig. 21b, c. α-Fe_2O_3, natural single crystal. (b) Magnetic phase diagram for applied magnetic fields $H_{c||}$ parallel to the trigonal axis as derived from anomalies in the attenuation of different acoustical modes indicated by the various symbols. The curve gives the magnitude of the magnetic field for the first order spin-flop transition of the spin from a direction parallel to the trigonal axis into a direction parallel to the trigonal plane. (c) Similar to (b), but now for magnetic fields perpendicular to the trigonal axis [69S1].

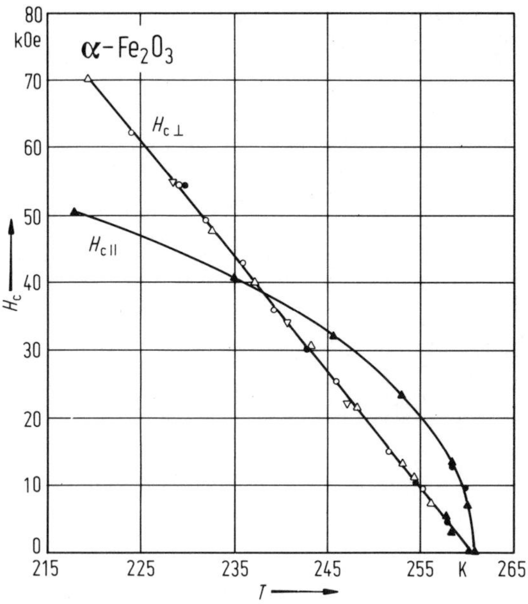

Fig. 22. α-Fe_2O_3. Enlarged parts of the Figs. 21 b, c near the Morin temperature of the crystal. $T_M = 260.6(5)$ K [69S1].

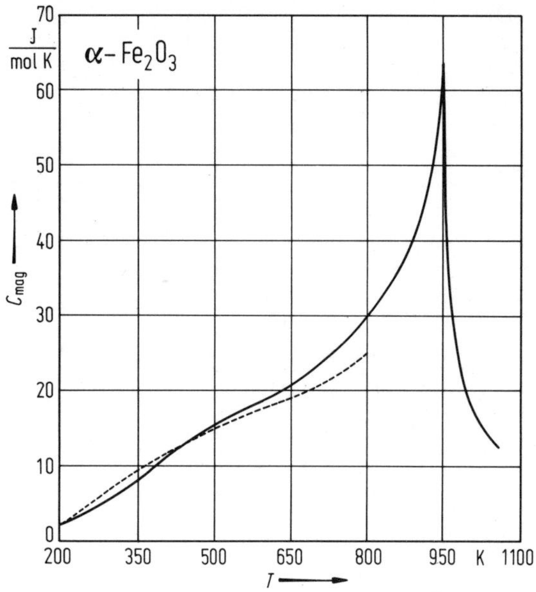

Fig. 25. α-Fe_2O_3. Molar magnetic heat capacity C_{mag} as derived from an experimental evaluation of the curves of Fig. 24. The dashed curve represents a calculated spin wave heat capacity [75G1].

For Fig. 24, see next page.

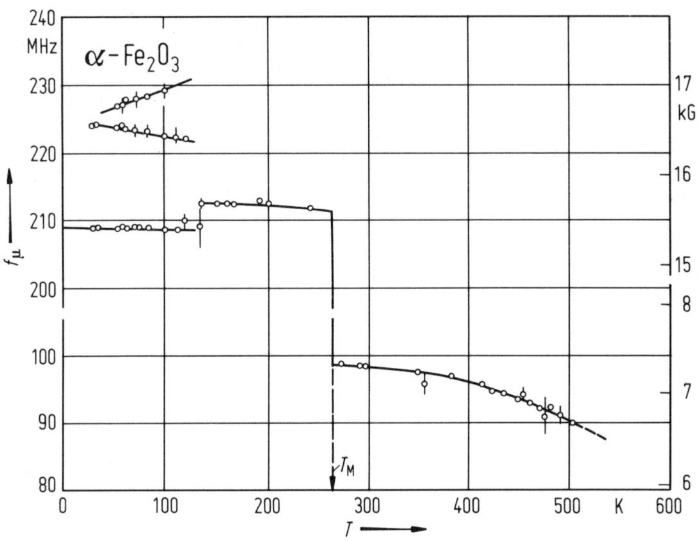

Fig. 23. α-Fe_2O_3. The muon spin rotation frequency f_μ vs. temperature T in a single crystal in zero applied magnetic field. Morin temperature T_M. Below $T = 120$ K three distinct frequencies are observed. Local muon field B_μ, which for $T < T_M$ is parallel to the c axis [80R1].

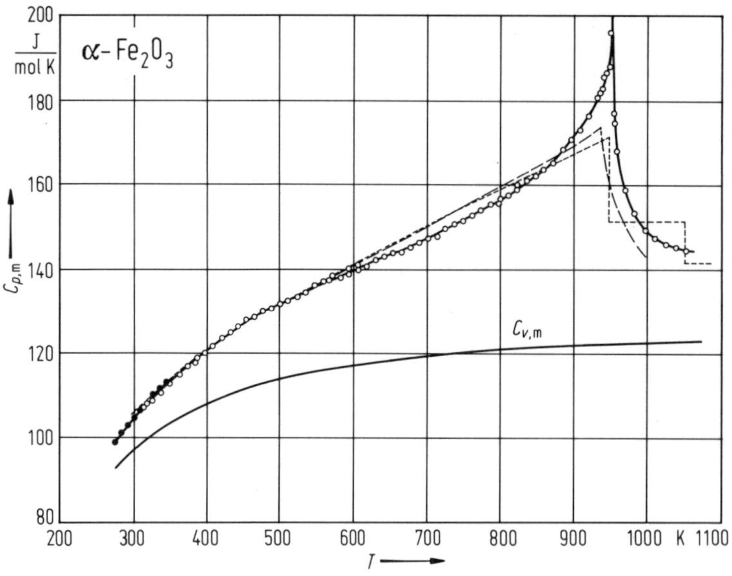

Fig. 24. α-Fe_2O_3. Heat capacity at constant pressure $C_{p,m}$ vs. temperature T. ○ [75G1]; ● low-temperature results [59G1]; short-dashed line [51C1]; long-dashed line [72R1]. The curve $C_{v,m}$ is the estimated lattice heat capacity at constant volume [75G1].

5.6.3.2 Pseudo-binary and ternary oxides $(M_{1-x}M'_x)_2O_3$; M, M′ = 3d transition elements

Ti_2O_3–V_2O_3 system

The compounds $(Ti_{1-x}V_x)_2O_3$ are paramagnetic for $0 < x < 0.075$ with a Curie-Weiss behavior and with effective magnetic moments per V ion decreasing from 3.8 μ_B to 1.8 μ_B when x is increased [76D1], see Figs. 26a, b, c and 27a, b. At low temperatures the position disorder of the V ions in the Ti_2O_3 matrix leads for $x > 0.005$ to spin glass properties [75D1], see Figs. 28, 29. The low-field ac susceptibility in the spin-glass freezing region depends on the measuring frequency (Fig. 30), the amplitude of the measuring field (Fig. 31) and the presence of a static magnetic field, Fig. 32. For an analysis of the spin-glass behavior see Fig. 33 and [80C1, 80M1, 80M2, 81C1, 81M1, 83S1, 85S1]. For the freezing temperature of the spin-glass system see Fig. 33b; for specific heat Fig. 34b; for phase diagram Fig. 41.

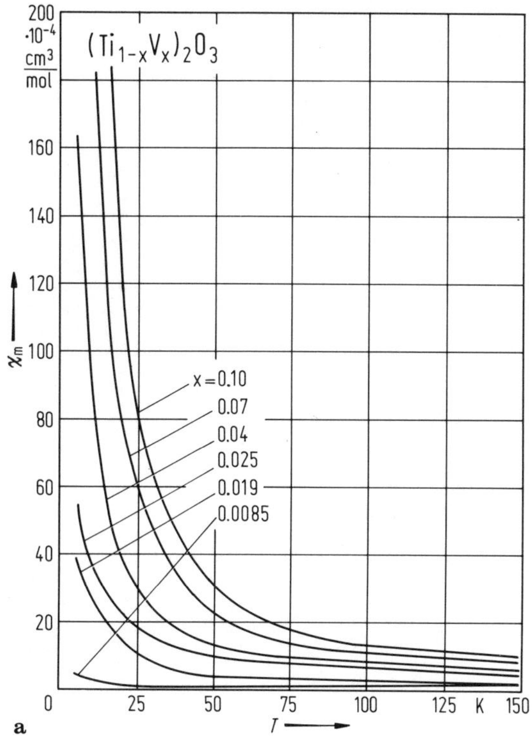

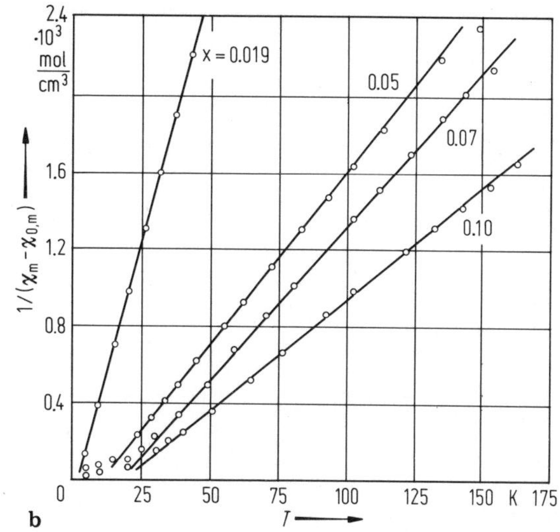

Fig. 26. $(Ti_{1-x}V_x)_2O_3$, powder. (a) Temperature dependence of the molar magnetic susceptibility χ_m measured in a field of 9.6 kOe. (b) If the Curie-Weiss behavior is fitted by $\chi_m = \chi_{o,m} + C_m/(T - \Theta_p)$, where $\chi_{o,m}$ is a temperature-independent term, then the straight lines for $(\chi_m - \chi_{o,m})^{-1}$ vs. T are obtained [76D1].

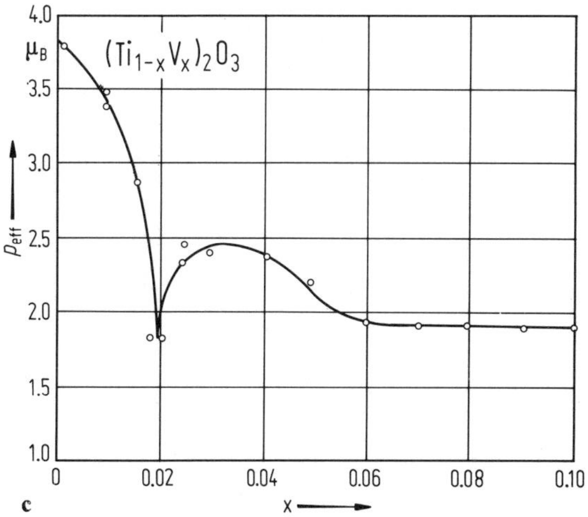

Fig. 26c. $(Ti_{1-x}V_x)_2O_3$, powder. Effective magnetic moments p_{eff} per V^{3+} ion as a function of the composition parameter x derived from the slopes of the straight lines in Fig. 26b [76D1].

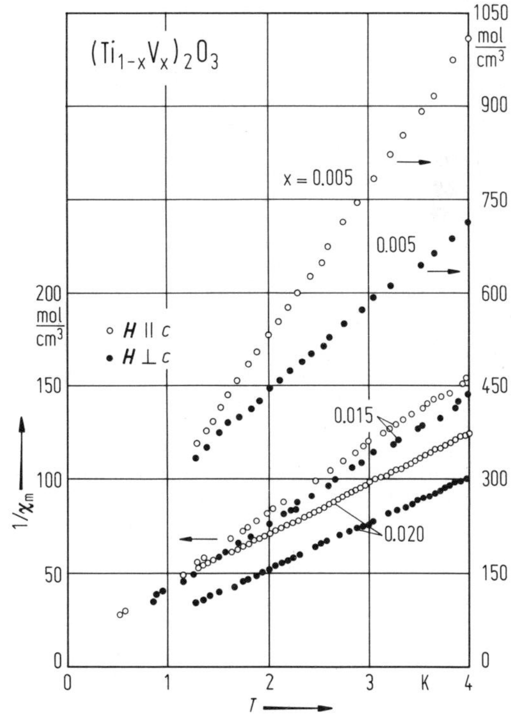

Fig. 27a. $(Ti_{1-x}V_x)_2O_3$, x = 5, 15, $20 \cdot 10^{-3}$, single crystals. The inverse molar magnetic susceptibility χ_m^{-1} vs. temperature T for magnetic field parallel (open circles) and perpendicular (full circles) to the crystallographic c axis [79C1].

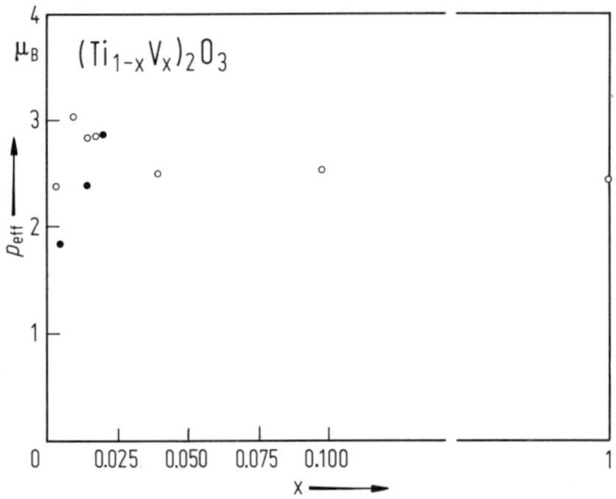

Fig. 27b. $(Ti_{1-x}V_x)_2O_3$. Composition dependence of the effective magnetic moment p_{eff} per V^{3+} ion, derived from the straight lines of Fig. 27a. Full circles and open circles denote values derived from measurements parallel and perpendicular to the c axis, respectively [79C1].

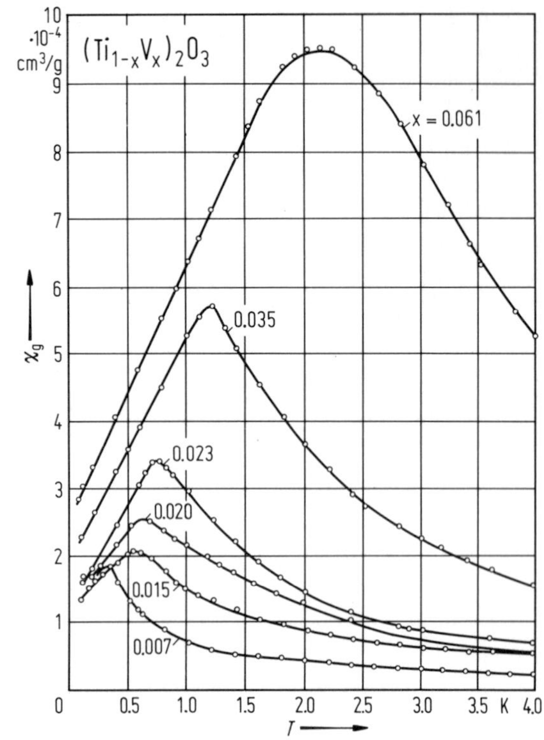

Fig. 28. $(Ti_{1-x}V_x)_2O_3$, single crystal. The initial reversible magnetic mass susceptibility χ_g measured in low dc fields (< 20 Oe and $\perp c$ axis) in the temperature range 0.05 ··· 4 K [76D1].

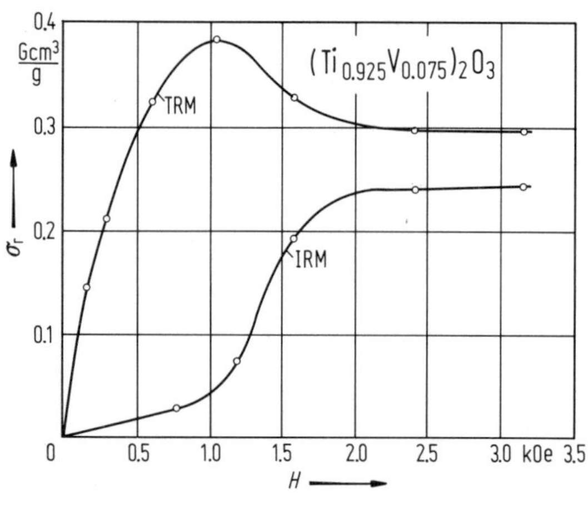

Fig. 29. $(Ti_{1-x}V_x)_2O_3$, x = 0.075, single crystal. Thermoremanent (TRM) and isothermal remanent magnetization (IRM) vs. magnetic field at $T = 0.11$ K [76D1].

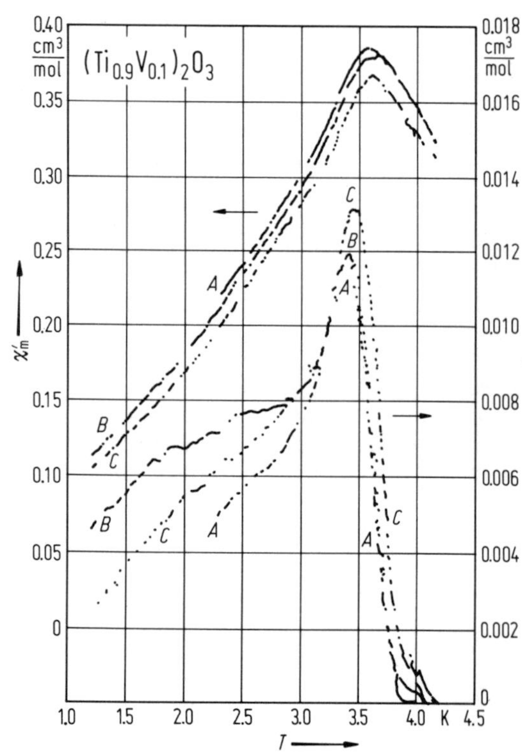

Fig. 30. $(Ti_{1-x}V_x)_2O_3$, x = 0.10, powder. Real and imaginary parts of the magnetic molar ac susceptibility, χ'_m and χ''_m, respectively, as a function of temperature T. The curves A, B and C were measured with a low magnetic field of amplitude 6 Oe and a frequency of 80, 240 and 480 Hz, respectively [80M1].

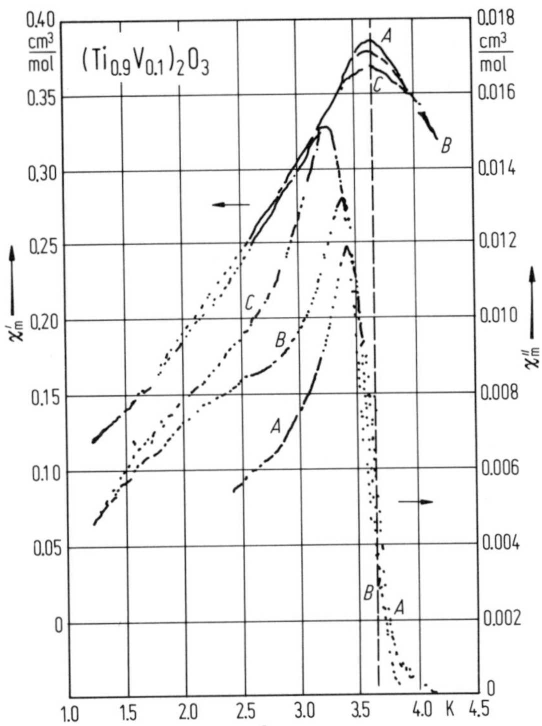

Fig. 31. $(Ti_{1-x}V_x)_2O_3$, x = 0.10, powder. Similar to Fig. 30, but now measured in ac fields of 80 Hz and amplitudes 6, 12 and 24 Oe corresponding to the curves *A*, *B* and *C*, respectively [80M2].

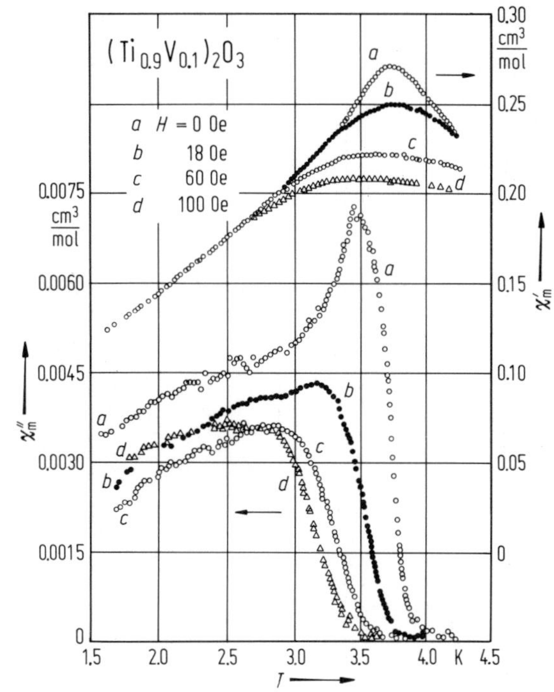

Fig. 32. $(Ti_{1-x}V_x)_2O_3$, x = 0.10, powder. Similar to Fig. 30, but now measured in ac fields of 80 Hz and an amplitude of 6 Oe in the presence of static magnetic fields of H = 0, 18, 60 and 100 Oe [83S1].

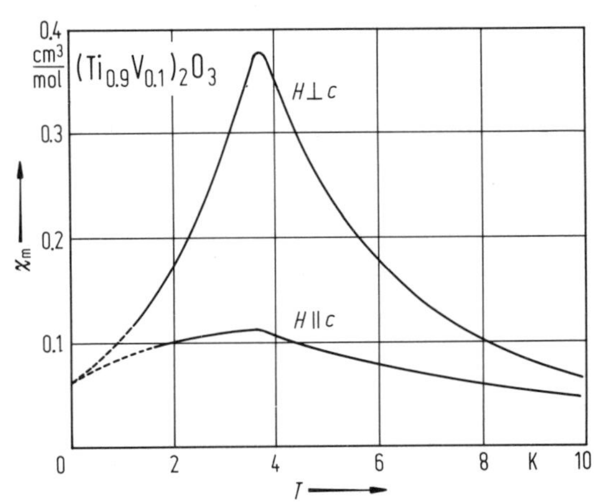

Fig. 33a. $(Ti_{1-x}V_x)_2O_3$, x = 0.10, single crystal. Anisotropic behavior of the molar magnetic susceptibility χ_m, measured parallel or perpendicular to the crystallographic *c* axis in a low frequency magnetic field of 80 Hz [81C1].

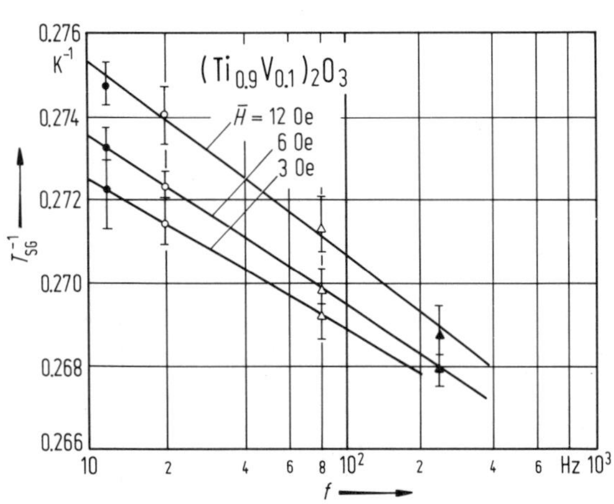

Fig. 33b. $(Ti_{1-x}V_x)_2O_3$, x = 0.10, powder. Frequency dependence of the inverse of the spin glass freezing temperature T_{SG} for various amplitudes of the measuring field $\bar{H}$ [81C1].

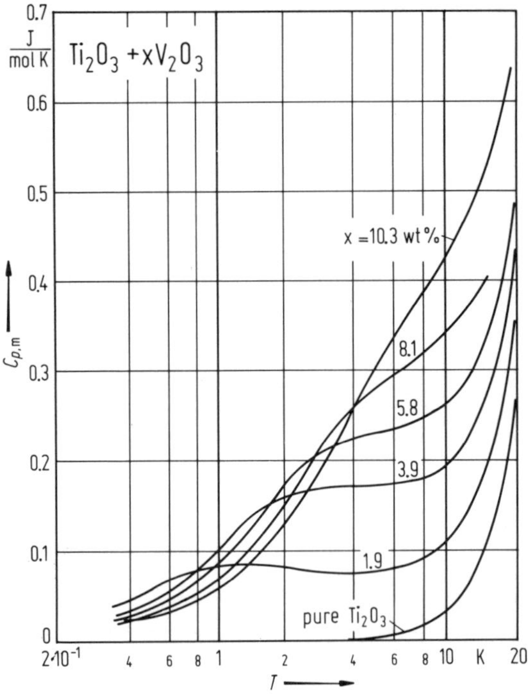

Fig. 34. Ti_2O_3 + xwt% V_2O_3. Low temperature molar specific heat $C_{p,m}$ vs. temperature T for metallic samples [73S2].

Ti_2O_3–Cr_2O_3 system, inclusive $CrTiO_3$

Between Cr_2O_3 and $CrTiO_3$ the sesquioxides form a homogeneous corundum-type solid solution. $CrTiO_3$ has the corundum structure below and above T_N with $a = 5.050(5)$ Å and $c = 13.630(5)$ Å at RT [72R2]. In $(Cr_xTi_{1-x})_2O_3$ the Cr_2O_3-type AF structure is found with the Ti^{3+} ions acting as a simple diluent. ESR gives a signal with line width ≈ 1600 G and $g = 2.00$ for both x = 0.9 and 1.0. For the magnetic susceptibility of $CrTiO_3$ see Fig. 35, $T_N = 135(3)$ K [72R2].

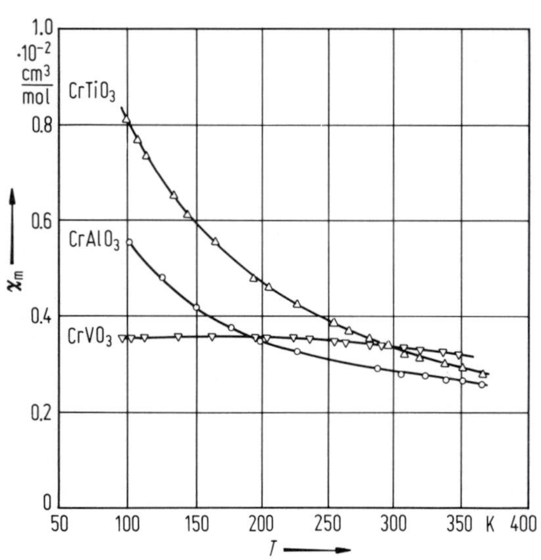

Fig. 35. $CrTiO_3$, $CrVO_3$, $CrAlO_3$. Molar magnetic susceptibilities χ_m vs. temperature T [72R2].

$MnTiO_3$ II

This is a high-pressure phase of $MnTiO_3$ with the corundum structure, see Fig. 36. Samples quenched under pressure to RT were analyzed [69S3]. For cell parameters, see Table 1. For the magnetic susceptibility see Figs. 37, 38. The g-factor determined from ESR data is 2.0 at RT and down to $T_N = 24(1)$ K, defined as the temperature obtained by extrapolating absorption and inverse line width of ESR to zero value. Mössbauer spectra of ^{57}Fe-doped samples at RT are similar to those of $FeTiO_3$. At RT is IS = 1.22 mm s^{-1} and $\Delta E_Q = 1.32$ mms^{-1}. At 4.2 K a magnetically ordered structure is found with an internal field $H_{hyp} \approx 150$ kOe.

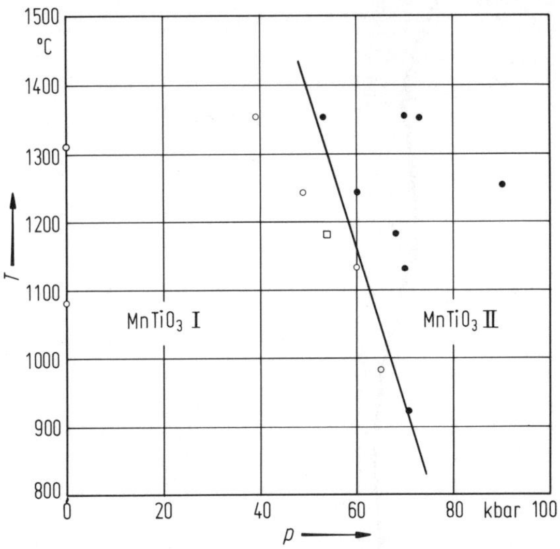

Fig. 36. $MnTiO_3$ I, II. Phase diagram at high temperatures and high pressures. Open circles $MnTiO_3$I, closed circles $MnTiO_3$II. The rectangle indicates a reverse reaction from II → I [69S3].

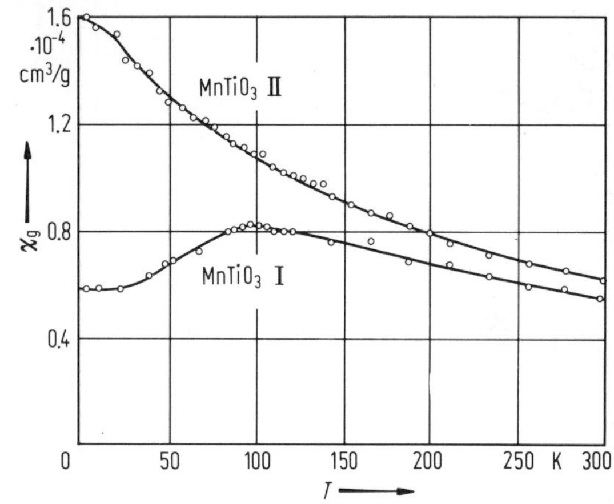

Fig. 37. $MnTiO_3$ I, II, powder. Magnetic mass susceptibility χ_g vs. temperature T [69S3].

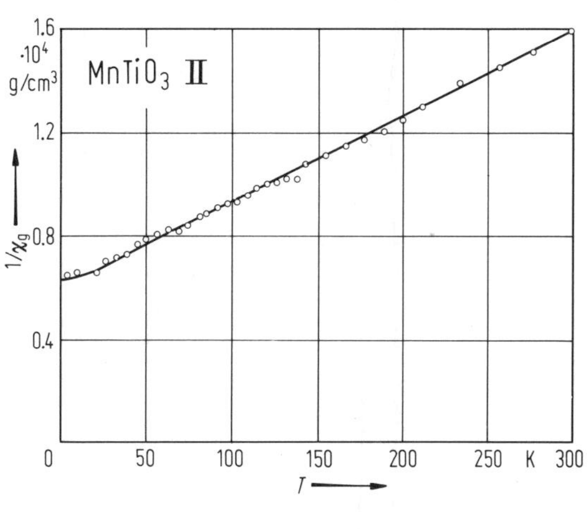

Fig. 38. $MnTiO_3$ II, powder. Inverse magnetic mass susceptibility χ_g^{-1} vs. temperature T. Effective magnetic moment per FU $p_{eff} = 5.95\,\mu_B$ and paramagnetic Curie temperature $\Theta_p = -175$ K [69S3].

V_2O_3–Cr_2O_3 system, inclusive $CrVO_3$

This system has the corundum structure at high temperature and it forms a continuous series of solid solutions [72R2]. Lattice parameters: Fig. 39. Phase diagram: Figs. 40–42. Electric properties: Table 7. Magnetic susceptibility data: Figs. 43–45 and Table 8.

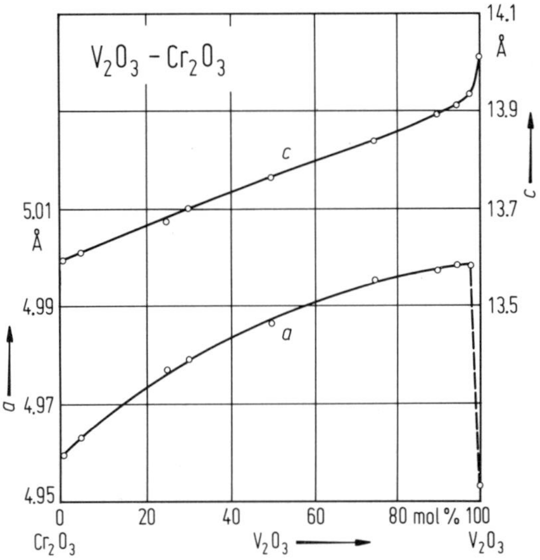

Fig. 39. V_2O_3–Cr_2O_3. Lattice parameters vs. composition at RT [72R2].

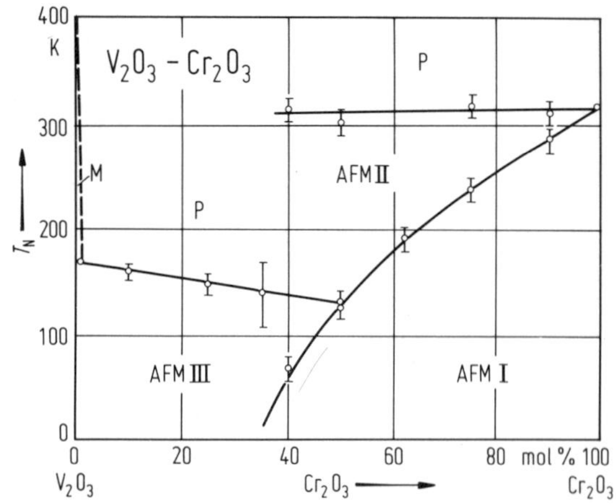

Fig. 40. V_2O_3–Cr_2O_3. Phase diagram defined by magnetic and crystallographic boundaries. Full lines by neutron diffraction, dashed line shows the metal-insulator boundary, which is also defined by crystallographic discontinuity. P: paramagnetic, AFMI: Cr_2O_3-type magnetic order, AFMII: new type of magnetic ordering, AFMIII: monoclinic V_2O_3, M: metallic conduction, corundum lattice [72R1]. For a similar phase diagram based on magnetic susceptibility data see [70M2].

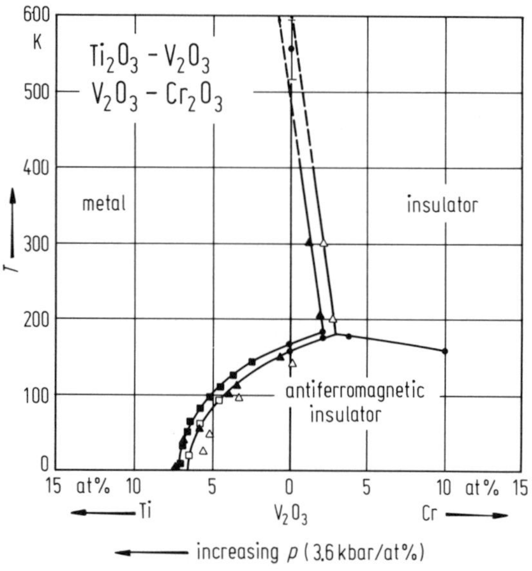

Fig. 41. Ti_2O_3–V_2O_3; V_2O_3–Cr_2O_3. Generalized phase diagram showing electric and magnetic transition temperatures vs. both pressure p and Cr or Ti content in V_2O_3. ● mixed oxides at 1 atm. ■, □ V_2O_3, increasing and decreasing pressure, respectively. ▲, △ $(V_{0.96}Cr_{0.04})_2O_3$, id. Below 10 kbar experimental uncertainties of ± 2 kbar in the pressure calibration [69M2, see also 71G1].

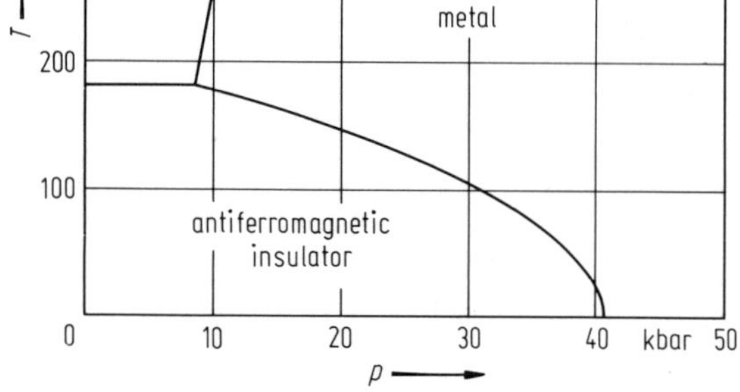

Fig. 42. $(V_{1-x}Cr_x)_2O_3$, $x = 0.0375$. $p - T$ phase diagram. I and M phases have corundum structure, the AF phase is monoclinic. For the critical point where the metal-insulator transition (M–I) terminates, see Table 7 [70J1].

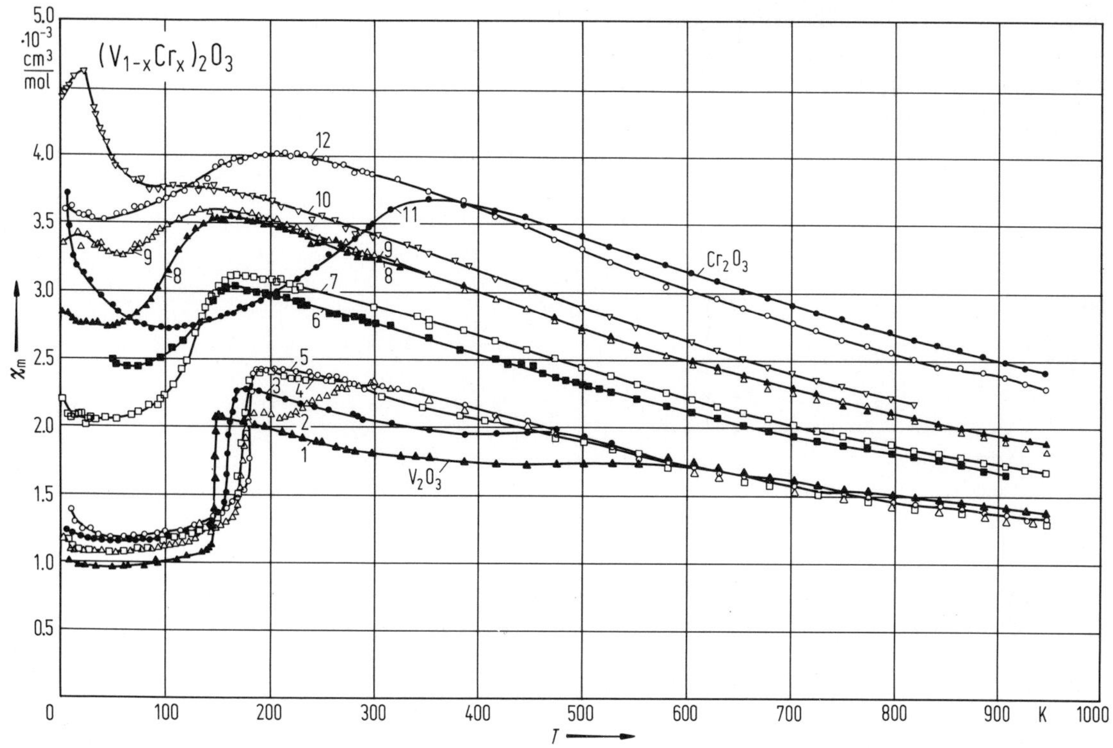

Fig. 43. $(V_{1-x}Cr_x)_2O_3$. Molar magnetic susceptibility χ_m vs. temperature T. For definition of the samples see Table 8 [70M2].

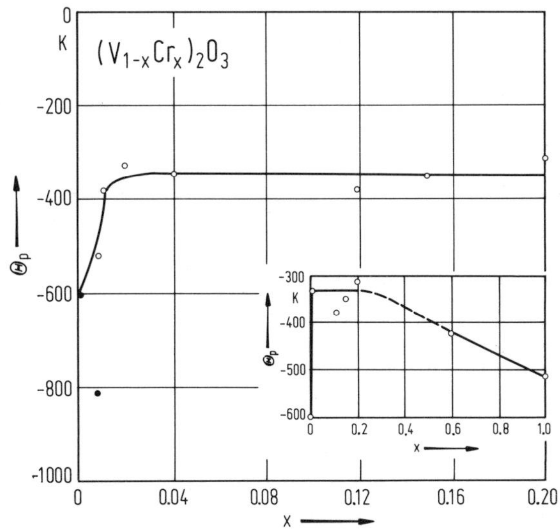

Fig. 44. $(V_{1-x}Cr_x)_2O_3$. Extrapolated paramagnetic Curie-temperatures Θ_p as a function of Cr content. Open circles correspond to the insulating high-temperature phase and closed circles to the metallic phase [70M2].

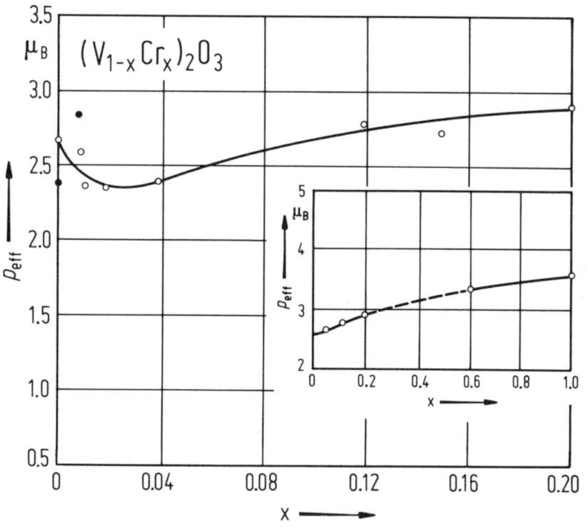

Fig. 45. $(V_{1-x}Cr_x)_2O_3$. Effective moment per transition metal ion [70M2]. Closed, open circles: see caption Fig. 44.

Table 7. $(V_{1-x}Cr_x)_2O_3$. The critical point in the $p-T$ phase diagram where the metal-insulator transition (M – I) terminates, depends on the composition. See also Fig. 42 [70J1].

x	p_c [kbar]	T_c [K]
0.0135	3.5	443
0.0187	5.5	433
0.0375	12.5	390
V_2O_3 [a])	−1.5	474

[a]) Extrapolated values.

Table 8. $(V_{1-x}Cr_x)_2O_3$. Definition of the samples corresponding to the susceptibility curves of Fig. 43 [70M2].

Material x	Susceptibility curve	Nature of sample
0	1	Single crystal
0.008	2	Single crystal
0.010	3	Single crystal
0.019	4	Single crystal
0.04	5	Single crystal
0.12	6	Ceramic
0.15	7	Ceramic
0.20	8	Ceramic
0.25	9	Ceramic
0.30	10	Ceramic
1	11	Powder
0.6	12	Ceramic

V_2O_3–Fe_2O_3 system

Almost complete solid solution exists at 1000°C. The magnetic behavior up to a V_2O_3 content of 40 mol% resembles that of α-Fe_2O_3, a weak spontaneous moment appearing at T_N. Between 40 and 80 mol% of V_2O_3, there is weak ferrimagnetism reaching a maximum of about 0.2 μ_B per FU at 65 mol% V_2O_3. Between 85 and 100% V_2O_3, the properties resemble those of V_2O_3 [62C1]. Lattice parameters: Fig. 46. Magnetization and magnetic susceptibility: Figs. 47–49. According to neutron diffraction data the hematite magnetic structure exists up to 60 mol% V_2O_3. Above this content the order is more short-range in nature [62C1].

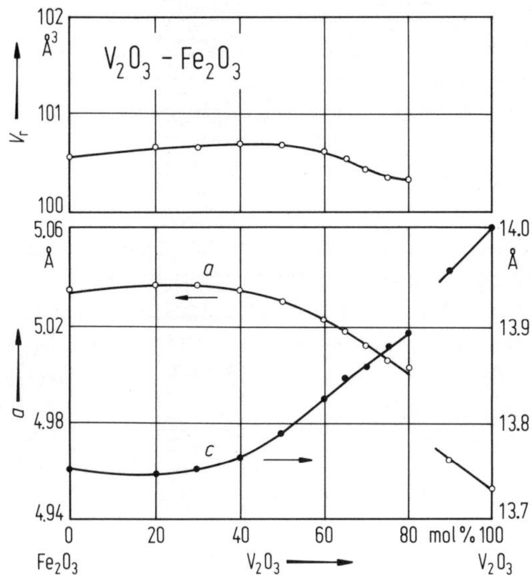

Fig. 46. V_2O_3–Fe_2O_3. Rhombohedral cell volume V_r and hexagonal lattice parameters a, c vs. composition [62C1, see also 60B1].

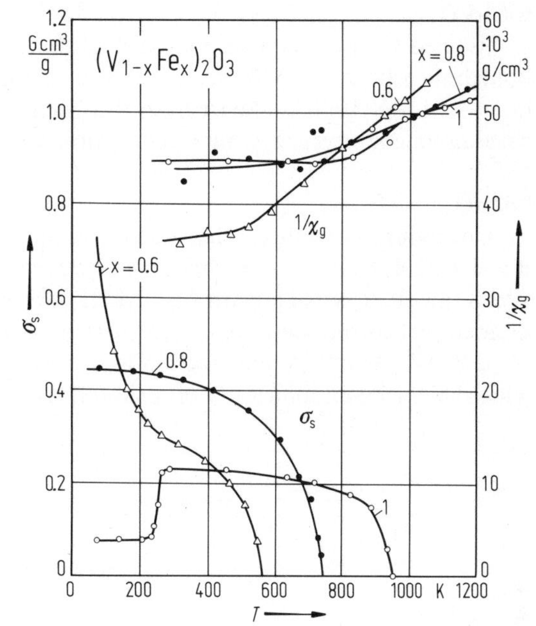

Fig. 47. $(V_{1-x}Fe_x)_2O_3$, $x \geq 0.6$. Temperature dependence of the spontaneous (parasitic) magnetization σ_s below the Curie temperature according to the relation $\sigma = \sigma_s + \chi_g H$, where χ_g is the magnetic mass susceptibility below and above the Curie point [62C1]. χ_g^{-1} vs. T is also shown.

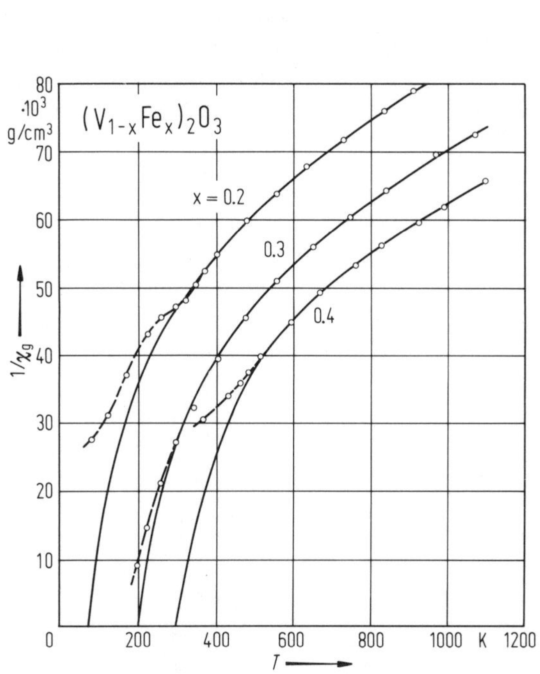

Fig. 48. $(V_{1-x}Fe_x)_2O_3$, $x \leq 0.4$. Reciprocal magnetic mass susceptibility χ_g^{-1} vs. temperature T [62C1].

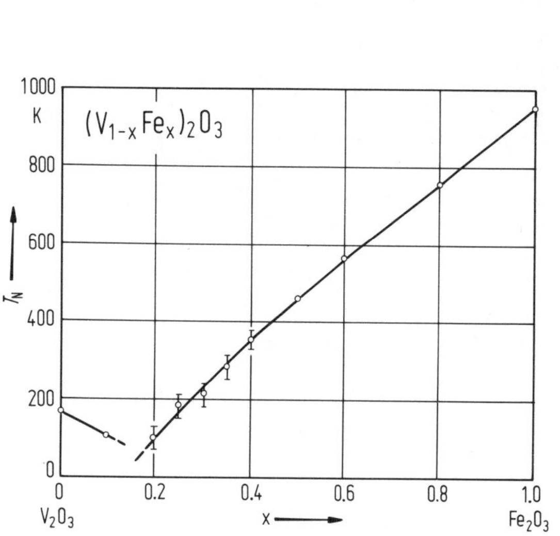

Fig. 49. $(V_{1-x}Fe_x)_2O_3$. Variation of magnetic ordering temperature T_N with composition [62C1].

α-$CuVO_3$

Refined Guinier data on the basis of an hexagonal cell $a = 12.859(2)$ Å, $c = 7.186(2)$ Å. Rhombohedral cell $a_r = 7.800(6)$ Å and $\alpha_r = 111.02(1)°$. Paramagnetic with $p_{eff} = 1.31\ \mu_B$/FU and $\Theta_p = -154$ K [70C1]. Apart from this phase, there have been prepared two high-pressure phases with an ilmenite type structure, see section 5.1.4.1.

$NiCrO_3$

Corundum structure. Hexagonal lattice parameters obtained from Guinier powder data: $a = 4.9252(4)$ Å and $c = 13.504(1)$ Å. Magnetic susceptibility: Fig. 50. A high-temperature Curie-Weiss law fit is possible with $C_m = 4.2(3)\ cm^3\ K\ mol^{-1}$ and $\Theta_p = -870(80)$ K. Effective magnetic moment per formula unit $p_{eff} = 5.8\ \mu_B$. It is suggested that an antiferromagnetic ordering occurs at $T_N \approx 250$ K. A small magnetization is found of 0.17 and 0.015 $G\ cm^3/g$ at 4.2 and 77 K, respectively, which might be an indication of a canted AF spin structure [69C1].

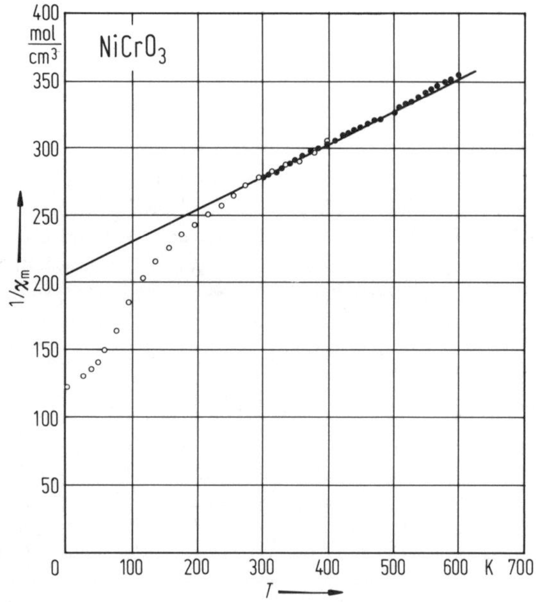

Fig. 50. $NiCrO_3$, powder. Reciprocal molar magnetic susceptibility χ_m^{-1} vs. temperature T [69C1]. Vibrating sample magnetometer and Faraday balance data are represented by open and closed circles, respectively.

Cr_2O_3–Fe_2O_3

Both Cr_2O_3 and α-Fe_2O_3 are antiferromagnets with identical crystalline structure. However, their magnetic structures are different. In Cr_2O_3, the spin couplings are antiferromagnetic both within a hexagonal *c* layer and between adjacent *c* layers. In Fe_2O_3, on the other hand, the spins are coupled ferromagnetically within a *c* layer and antiferromagnetically between adjacent *c* layers. The difference of the antiferromagnetic structures between Cr_2O_3 and Fe_2O_3 produces a cone spiral structure in $(1-x)Cr_2O_3 \cdot x\,Fe_2O_3$. In order to describe the cone spiral structure conventionally, the magnetic structure is divided into two sublattices as shown in Fig. 51. In samples with $x \le 0.15$, a cone spiral arrangement is established within each sublattice, where both the cone axis and the spiral wave vector align in the *c* axis. The spin components along the cone axis have a Cr_2O_3 type of magnetic structure. The cone angle increases with the increase of x up to $x = 0.15$. In samples with $0.2 \le x \le 0.35$, the cone axis is perpendicular to the *c* axis. However, the spiral wave vector remains parallel to the *c* axis, which makes a cycloidal arrangement. The spin components along the cone axis have an Fe_2O_3-type of magnetic structure. As the x increases above 0.2, the cone angle decreases. Then it becomes zero and an Fe_2O_3-type of magnetic structure is attained.

Spontaneous magnetic moment and susceptibility: Figs. 52, 53. Magnetic transition temperatures: Fig. 54. For neutron diffraction and Mössbauer data see Tables 9, 10.

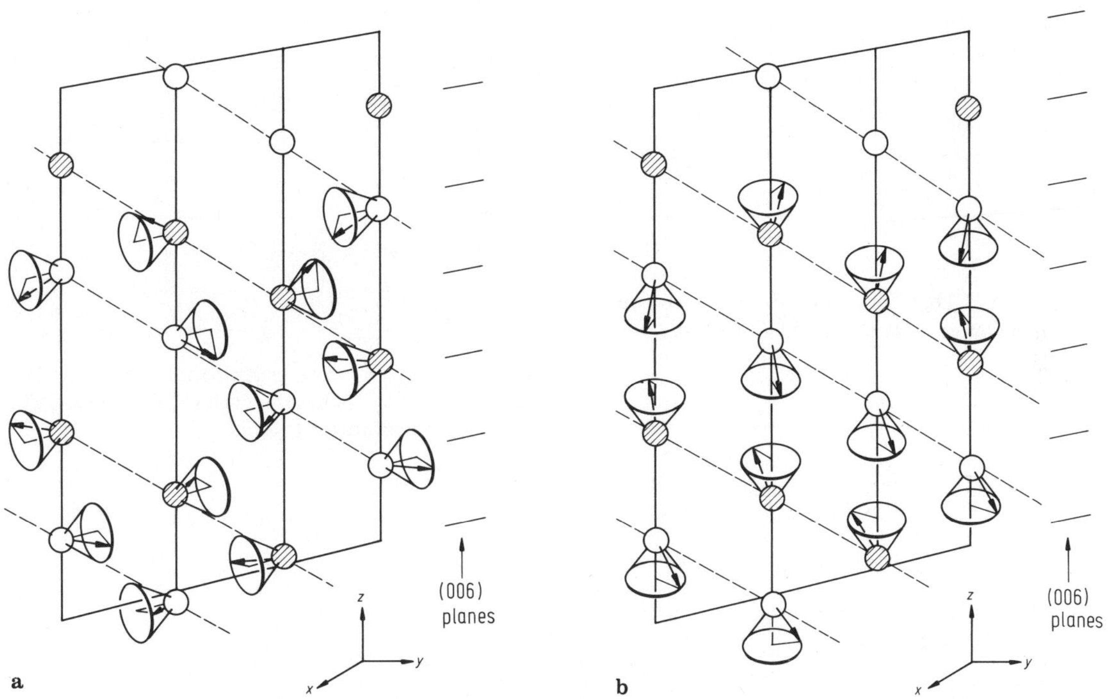

Fig. 51. Cr_2O_3–Fe_2O_3. A (110) section of the corundum lattice showing cone spirals based upon the Fe_2O_3 (a) and Cr_2O_3 (b) magnetic structures. The heavy arrows indicate the spin directions [63C1]. Shaded and open circles indicate different sublattices.

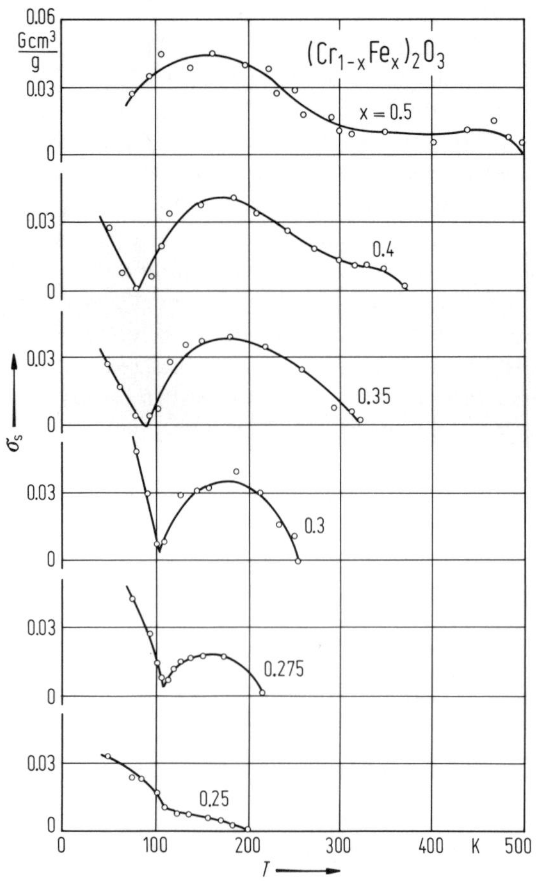

Fig. 52. $(Cr_{1-x}Fe_x)_2O_3$. Parasitic spontaneous magnetic moment σ_s vs. temperature T [63C1].

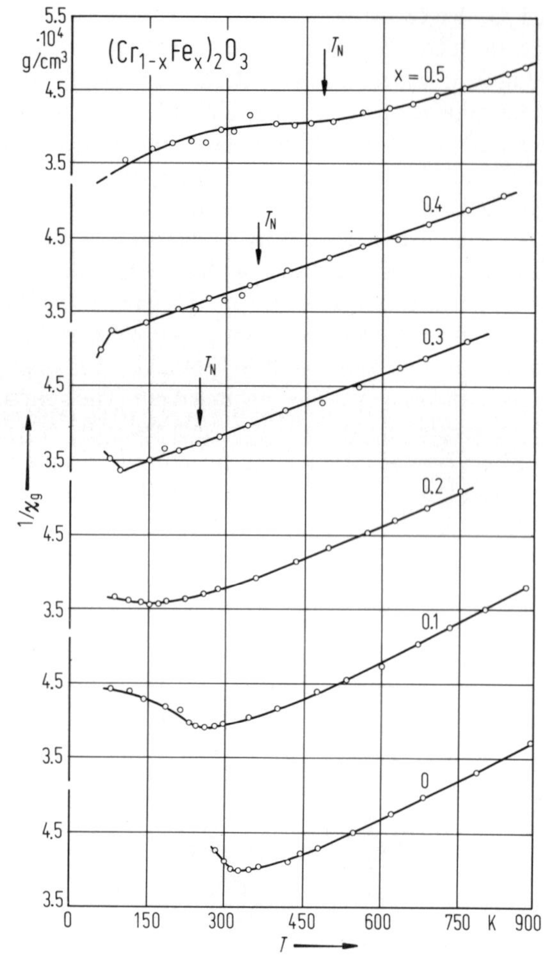

Fig. 53. $(Cr_{1-x}Fe_x)_2O_3$. Reciprocal magnetic mass susceptibility χ_g^{-1} vs. temperature T. T_N magnetic ordering temperature [63C1].

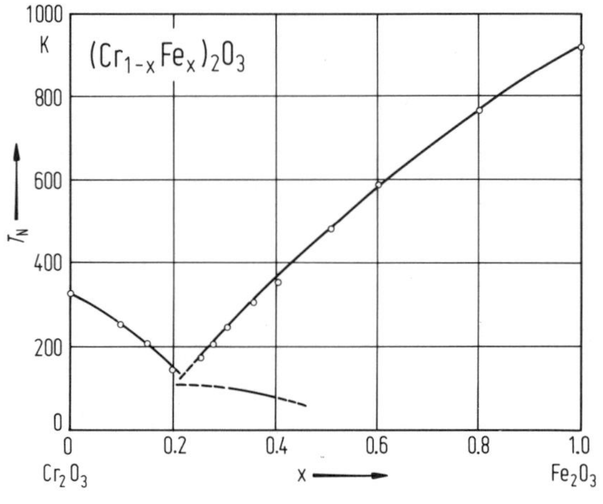

Fig. 54. $(Cr_{1-x}Fe_x)_2O_3$. Magnetic transition temperatures T_N as a function of composition. The lowest line corresponds to a low temperature discontinuity in the magnetic moment [63C1], similar results obtained from Mössbauer spectra [75B1], and from ESR measurements [63K1].

Table 9. Cr_2O_3–Fe_2O_3 system. Mössbauer parameters [75B1].

$\%Fe_2O_3$	T [K]	ΔE_Q a) [mm/s]	ε b) [mm/s]	H_{hyp} c) [kOe]	$\%Fe_2O_3$	T [K]	ΔE_Q a) [mm/s]	ε b) [mm/s]	H_{hyp} c) [kOe]
100 d)	0			544	30	255		–	T_N
	100 ··· 220		+ 0.201			295	0.38		0
	260		0.000	T_M	25	4.2		0.0	532
	295		– 0.110 e)			77		0.0	456
	310 ··· 900		– 0.103			85		–	417
	956			T_N		111		–	340
	~ 1000	0.465				122		–	250
	1050	0.49 f)				149			T_N
75	4.2		– 0.12	525		160	0.36		0
	77		– 0.11	528	20	4.2		0.0	498
	200		– 0.11	515		77		0.0	422
	295		– 0.11	494		90		–	392
	310		– 0.11	490		117		–	322
50 g)	77		– 0.10	512		162		–	142
	295		– 0.11	429		175			T_N
	325		– 0.10	421		232	0.30		0
	347		(– 0.08)	405	10	77		+ 0.08	410
	376		(– 0.06)	384		235			T_N
40 g)	4.2		– 0.09	510		295	0.30		0
	77		– 0.10	495	5	77		– i)	415
	115		– 0.10	485		265			T_N
	195		– 0.10	455		295	0.27		0
	222		(– 0.08)	439	0.4	77		+ 0.11	438
	273		(– 0.05)	393		195		+ 0.12	283
	295		(– 0.08)	371		295		+ 0.12	89
	310		– h)	355		305		–	T_N
	359		–	266		310	0.27	–	0
	371		–	222	0.2	4.2		+ 0.10	521
	390		–	T_N		77		+ 0.12	440
30	4.2		– 0.07	493		195		+ 0.12	284
	77		– 0.07	466		273		+ 0.13	153
	181		–	380		295		–	93
	200		–	336		305		–	T_N
	220		–	296		330	0.26	–	0

a) Quadrupole splitting above T_N.
b) Quadrupole interaction below T_N. Estimated standard deviations $\pm$ 0.01 mm/s except for values in parentheses.
c) Hyperfine field at the Fe nucleus.
d) Data for α-Fe_2O_3 from van der Woude [66W1] except where indicated. T_M is the Morin transition temperature, T_N the Néel temperature.
e) Data of Artman et al. [68A1] for high purity synthetic α-Fe_2O_3.
f) Data of Ono and Ito [62O1] for α-Fe_2O_3 enriched to 60% in ^{57}Fe.
g) For these compositions, values of ε in parentheses indicate uncertain values obtained from broadened spectra.
h) Poor counting statistics from 5% of natural Fe prevented the measurement of ε.
i) Dashed entries indicate spectra too broadened to allow adequate measurements.

Table 10. Cr_2O_3–Fe_2O_3 system. Quadrupole shifts ε and spin configurations. The latter defined by the cone angle θ for Fe^{3+} spin vectors in the spin-spiral arrangements for intermediate compositions, see footnote [b]) [75B1].

%Fe_2O_3	ε[a]) [mms^{-1}]	θ[b]) [deg] Mössbauer	θ[c]) [deg] Neutrons
0.2	0.113(10)	0, ∥	0, ∥
10	0.08(1)	26(4), ∥	40, ∥
15	–	–	66, ∥
20	0.00(1)	55(4), ∥	71, ⊥
25	0.00(1)	55(4), ⊥	51, ⊥
30	– 0.07(1)	40(6), ⊥	35, ⊥
35	–	–	19, ⊥
45[d])	– 0.100(5)	14(4), ⊥	–
75	– 0.110(5)	0(10), ⊥	–
100[e])	– 0.110(3)	0, ⊥	0, ⊥

[a]) From Table 9.
[b]) θ is the cone half-angle. For cone-axis parallel to [111], e^2qQ is assumed to be that for the 0.2% Fe_2O_3 composition, + 0.448 mm s^{-1}. For cone axis perpendicular to [111], e^2qQ is assumed to be that for α-Fe_2O_3, + 0.880 mm s^{-1}.
[c]) [63C1].
[d]) Average of 40 and 50% values.
[e]) [68A1].

α-Fe_2O_3–Rh_2O_3 system

In this system all the compounds have the hematite corundum structure [65K1]. For the magnetic phase diagram see Fig. 55. For Mössbauer data see Table 11.

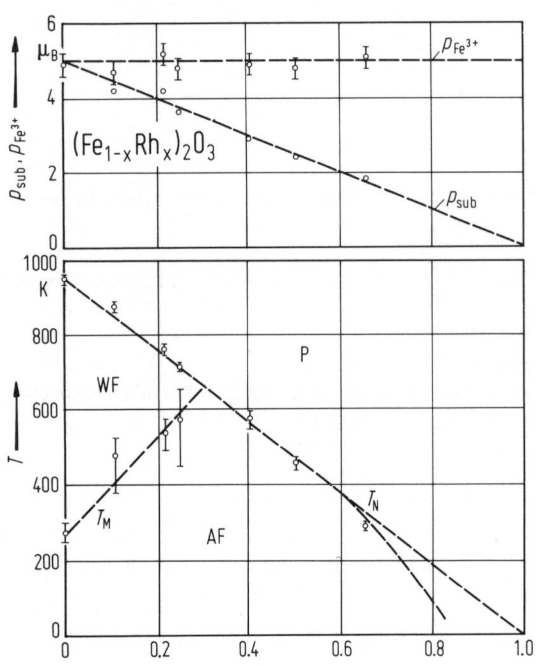

Fig. 55. $(Fe_{1-x}Rh_x)_2O_3$. Néel temperature T_N and Morin temperature T_M as dependent on composition parameter x. The average magnetic moment per Fe-ion, $p_{Fe^{3+}}$ and the average magnetic moment of the sublattice, p_{sub} are also determined from neutron diffraction measurements [65K1].

Table 11. $(Fe_{1-x}Rh_x)_2O_3$. Morin temperature T_M, Néel temperature T_N and magnetic hyperfine field H_{hyp} at 82 K, as derived from Mössbauer spectra [67D1].

x	T_N[K]	T_M[K]	H_{hyp}[kOe]
0	958(3)	257(5)	540(10)
0.11	823(10)	507(50)	530(10)
0.22	805(10)	544(25)	530(10)
0.25	783(10)		530(10)
0.41	628(10)		520(10)
0.81	< 82		

5.6.4 Oxides with ilmenite structure

5.6.4.1 Single ternary oxides $MM'O_3$; M,M' = 3d transition elements

Ilmenites, general

The rhombohedral ilmenite structure C_{3i}^2–$R\bar{3}$ of the compounds $M^{2+}M'^{4+}O_3$ can be considered as an ordered superstructure of the corundum structure. A crystallographic order of the M and M' ions occurs on the iron sites of the α-Fe_2O_3 structure. Table 12a gives a survey of crystallographic parameters of some ilmenites. The crystallographic parameters of the distorted ilmenites with triclinic crystal symmetry synthesized at high temperature are given in Table 12b. The magnetic structure of the non-distorted ilmenites depends upon whether both types of ions, M and M' (case A) or only the M ions (case B) are magnetic. In the following, the general properties of the two categories are described and experimental results on prominent compounds are presented.

Case A: Both M and M' ions are magnetic

Very alike to the corundum case, the magnetic structure is mainly determined by the exchange between the M and M' sublattices. Since this exchange is antiferromagnetic, the spins in the neighbouring planes order antiparallel and the difference in the magnetic moments determines the ferrimagnetism. $NiMnO_3$ and $CoMnO_3$ are examples of this category.

In $NiMnO_3$ the magnetization value deviates from the theoretical spin only contribution because the orbital moments on both ions are not fully quenched. In $CoMnO_3$ the saturation magnetization is determined purely by the orbital moments, because Co^{2+} and Mn^{4+} ions possess the same spin value. This is referred to as 'orbital ferrimagnetism'.

Case B: Only the M ions are magnetic

This case includes the far most intensively studied ilmenite compounds with Ti^{4+} as M'-ions. The reason for the strong interests in these compounds in the years of 1950 to 1970 was the understanding of the magnetic exchange interaction and anisotropy in the ionic materials based on crystal field, superexchange and molecular field considerations [56L1, 56I1, 64O1, 67G1].

When the M′-ions are non-magnetic, the M-sublattice can be subdivided into four sublattices. The magnetic structure is then determined by the intra-planar exchange, i.e. exchange between the ions in the same plane, and the interplanar exchange via two oxygen ions, i.e. exchange between the ions in the neighbouring planes. The three possible magnetic structures are shown in Fig. 56. $FeTiO_3$, $CoTiO_3$ and $NiTiO_3$ are examples of the structure a), the structure c) is observed in $MnTiO_3$ and the structure b) in $MnGeO_3$.

In Fig. 57 the temperature dependences of the magnetic susceptibility of powder samples are shown [67S1]. The anomaly due to the 3-D magnetic ordering in the structure a) is clearly seen, where the ferromagnetic sheets are antiferromagnetically coupled by the interplanar exchange. It is interesting that the interplanar exchange going over two oxygen ions results in a Néel temperature between 20 and 60 K.

On the other hand because of the antiferromagnetic coupling in the plane in the structure c) there is a cancellation of the interplanar couplings in $MnTiO_3$ which results in a weak net interplanar interaction. The susceptibility therefore shows a broad maximum well above the Néel temperature, a typical feature of the low dimensional magnetism. In the high-temperature region ($T > T_N$), the susceptibilities appear to obey a Curie-Weiss law with the paramagnetic parameters of Table 13. Mössbauer data Figs. 58–60 and Table 14.

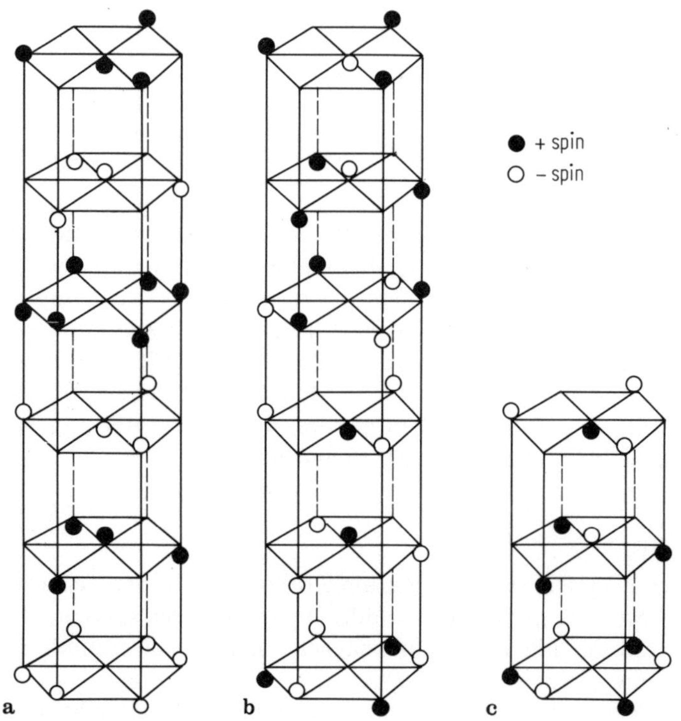

Fig. 56. The three possible magnetic structures of $MM'O_3$ ilmenite compounds for the case that M′ is a non-magnetic ion. The open and closed circles represent the magnetic M ions with antiparallel oriented magnetic moments. The M′ ions and the oxygen ions are omitted.

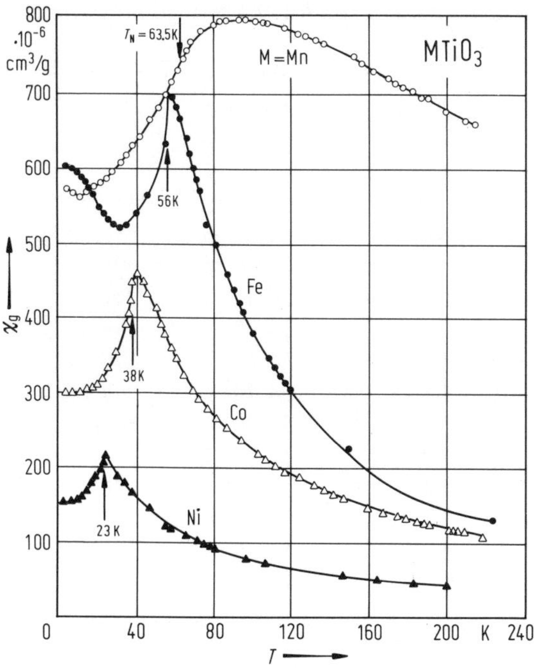

Fig. 57. $MTiO_3$, M=Mn, Fe, Co or Ni; powders. Magnetic mass susceptibilities χ_g vs. temperature T. T_N is Néel temperature. The susceptibilities of $MnTiO_3$ have been multiplied by a factor 10. Applied magnetic field 7.32 kOe. Since the χ_g vs. T curve for $MnTiO_3$ exhibits a broad peak at the ordering temperature, the value $T_N = 63.5$ K is given, which is obtained from resonance and specific heat data [67S1].

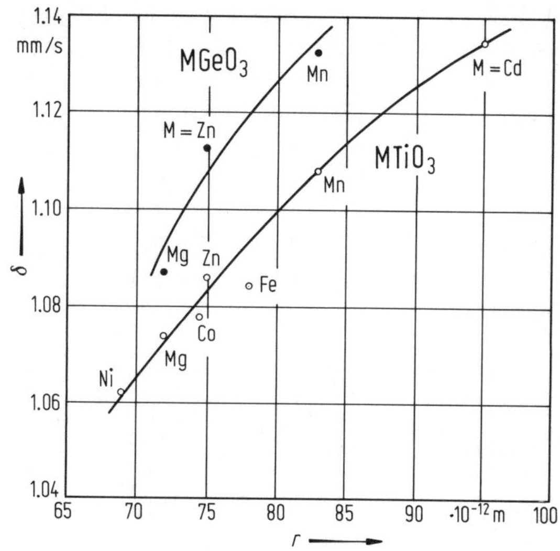

Fig. 58. $MTiO_3$, M=Mn, Co, Zn, Cd, Mg, Ni, Fe; $MGeO_3$, M=Mn, Zn, Mg. Fe^{2+} doped ilmenites. Isomer shift δ of Mössbauer resonance line at RT vs. ionic radius r. The center shift of iron is taken to be zero [81S1].

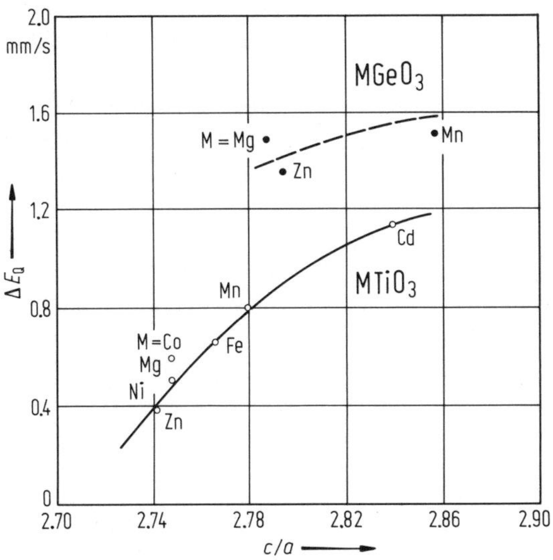

Fig. 59. Similar to Fig. 58, but now for the dependence of the quadrupole splitting $\Delta E_Q = (1/2)e^2qQ$ on the ratio of the unit cell parameters, c/a. The splitting of the innermost pair of the hyperfine spectrum of α-Fe_2O_3 was taken to be 2.625 $mm\,s^{-1}$ [81S1].

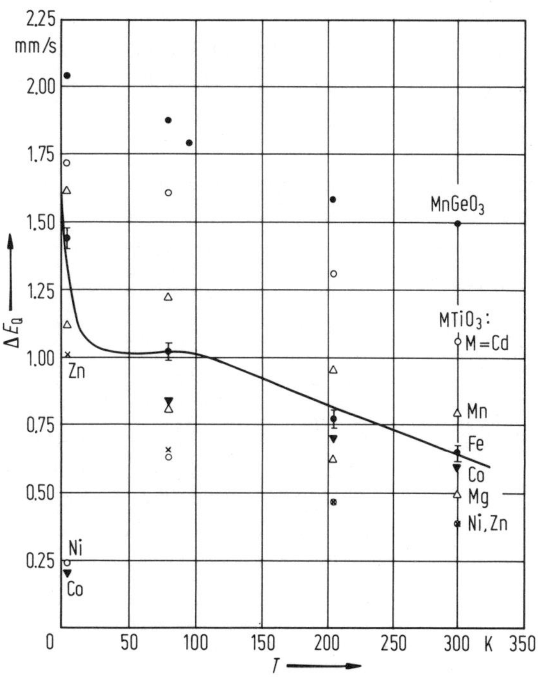

Fig. 60. $MTiO_3$ and $MnGeO_3$. Temperature dependence of the quadrupole splitting ΔE_Q. See also caption Fig. 58. The curve is the result of an analysis on the basis of a suggested Hamiltonian [81S1].

Table 12a. Ilmenites, space group $R\bar{3}$. RT crystallographic parameters for the hexagonal symmetry (a, c) and the rhombohedral symmetry (a_r, α_r).

Compound	Hexagonal system a [Å]	c [Å]	c/a	Rhombohedral system a_r [Å]	α_r	Ref.
$MnTiO_3$ I	5.138	14.283	2.780	5.609	54°30′	58I1, 59S1[a])
	5.1374(2)	14.284(1)	2.7803(40)	5.6096(5)	54°30′(1)	69S3
$FeTiO_3$				5.538	54°41′	34B1
				5.610	54°30′	59S1[a])
	5.088	14.073	2.766	5.535	54°44′	69S3
	5.088	14.18	2.787	5.565	54°24′	84K1[a])
$CoTiO_3$	5.066	13.920	2.748	5.485	55°00′	58I1
				5.4846(7)	55°01′(2)	64N1[a])
$NiTiO_3$	5.031	13.788	2.741	5.437	55°07′	58I1
						59S1[a])
$MgTiO_3$	5.052	13.893	2.748	5.471	54°58′	58I1
$CuVO_3$ II[b])	4.9547(6)	14.066(4)				70C1
$CoMnO_3$	4.933	13.71				58S1
$NiMnO_3$	4.901	13.601				58B2
	4.905	13.59				58S1
	–	–		5.343	54°39′	75P1
$ZnMnO_3$	4.965	13.80				70C3
$MgMnO_3$	4.945	13.73				70C3
$MnGeO_3$				5.576	53°25′	74T1[a])

[a]) Neutron diffraction results. Detailed information on atomic positions are given in these refs.
[b]) Ilmenite-type structure, space group $R\bar{3}m$ [70C1].

Table 12b. Distorted ilmenites with triclinic crystal symmetry, prepared at high pressure. RT crystallographic parameters.

Compound	a [Å]	b [Å]	c [Å]	α	β	γ	Ref.
$MnVO_3$ I	5.072(5)	5.550(5)	5.023(4)	90°0′(6)	118°38′(4)	63°4′(4)	71S2
$CoVO_3$	4.980(1)	5.411(1)	4.929(1)	90°2′(1)	118°47′(1)	63°26′(1)	70C1
$NiVO_3$	4.935(1)	5.377(1)	4.903(1)	90°43′(1)	119°8′(1)	63°0′(1)	70C2
$CuVO_3$ I	4.967(3)	5.404(3)	4.914(2)	90°19′(1)	119°3′(1)	65°0′(1)	70C1
$MnCrO_3$	5.113(9)	5.385(7)	4.991(7)	89°57′(9)	118°12′(7)	62°33′(6)	73S1

Table 13. Ilmenites with rhombohedral (R) and with triclinic crystal symmetry (T). Néel temperature T_N, Curie temperature T_C, paramagnetic Curie temperature Θ_p, effective magnetic moment per FU, p_{eff}, and effective spectroscopic splitting factor g_{eff} derived from high-temperature dc susceptibilities.

Compound	R or T	T_N, T_C [K]	C_m [cm^3K/mol]	Θ_p [K]	p_{eff} [μ_B/FU]	g_{eff}	Ref.
$MnTiO_3$ I	R	41	4.15	− 210	5.80		58I2
		65(2)[f])	4.36	− 219	5.93	2.01	67S1
		63.6	4.31	− 220	5.90		83Y1
$FeTiO_3$[e])	R	55	3.92	+ 17	5.62	–	58I2
		56	3.40	+ 59	5.23	2.14	67S1
$CoTiO_3$[c])	R	–	3.30	− 15	5.17	–	58I2
		42	–	− 4.5	4.91	–	64U1
		38	–	− 15	5.3	–	64N1
		36	3.61	− 3	5.46	2.79	67S1
$NiTiO_3$	R	23.2	1.30	− 55	3.24	–	58I2
		25	–	− 36	3.73	–	64U1
		22	1.26	− 13	3.16	2.21	67S1
		–	2.00	− 58	4.01	–	80W1[b])
$MnVO_3$ I	T[d])	70 (T_C)	–	− 430	8.0	–	71S1
$CoVO_3$ I	T[d])	142	–	− 22	5.53	2.71	70C2
$NiVO_3$	T[d])	153	–	− 129.5	3.36	–	70C2
$CuVO_3$ I	T[d])	–	–	− 41	1.12	–	70C1
$MnCrO_3$	T[d])	–	–	− 380(20)	4.00(15)	–	73S1
$CoMnO_3$	R	391(T_C)	–	–	–	–	58S1
$NiMnO_3$	R	437(T_C)	–	–	–	–	75P1
$ZnMnO_3$	R	18	–	+ 6	3.90	–	70C3
$MgMnO_3$	R	≈ 40	–	− 36	3.48	–	70C3
$MnGeO_3$	R	94[a])	–	− 245	5.86	–	74T1

[a]) From Mössbauer data [81S1].
[b]) Single crystal measurement, see Fig. 72.
[c]) For single crystal data, see Table 17.
[d]) Distorted structure.
[e]) See also Fig. 65.
[f]) See also caption Fig. 57.

Table 14. Ilmenites doped with Fe^{2+}. Mössbauer data at 4.2 K. Quadrupole splitting $\Delta E_Q = (1/2)e^2qQ$ and internal field H_{int}. Also the spin direction with respect to the rhombohedral crystal axes is given [81S1].

Compound	Direction of spins parallel to	ΔE_Q [mms^{-1}]	H_{int} [kOe]
$MnTiO_3$	[111]	1.62	33
$FeTiO_3$[a])	[111]	1.44	56
$CoTiO_3$	(111)	0.20	112
$NiTiO_3$	(111)	0.24	93
$MnGeO_3$	[111]	2.04	100

[a]) At 5 K is found $\Delta E_Q = +1.44(1)\ mms^{-1}$ and $H_{int} = -43(3)$ kOe [72G1].

$MnTiO_3$

With $MnTiO_3$ we denote the low-temperature phase $MnTiO_3$ I with ilmenite structure, see Fig. 36. Lattice parameters: Table 12a. Magnetic susceptibility: Figs. 57, 61 (and Fig. 37), Table 13. The compound orders at $T_N = 63.5$ K [67S1]. Its magnetic structure was determined by neutron powder diffraction to be of type c), Fig. 56, with the spin direction parallel to the *c* axis [59S1], see Fig. 62. Magnetization measurements show a spin flop field of 50.8 kOe at 4.2 K, indicating a small easy axis anisotropy, Fig. 63. The spin wave dispersion relations in the *c* plane and perpendicular to the *c* plane, Fig. 64, are studied by inelastic neutron scattering [86T1]. The smaller dispersion perpendicular to the plane reflects the two-dimensional nature of the magnetic system. The exchange parameters J_i (see Fig. 62) derived from these curves are listed in Table 15 together with the effective uniaxial anisotropy field H_A. The two-dimensional antiferromagnetic properties are realized by an accidental cancellation of the interlayer exchange interactions [70A1, 86T1]. Antiferromagnetic and paramagnetic resonance [67S1].

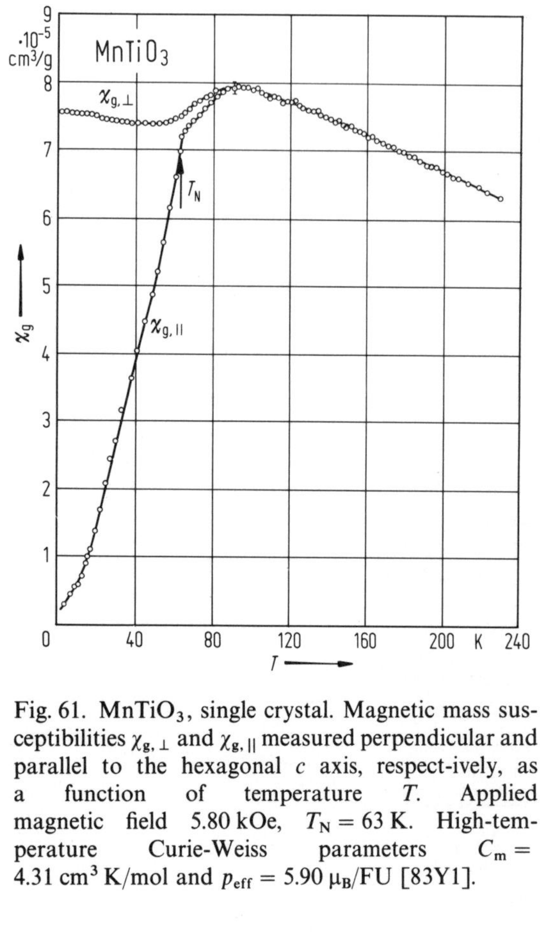

Fig. 61. $MnTiO_3$, single crystal. Magnetic mass susceptibilities $\chi_{g,\perp}$ and $\chi_{g,\parallel}$ measured perpendicular and parallel to the hexagonal *c* axis, respect-ively, as a function of temperature *T*. Applied magnetic field 5.80 kOe, $T_N = 63$ K. High-temperature Curie-Weiss parameters $C_m = 4.31\ cm^3$ K/mol and $p_{eff} = 5.90\ \mu_B$/FU [83Y1].

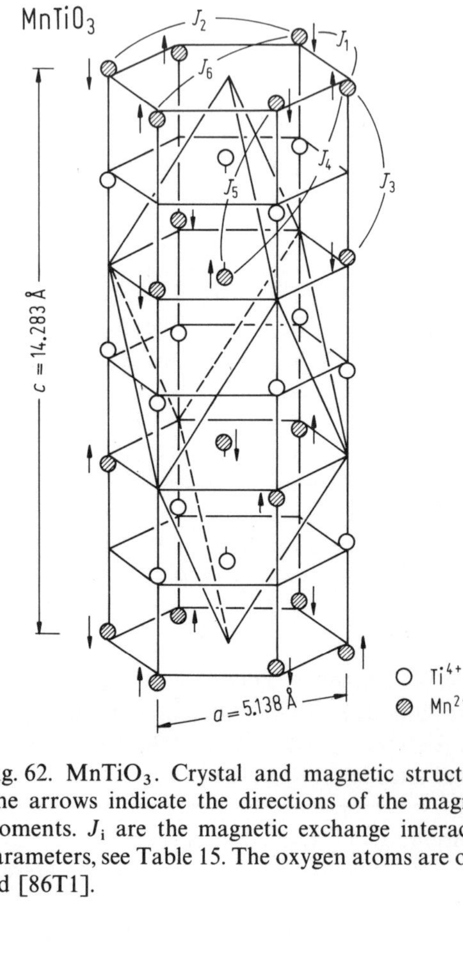

Fig. 62. $MnTiO_3$. Crystal and magnetic structures. The arrows indicate the directions of the magnetic moments. J_i are the magnetic exchange interaction parameters, see Table 15. The oxygen atoms are omitted [86T1].

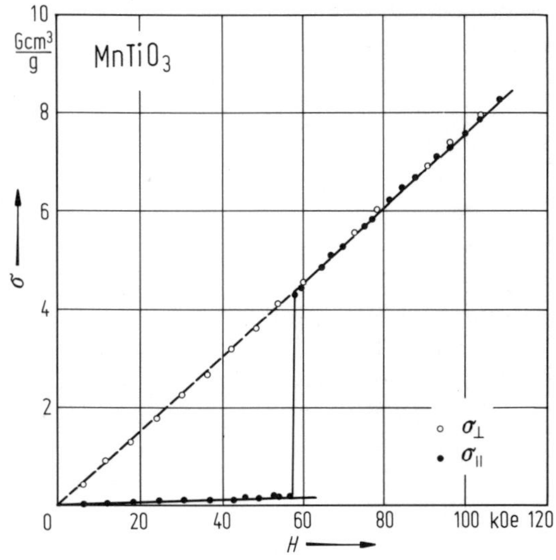

Fig. 63. $MnTiO_3$, single crystal. Magnetization curves at 4.2 K parallel ($\sigma_{\parallel}$) and perpendicular ($\sigma_{\perp}$) to the hexagonal axis [83Y1].

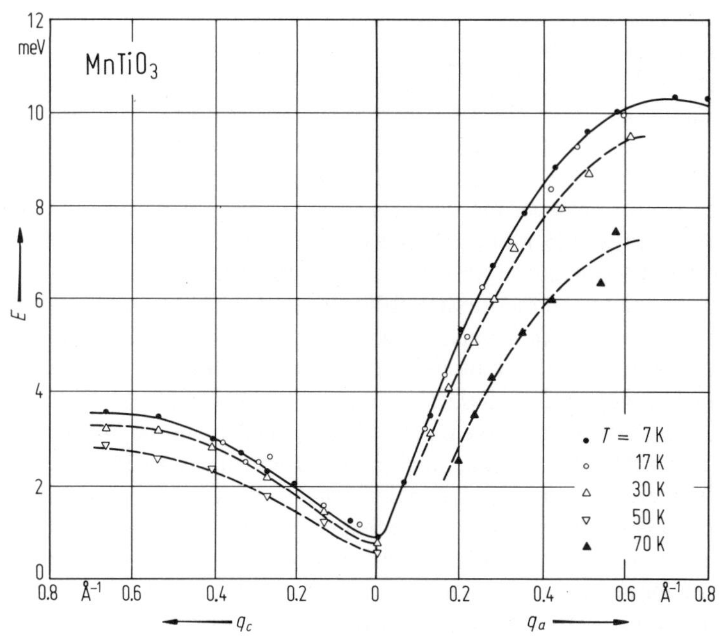

Fig. 64. $MnTiO_3$. Spin wave dispersions curves for the [100] and [001] directions at various temperatures as derived from INS experiments. Solid line calculated with parameters of Table 15. Broken lines are guides to the eye [86T1].

Table 15. $MnTiO_3$. Magnetic exchange interaction parameters J_i defined in Fig. 62, and effective anisotropy field H_A as determined from the spin-wave dispersion relations of Fig. 64 [86T1].

Parameter J_i/k	Number of neighbours	Mn–Mn distance [Å]	Parameter value [K]
J_1/k	3	3.04	– 7.28(20)
J_2/k	6	5.14	– 0.11(5)
J_3/k interlayer	1	4.09	– 2.20(15)
J_4/k interlayer	6	5.61	– 1.12(4)
J_5/k interlayer	3	6.19	– 1.83(10)
J_6/k	3	5.97	– 0.53(10)
$g\mu_B H_A$/k			0.065(10)

$FeTiO_3$

Earlier work reported in Landolt-Börnstein, New Series, subvolume III/4a p. 29, 30. Lattice parameters: Table 12a. Magnetic susceptibility: Figs. 57, 65 and Table 13. $FeTiO_3$ orders at $T_N = 56$ K [67S1]. Its magnetic structure was determined by neutron powder diffraction to be of type a), Fig. 56, [59S1] with the spin direction almost parallel to the c axis, see Fig. 66. The collinear spins make an angle φ with the hexagonal axis. $\varphi = 2.2(2)°$ at 4.3 K and is almost insensitive to temperature variations up to the Néel temperature $T_N = 57.35(5)$ K [83K1, 86K1]. If multiple scattering effects are taken into account $\varphi = 1.6(1)°$ [86Y1]. A theoretical explanation for this deviation angle φ is still missing [86K1]. The inelastic neutron scattering experiment [86K1], Fig. 67, mapping the spin wave dispersion in the plane and perpendicular to the plane indicates the existence of an easy plane anisotropy as the reason for the gapless acoustic mode observed at the zone center. The easy plane like anisotropy would be consistent with the above mentioned obliqueness of the moment direction. But the obliqueness and the high-field metamagnetic behaviour with $\boldsymbol{H}$ parallel to a axis cannot be consistently explained within the single-ion anisotropy model even when an anisotropic exchange interaction is taken into account. For the magnetic exchange parameters see Table 16. In this table also the magnetic anisotropy constants K_x and K_y are given, defined as coefficients of the anisotropy energy Hamiltonian $\mathscr{H}_{an} = K_x S_x^2 + K_y S_y^2$. Results of magnetization measurements up to 300 kOe are shown in Fig. 68. A spin-flop phase transition for the applied field parallel to the c axis is observed at $H_c = 80.3(1)$ kOe, due to the easy-axis-anisotropy inherent to the magnetic structure. The saturation moment of 4.6 μ_B per Fe ion suggests a considerable orbital contribution. Another metamagnetic behaviour is observed around a field of about 235 kOe, when this field is applied parallel to the a axis. However, no anomalous change in magnetization is found along the a^* axis, which is taken to be perpendicular to both the a and c axes [90K1].

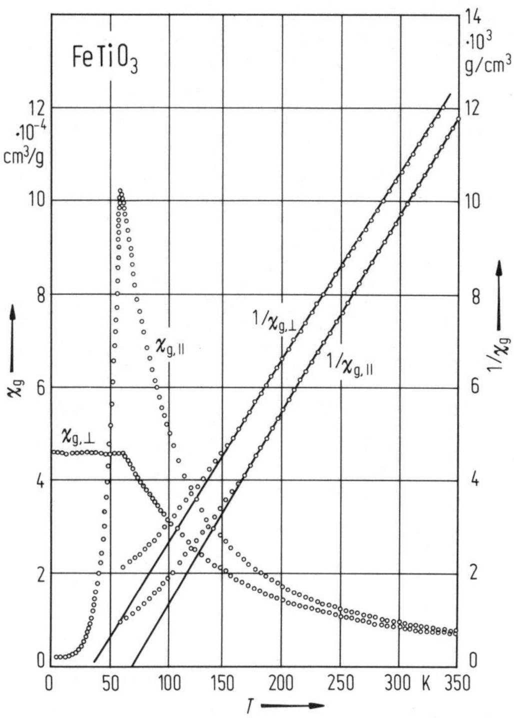

Fig. 65. $FeTiO_3$, single crystal. Magnetic mass susceptibilities $\chi_{g,\perp}$ and $\chi_{g,\parallel}$ and their inverses, measured perpendicular and parallel to the hexagonal c axis, respectively, as a function of temperature T. Applied magnetic field 13.8 kOe. $T_N = 58.0(3)$K. Extrapolated paramagnetic Curie temperatures $\Theta_p = 70$ K and 36.5 K [82K1].

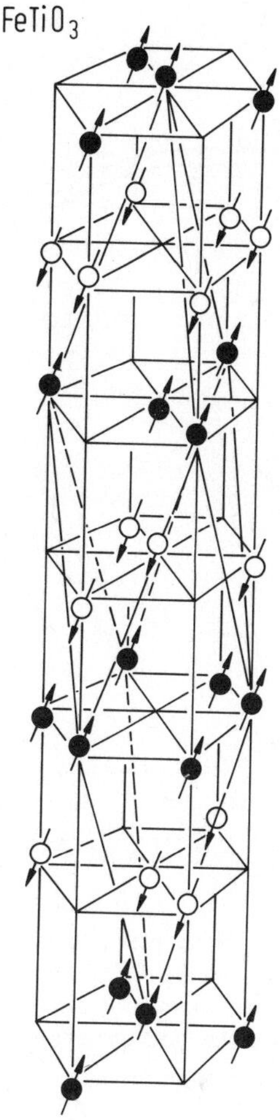

Fig. 66. $FeTiO_3$. Crystal and magnetic structures. The arrows indicate the directions of the magnetic moments of the Fe^{2+} ions with nearly up and down spins, respectively. The Ti and O ions are omitted [83K2].

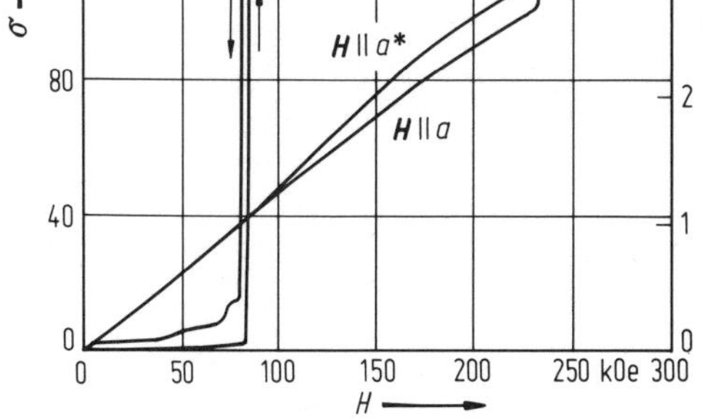

For Fig. 67, see next page.

Fig. 68. $FeTiO_3$, single crystal. Magnetization curves along three principal crystallographic axes measured at 4.2 K [90K1, see also 82K1]. a^*: axis perpendicular to both a and c axes.

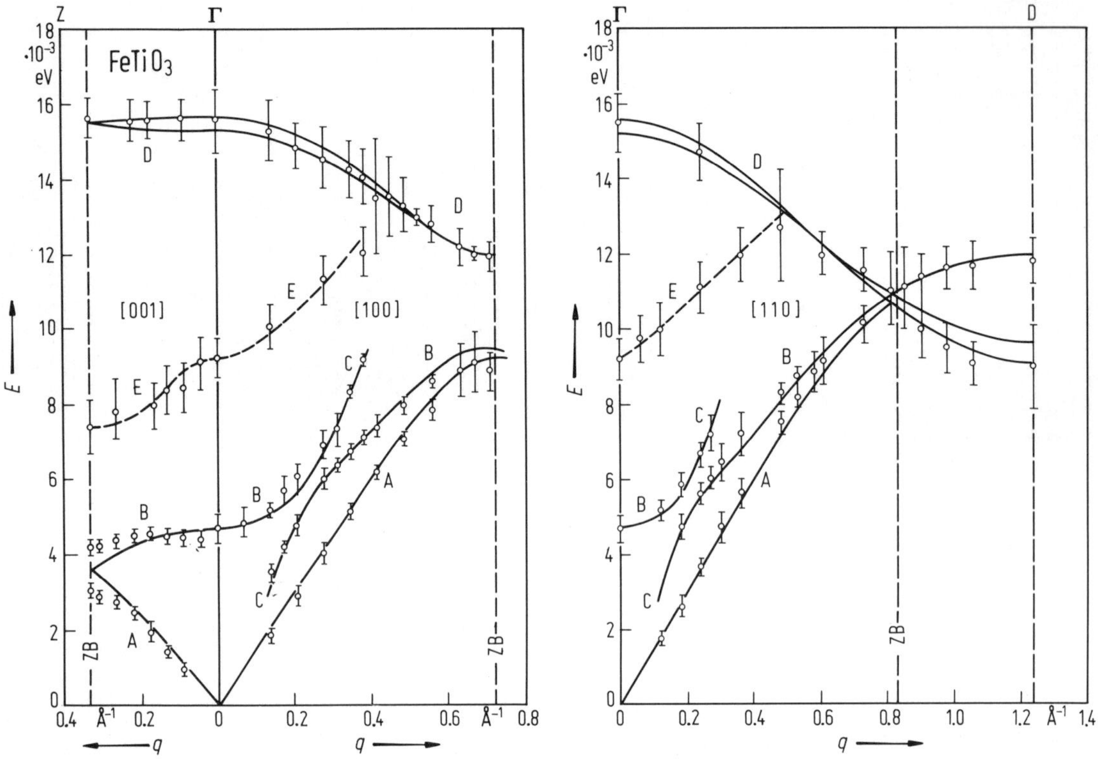

Fig. 67. $FeTiO_3$. Spin wave dispersions curves at $T = 12$ K for the [001], [100] and [110] crystallographic directions, derived from INS experiments. The various branches are designated A until E. The full curves show the results of phenomenological calculations based on the magnetic exchange interaction parameters of Table 16; broken curve is only guide to the eye [86K1]. ZB: zone boundary.

Table 16. $FeTiO_3$. Magnetic exchange interaction parameters J_i as defined in Fig. 62, and anisotropy energy constants K_x, K_y derived from the spin-wave dispersion relations of Fig. 67. Parameters expressed in K [86K1, 82K1, 83K2, see also 85I1].

J_1	J_2	J_3	J_4	J_5	J_6	K_x	K_y
4.77	− 0.90	0.28	− 0.55	0.38	− 0.04	21.8	0

CoTiO$_3$

For crystallographic properties see Table 12a. Each Ti and Co ion is octahedrally coordinated. Each Ti ion has three oxygen neighbours at 1.92 Å and three more at 2.06 Å distance. The corresponding Co-O distances are 2.08 Å and 2.16 Å, see [64N1], in which also their positional parameters are given. For magnetic susceptibilities and Néel temperature see Figs. 57, 69, 70 and Table 17. Magnetization curves: Fig. 71. Powder neutron diffraction reveals an antiferromagnetic structure of type a), Fig. 56. The Co^{2+} spins are mutually parallel in each hexagonal layer but antiparallel oriented between adjacent layers. The spins are parallel to the layers [59S2, 64N1].

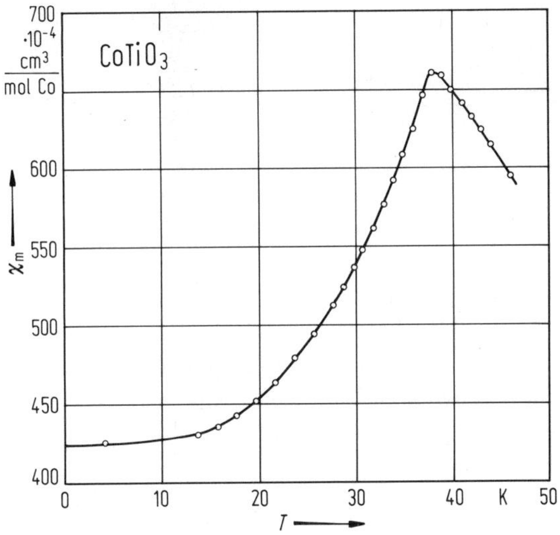

Fig. 69. CoTiO$_3$, polycrystalline sample. Molar magnetic susceptibility χ_m vs. temperature T, measured up to a magnetic field of 12 kOe. χ_m is independent of the magnitude of this field. T_N = 38(1) K [64N1].

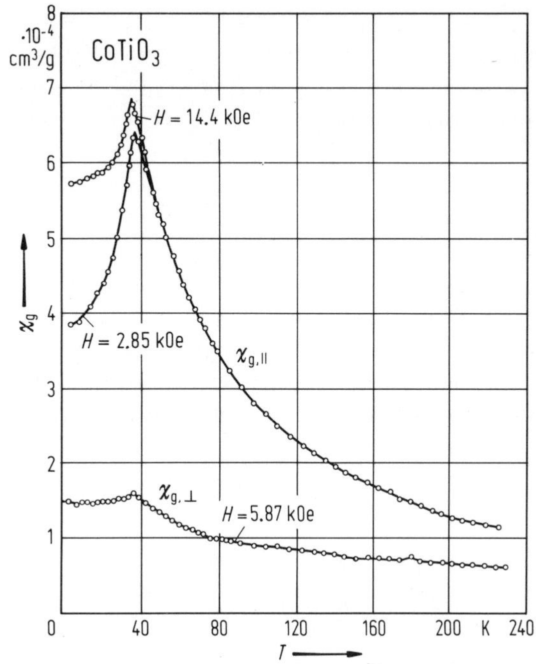

Fig. 70. CoTiO$_3$, single crystal. Magnetic mass susceptibility χ_g vs. temperature T. The indices $\parallel$ and $\perp$ denote the properties parallel and perpendicular to the c plane, respectively. The measuring fields H are indicated. T_N = 35.6 K, see also Table 17 [80W1].

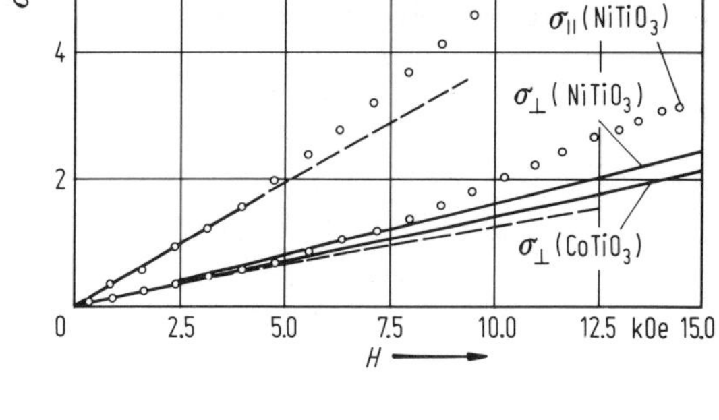

Fig. 71. CoTiO$_3$; NiTiO$_3$, single crystals. Magnetization curves parallel and perpendicular to the c plane at 4.2 K. The slopes of the linear parts for $\sigma_\parallel$ at lower fields are indicated by dashed straight lines [80W1].

Table 17. $CoTiO_3$; $NiTiO_3$. Single crystal Curie-Weiss parameters and Néel temperature T_N [80W1].

	$CoTiO_3$		$NiTiO_3$[a])
	$\boldsymbol{H} \parallel c$ plane	$\boldsymbol{H} \perp c$ plane	
T_N[K]	35.6		23.2
Θ_p[K]	33	−174	−58
C_m [cm^3 K/mol]	3.58	3.93	2.00
p_{eff} [μ_B/FU]	5.28	5.62	4.01

[a]) Same values for $\boldsymbol{H} \parallel c$ or $\boldsymbol{H} \perp c$.

$NiTiO_3$

Lattice parameters: Table 12a. Magnetic susceptibilities and Néel temperature: Figs. 57, 72 and Table 17. Magnetization curves: Fig. 71. One should note that the effective magnetic moment per formula unit derived from single crystal measurement is $p_{eff} = 4.01\ \mu_B$, which is considerably larger than the moment (3.16 μ_B) determined from powder experiments [67S1]. The magnetic structure is similar to that of $CoTiO_3$, the spins are directed perpendicular to the [111] axis [59S1].

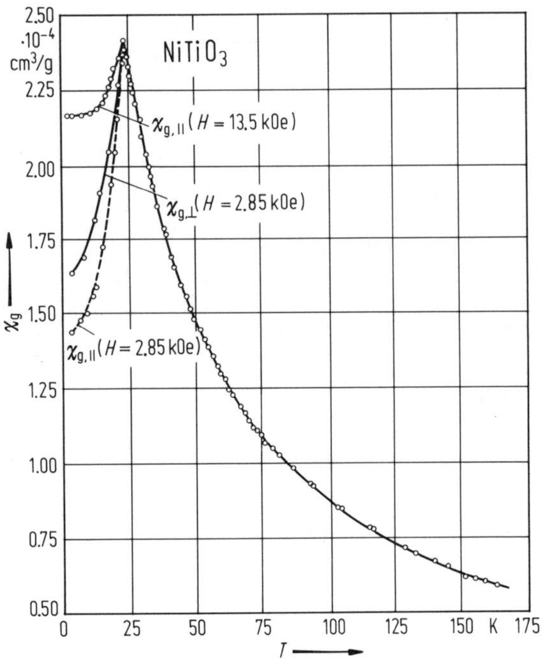

Fig. 72. $NiTiO_3$, single crystal. Magnetic mass susceptibility χ_g vs. temperature T. The indices $\parallel$ and $\perp$ denote measurements parallel and perpendicular to the c plane, respectively. The measuring fields H are indicated. $T_N = 23.2$ K [80W1].

$MnVO_3$

The low-pressure triclinic ilmenite-type $MnVO_3$ I has lattice parameters as given in Table 12b. The high-pressure form of $MnVO_3$ II has the orthorhombic perovskite structure [71S1]. Phase diagram: Fig. 73. Magnetic susceptibility and magnetization: Fig. 74a. For comparison the properties of type II are also given in Fig. 74b. The compounds are antiferromagnetic with some not compensated magnetism at low temperature.

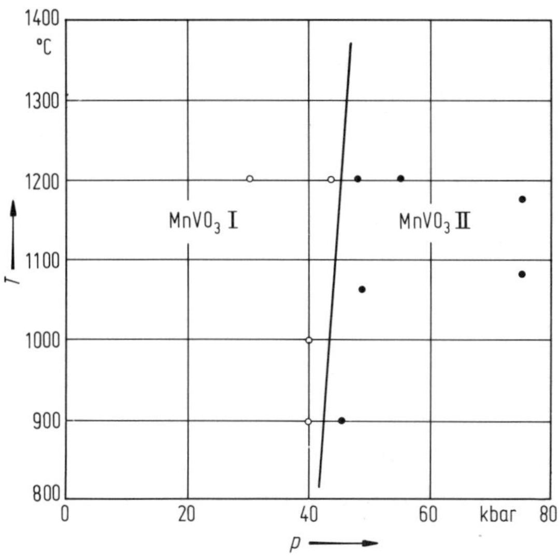

Fig. 73. $MnVO_3$ I, II. Temperature-pressure phase diagram for the ilmenite type I and the perovskite type II structures [71S2].

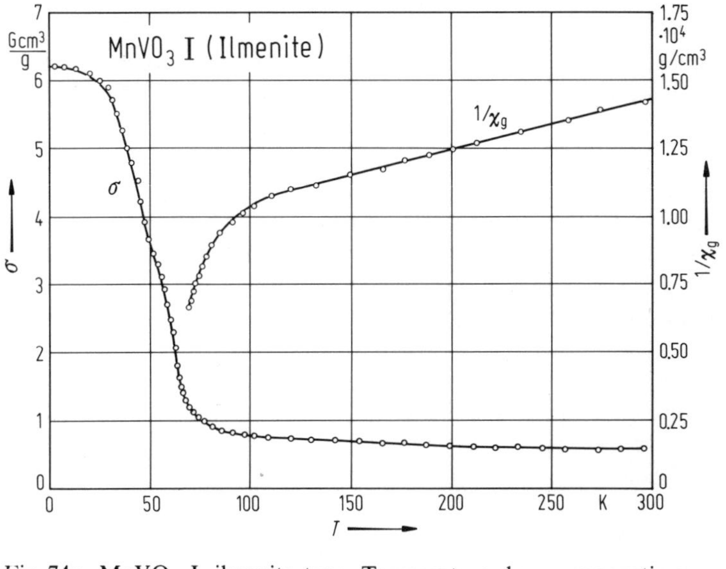

Fig. 74a. $MnVO_3$ I, ilmenite type. Temperature dependence of the mass magnetization σ and the inverse magnetic mass susceptibility χ_g^{-1}, measured in a field of 8.35 kOe, $T_N = 70$ K [71S2].

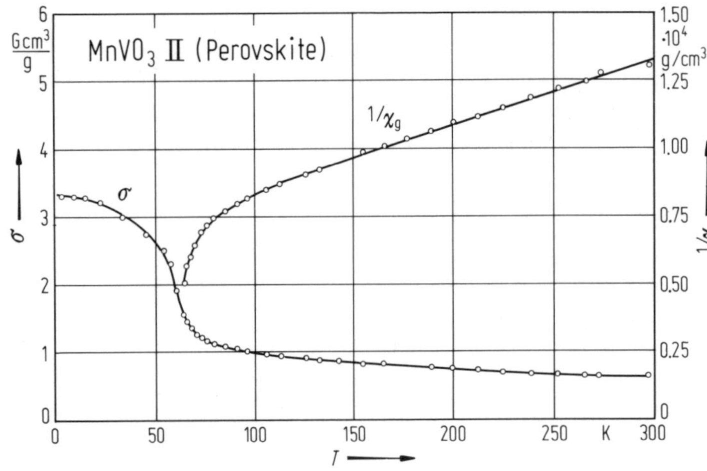

Fig. 74b. $MnVO_3$ II, perovskite type. Similar to Fig. 74a; $T_N = 65$ K [71S2].

$FeVO_3$

See V_2O_3–Fe_2O_3 system in Section 5.6.3.2.

$CoVO_3$

Triclinic, ilmenite type $CoVO_3$ was prepared [70C1] at high temperatures at 65 kbar pressure. Crystallographic parameters: Table 12b. The compound is antiferromagnetically ordered below $T_N = 142$ K. The Curie-Weiss analysis gives $p_{eff} = 5.3\ \mu_B$/FU and $\Theta_p = -9$ K. However a reevaluation and correction for atomic diamagnetism gives $p_{eff} = 5.53\ \mu_B$/FU, $\Theta_p = -22$ K with $g_{Co^{2+}} = 2.71$, assuming the "g factor" for V^{4+} to be 2.00 [70C2].

$NiVO_3$

X-ray powder diffraction data indicate a triclinic ilmenite structure with lattice parameters as given in Table 12b. The compound is antiferromagnetic with $T_N = 153$ K, $\Theta_p = -129.5$ K and $p_{eff} = 3.36\ \mu_B$/FU [70C2].

$CuVO_3$

There have been prepared two high-pressure phases $CuVO_3$ I and $CuVO_3$ II with triclinic, distorted and with rhombohedral ilmenite structure, respectively. Lattice parameters: Table 12a, b. The compound I is paramagnetic with $p_{eff} = 1.12\ \mu_B$/FU (see also Table 13) and $\Theta_p = -41$ K. The compound II shows van Vleck paramagnetism [70C1].

$MnCrO_3$

High-pressure synthesization, triclinic ilmenite structure [73S1]. Lattice parameters: Table 12b. Magnetic susceptibility: Fig. 75. Curie-Weiss law above 130 K with $\Theta_p = -380(20)$ K and $p_{eff} = 4.00(15)\ \mu_B$/FU and probably AF below 130 K [73S1].

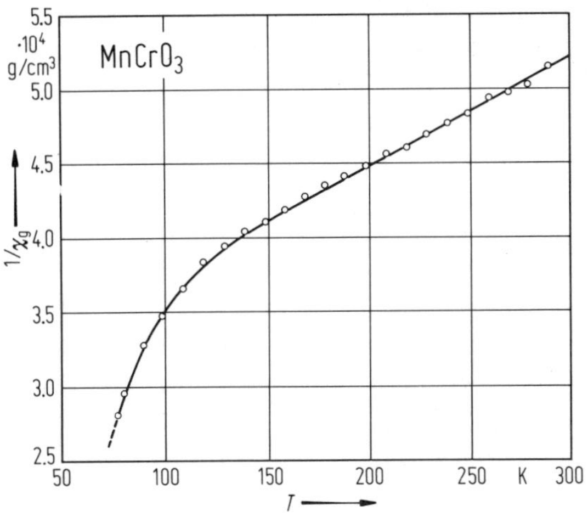

Fig. 75. $MnCrO_3$. Inverse magnetic mass susceptibility χ_g^{-1} vs. temperature T, measured in a field of 1 kOe. The high-temperature Curie-Weiss approximation yields $\Theta_p = -380(20)$ K and $p_{eff} = 4.00(15)\,\mu_B$/FU [73S1].

$CoMnO_3$

Lattice parameters: Table 12a. Orbital ferrimagnet below $T_C = 118$°C, spin direction perpendicular to c axis [58S1]. Magnetic moment extrapolated to 0 K and infinite field strength is 0.72 μ_B per FU. This is interpreted as a consequence of antiferromagnetic coupling between layers of cobalt and manganese ions, the difference in moment being attributed to the orbital moment of the cobalt ion [58B1].

$NiMnO_3$

Lattice parameters: Table 12a. Magnetic susceptibility and magnetization: Fig. 76. $T_C = 437$ K [75P1]. Ferrimagnetic compound. Below T_C the magnetic structure consists of an alternation of planes of ferromagnetically coupled Ni^{2+} or Mn^{4+} ions with the spin direction perpendicular to the [111] axis. These planes are antiferromagnetically coupled. Magnetic moments $p_{Ni^{2+}} = 1.8\,\mu_B$ and $p_{Mn^{4+}} = 2.2\,\mu_B$ as derived from neutron diffraction data [75P1]. A magnetization curve is given in Fig. 77. RT value of the saturation magnetization 15.3 $Gcm^3\,g^{-1}$ corresponding to 0.44 μ_B per FU. At 0 K is found 0.76 μ_B/FU [58B2, 78B1].

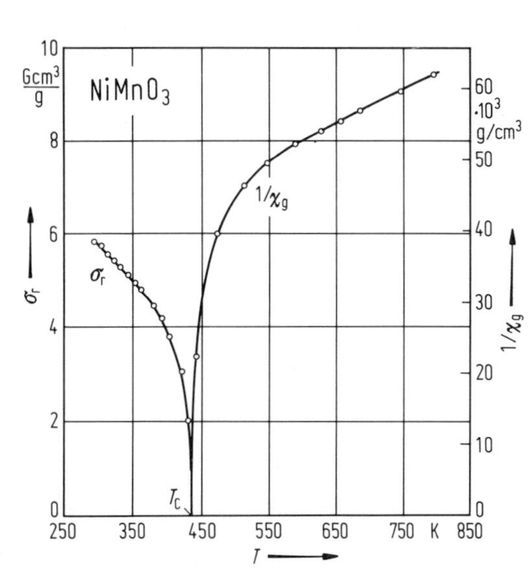

Fig. 76. $NiMnO_3$. Inverse magnetic mass susceptibility χ_g^{-1} vs. temperature T. Also the remanent mass magnetization σ_r is shown [75P1].

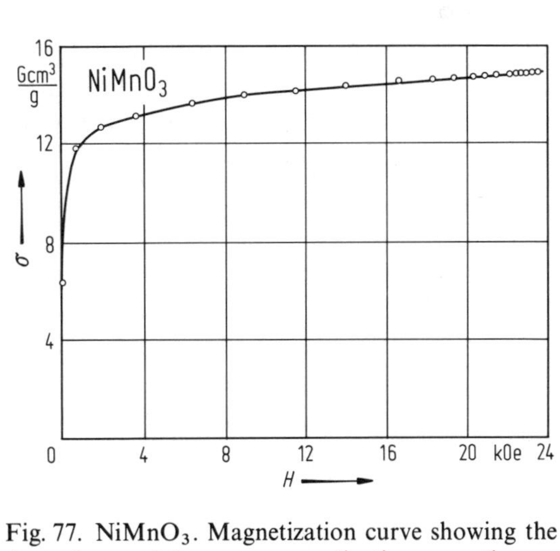

Fig. 77. $NiMnO_3$. Magnetization curve showing the dependence of the mass magnetization σ on the applied field H at 294 K [75P1].

$ZnMnO_3$

High-pressure phase. Lattice parameters: Table 12a. Magnetic susceptibility: Fig. 78, with $p_{eff} = 3.90\ \mu_B/FU$, $\Theta_p = 6$ K and $T_N = 18$ K [70C3].

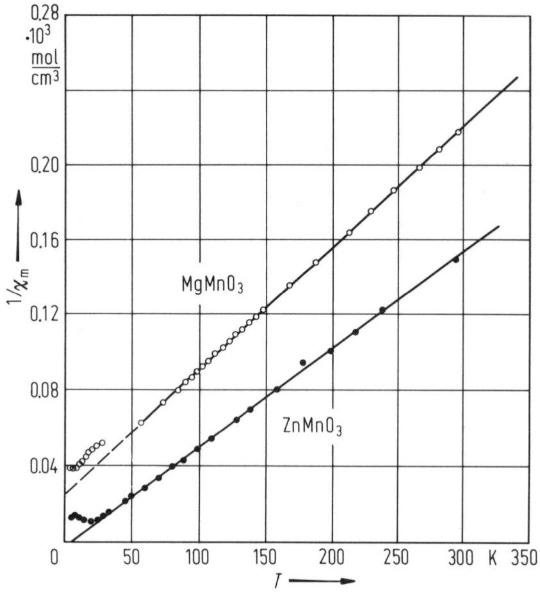

Fig. 78. $ZnMnO_3$; $MgMnO_3$. Reciprocal molar magnetic susceptibility χ_m^{-1} vs. temperature T. The high-temperature Curie-Weiss approximation yields for $ZnMnO_3$: $\Theta_p = +6$ K, $p_{eff} = 3.90\ \mu_B/FU$, and for $MgMnO_3$: $\Theta_p = -36$ K, $p_{eff} = 3.48\ \mu_B/FU$ [70C3].

$MgMnO_3$

High-pressure phase. Lattice parameters: Table 12a. Magnetic susceptibility: Fig. 78, with $p_{eff} = 3.48\ \mu_B/FU$ and $\Theta_p = -36$ K [70C3].

$MnGeO_3$

High-pressure phase. Lattice parameters: Table 12a. Magnetic susceptibility: Fig. 79. $p_{eff} = 5.86\ \mu_B/FU$ and $\Theta_p = -245$ K. Saturation magnetic moment is 4.6 μ_B. Neutron diffraction data reveal an antiferromagnetic structure similar to $MnTiO_3$; the spins are directed along the [111] axis. An analysis of these properties leads to the following exchange parameter: $J_1 = -0.84(1)$ meV (definition see Fig. 62) [74T1].

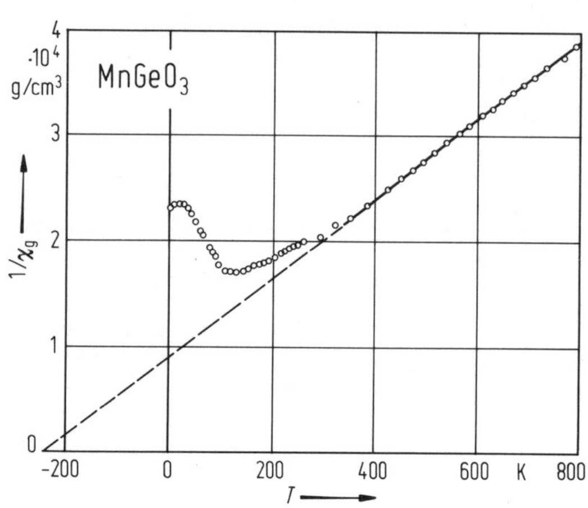

Fig. 79. $MnGeO_3$. Reciprocal magnetic mass susceptibility χ_g^{-1} vs. temperature T measured in a field of 4 kOe. The high-temperature Curie-Weiss approximation yields: $\Theta_p = -245$ K and $p_{eff} = 5.86\ \mu_B/FU$ [74T1, see also 66S1].

5.6.4.2 Mixed pseudo-ternary Ti oxides ($(M_xM'_{1-x})TiO_3$; M,M′ = Mn,Fe,Co or Ni

$Fe_xMn_{1-x}TiO_3$

The $Fe_xMn_{1-x}TiO_3$ is a mixture of two ilmenite systems $FeTiO_3$ and $MnTiO_3$. Both are antiferromagnets having an easy axis of spins along the hexagonal *c* axis of the ilmenite structure. The inter-layer couplings of spins in both compounds are antiferromagnetic. Within the *c* layer, the Fe^{2+} spins in $FeTiO_3$ are coupled ferromagnetically, whereas the Mn^{2+} spins in $MnTiO_3$ are coupled antiferromagnetically. In the mixed compound, therefore, the competition between the ferromagnetic and antiferromagnetic interactions occurs within each *c* layer, which brings about frustration among spins. The compounds are typical Ising systems with exchange frustrations.

The temperature versus Fe concentration phase diagram of $Fe_xMn_{1-x}TiO_3$ is shown in Fig. 80, which is determined by the low-field dc magnetization measurement and the neutron scattering measurement. Spin-glass transitions are observed in many samples which have the intermediate Fe concentrations. The samples with the Fe-rich and the Mn-rich concentrations undergo reentrant transitions from the antiferromagnetic phase to the reentrant spin-glass phase at low temperatures. The samples with the concentration close to 0 and 1 are antiferromagnets like $MnTiO_3$ and $FeTiO_3$, respectively. The typical results of the low-field dc magnetization measurements are shown in Fig. 81.

In the low-field dc magnetization measurements, the zero-field-cooled magnetization of the samples with intermediate concentrations shows a sharp cusp when the magnetic field is applied parallel to the *c* axis, which is shown in Fig. 82. On the other hand, applying the field perpendicular to the *c* axis, no evidence of the spin freezing appears in the temperature dependence of the magnetization. As the measuring field increases, the cusp is broadened and the spin-glass transition temperature, which is defined as the temperature where the irreversibility appears between the zero-field-cooled and the field-cooled magnetization, shifts to lower temperatures as shown in Fig. 83. This field H dependence of the spin-glass transition temperature $T_{SG}(H)$ obeys the following formula predicted by de Almeida and Thouless [58A1].

$$H \propto \left[1 - \frac{T_{SG}(H)}{T_{SG}(0)}\right]^{3/2} .$$

This means that there is a spin-glass transition even in an external magnetic field.

In the ac susceptibility measurements, the cusp indicating the spin-glass transition becomes broad and shifts to higher temperatures as the measuring frequency increases as seen in Fig. 84. This frequency f dependence of the cusp-temperature $T_{SG}(f)$ obeys the following simple power law,

$$f = f_0 \left[\frac{T_{SG}(f) - T^*}{T_{SG}(f)}\right]^{z\nu} ,$$

where, $f_0 = 2.4 \cdot 10^{13}$ Hz, $z\nu = 11.5$ and $T^* = 20.5$ K [88A1]. This critical behavior implies that the present spin-glass transition is a second-order phase transition.

The results of the neutron scattering measurements for the reentrant spin-glasses are shown in Figs. 85a, b. A decrease of the intensity of the magnetic Bragg reflection and an increase of the intensity of the magnetic diffuse scattering are observed when the sample undergoes the reentrant transitions. The inverse correlation length has two minima at the antiferromagnetic transition temperature and around the reentrant transition temperature.

This means that some spins which first belong to the antiferromagnetic ordered network extending over the system, start to construct a new order of spin-glass. However, the intensity of the Bragg reflection has a finite value even in lowest temperature examined. This result indicates that the long-range order coexists with the spin-glass state in the reentrant spin-glass phase in contrast with the ordinary spin-glass phase for which only a short-range order is observed.

The field variations of the magnetization for the reentrant spin-glasses with the Fe-rich concentrations are shown in Fig. 86. The field variations of the transition temperatures are shown in Fig. 87. The reentrant spin-glass transition temperature rises as the magnetic field increases, until it coincides with the antiferromagnetic transition temperature. Then, it decreases with the increase of the magnetic field. Such a field dependence of the reentrant transition temperature has not been found in the systems showing reentrant transitions from the ferromagnetic phase. Therefore, this behavior is considered to be characteristic of the systems undergoing reentrant transitions from the antiferromagnetic phase. Recently, such an enhancement of the reentrant transition temperature is derived in some theoretical models, where the antiferromagnetic phase plays an important role [88T1].

From the results obtained by the low-field magnetization and the neutron scattering measurements, this system is regarded as an ideal Ising system. However, the Mössbauer spectra shown in Fig. 88 indicate that the hyperfine field is tilted from the *c* axis. This implies the existence of the transverse spin component. By combining these experimental results, it is considered that the transverse spin components exist and fluctuate rather incoherently with characteristic time scales longer than the Mössbauer time scale of 10^{-7} s [90I2].

For the behavior of the magnetization in strong magnetic fields see Figs. 89, 90. Spin wave excitations [92Y1].

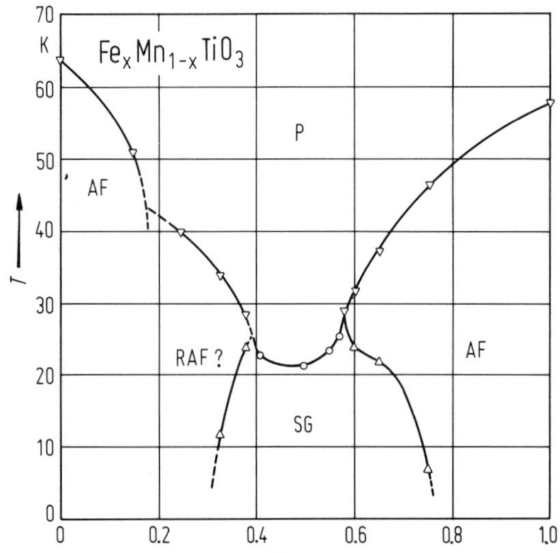

Fig. 80. $Fe_xMn_{1-x}TiO_3$. Magnetic phase diagram determined by low-field dc magnetization measurements. RAF ? indicates a suggested random-antiferromagnetic state [88I1, see also 89Y1].

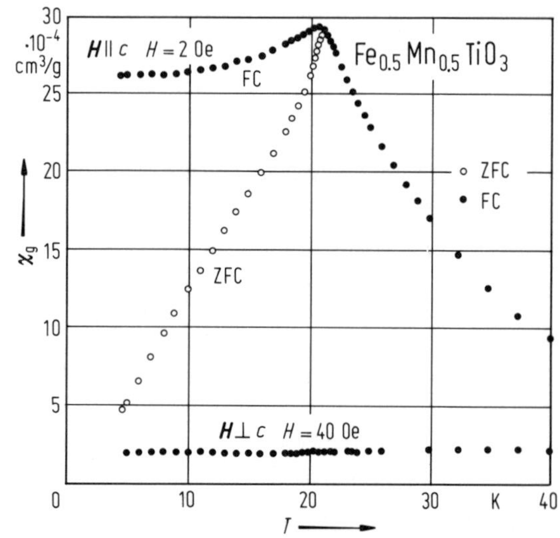

Fig. 82. $Fe_{0.5}Mn_{0.5}TiO_3$, zero-field cooled (ZFC) and field cooled (FC). Susceptibilities χ_g measured in a field $H = 2$ Oe parallel to the *c* axis and a field of 40 Oe perpendicular to this axis [90I1].

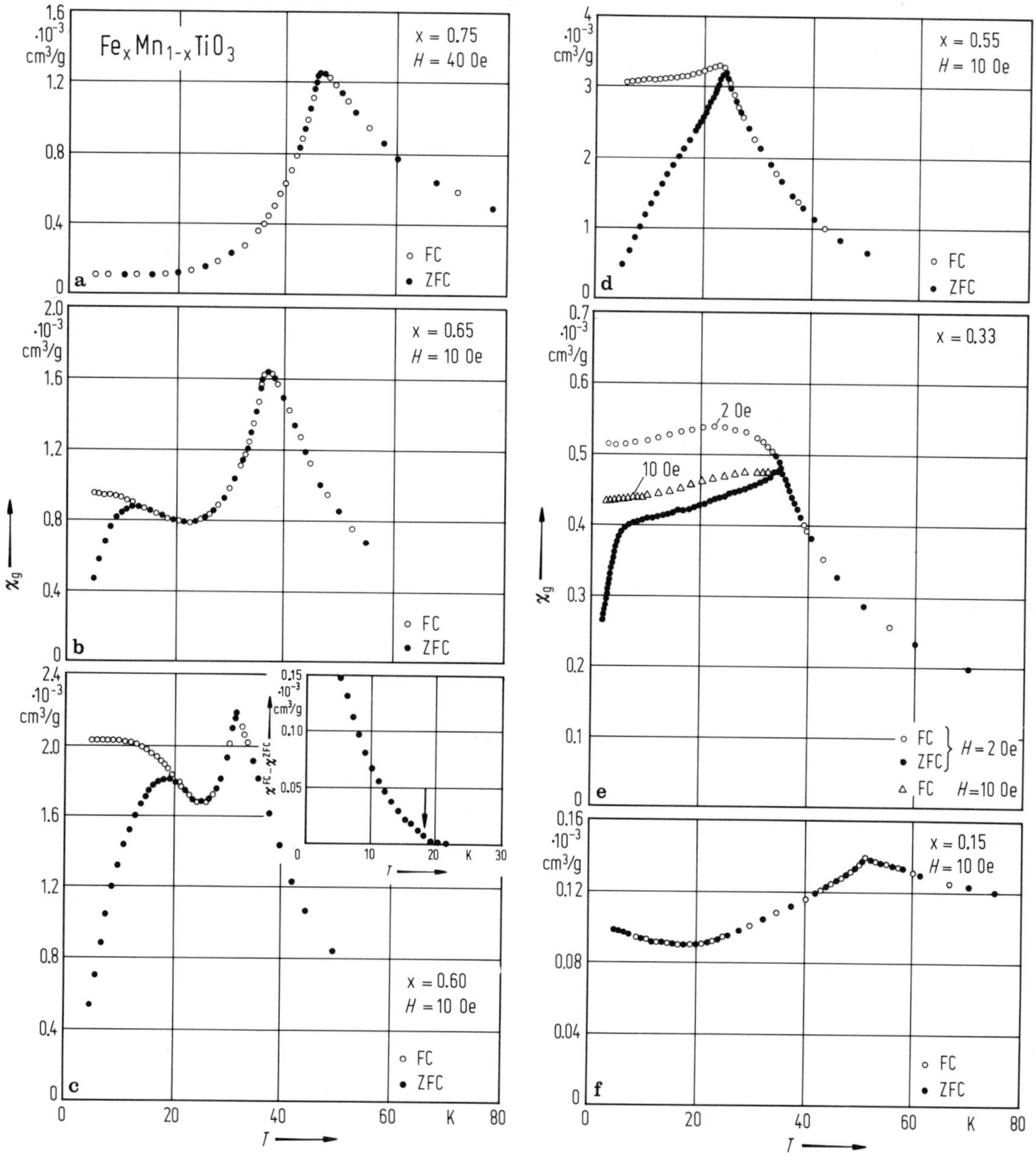

Fig. 81. $Fe_xMn_{1-x}TiO_3$. Low dc field susceptibility curves measured along the *c* axis for various values of the composition parameter x. FC: field cooled, ZFC: zero-field cooled. The inset of Fig. 81c shows the difference of both susceptibilities [88I1].

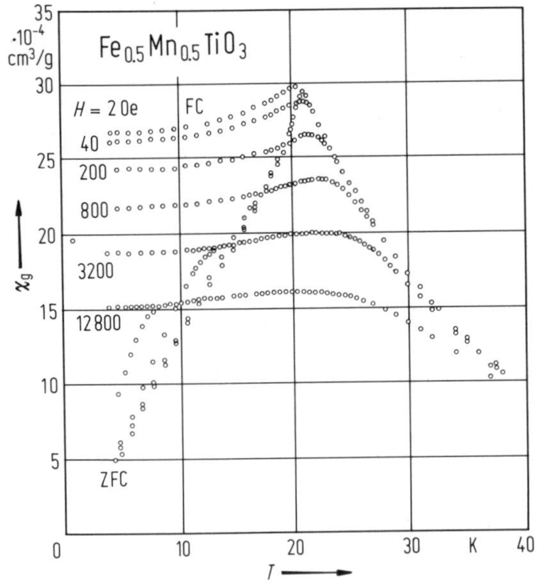

Fig. 83. $Fe_{0.5}Mn_{0.5}TiO_3$. Temperature dependence of the susceptibilities measured for various magnitudes of the dc field H parallel to the c axis for field cooled (FC) and zero-field cooled (ZFC) samples [86I1].

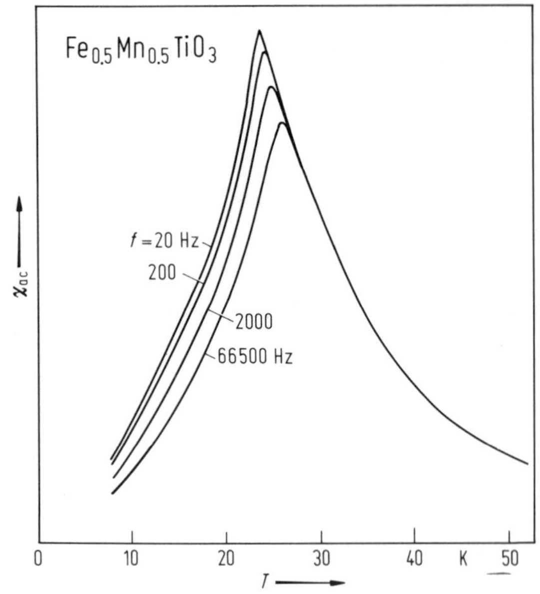

Fig. 84. $Fe_{0.5}Mn_{0.5}TiO_3$. Temperature dependence of the ac susceptibility χ_{ac} parallel to the c axis for various frequencies f of the applied magnetic field [88A1].

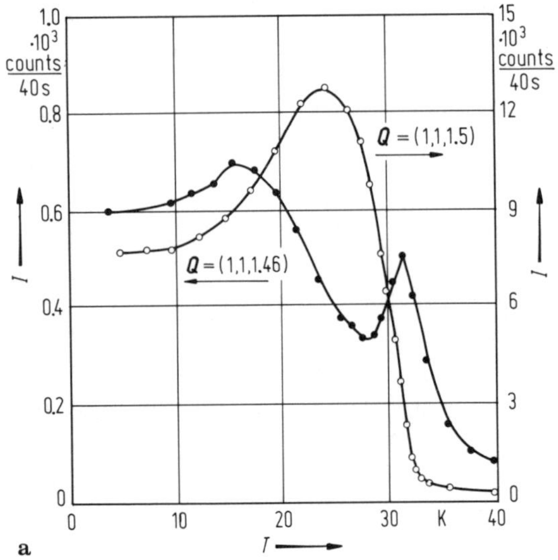

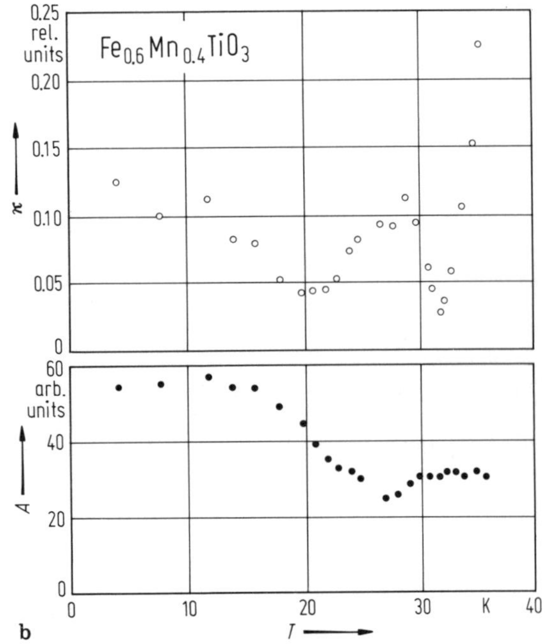

Fig. 85. $Fe_{0.6}Mn_{0.4}TiO_3$. (a) Temperature variations of the magnetic Bragg and the diffuse neutron scatterings near the reciprocal lattice positions indicated by the two Q values, respectively [88I2].

(b) Inverse correlation length κ and amplitude A (in arbitrary units) of a Lorentzian fit for the diffuse neutron scattering [89Y1].

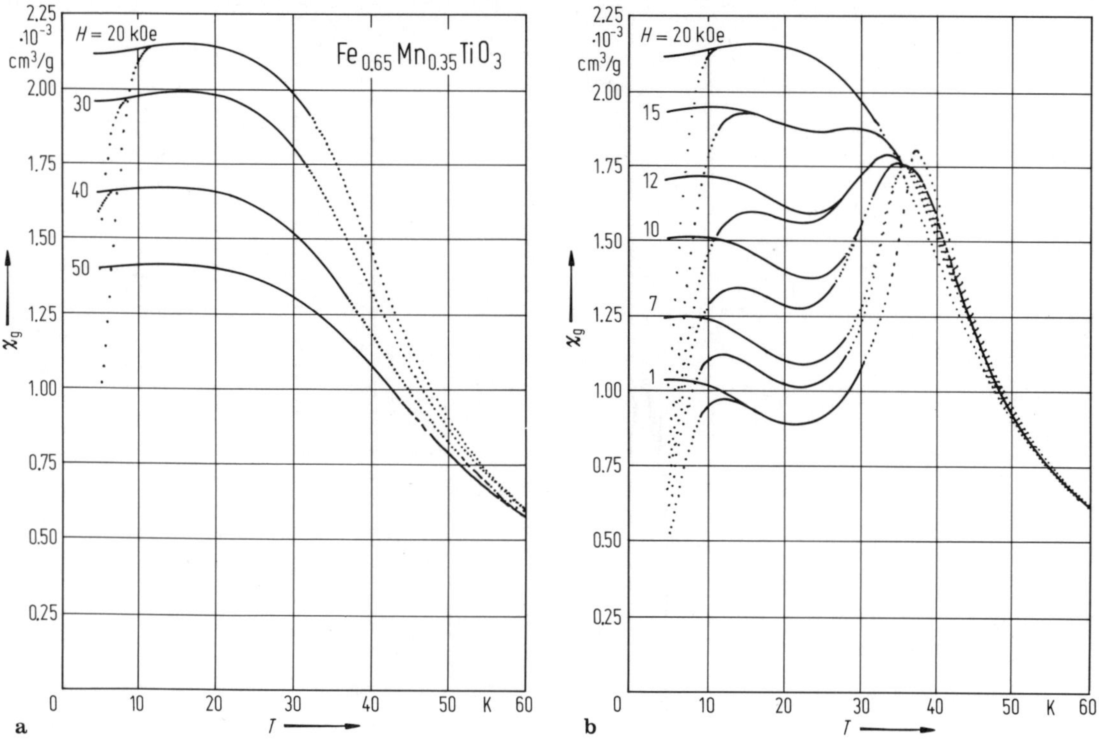

Fig. 86. $Fe_{0.65}Mn_{0.35}TiO_3$, zero-field-cooled (ZFC) and field-cooled (FC). Temperature dependence of the susceptibility χ_g parallel to the c axis, for various magnitudes of the magnetic measuring field H. The measuring results coincide at higher temperatures but they deviate from each other at lower temperatures where the susceptibility in the ZFC state is lower than in the FC state [88A2].

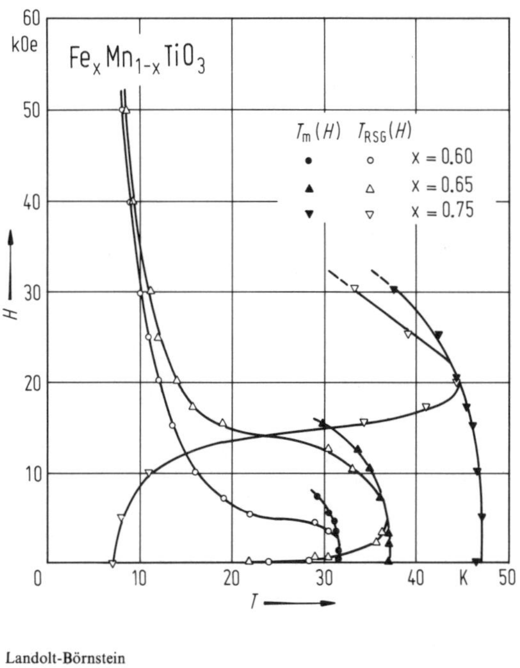

Fig. 87. $Fe_xMn_{1-x}TiO_3$; x = 0.60, 0.65 and 0.75. Field dependence of the temperatures $T_m(H)$: the temperature for maximum χ_g value. $T_{RSG}(H)$: the temperature below which the irreversibility appears between $\chi_g^{ZFC}(T)$ and $\chi_g^{FC}(T)$, see for instance the inset of Fig. 81c [88A2]. $T_m(H)$ and $T_{RSG}(H)$ are the temperatures characterizing the paramagnetic to antiferromagnetic transition and the reentrant spin glass transition, respectively, as detected by neutron scattering measurements [87Y1].

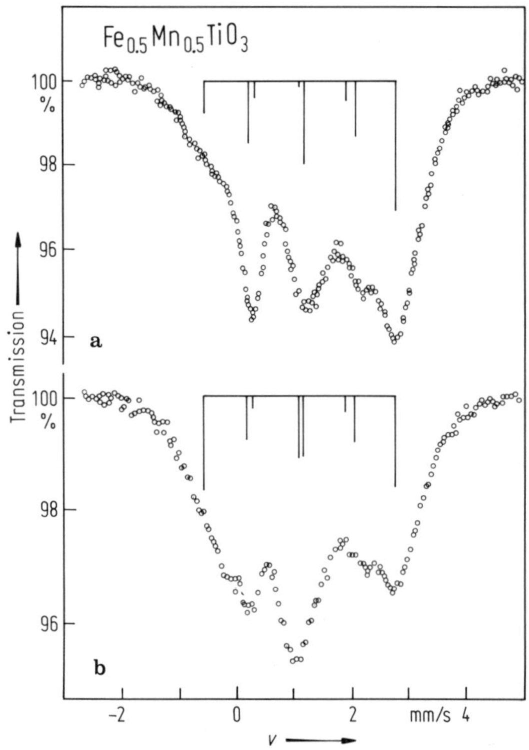

Fig. 88. $Fe_{0.5}Mn_{0.5}TiO_3$. Mössbauer spectra at 4.2 K in zero field. (a) Single crystal absorber with γ rays incident along the *c* axis. (b) Powder absorber. The vertical lines have been calculated using the values $H_{hyp} = 75$ kOe, $\Delta E_Q = (1/2)\ e^2qQ = 1.5\ \mathrm{mm\,s^{-1}}$, $\eta = 0.2$, $\theta_H = 60°$, $\phi_H = 0°$, $\theta_\gamma \approx 0°$ and $\phi_\gamma \approx 0°$, where (θ_H, ϕ_H) and $(\theta_\gamma, \phi_\gamma)$ are the polar and azimuthal angles defining the orientation of H_{hyp} and the incident rays, respectively, relative to the principal axes of the electric field gradient tensor EFG [90I1].

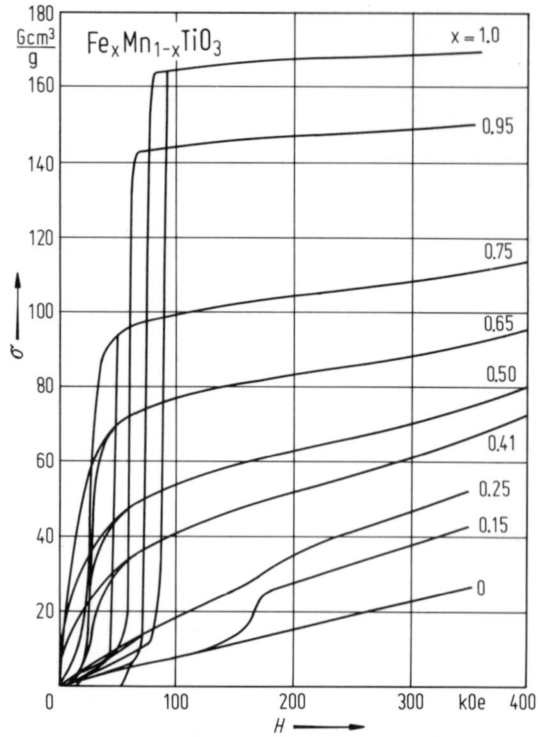

Fig. 89. $Fe_xMn_{1-x}TiO_3$, single crystals. Magnetization σ vs. applied magnetic field measured along the *c* axis at 4.2 K [92I1].

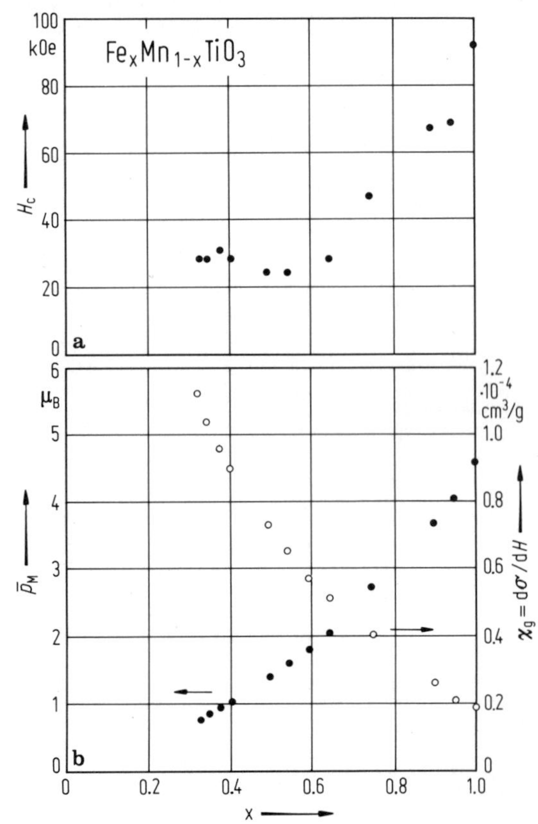

Fig. 90. $Fe_xMn_{1-x}TiO_3$, single crystals. Analyses of the curves of Fig. 89. (a) Critical field H_c of the spin-flip like transition, (b) average magnetic moment $\bar{p}_M$ per magnetic ion just after this transition, and high-field susceptibility $d\sigma/dH$ in the region where σ is proportional to *H* [92I1, see also 92A1].

$Co_xMn_{1-x}TiO_3$

In the compounds $Co_xMn_{1-x}TiO_3$ there exist competing exchange interactions and competing magnetic anisotropies. A phase diagram derived from low-field dc magnetization measurements is shown in Fig. 91.

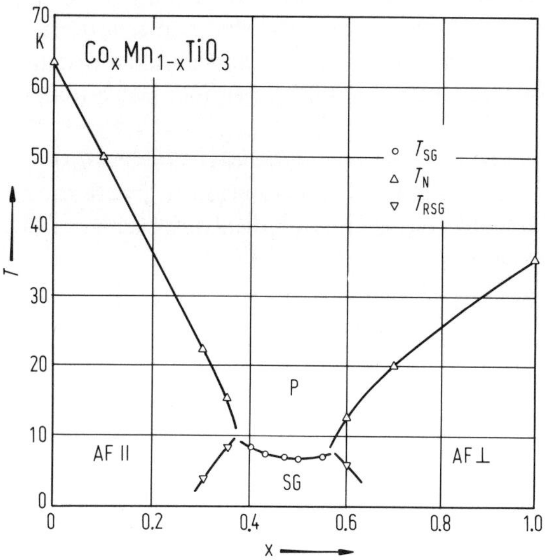

Fig. 91. $Co_xMn_{1-x}TiO_3$. Magnetic phase diagram. The symbols AF|| and AF⊥ correspond to $MnTiO_3$-type or $CoTiO_3$-type order with spin axis parallel or perpendicular to the *c* axis, respectively [91K1].

$Ni_xMn_{1-x}TiO_3$

In the system $Ni_xMn_{1-x}TiO_3$ the same frustrations occur as in the system $Co_xMn_{1-x}TiO_3$. However for a system with a weak single-ion anisotropy these frustrations give rise to new successive phase transitions near the reentrant spin glass phase, see Fig. 92.

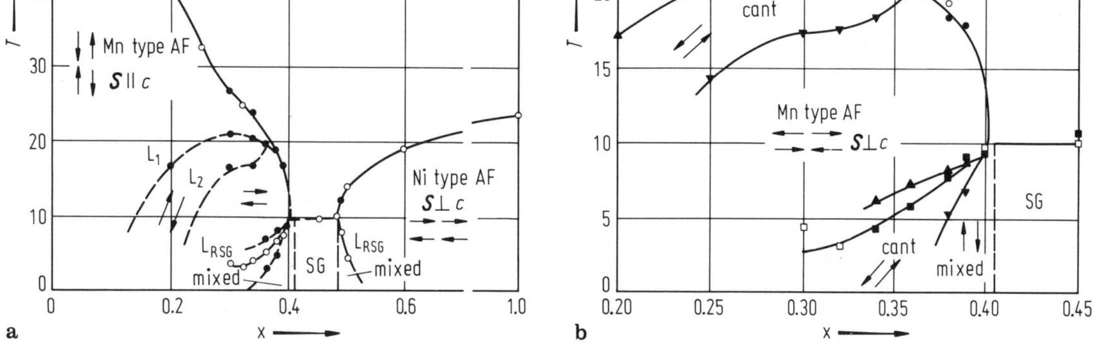

Fig. 92. $Ni_xMn_{1-x}TiO_3$. Magnetic phase diagram derived from magnetization measurements (open symbols) and from neutron scattering data (closed symbols). (a) Full diagram [92I2]. The lines L_1, L_2 indicate starting temperature and end of spin rotation, respectively. (b) Enlarged part of the diagram [92Y1].

$Fe_xCo_{1-x}TiO_3$

The $Fe_xCo_{1-x}TiO_3$ is a mixture of two ilmenite systems $FeTiO_3$ and $CoTiO_3$. The spin structures of both systems are identical except for the spin directions; the easy axis of $FeTiO_3$ is along the hexagonal *c* axis and that of $CoTiO_3$ is in the *c* plane. Therefore, $Fe_xCo_{1-x}TiO_3$ is a random mixture with competing spin anisotropy. The temperature versus Fe concentration phase diagram of this mixture is shown in Fig. 93, which is determined by neutron scattering and Mössbauer measurements. It shows the existence of three ordered phases: two antiferromagnetic phases, and a mixed ordering phase (named by Aharony and Fishman [76A1, 78F1]) or an oblique antiferromagnetic phase (named by Matsubara and Inawashiro [77M1]).

For the neutron scattering measurements, the temperature variations of the intensities at (0, 0, 1.5) and (1, 1, 1.5) magnetic reciprocal-lattice positions have been measured. The typical results are shown in Fig. 94. The intensities at these positions are related to the squares of the components of the sub-lattice magnetization, $|M_\perp|^2$ and $|M_\parallel|^2$, as follows [82I1],

$$I_{0,0,1.5} \propto |\bar{M}_\perp|^2 \ ,$$

$$I_{1,1,1.5} \propto 0.931\,|\bar{M}_\parallel|^2 + 0.534\,|\bar{M}_\perp|^2 \ ,$$

where

$$|\bar{M}_{\parallel(\perp)}| \equiv \mathrm{x}\,|M^{\mathrm{Fe}}_{\parallel(\perp)}| + (1-\mathrm{x})\,|M^{\mathrm{Co}}_{\parallel(\perp)}| \ .$$

The results for the samples with intermediate concentrations indicate that the long-range order of one spin component $S_\parallel(S_\perp)$ is established first at T_N, and then that of another one $S_\perp(S_\parallel)$ is done at T_L ($< T_N$) as the temperature decreases. Therefore, the orthogonal spin components are decoupled with each other.

From the Mössbauer spectra of $Fe_xCo_{1-x}TiO_3$ at 4.2 K, on the other hand, it is observed that, except for the samples with $x \leq 0.02$ and $x = 1.0$, the angles between the direction of Fe^{2+} spin and the *c* axis are not equal to 90° nor 0°. Moreover, the results of the temperature variations of the Mössbauer spectra indicate that the directions of Fe^{2+} spins for all samples except for the samples with $x \leq 0.02$ and $x = 1.0$ are oblique to the spin axis of pure compounds at all temperatures below T_N. From the Mössbauer spectra of the samples with intermediate concentrations, it is found that the direction of Fe^{2+} spin varies gradually with temperature without any abrupt change at T_L determined by the neutron scattering measurements [82I1]. These results mean that the spin components $S_\parallel$ and $S_\perp$ are closely coupled below T_N at least in the time scale of the Larmor precession time $\approx 10^{-7}$ s of ^{57}Fe.

The neutron scattering gives information concerning the site-averaged value of spins. On the other hand, the Mössbauer spectroscopy gives information concerning the time-averaged value of a single spin within the nuclear precession time. Therefore, the discrepancy between the result of the neutron scattering measurement and that of the Mössbauer measurement is interpreted as follows. In the $S_\parallel(S_\perp)$-ordered antiferromagnetic phase, the orientation of $S_\perp(S_\parallel)$ distributes, which does not produce the coherent scattering of neutron. But $S_\perp(S_\parallel)$ of each spin exists statically at least in the time scale of the Mössbauer time scale of 10^{-7} s. For Mössbauer data see Table 18.

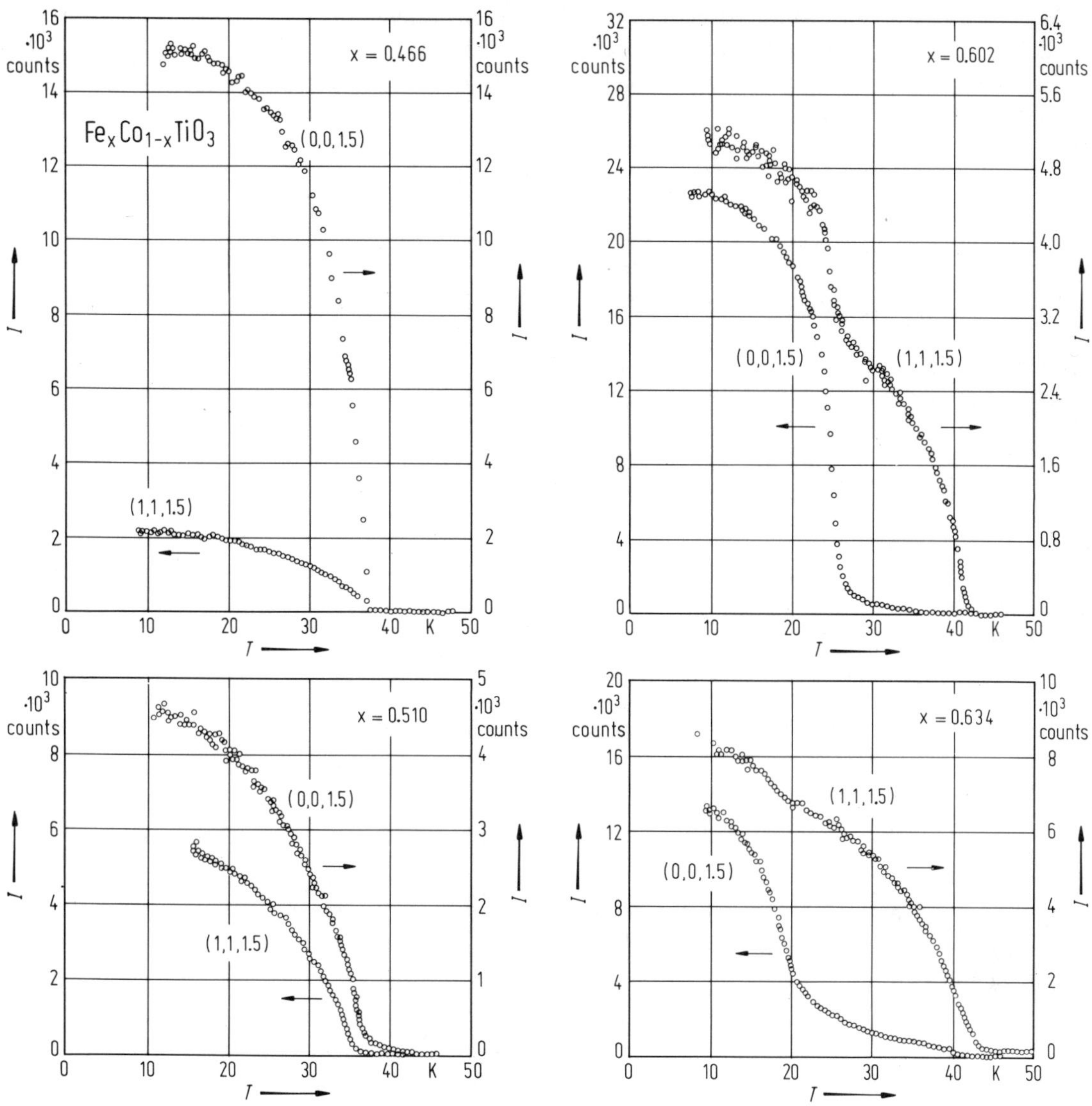

Fig. 94. $Fe_xCo_{1-x}TiO_3$, single crystals. Temperature variation of the neutron scattering intensities at the (0, 0, 1.5) and (1, 1, 1.5) magnetic reciprocal lattice positions for various values of the composition parameter x [82I1].

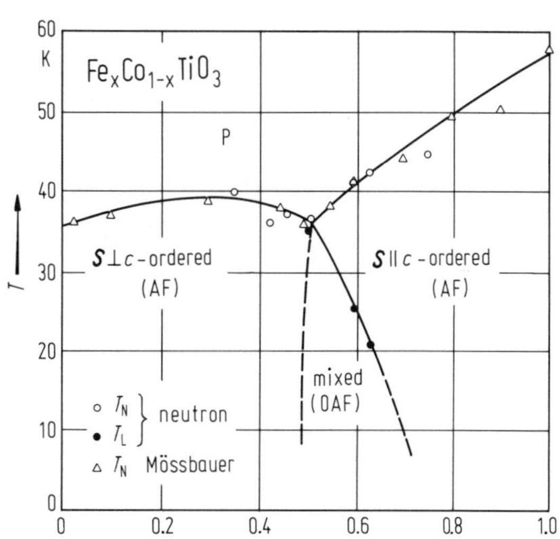

Fig. 93. $Fe_xCo_{1-x}TiO_3$. Magnetic phase diagram as derived from neutron scattering (circles) and Mössbauer spectra (triangles) [82I1]. T_N denotes the Néel temperature, below which a long range order of the spin components parallel or perpendicular to the c axis occurs. T_L denotes the transition to the mixed order phase OAF where also an ordering of the perpendicular components of the spins occurs.

Table 18. $Fe_xCo_{1-x}TiO_3$, enriched with ^{57}Fe. Mössbauer parameters at 4.2 K. The angle θ_H is the polar angle defining the orientation of H_{int} relative to the principal axis of the EFG tensor. It is assumed that the azimuthal angle is zero [82I1].

Compound x	θ_H [deg]	η	H_{int} [kOe]	$e^2qQ/2$ [mm/s]
0.02	90	0.6	106(2)	0.71
0.10	75	0.6	103	0.8
0.30	65	0.6	95	1.1
0.50	55	0.6	93	1.3
0.60	50	0.6	80	1.3
0.75	40	0.6	70	1.35
0.80	25	0.4	60	1.4
1.00	0	0.0	45(2)	1.42

$Fe_xMg_{1-x}TiO_3$

Néel temperatures: Fig. 95a. Susceptibility and magnetization curves: Fig. 95b, c. For low Fe concentration around the percolation concentration $x_c \approx 0.25$ a spin glass or a reentrant spin glass like behavior is found [92I3]. Mössbauer spectra: Fig. 95d.

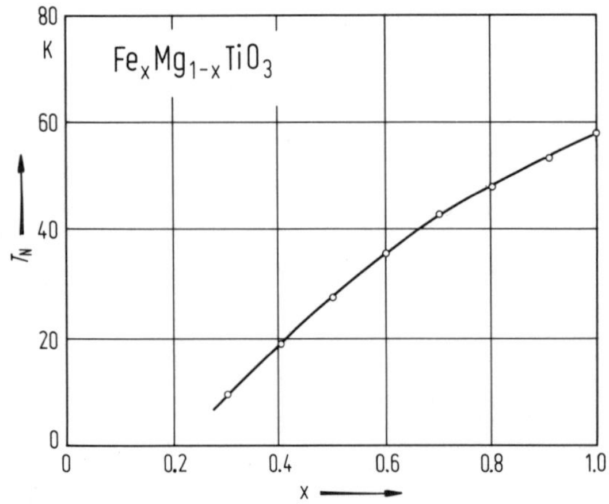

Fig. 95a. $Fe_xMg_{1-x}TiO_3$, single crystals. Néel temperature T_N vs. composition parameter x [91K1].

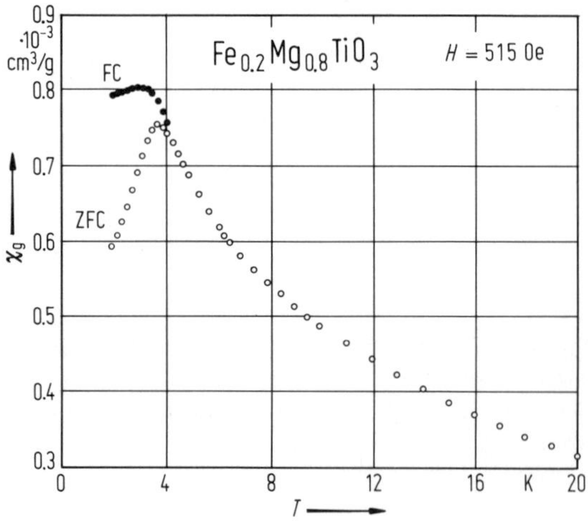

Fig. 95c. $Fe_{0.2}Mg_{0.8}TiO_3$. Temperature variation of FC and ZFC low amplitude dc susceptibilities. $T_{SG} \simeq 4.6$ K [92I3].

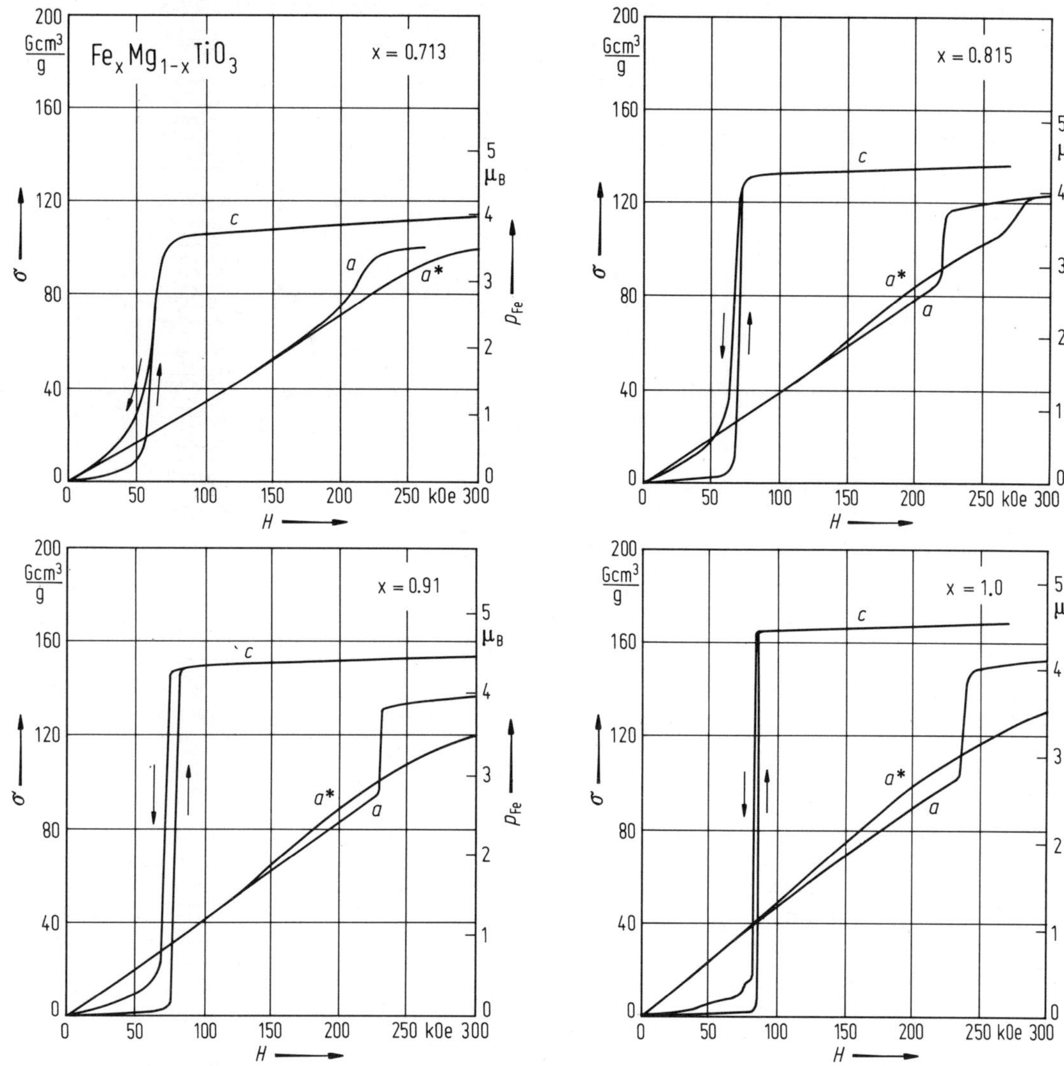

Fig. 95b. $Fe_xMg_{1-x}TiO_3$, single crystals, Magnetization curves in three orthogonal directions at 4.2 K. The right hand scale gives the magnetic moment per Fe^{2+} ion [91K1]. a*: axis perpendicular to both a and c axes.

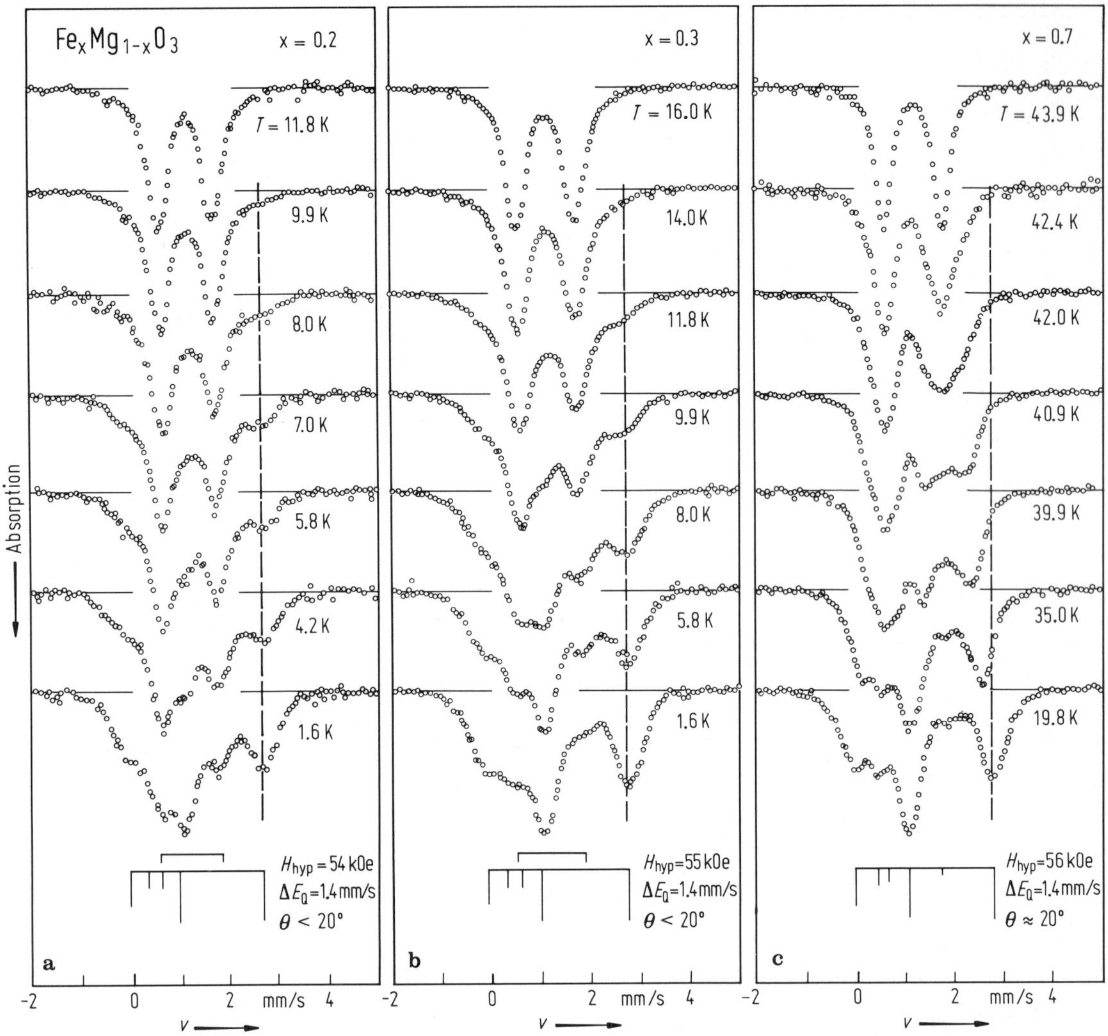

Fig. 95d. $Fe_xMg_{1-x}TiO_3$, powder. Mössbauer spectra, the Mössbauer parameters H_{hyp}, $\Delta E_Q = e^2qQ/2$ and the angle θ between the direction of H_{hyp} and the c axis are also given. The positions of the absorption lines at the lowest temperature examined are shown at the bottom [92I3].

5.6.5 Solid solutions between oxides with corundum and ilmenite structure

$FeTiO_3$–Fe_2O_3 system

Earlier work reported in Landolt-Börnstein, New Series, subvolume III/4a, p. 31–33.

The $(1-x)FeTiO_3 \cdot xFe_2O_3$ is a solid solution of ilmenite and hematite. At low temperatures, both $FeTiO_3$ and α-Fe_2O_3 become antiferromagnetic with slightly different crystal symmetries of $R\bar{3}$ and $R\bar{3}c$, respectively. In $FeTiO_3$, the Fe^{2+} spins are aligned along the *c* axis, while the Fe^{3+} spins in Fe_2O_3 are parallel to the *c* plane. When the Fe_2O_3 concentration x in $(1-x)FeTiO_3 \cdot xFe_2O_3$ is small, the Fe^{3+} ions in the Ti^{4+} layers align the moments of the adjacent Fe^{2+} ions in parallel by the strong nearest-neighbour interactions. It brings about an isolated ferrimagnetic cluster, which behaves superparamagnetically at high temperatures. In samples with x larger than the percolation threshold $x \approx 0.2$, the ferrimagnetic long-range order is established at high temperatures. The behavior of the cluster spin-glass appears at low temperatures in samples with near $x \approx 0.2$. This is because there exists the competition between the nearest-neighbour and the next-nearest-neighbour interactions and it creates the frustrated spins located at the boundary of the ferrimagnetic clusters.

The phase diagram of $(1-x)FeTiO_3 \cdot xFe_2O_3$ around the percolation threshold is shown in Fig. 96, which is determined by the magnetization measurement. The typical results of the temperature variations of the magnetization are shown in Figs. 97 and 98. For the sample with $x = 0.1$, a sharp cusp indicating the spin-glass transition appears in a low field parallel to the *c* axis. This cusp is broadened as the measuring field increases. By the neutron scattering measurements, it is found that the spin axis inside the cluster aligns in the *c* axis and the antiferromagnetic modulation starts to occur inside the cluster by the next-nearest-neighbour interactions at the spin-glass transition temperature. On the other hand, for the sample with $x = 0.21$, the magnetization in a low field perpendicular to the *c* axis shows a typical behavior of the reentrant spin-glass. By the neutron scattering measurements, the squared-Lorentzian-type of spin correlations between transverse spin components have been found to develop below the ferrimagnetic transition temperature. At the reentrant spin-glass transition temperature, on the other hand, the correlations between parallel components appear as well as the correlations between transverse components, which is attributed to the onset of the large domains with the canted spin structure [85I1, 85A1, 85A2].

Muon spin relaxation properties [83U1] and Fig. 99.

For Figs 96 ··· 98, see next page.

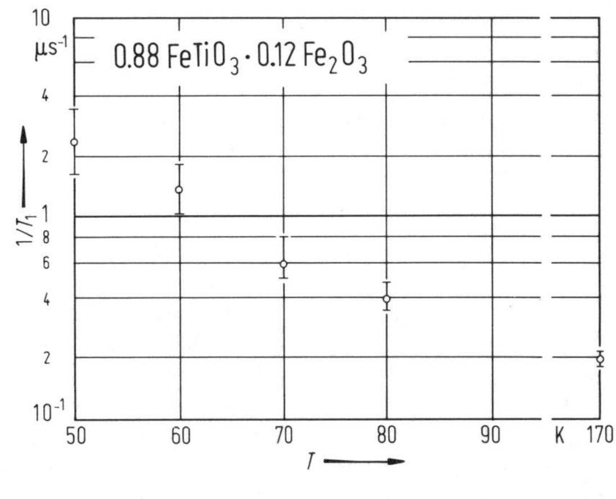

Fig. 99. $0.88\ FeTiO_3 \cdot 0.12\ Fe_2O_3$, powder. Zero-field muon spin relaxation rate T_1^{-1} vs. temperature T [83U1].

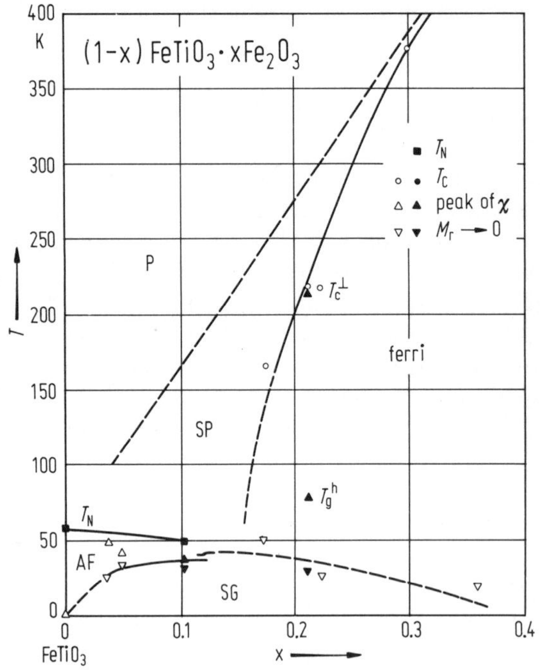

Fig. 96. (1 − x) $FeTiO_3 \cdot xFe_2O_3$. Magnetic phase diagram based on magnetization measurements and on neutron scattering. Temperature where the thermo-remanent magnetization disappears TRM (▽, ▼), temperature of the low-field susceptibility peak (△, ▲), closed symbols: single crystals; open symbols: powder measurements. SP: superparamagnetism; T_g^h : see caption Fig. 98 [85I1].

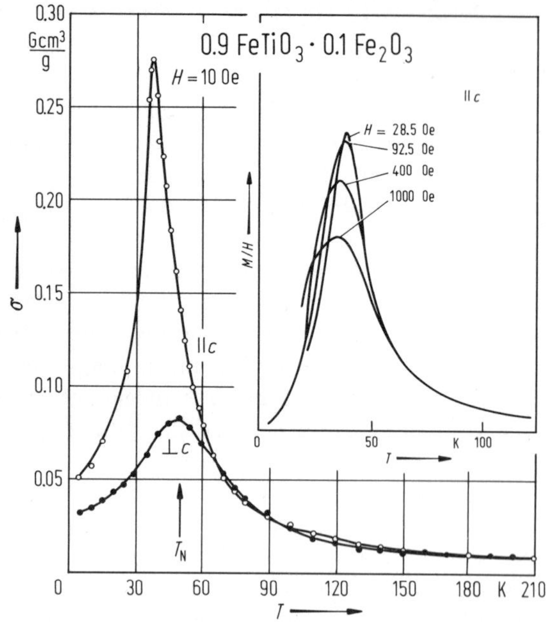

Fig. 97. 0.9 $FeTiO_3 \cdot 0.1\ Fe_2O_3$, single crystal. Temperature dependence of the mass magnetization σ measured with increasing T in a static magnetic field of 10 Oe parallel and perpendicular to the c axis. The influence of the magnitude of the field is shown in the inset [83I1].

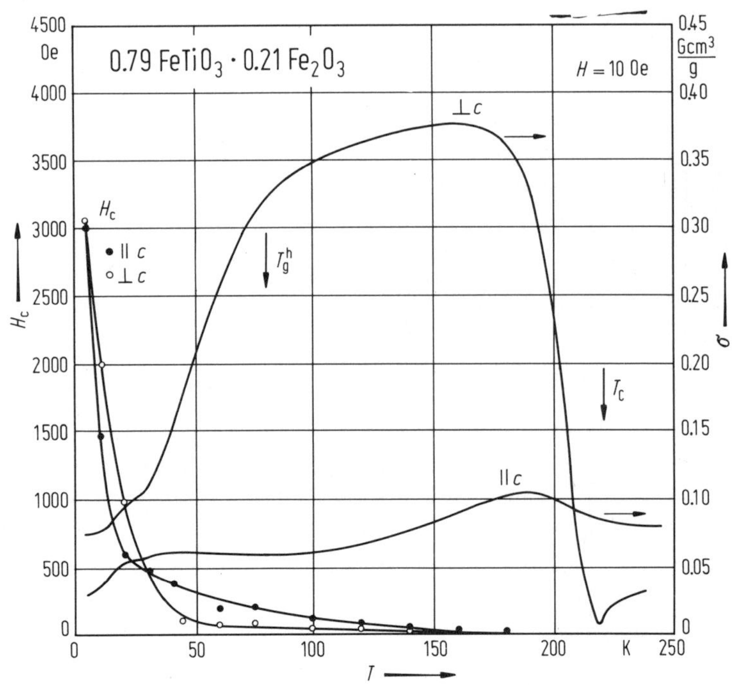

Fig. 98. 0.79 $FeTiO_3 \cdot 0.21\ Fe_2O_3$, single crystal. Mass magnetization σ measured in a static field of 10 Oe at increasing temperatures T, parallel and perpendicular to the c axis. The value of the coercive force H_c is also given. T_g^h = 80 K, the temperature where there is a peak in the short range correlations between the spin components perpendicular to the c axis [85I1, 85A3].

$MnTiO_3$–Fe_2O_3 system

$CoTiO_3$–Fe_2O_3 system

$NiTiO_3$–Fe_2O_3 system

$MgTiO_3$–Fe_2O_3 system

These systems have been dealt with in Landolt-Börnstein, subvolume III/4a, pages 33–36.

5.6.6. References for 5.6

34B1 Barth, F. W., Posnjak, E.: Z. Kristallogr. **88** (1934) 265.
39F1 Foëx, G., Graff, M.: C. R. Acad. Sci. (Paris) **209** (1939) 16c.
50T1 Trounson, E. P., Bleil, D. F., Wangsness, R. K., Maxwell, L. R.: Phys. Rev. **79** (1950) 542.
51C1 Coughlin, J. P., King, E. G., Bonnickson, K. R.: J. Am. Chem. Soc. **73** (1951) 3891.
51S1 Shull, C. G., Strauser, W. A., Wollan, E. O.: Phys. Rev. **83** (1951) 333.
53M1 Maxwell, L. R., McGuire, T. R.: Rev. Mod. Phys. **25** (1953) 279.
54A1 Adler, S. F., Selwood, P. W.: J. Am. Chem. Soc. **76** (1954) 346.
55F1 Foëx, M., Wucher, J.: C. R. Acad. Sci. (Paris) **241** (1955) 184.
56I1 Iida, S.: J. Phys. Soc. Jpn. **11** (1956) 1300.
56L1 Li, Y.: Phys. Rev. **4** (1956) 1015.
58A1 Almeida, J. R. L. de, Thouless, D. J.: J. Phys. **A11** (1958) 983.
58B1 Bozorth, R. M., Walsh, D. E.: J. Phys. Chem. Solids **5** (1958) 299.
58B2 Bertaut, E. F., Forrat, F.: J. Appl. Phys. **29** (1958) 247.
58I1 Ishikawa, Y., Akimoto, S.: J. Phys. Soc. Jpn. **13** (1958) 1110.
58I2 Ishikawa, Y., Akimoto, S.: J. Phys. Soc. Jpn. **13** (1958) 1298.
58P1 Pearson, A. D.: J. Phys. Chem. Solids **5** (1958) 316.
58S1 Swoboda, T. J., Toole, R. C., Vaughan, J. D.: J. Phys. Chem. Solids **5** (1958) 293.
59G1 Grønvold, F., Westrum, E. F., jr.: J. Am. Chem. Soc. **81** (1959) 1780.
59S1 Shirane, G., Pickart, S. J., Ishikawa, Y.: J. Phys. Soc. Jpn. **14** (1959) 1352.
59S2 Shirane, G., Pickart, S, J., Nathans, R., Ishikawa, Y.: J. Phys. Chem. Solids **10** (1959) 35.
60B1 Berry, C. R., Combs, C. M.: J. Appl. Phys. **31** (1960) 1130.
60C1 Carr, P. H., Foner, S.: J. Appl. Phys. **31S** (1960) 344S.
61F1 Folen, V. J., Rado, G. T., Stalder, E. W.: Phys. Rev. Lett. **6** (1961) 607.
62C1 Cox, D. E., Takei, W. J., Miller, R. C., Shirane, G.: J. Phys. Chem. Solids **23** (1962) 863.
62N1 Newnham, R. E., de Haan, Y. M.: Z. Kristallogr. **117** (1962) 235.
62O1 Ono, K., Ito, A.: J. Phys. Soc. Jpn. **17** (1962) 1012.
62R1 Rado, G. T., Folen, V. J.: J. Appl. Phys. **33S** (1962) 1126.
63A1 Abrahams, S. C.: Phys. Rev. **130** (1963) 2230.
63C1 Cox, D. E., Takei, W. J., Shirane, G.: J. Phys. Chem. Solids **24** (1963) 405.
63F1 Foner, S.: Phys. Rev. **130** (1963) 183.
63K1 Knappwost, A., Gunsser, W.: Ber. Bunsenges. Phys. Chem. **67** (1963) 588.
64C1 Corliss, L. M., Hastings, J. M.: J. Phys. (Paris) **25** (1964) 557.
64N1 Newnham, R. E., Fang, J. H., Santoro, R. P.: Acta Crystallogr. **17** (1964) 240.
64O1 Osmond, W. P.: Brit. J. Appl. Phys. **15** (1964) 1377.
64U1 Uitert, L. G. van, Sherwood, R. C., Williams, H. J., Rubin, J. J., Bonner, W. A.: J. Phys. Chem. Solids **25** (1964) 1447.
65A1 Artman, J. O., Murphy, J. C., Foner, S.: J. Appl. Phys. **36** (1965) 986.

65C1 Corliss, L. M., Hastings, J. M., Nathans, R., Shirane, G.: J. Appl. Phys. **36** (1965) 1099.
65J1 Jones, E. D.: Phys. Rev. **137** (1965) A978.
65K1 Krén, E., Szabó, P., Konczos, G.: Phys. Lett. **19** (1965) 103.
65M1 Møglestue, K. T.: Thesis, Oslo University 1965.
65S1 Shinjo, T., Kosuge, K., Shiga, M., Nakamura, Y., Kachi, S., Takaki, H.: Phys. Lett. **19** (1965) 91.
66S1 Sawaoka, A., Miyahara, S., Akimoto, S., Fujisawa, H.: J. Phys. Soc. Jpn. **21** (1966) 185.
66W1 Woude, F. van der: Phys. Status Solidi **17** (1966) 417.
67D1 Dézsi, I., Erlaki, G., Keszthelyi, L.: Phys. Status Solidi **21** (1967) K121.
67G1 Goodenough, J. B., Stickler, J. J.: Phys. Rev. **164** (1967) 768.
67S1 Stickler, J. J., Kern, S., Wold, A., Heller, G. S.: Phys. Rev. **164** (1967) 765.
67S2 Steinwehr, H. F. von: Z. Kristallogr. **125** (1967) 377.
67V1 Vaughan, R. W., Drickamer, H. G.: J. Chem. Phys. **47** (1967) 1530.
67W1 Worlton, T. G., Bennion, R. B., Brugger, R. M.: Phys. Lett. **24A** (1967) 653.
68A1 Arnold, D. J., Mires, R. W.: J. Chem. Phys. **48** (1968) 2231.
68K1 Kendrick, H., Arrott, A., Werner, S. A.: J. Appl. Phys. **39** (1968) 585.
68S1 Shaked, H., Shtrikman, S.: Solid State Commun. **6** (1968) 425.
68W1 Worlton, T. G., Brugger, R. M., Bennion, R. B.: J. Phys. Chem. Solids **29** (1968) 435.
69A1 Alikhanov, R. A., Dimitrijevic, Z., Kowalska, A., Krasnicki, S., Rzany, H., Todorovic, J., Wanic, A.: Phys. Status Solidi **32** (1969) 41.
69C1 Chamberland, B. L., Cloud, W. H.: J. Appl. Phys. **40** (1969) 434.
69M1 Moon, R. M., Riste, T., Koehler, W. C.: J. Appl. Phys. **40** (1969) 1445.
69M2 McWhan, D. B., Rice, T. M., Remeika, J. P.: Phys. Rev. Lett. **23** (1969) 1384.
69S1 Shapira, Y.: Phys. Rev. **184** (1969) 589.
69S2 Shapira, Y.: Phys. Rev. **187** (1969) 734.
69S3 Syono, Y., Akimoto, S., Ishikawa, Y., Endoh, Y.: J. Phys. Chem. Solids **30** (1969) 1665.
70A1 Akimitsu, J., Ishikawa, Y., Endoh, Y.: Solid State Commun. **8** (1970) 87.
70A2 Andres, K.: Phys. Rev. **B2** (1970) 3768.
70C1 Chamberland, B. L.: J. Solid State Chem. **1** (1970) 138.
70C2 Chamberland, B. L.: J. Solid State Chem. **2** (1970) 521.
70C3 Chamberland, B. L., Sleight, A. W., Weiher, J. F.: J. Solid State Chem. **1** (1970) 512.
70H1 Heidemann, A.: Z. Physik **238** (1970) 208.
70J1 Jayaraman, A., McWhan, D. B., Remeika, J. P., Dernier, P. D.: Phys. Rev. **B2** (1970) 3751.
70M1 Moon, R. M.: Phys. Rev. Lett. **25** (1970) 527.
70M2 Menth, A., Remeika, J. P.: Phys. Rev. **B2** (1970) 3756.
70S1 Samuelsen, E. J., Shirane, G.: Phys. Status Solidi **42** (1970) 241.
70S2 Samuelsen, E. J., Hutchings, M. T., Shirane, G.: Physica **48** (1970) 13.
71D1 Dudko, K. L., Eremenko, V. V., Semenenko, L. M.: Phys. Status Solidi **B43** (1971) 471.
71G1 Gossard, A. C., Menth, A., Warren, W. W., Remeika, J. P.: Phys. Rev. **B3** (1971) 3993.
71S1 Shapira, Y.: J. Appl. Phys. **42** (1971) 1588.
71S2 Syono, Y., Akimoto, S., Endoh, Y.: J. Phys. Chem. Solids **32** (1971) 243.
71Y1 Yasuoka, H., Nishihara, H., Nakamura, Y., Remeika, J. P.: Phys. Lett. **A37** (1971) 299.
72G1 Graut, R. W., Housley, R. M., Geller, S.: Phys. Rev. **B5** (1972) 1700.
72R1 Reznitskii, L. A., Filippora, S. E.: Inorg. Mater. **8** (1972) 419.
72R2 Reid, A. F., Sabine, T. M., Wheeler, D. A.: J. Solid State Chem. **4** (1972) 400.
73G1 Gorodetsky, G., Hornreich, R. M., Shtrikman, S.: Phys. Rev. Lett. **31** (1973) 938.
73S1 Sawamoto, H.: Mater. Res. Bull. **8** (1973) 767.
73S2 Sjöstrand, M. E., Keesom, P. H.: Phys. Rev. **B7** (1973) 3558.
74B1 Bachellerie, A., Frenois, Ch.: J. Phys. (Paris) **35** (1974) 437.
74T1 Tsuzuki, K., Ishikawa, Y., Watanabe, N., Akimoto, S.: J. Phys. Soc. Jpn. **37** (1974) 1242.
75B1 Birchall, T., Freid, A. F.: J. Solid State Chem. **13** (1975) 351.
75D1 Dumas, J., Schlenker, C., Tholence, J. L., Tournier, R.: Solid State Commun. **17** (1975) 1215.

75G1 Grønvold, F., Samuelsen, E. J.: J. Phys. Chem. Solids **36** (1975) 249.
75M1 Motoya, K., Yasuoka, H., Nakamura, Y., Kosuge, K., Kachi, S.: J. Phys. Soc. Jpn. **39** (1975) 1137.
75P1 Pernet, M., Joubert, J. C., Ferrand, B.: Solid State Commun. **16** (1975) 503.
76A1 Aharony, A., Fishman, S.: Phys. Rev. Lett. **37** (1976) 1587.
76D1 Dumas, J., Schlenker, C.: J. Phys. Coll. **37** (1976) C4–41.
77B1 Bruce, R. H., Cannell, D. S.: Phys. Rev. **B15** (1977) 4451.
77M1 Matsubara, F., Inawahiro, S.: J. Phys. Soc. Jpn. **42** (1977) 1529.
77P1 Parette, G., Madhav Rao, L.: Solid State Commun. **23** (1977) 179.
77Y1 Yacovitch, R. D., Shapira, Y.: Physica **86–88B** (1977) 1126.
78B1 Belbéoch, R., Kleinberger, R., Roulliay, M.: Solid State Commun. **25** (1978) 1043.
78B2 Belbéoch, R., Kleinberger, R., Roulliay, M.: J. Solid State Chem. **39** (1978) 1007.
78F1 Fishman, S., Aharony, A.: Phys. Rev. **B18** (1978) 3507.
78K1 Kowalska, A., Sloniowska, B.: Acta Phys. Pol. **A 54** (1978) 679.
78N1 Napijalo, M. L., Srećković, A., Novaković, L.: Fizika (Zagreb) **10** Suppl. (1978) 169.
78U1 Ueda, Y., Kosuge, K., Kachi, S., Yasuoka, H., Nishihara, H., Heidemann, A.: J. Phys. Chem. Solids **39** (1978) 1281.
79C1 Chikazawa, S., Sato, T., Miyako, Y., Yuochunas, Y. G.: Phys. Lett. **A72** (1979) 470.
80C1 Chikazawa, S., Yuochunas, Y. G., Miyako, Y.: J. Phys. Soc. Jpn. **49** (1980) 1276.
80M1 Miyako, Y., Chikazawa, S., Saito, T., Yuochunas, Y. G.: J. Phys. Soc. Jpn. **48** (1980) 329.
80M2 Miyako, Y., Saito, T., Chikazawa, S., Yuochunas, Y. G.: J. Phys. Soc. Jpn. **49** (1980) 1193.
80R1 Rüegg, K., Boekema C., Denison, A., Hofmann, W., Kündig, W.: J. Magn. Magn. Mater. **15–18** (1980) 669.
80U1 Ueda, Y., Kosuge, K., Kachi, S.: J. Solid State Chem. **31** (1980) 171.
80W1 Watanabe, H., Yamauchi, H., Takei, H.: J. Magn. Magn. Mater. **15–18** (1980) 549.
81C1 Chikazawa, S., Sandberg, C. J., Miyako, Y.: J. Phys. Soc. Jpn. **50** (1981) 2884.
81M1 Miyako, Y., Chikazawa, S., Saito, T., Yuochunas, Y. G.: J. Appl. Phys. **52** (1981) 1779.
81S1 Syono, Y., Ito, A., Morimoto, S.: J. Phys. Chem. Solids **42** (1981) 483.
81W1 Word, R. E., Werner, S. A., Yelon, W. B., Honig, J. M., Shivashankar, S.: Phys. Rev. **B 23** (1981) 3533.
81Y1 Yelon, W. B., Werner, S. A., Shivashankar, S., Honig, J. M.: Phys. Rev. **B24** (1981) 1818.
82I1 Ito, A., Morimoto, S., Someya, Y., Syono, Y., Takei, H.: J. Phys. Soc. Jpn. **51** (1982) 3173.
82K1 Kato, H., Yamada, M., Yamauchi, H., Hiroyoshi, H., Takei, H., Watanabe, H.: J. Phys. Soc. Jpn. **51** (1982) 1769.
83B1 Bruzzone, C. L., Ingalls, R.: Phys. Rev. **B28** (1983) 2430.
83I1 Ishikawa, Y., Arai, M., Saito, N., Kohgi, M.: J. Magn. Magn. Mater. **31–34** (1983) 1381.
83K1 Kato, H., Yamaguchi, Y., Ohashi, M., Yamada, M., Takei, H.: Solid State Commun. **45** (1983) 669.
83K2 Kato, H., Funahashi, S., Yamaguchi, Y., Yamada, M., Takei, H.: J. Magn. Magn. Mater. **31–34** (1983) 617.
83S1 Saito, T., Miyako, Y., Sandberg, C. J.: J. Magn. Magn. Mater. **31–34** (1983) 1387.
83U1 Uemura, Y. J., Nishiyama, K., Kadono, R., Imazato, J., Kuno, Y., Nagamine, K., Yamazaki, T., Ishikawa, Y.: J. Magn. Magn. Mater. **31–34** (1983) 1379.
83Y1 Yamaguchi, H., Hiroyoshi, H., Yamada, M., Watanabe, H., Takei, H.: J. Magn. Magn. Mater. **31–34** (1983) 1071.
85A1 Arai, M., Ishikawa, Y., Saito, N., Takei, H.: J. Phys. Soc. Jpn. **54** (1985) 781.
85A2 Arai, M., Ishikawa, Y.: J. Phys. Soc. Jpn. **54** (1985) 795.
85A3 Arai, M., Ishikawa, Y., Takei, H.: J. Phys. Soc. Jpn. **54** (1985) 2279.
85I1 Ishikawa, Y., Saito, N., Arai, M., Watanabe, Y., Takei, H.: J. Phys. Soc. Jpn. **54** (1985) 312.
85S1 Saito, T., Sandberg, C. J., Miyako, Y.: J. Phys. Soc. Jpn. **54** (1985) 231.
86I1 Ito, A., Aruga, H., Torikai, E., Kikuchi, M., Syono, Y., Takai, H.: Phys. Rev. Letters **57** (1986) 483.

86K1 Kato, H., Yamaguchi, Y., Yamada, M., Funahashi, S., Nakagawa, Y., Takei, H.: J. Phys. C: Solid State Phys. **19** (1986) 6993.

86T1 Todate, Y., Ishikawa, Y., Tajima, K., Tomiyoshi, S., Takei, H.: J. Phys. Soc. Jpn. **55** (1986) 4464.

86Y1 Yamaguchi, Y., Kato, H., Takei, H.: Solid State Commun. **59** (1986) 865.

87Y1 Yoshizawa, H., Mitsuda, S., Aruga, H., Ito, A.: Phys. Rev. Letters **59** (1987) 2364.

88A1 Aruga, H., Tokoro, T., Ito, A.: J. Phys. Soc. Jpn. **57** (1988) 261.

88A2 Aruga, H., Ito, H., Wakabayashi, H., Goto, T.: J. Phys. Soc. Jpn. **57** (1988) 2636.

88I1 Ito, A., Aruga, H., Kikuchi, M., Syono, Y., Takei, H.: Solid State Commun. **66** (1988) 475.

88I2 Ito, A., Aruga, H., Morimoto, S., Yoshizawa, H.: J. Phys. (Paris) Suppl. **49** (1988) C8–1129.

88T1 Takayama, H.; Prog. Theor. Phys. 80 (1988) 827.

89Y1 Yoshizawa, H., Mitsuda, S., Aruga, H., Ito, A.: J. Phys. Soc. Jpn. **58** (1989) 1416.

90I1 Ito, A., Torikai, E., Morimoto, S., Aruga, H., Kikuchi, M., Syono, Y., Takei, H.: J. Phys. Soc. Jpn. **59** (1990) 829.

90I2 Ito, A., Morimoto, S., Aruga, H.: Hyperfine Interact. **54** (1990) 567.

90K1 Kato, H., Yamada, M., Nakagawa, M., Kido, G., Nakagawa Y.: J. Magn. Magn. Mater. **90, 91** (1990) 67.

91K1 Kato, H., Nakagawa, Y., Hosoya, S., Kido, G., Nakagawa, M., Fukuda, T.: J. Appl. Phys. **69** (1991) 4819.

91K2 Kurihara, T., Komai, T., Ito, A., Goto, T.: J. Phys. Soc. Jpn. **60** (1991) 2057.

92A1 Aruga Katori, H., Goto, T., Ebii, S., Ito, A.: J. Magn. Magn. Mater. **104–107** (1992) 1639.

92B1 Bakkaloglu, Ö. F., Thomas, M. F.: J. Magn. Magn. Mater. **104–107** (1992) 1921.

92I1 Ito, A., Ebii, S., Aruga Katori, H., Goto, T.: J. Magn. Magn. Mater. **104–107** (1992) 1635.

92I2 Ito, A., Kawano, H., Yoshizawa, H., Motoya, K.: J. Magn. Magn. Mater. **104–107** (1992) 1637.

92I3 Ito, A., Iwai, K., Oyama, C., Morimoto, S., Kato, H.: Hyperfine Interact. **71** (1992) 1371.

92Y1 Yoshizawa, H., Mori, H., Mitsuda, S., Aruga, H., Kawano, H., Ebii, S., Ito, A.: Physica **B180, 181** (1992) 209.

92Y2 Yoshizawa, H., Kawano, H., Mori, H., Mitsuda, S., Ito, A.: Physica **B180, 181** (1992) 94.

5.7 Amorphous oxides

5.7.1 Introduction

5.7.1.0 Scope of the review

In this review, we have collected the experimental results and summarized theoretical understanding on the magnetic properties of amorphous oxides of the 3d transition elements, viz., Sc, Ti, V, Cr, Mn, Fe, Co, Ni, and Cu. By amorphous, we mean materials which are not spatially periodic, i.e., they are non crystalline and do not possess long range crystalline order as determined by X-ray, neutron and electron diffraction experiments. This review is limited to amorphous oxides which in general tend to be electrical insulators and it does not include metallic alloy glasses and glassy metals which are usually good electrical conductors. All amorphous materials which contain at least one 3d element and oxygen are included. The subchapter 5.7.2 deals with substances for which the amorphous structure is the consequence of the chemistry of glass formers. For the substances of subchapter 5.7.3 the amorphous structure is the consequence of a certain special manufacturing technique. The following books and reviews on amorphous magnets have appeared in recent years.

73 H1	"Amorphous magnetism," edited by H. O. Hooper and A. M. de Graaf (Plenum Press, New York, 1973).
75 C	"Structure of metallic alloy glasses," by G. S. Cargill in Solid State Physics **30** (1975) 227 (Academic Press, New York), edited by F. Seitz, D. Turnbull and S. Ehrenreich.
77 L1	"Amorphous magnetism II," edited by R. A. Levy and R. Hasegawa (Plenum Press, New York, 1977).
78 C	"Amorphous magnetic order," by J. M. D. Coey, J. Appl. Phys. **49** (1978) 1646.
80 C	"Glassy metals," by H. S. Chen, Rep. Prog. Phys. **43** (1980) 353.
83 H	"Glassy metals," edited by R. Hasegawa (CRC Press, New York, 1983).
84 C	"Current trends in amorphous magnetism," by J. M. D. Coey and D. H. Ryan, IEEE Trans. Magn. MAG-**20** (1984) 1278.
84 K	"Amorphous Magnetism," by T. Kaneyoshi (CRC Press, Boca Raton, Florida 1984).
84 M	"Magnetic glasses," by K. Moorjani and J. M. D. Coey (Elsevier, Amsterdam, 1984).
93 S3	"Magnetic ordering in amorphous oxides," by G. Srinivasan, B. Uma Maheshwar Rao, and J. Chen, IEEE. Trans. Magn. **29** (1993) 3078.

Of all the above references, [73H1, 77L1, 78C, 84C, 84K, 84M, and 93S3] contain material relevant to amorphous oxides.

5.7.1.1 Symbols and abbreviations

Symbols

Symbol	Unit	Property
C	$cm^3\ K\ g^{-1}$	Curie constant
C	$J\ g^{-1}\ K^{-1}$	specific heat
d	Å	separation between lattice planes
D	$C\ cm^{-2}$	electric displacement
E	$V\ cm^{-1}$	electric field strength
E_a	eV	activation energy
f	Hz	frequency
g	–	spectroscopic splitting factor
H	Oe	static magnetic field
H_A	Oe	magnetic anisotropy field
H_c	Oe	coercive field
H_{int}	Oe	hyperfine field
H_{res}	Oe	magnetic resonance field
ΔH	Oe	ESR or FMR linewidth
I	–	intensity
J	$eV,\ cm^{-1}$	exchange constant (J/k in K)
k	$J\ K^{-1}$	Boltzmann constant
M	$emu\ cm^{-3} = G$	magnetic moment per unit volume = magnetization
$p(O_2)$	–	oxygen partial pressure (as percentage of total pressure)
p_{eff}	μ_B	effective paramagnetic moment
p_{FU}	μ_B	magnetic moment per formula unit
P	–	probability
T	K, °C	temperature
ΔT	%	temperature change
T_a	K, °C	annealing temperature
T_C	K, °C	Curie temperature
T_{crys}	K, °C	crystallization temperature
T_f	K	spin freezing temperature, spin glass transition temperature
T_g	K	glass transition temperature
T_N	K	Néel temperature
T_s	K	substrate temperature
v	$mm\ s^{-1}$	velocity for Mössbauer spectroscopy, sometimes used for sound velocity
ΔW	%	weight change
γ	$Hz\ Oe^{-1}$	gyromagnetic ratio
δ	$mm\ s^{-1}$	isomer shift
Δ	$mm\ s^{-1}$	quadrupole splitting
θ	deg	diffraction angle
θ_F	$deg\ cm^{-1}$	Faraday rotation
Θ_p	K	paramagnetic Curie temperature
λ	nm	wavelength
μ_B	$J\ T^{-1}$	Bohr magneton

Symbol	Unit	Property
$\overline{\nu}$	cm^{-1}	wavenumber
ρ	Ω cm	electrical resistivity
$\sigma_{(s)}$	emu g^{-1} = G cm^3 g^{-1}	(saturation) magnetic moment per unit mass = specific magnetization
χ	emu	magnetic susceptibility (sometimes given in MKS units)
χ_g	emu g^{-1} = cm^3 g^{-1}	specific magnetic susceptibility
χ_v	emu cm^{-3}	volume susceptibility

Abbreviations

ac	alternating current
AF	antiferromagnetic order
Am, a	amorphous
cryst	crystalline
dc	direct current
DTA	differential thermal analysis
endo	endothermal
eff	effective
exo	exothermal
ESR	electron spin resonance
F	ferromagnetic order
FC	field cooled
Ferri	ferrimagnetic order
FMR	ferromagnetic resonance
M	metal
Mag.	magnetization studies
Möss.	Mössbauer spectroscopy
Neu.	neutron diffraction
P	paramagnetic
rf	radio frequency
R	rare earth (element)
RT	room temperature
S	random antiferromagnetic or speromagnet
SG	spin glass
Sus.	susceptibility data
TG	thermogravimetry
TGA	thermogravimetric analysis
XRD	X-ray diffraction
ZFC	zero field cooled

5.7.2 Oxide glasses

5.7.2.0 General remarks

Most glasses contain two constituents, a glass former and a network modifier. Materials such as P_2O_5, B_2O_3, or SiO_2 provide the skeleton of the amorphous structure and are, therefore, called glass formers. Network modifiers, usually either an alkali or alkali-earth oxide, prevent crystallization by occupying larger interstices in the amorphous network. Glasses are most often prepared by rapidly cooling the molten mixture of constituents with a water cooled copper block or by pouring the mixture onto a high speed roller [84M]. Oxide glasses containing 3d transition metal ions are grouped into three categories, based on the type of glass formers: phosphate, borate, or silicate glasses.

5.7.2.1 Phosphate glasses

5.7.2.1.0 General remarks

A long range magnetic order is observed in glasses containing V^{4+}, Fe^{3+} and amorphous ferrites. (Refer to Tables 1 and 3).

Table 1. Type of magnetic order and ordering temperatures for amorphous phosphate glasses containing $3d^n$-transition metal ions. Refer to Table 3 for similar data for amorphous spinel ferrites in the glasses.

n	Ion	Magnetic order	Ordering temp. [K]	Remarks	Ref.	Figs.
1	Ti^{3+}	–	–	ESR, Sus.	77P, 79P	1···3
	V^{4+}	AF	203 $[(V_2O_5)_{0.6\cdots0.9}(P_2O_5)_{0.4\cdots0.1}]$	ESR	72F	4
3	Cr^{3+}	–	–	ESR	67L	5
5	Mn^{2+}	–	–	Sus.	73E, 73W, 74F1	6
	Fe^{3+}	AF	7 $[(Fe_2O_3)_{0.79}P_2O_5]$	Sus., Neu.	72E, 73E	7,8
6	Fe^{2+}	–	–	Sus.	73W	8
7	Co^{2+}	–	–	Sus.	73E	9
8	Ni^{2+}	–	–	Sus., Neu.	73E	
9	Cu^{2+}	–	–	Sus.	73E	

Table 2. The paramagnetic Curie temperature Θ_p, Curie constant C and the effective magnetic moment p_{eff} for some amorphous transition metal oxide-phosphate glasses determined from the Curie-Weiss law $\chi = C/(T-\Theta_p)$. Refer to Fig. 2 in this section for similar data for titanium-phosphate glasses.

Compound	Θ_p [K]	C [cm³ K / g]	p_{eff} [μ_B]	Ref.
$(V_2O_5)_{0.6}(P_2O_5)_{0.4}$	−117	0.0015	–	73W
$(V_2O_5)_{0.75}(P_2O_5)_{0.25}$	−274	0.0009	–	73W
$(MnO)_{1.3}P_2O_5$ [a])	− 26	–	5.97	73E
$(MnO)_{0.3}(P_2O_5)_{0.7}$	− 48	0.0205	–	73W
$(MnO)_{0.55}(P_2O_5)_{0.45}$	− 31	0.028	–	73W
$(Fe_2O_3)_{0.63}P_2O_5$ [a])	− 58	–	6.01	73E
$(Fe_2O_3)_{0.79}P_2O_5$ [a])	− 58	–	5.85	73E
$(FeO)_{0.55}(P_2O_5)_{0.45}$	−105	0.0288	–	73W
$(FeO)_{0.8}(P_2O_5)_{0.2}$	−289	0.0357	–	73W
$(CoO)_{1.4}P_2O_5$ [a])	− 8	–	4.51	72E

[a]) Phosphate glasses of the composition $(M,O)_x$ P_2O_5 where x is the mole ratio of total M oxide to P_2O_5.

5.7.2.1.1 Ti - based phosphates

In phosphate glasses, Ti^{3+} ($3d^1$) ions occupy distorted octahedral sites [62Y]. Two such crystallographic sites for titanium ions are evident from ESR measurements [77P]: isolated Ti^{3+} ions at octahedral sites of very low symmetry that give a narrow line and ions at sites of higher symmetry that result in a broad component (Fig. 1). Even though magnetic susceptibility data indicate strong antiferromagnetic interactions between trivalent Ti ions [77P], [79P], the data do not show a magnetic phase transition down to 1K (Figs. 2 and 3).

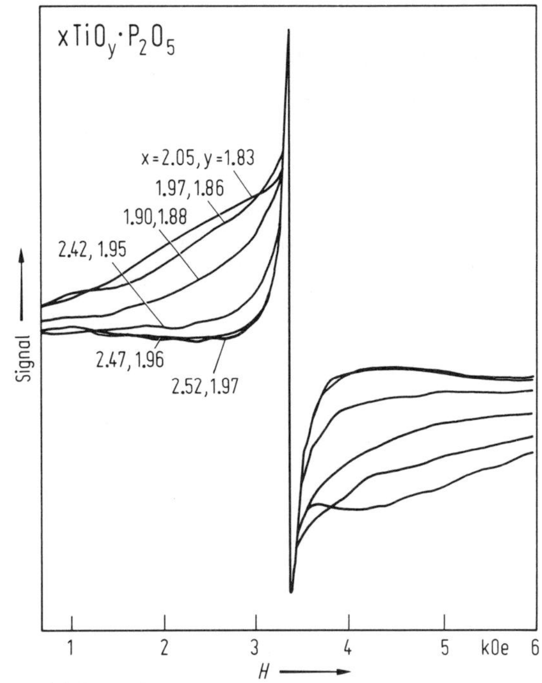

Fig. 1. $xTiO_y{\cdot}P_2O_5$ glass. ESR absorption (g = 1.93(1)) at RT. x is the mole ratio of total titanium oxide to P_2O_5 and y is the titanium oxide stoichiometry. The broad and narrow ESR components are due to Ti^{3+} ions occupying different distorted octahedral crystallographic sites [77P].

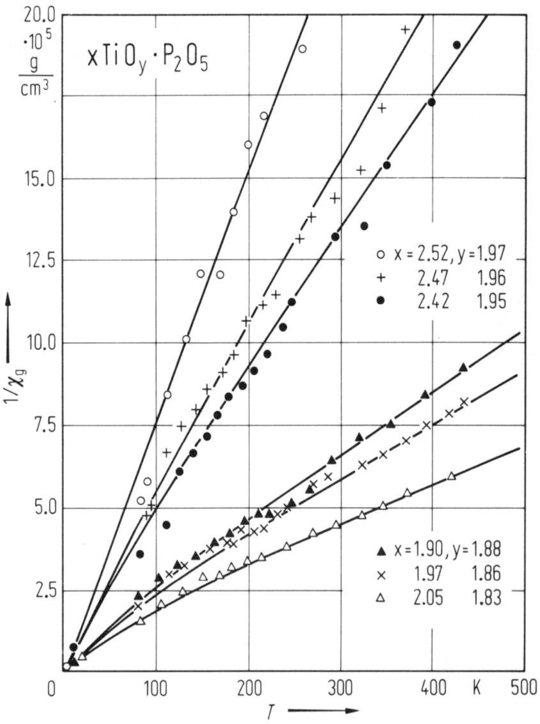

Fig. 2. $xTiO_y·P_2O_5$ glass. Inverse susceptibility vs. temperature. Smooth curves are theoretical plots [79P].

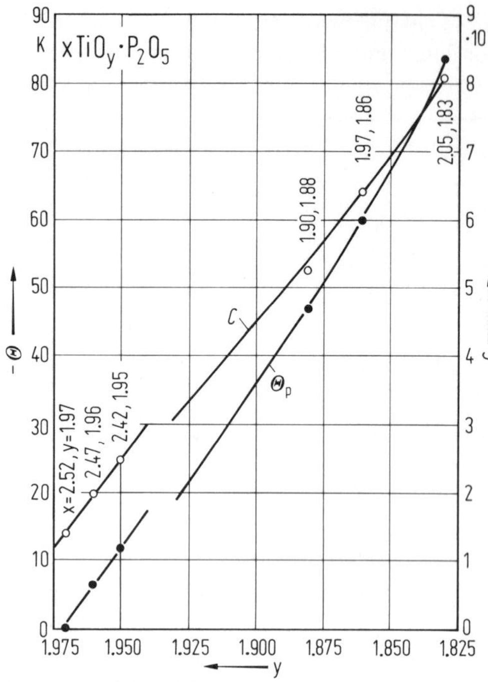

Fig. 3. $xTiO_y·P_2O_5$ glass. Variation of negative paramagnetic Curie temperature Θ_p and the Curie constant C with titanium oxide stoichiometry y [77P]. C is supposed to be in CGS units (cm^3Kg^{-1}).

5.7.2.1.2 V - based phosphates

Studies on 60···90 mol% of V_2O_5 in phosphate glasses reveal an antiferromagnetic exchange interaction between V^{4+} ions, leading to a paramagnetic to antiferromagnetic transition at 203 K (Fig. 4) [72F], [73W]. The magnetic transition is identified from data on the intensity of ESR absorption corresponding to g = 1.960(5) [72F]. High temperature susceptibility data for the oxides obey the Curie-Weiss law [73W]. The magnetic transitions were not confirmed with any other measurements on the system.

5.7.2.1.3 Cr - based phosphates

An octahedral crystallographic environment for Cr^{3+} is inferred from optical absorption and ESR studies on $Cr_2O_3·P_2O_5$ glasses [67L], [71F]. The ESR spectra (Fig. 5) show absorption corresponding to g = 5.25 due to isolated Cr^{3+} and g = 1.96 due to exchange coupled pairs. The antiferromagnetic coupling constant J between chromium ions in such pairs is about 24(2) cm^{-1} [67L].

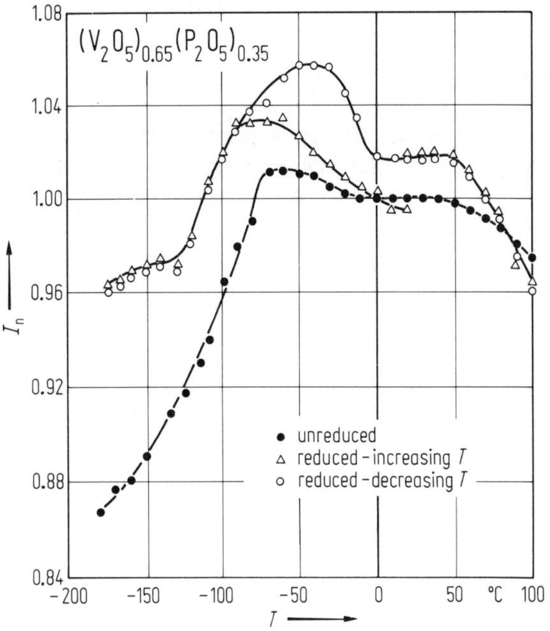

Fig. 4. $(V_2O_5)_{0.65}(P_2O_5)_{0.35}$ glass. Intensity of ESR absorption at g = 1.960(5) vs. temperature data for as-cast and reduced samples. The reduction of V^{5+} was accomplished by adding dextrose to as-cast samples. Notice the hysteresis in the data for the reduced sample for decreasing and increasing temperatures. A magnetic transition is observed at –70°C for the as-cast sample, due to antiferromagnetic ordering of V^{4+}. Additional transitions occur at 70°C and –120°C for the reduced sample [72F]. I_n: normalized intensity.

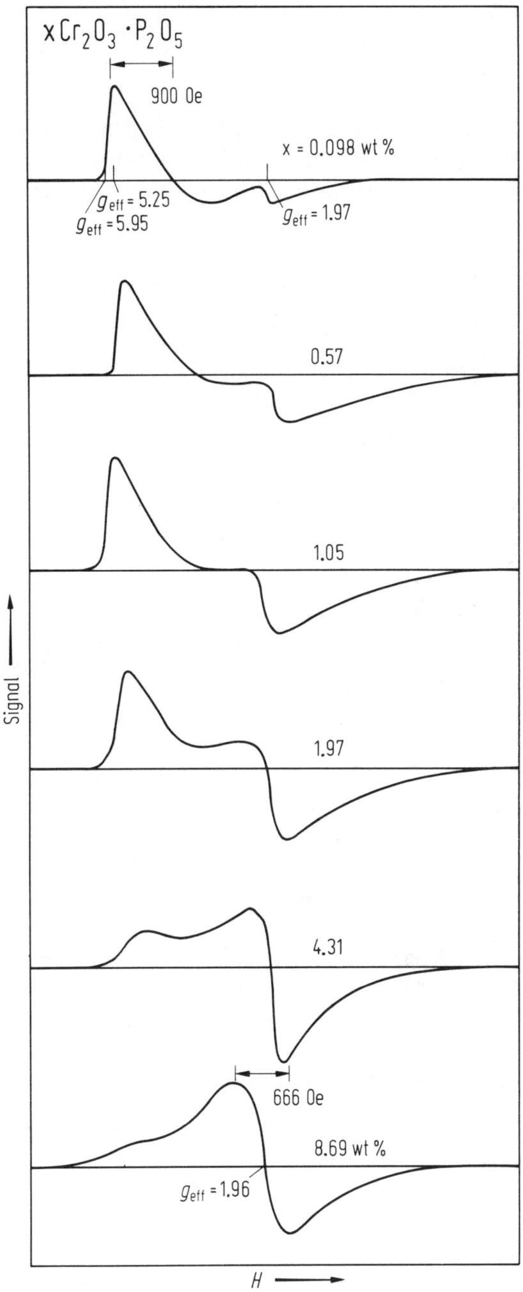

Fig. 5. $xCr_2O_3{\cdot}P_2O_5$ glass. ESR absorption profile for a series of concentration of Cr_2O_3 at T = 77 K. The low and high field absorption are due to isolated and pairs of Cr^{3+}, respectively [67L].

5.7.2.1.4 Mn - based phosphates

From susceptibility studies divalent Mn ions are found to be antiferromagnetically coupled in $MnO{\cdot}P_2O_5$ glasses (Fig. 6) [73W]. However, a paramagnetic to antiferromagnetic transition is not observed in the system [73 E], [74 F1], [74 F2].

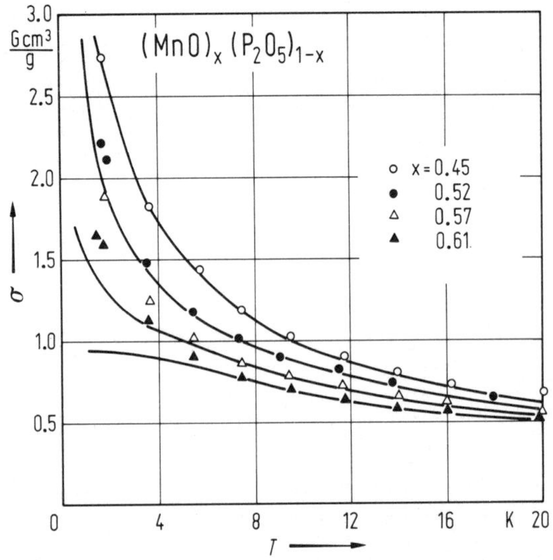

Fig. 6. $(MnO)_x(P_2O_5)_{1-x}$ glass. The magnetization at 1 kOe as a function of temperature [74 F1].

5.7.2.1.5 Fe - based phosphates

Phosphate glasses containing trivalent iron show antiferromagnetic order at low temperatures. Fig. 7 shows the temperature dependence of the inverse susceptibility for such amorphous oxides [72E], [73E]. Data for the high iron content glass, $(Fe_2O_3)_{0.79}P_2O_5$, show a transition at 7 K.

A somewhat different magnetic character is observed in glasses containing both divalent and trivalent iron ions [73W]. Susceptibility data in Fig. 8 for such a sample, $(FeO)_{0.55}(P_2O_5)_{0.45}$, show a constant susceptibility at low temperatures [73W].

For Fig. 7 see next page.

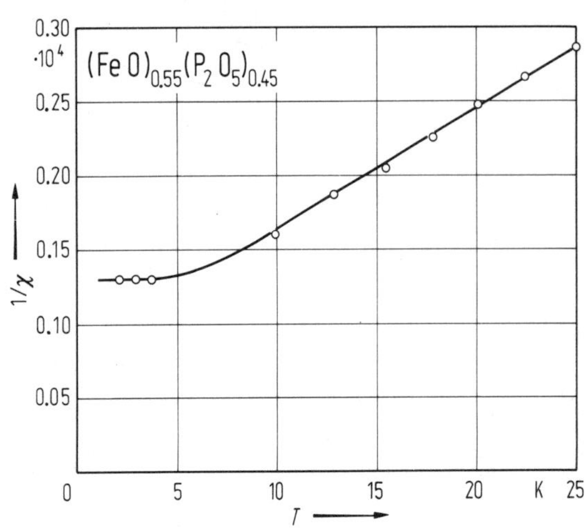

Fig. 8. $(FeO)_{0.55}(P_2O_5)_{0.45}$ glass. Low temperature inverse susceptibility data [73W]. χ in cm^3g^{-1}.

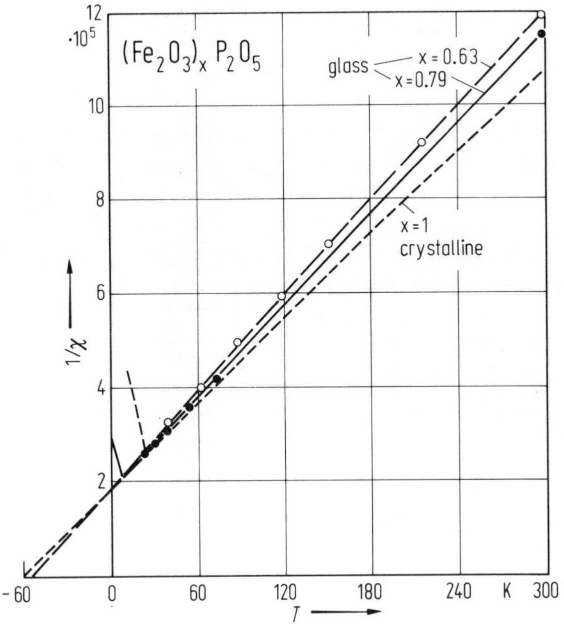

Fig. 7. $(Fe_2O_3)_xP_2O_5$ glass. Inverse susceptibility vs. temperature data for amorphous oxides with x = 0.63 and 0.79 and for crystalline $Fe_2O_3{\cdot}P_2O_5$. Notice the magnetic transition at 7 K for the amorphous sample with x = 0.79 [73E]. χ in $10^{-3}\ cm^3g^{-1}$.

5.7.2.1.6 Co - based phosphates

In phosphorus pentoxide glasses containing high concentrations of Co^{2+}, Ni^{2+} or Cu^{2+}, susceptibility measurements indicate antiferromagnetic interactions between the 3d-ions (Fig. 9). However, antiferromagnetic transitions are not observed in any of these materials [72E], [73E].

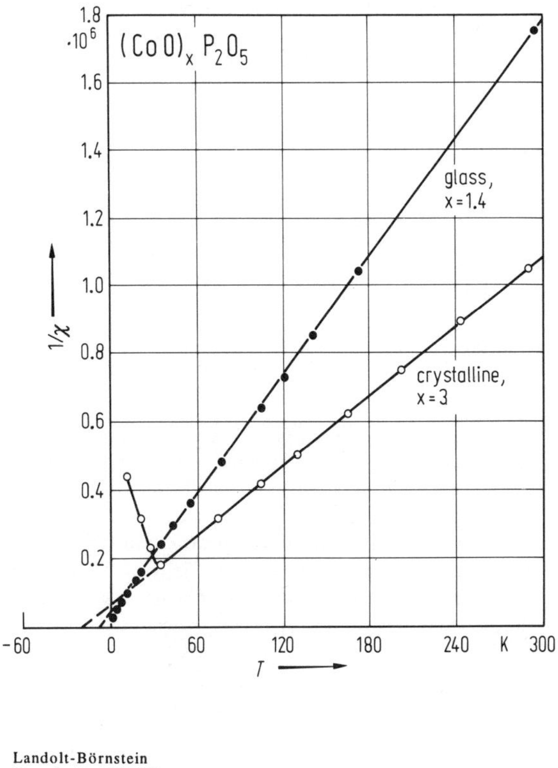

Fig. 9. $(CoO)_xP_2O_5$. The temperature variation of the reciprocal susceptibility for crystalline cobalt orthophosphate and for a cobalt phosphate glass [73E]. χ in $10^{-3}\ cm^3g^{-1}$.

5.7.2.1.7 Spinel ferrite based phosphates

Non-crystalline spinels, MFe_2O_4 (M = Mn, Fe, Co, Ni, Cu), in phosphate glasses were prepared by rapidly quenching a molten mixture of the ferrite and P_2O_5 [82S]. Quenching the melt from temperatures lower than 1300°C resulted in crystalline precipitates in the samples. Differential thermal analysis shows two crystallization temperatures for the amorphous samples, depending on the nature of M ions (Table 3). The amorphous samples show a relatively small magnetization (Figs. 10 and 11) and Mössbauer data at room temperature provide evidence for a long range magnetic order (Fig. 12) [82S], [83S].

Table 3. $(MFe_2O_4)_x(P_2O_5)_{1-x}$ glasses. Crystallization temperatures T_{crys} and ferrimagnetic Curie temperature T_C for amorphous ferrites MFe_2O_4 in phosphate glasses and T_C for crystalline ferrites [83S].

M [1])	T_{crys} [°C] [2])	T_C [°C] Amorphous	T_C [°C] Crystalline
Mn	618, 660	519	300
Fe	592, 632	585	585
Co	624, 662	550	520
Ni	664, 686	585	585
Cu	580, 628	457	455

[1]) The P_2O_5 content in the amorphous oxides varied as follows: $CuFe_2O_4 < Fe_3O_4 < CoFe_2O_4 < MnFe_2O_4 < NiFe_2O_4$.

[2]) Two crystallization temperatures were measured according to a two-step process of crystallization.

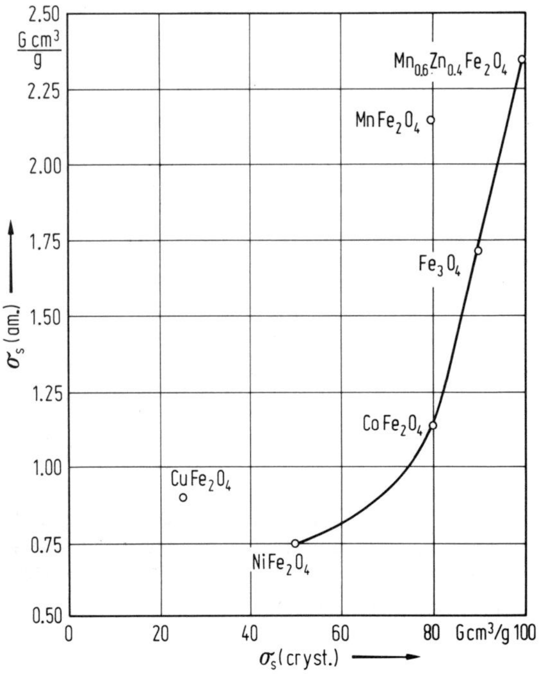

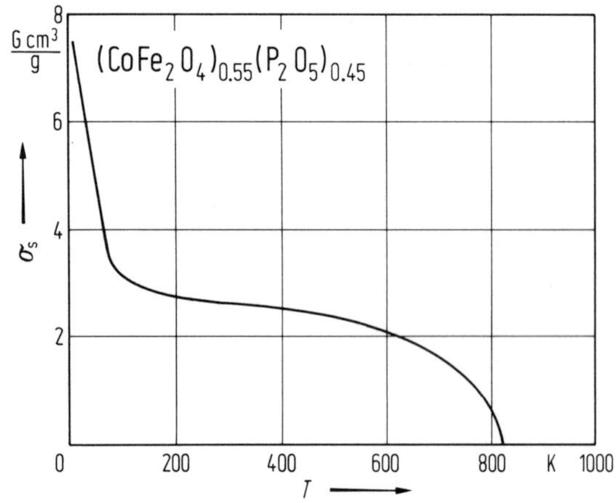

Fig. 11. $(CoFe_2O_4)_{0.55}(P_2O_5)_{0.45}$ glass. Specific saturation magnetization as a function of temperature [82S].

←

Fig. 10. Room temperature specific saturation magnetization measured with a static field of 18 kOe for amorphous ferrites in phosphate glasses and for crystalline spinel ferrites [83S].

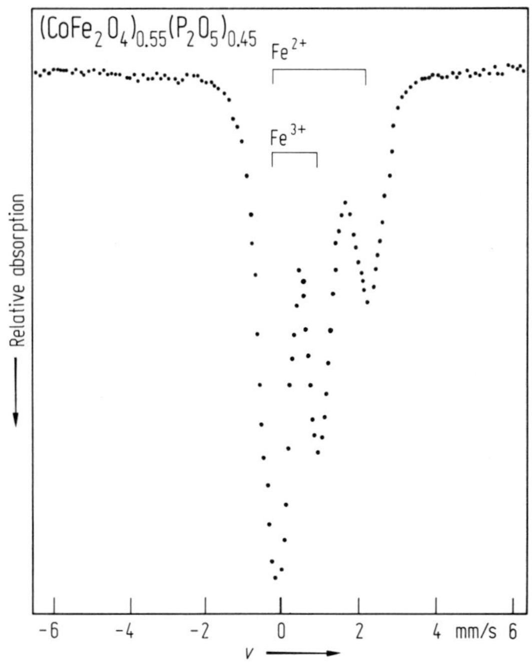

Fig. 12. $(CoFe_2O_4)_{0.55}(P_2O_5)_{0.45}$ glass. Mössbauer spectra at room temperature. A broad weak hyperfine splitting is evident in the spectra [82S].

5.7.2.2 Borate glasses

5.7.2.2.0 General remarks

Susceptibility, magnetization, and Mössbauer studies on amorphous alkali or alkali-earth borate glasses containing manganese or iron show evidence for the onset of a magnetic order in the systems. In Landolt-Börnstein, Subvolume III/27h, p. 205ff a comprehensive study of the structure of boron glasses is given together with additional data of their magnetic properties. Details on the sample composition, the nature of magnetic order and the ordering temperature of some borate glasses are given in Table 4.

Table 4. Magnetic order and ordering temperatures for borate glasses with manganese or iron.

Composition	Order	Ordering temp. [K]	Remarks	Ref.	Figs.
$(La_{1-x}Sr_xMnO_3)_{0.5}\,(B_2O_3)_{0.5}$ ($x = 0.2\cdots1.0$)	F	340···390	Mag., Möss.	88I,90I 90M	13···17
26 wt% $Fe_2O_3{\cdot}Na_2O{\cdot}B_2O_3$ [1])	P	–	Sus., Möss.	73H2	18,19
14.2···25 wt% $Fe_2O_3{\cdot}Li_2O{\cdot}B_2O_3{\cdot}Na_2O$	P	–	Sus., Möss.	73H2	
$(Fe_2O_3)_x[0.2BaO{\cdot}0.8B_2O_3]_{1-x}$ ($x \le 0.1$)	Ṗ	–	Sus., Möss. ESR	77K	20,21
$(Fe_2O_3)_{0.3}(BaO)_{0.45}(B_2O_3)_{0.25}$	SG?	12(1) 44(1)	Sus. Möss.	80L	22···24,26,27

continued

Table 4 (*continued*)

Composition	Order	Ordering temp. [K]	Remarks	Ref.	Figs.
$(Fe_2O_3)_{0.39}(BaO)_{0.48}(B_2O_3)_{0.13}$	S	< 77	Möss.	78H1	
$(Fe_2O_3)_{0.54}(BaO)_{0.39}(B_2O_3)_{0.07}$	S	100(10)	Möss.	78H1	
$(Fe_2O_3)_{0.63}(BaO)_{0.32}(B_2O_3)_{0.05}$	S	130(10)	Möss.	78H1	25
$(Fe_2O_3)_{0.55}(CaO)_{0.15}(B_2O_3)_{0.3}$	F?	723	Mag., Möss.	87N1	28···32
$(Fe_2O_3)_{0.6}(BaO)_{0.1}(Na_2O)_{0.3}$	S?	100(10) 125(5)	Sus. Möss.	78L	33,34

[1]) Base glass composition is 1 part Na_2O and 2 parts B_2O_3. To this was added an amount of 26 wt% Fe_2O_3.

5.7.2.2.1 Mn - borate glasses

A long range order with a ferromagnetic character is observed in borate glasses containing Mn. Amorphous samples $(La_{1-x}Sr_xMnO_3)_{0.5}(B_2O_3)_{0.5}$ for $0.2 \le x \le 1.0$ were prepared by the twin-roller quench technique [88I]. The samples which were 10···30 μm thick, brown colored, and transparent were confirmed to be amorphous (Fig. 13) by X-ray diffraction and transmission electron microscopy. The crystallization temperature (Fig. 14) is about 920 K and is much higher than the Curie temperature of 340···390 K (Figs. 15···17). Anomalous X-ray scattering studies provide evidence for the presence of clusters having a perovskite (La,Sr)MnO_3 -like structure in an amorphous B_2O_3 network. The spontaneous moment is attributed to the ferromagnetic double exchange interaction between Mn^{3+} and Mn^{4+} ions [88I], [90M], [90I].

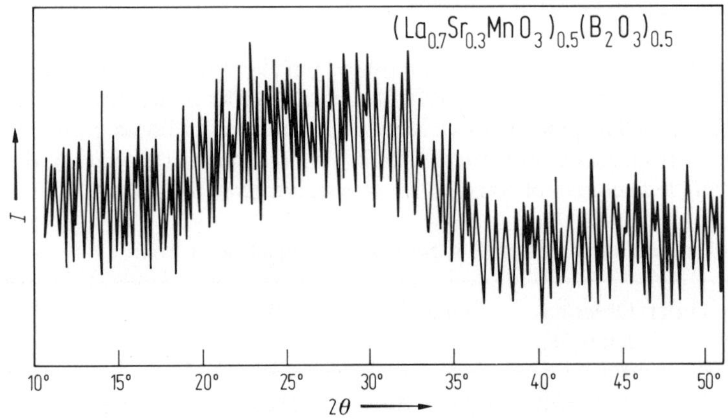

Fig. 13. $(La_{0.7}Sr_{0.3}MnO_3)_{0.5}(B_2O_3)_{0.5}$ glass. X-ray diffraction pattern for a sample prepared by melt quenching [88I].

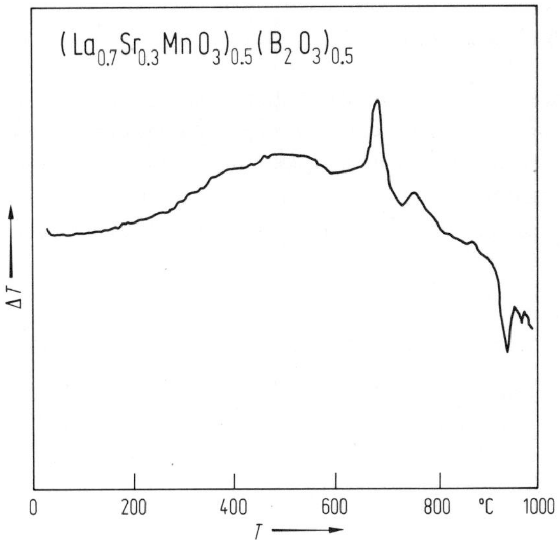

Fig. 14. $(La_{0.7}Sr_{0.3}MnO_3)_{0.5}(B_2O_3)_{0.5}$ glass. Data on differential thermal analysis for a sample prepared by melt quenching. The crystallization temperature is 660°C [88I].

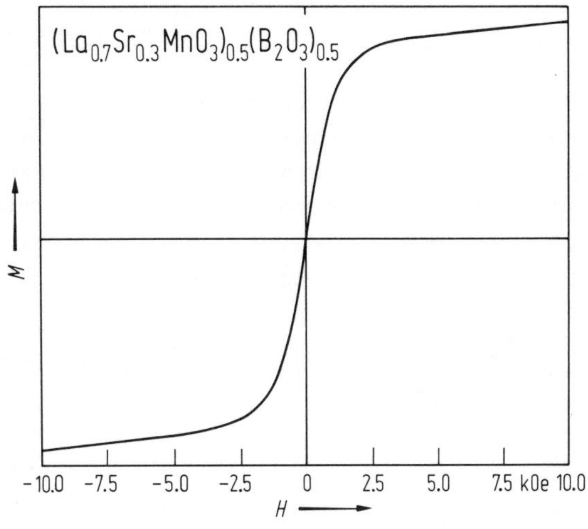

Fig. 15. $(La_{0.7}Sr_{0.3}MnO_3)_{0.5}(B_2O_3)_{0.5}$ glass. Magnetization M vs. static field H at room temperature. The magnetization at 10 kOe is 3.2 G cm^3/g[88I].

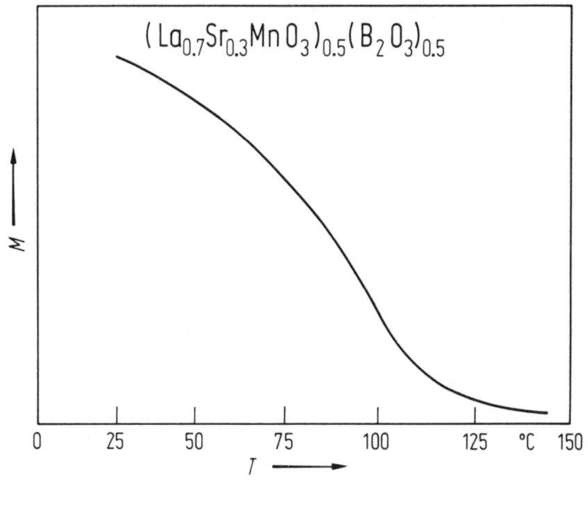

Fig. 16. $(La_{0.7}Sr_{0.3}MnO_3)_{0.5}(B_2O_3)_{0.5}$ glass. Temperature dependence of the magnetization at 10 kOe. The Curie temperature is 107°C [88I].

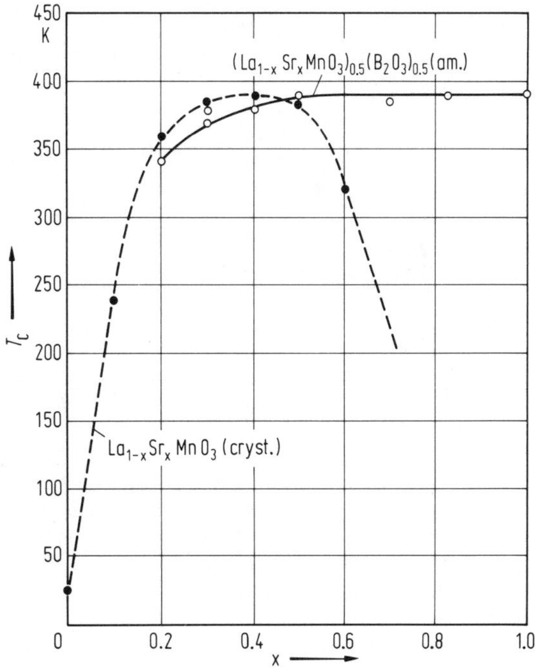

Fig. 17. $(La_{1-x}Sr_xMnO_3)_{0.5}(B_2O_3)_{0.5}$ glass, $La_{1-x}Sr_xMnO_3$. Variation of the Curie temperature T_C with the concentration of Sr, x, for amorphous (solid line) and crystalline (dotted line) material [88I].

5.7.2.2.2 Fe - alkali - borate glasses

Iron ions in alkali borate glasses, $Li_2O{\cdot}B_2O_3$ or $Na_2O{\cdot}B_2O_3$, are coupled antiferromagnetically [69B], [73H2], [80I]. ESR spectra (Fig. 18) indicate the presence of isolated Fe^{3+} in distorted tetrahedral sites and exchange coupled iron pairs [73H2]. For samples containing less than 26 wt% of Fe_2O_3, χ^{-1} vs. T is a straight line with a negative paramagnetic Curie temperature (Fig. 19). For higher iron content, a complex χ vs. T behavior is observed possibly due to microcrystals of Fe_2O_3 [73H2], and their presence is confirmed by Mössbauer studies [69B].

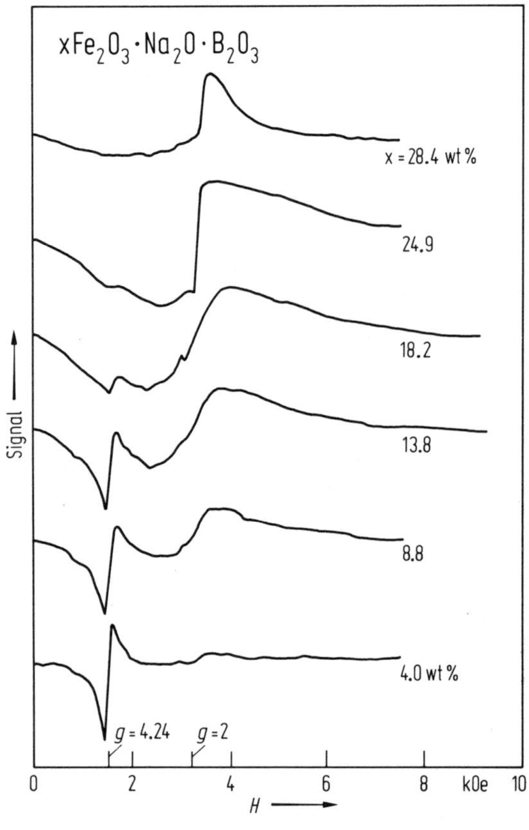

Fig. 18. $xFe_2O_3{\cdot}Na_2O{\cdot}B_2O_3$ glass. X-band ESR spectra for a series of Fe_2O_3 content in wt%. (Base glass composition is one part of Na_2O and two parts of B_2O_3.) The absorption at low fields ($g = 4.24$) is due to isolated Fe^{3+} in distorted tetrahedral sites and at high fields (g = 2.0), resonance due to exchange coupled iron ions is observed [73H2].

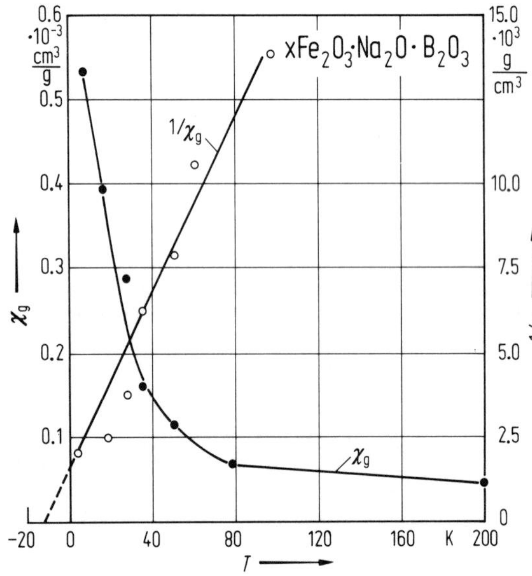

Fig. 19. x $Fe_2O_3{\cdot}Na_2O{\cdot}B_2O_3$ glass. Low-field ac susceptibility vs. temperature for sodium borate glasses containing 26 wt% of Fe_2O_3 [73H2]. (Base glass composition is one part of Na_2O and two parts of B_2O_3.)

5.7.2.2.3 Fe - BaO - borate glasses

In borate glasses with Ba and Fe, for Fe_2O_3 concentration less than 10 mol%, iron exists as isolated ions and as 1.5 nm clusters of two or three ions. Susceptibility data and ESR spectra for the amorphous sample are shown in Figs. 20 and 21 [77K]. Such samples when annealed show superparamagnetism due to the precipitation of Fe_2O_3 and Fe_3O_4 microcrystals [73M]. When the content of Fe_2O_3 in the glass is increased to 30 mol% or more, a long range magnetic order with the ordering temperature ranging from 12 K to 130 K is observed through magnetization (Table 4, Figs. 22 and 23) and Mössbauer studies (Figs. 24 and 25) [80L], [78H1], [79S2], [82B]. Samples containing 30 mol% of Fe_2O_3 when annealed at progressively higher temperatures show paramagnetism, superparamagnetism and then a ferrimagnetic character at room temperature (Figs. 26 and 27).

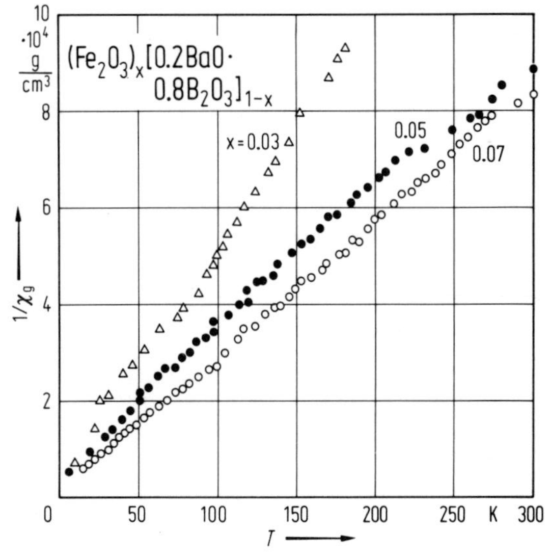

Fig. 20. $(Fe_2O_3)_x[0.2BaO{\cdot}0.8B_2O_3]_{1-x}$ glass. Inverse susceptibility data for x = 0.03, 0.05 and 0.07. The samples show a paramagnetic character down to 1 K [77K].
←

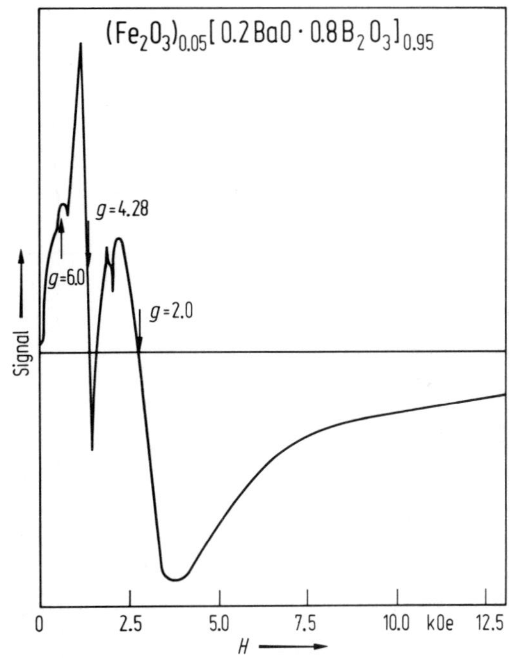

Fig. 21. $(Fe_2O_3)_{0.05}[0.2BaO{\cdot}0.8B_2O_3]_{0.95}$ glass. X-band ESR spectra at 292 K for an amorphous sample of 5 mol% of Fe_2O_3 in $BaO{\cdot}4B_2O_3$. Resonances at g = 2.0 and 6.0 are due to clusters of iron ions (pairs or triads), and at g = 4.28 is due to isolated iron ions [77K].

For Fig. 22 see next page.

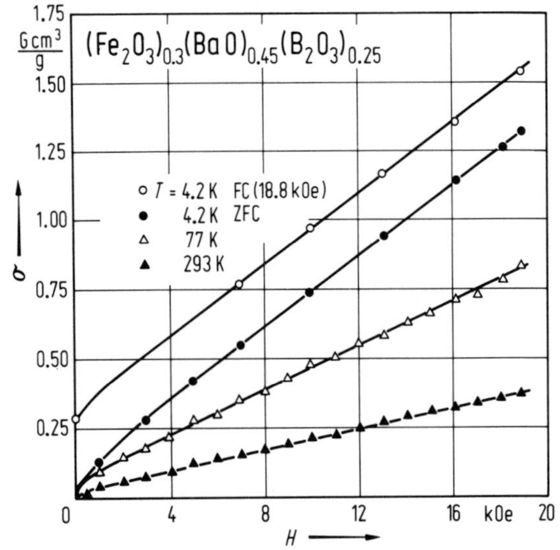

←
Fig. 23. $(Fe_2O_3)_{0.3}(BaO)_{0.45}(B_2O_3)_{0.25}$ glass. Static field H dependence of the specific magnetization for temperatures above and below the magnetic phase transition at 12 K [80L].

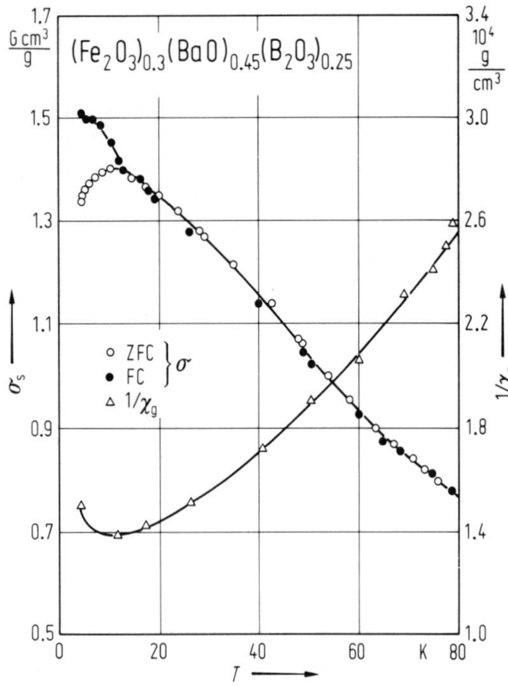

Fig. 22. $(Fe_2O_3)_{0.3}(BaO)_{0.45}(B_2O_3)_{0.25}$ glass. Temperature variation of the specific saturation magnetization σ_s and the inverse susceptibility χ_g^{-1}. Magnetization data are for zero field cooled and field cooled (H = 18.8 kOe) conditions. Notice the magnetic transition at 12 K in the susceptibility data [80L].

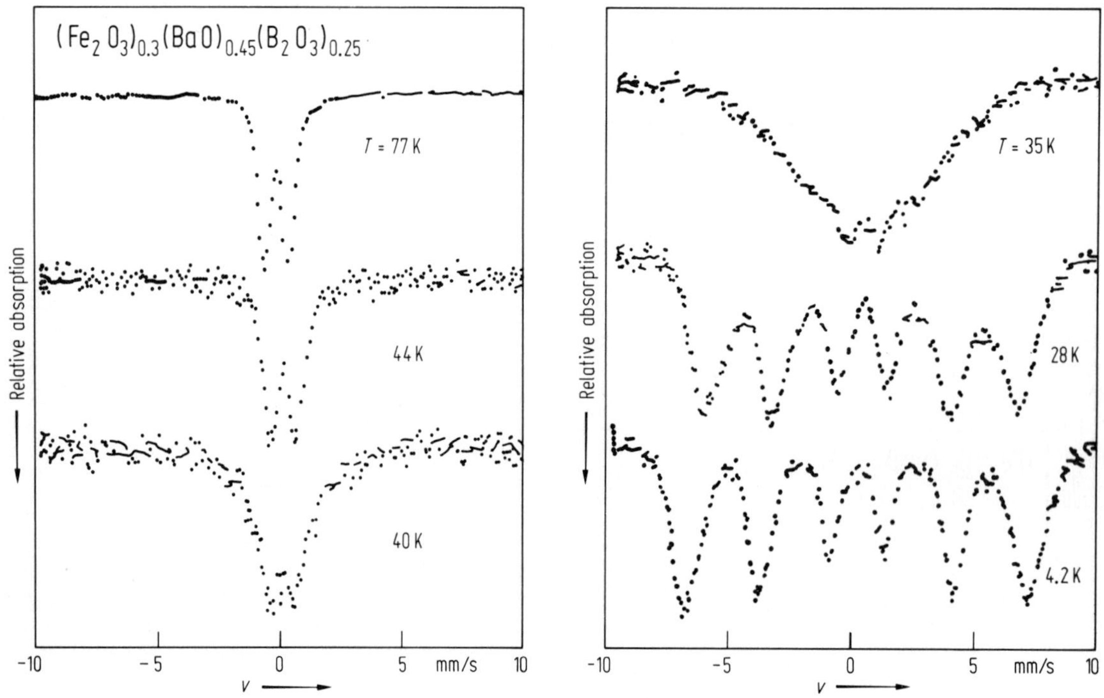

Fig. 24. $(Fe_2O_3)_{0.3}(BaO)_{0.45}(B_2O_3)_{0.25}$ glass. Mössbauer spectra. The onset of a magnetic hyperfine splitting occurs at 44 K [80L].

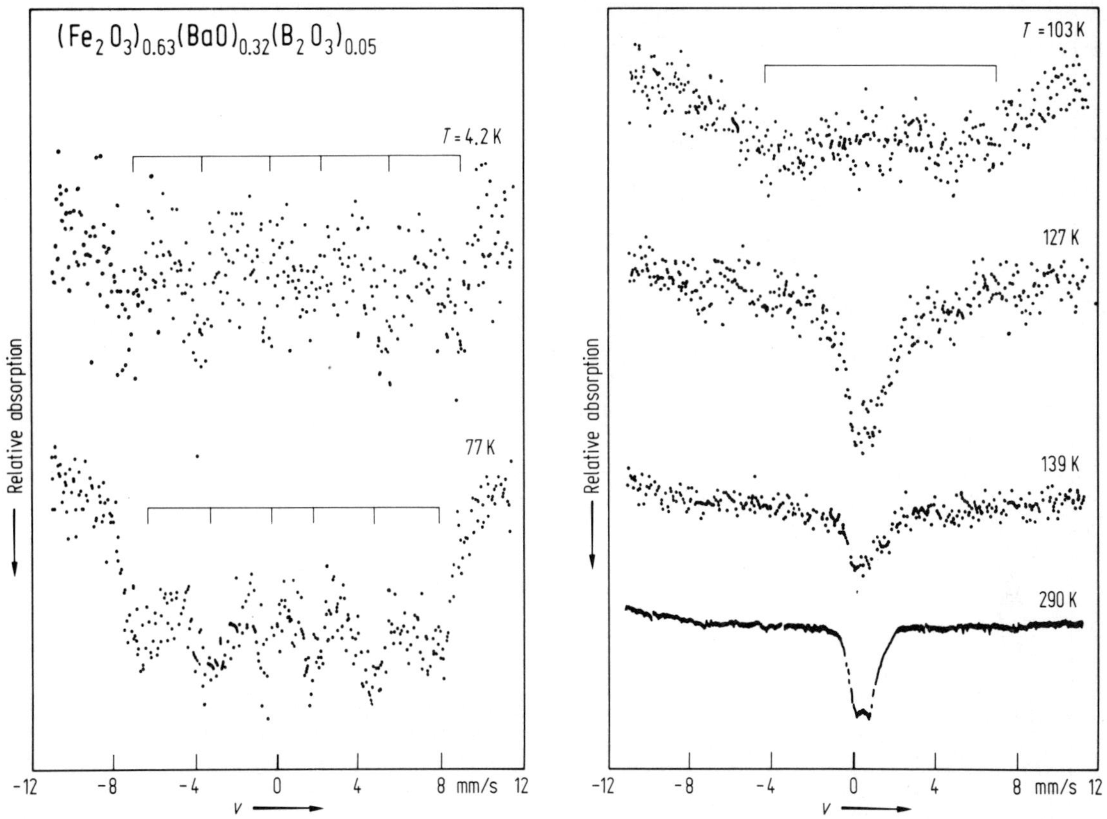

Fig. 25. $(Fe_2O_3)_{0.63}(BaO)_{0.32}(B_2O_3)_{0.05}$ glass. Similar data as in Fig. 24. The magnetic transition temperature is 130 K for the sample [78H1].

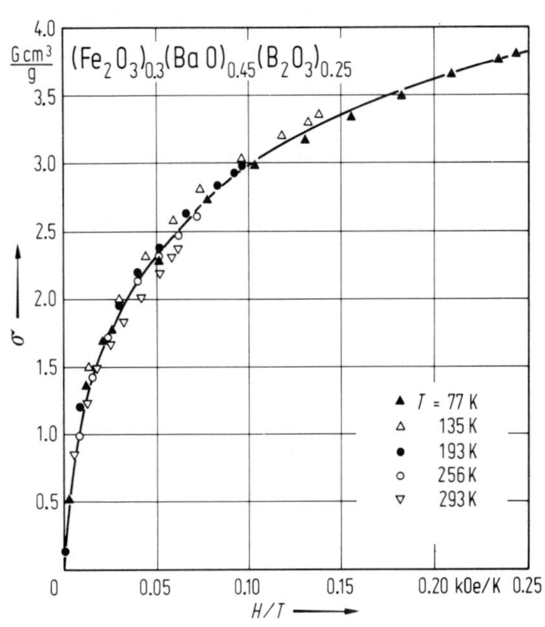

Fig. 26. $(Fe_2O_3)_{0.3}(BaO)_{0.45}(B_2O_3)_{0.25}$ glass. Specific magnetization vs. *H*/*T* data for a sample annealed at 845 K in air. The data show superparamagnetism above the blocking temperature of 65 K [80L].

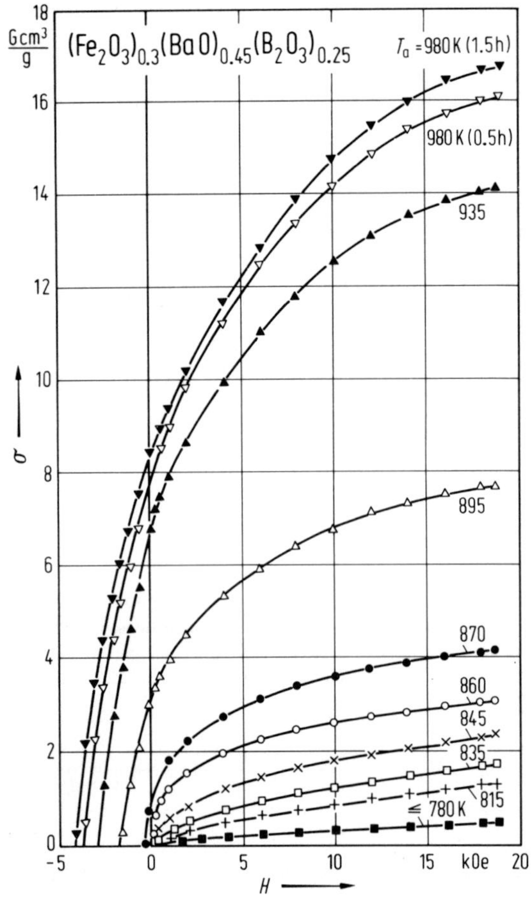

Fig. 27. $(Fe_2O_3)_{0.3}(BaO)_{0.45}(B_2O_3)_{0.25}$ glass. Specific magnetization data at room temperature for samples annealed at a series of high temperatures in air. With increasing annealing temperature, the sample shows paramagnetism, superparamagnetism or ferrimagnetism at room temperature [80L].

5.7.2.2.4 Fe - CaO - borate glasses

A ferromagnetic character is also observed in Fe based borate glasses containing Ca [87N1]. The phase diagram in Fig. 28 shows the range of composition over which the amorphous phase is stable. The crystallization temperature, measured by DTA, is 650°C and is dependent slightly on the concentration of B_2O_3. X-ray analysis shows very small amount of Fe_3O_4, $CaFe_2O_4$ or $FeBO_3$. The magnetization is dependent on the concentration of B_2O_3 (Figs. 29 and 30) and T_C is smaller than the crystallization temperature (Fig. 31). A long range magnetic order is evident from Mössbauer data (Fig. 32).

Finally, for comparison data on the amorphous glasses without borate, $Fe_2O_3{\cdot}BaO{\cdot}Na_2O$ are given. $(Fe_2O_3)_{0.6}(BaO)_{0.1}(Na_2O)_{0.3}$ orders magnetically at 100(10) K (Fig. 33) [78L]. Mössbauer studies reveal a distorted tetrahedral environment for Fe^{3+} ions and a magnetic transition at 125(5) K (Fig. 34) for this substance.

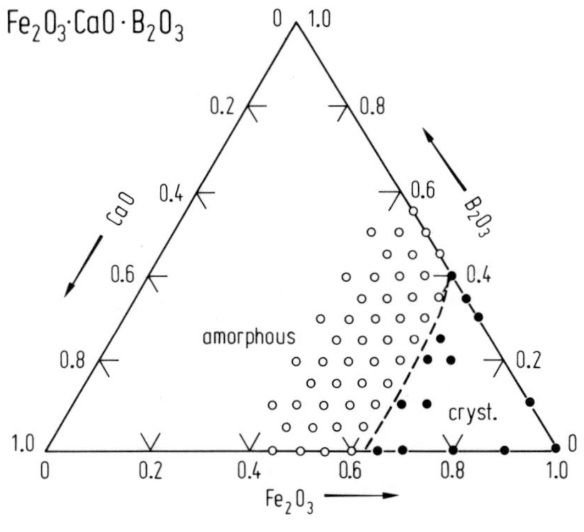

Fig. 28. Fe_2O_3·CaO·B_2O_3. Phase diagram at RT for the ternary system showing the region of amorphous and crystalline phases [87N1].

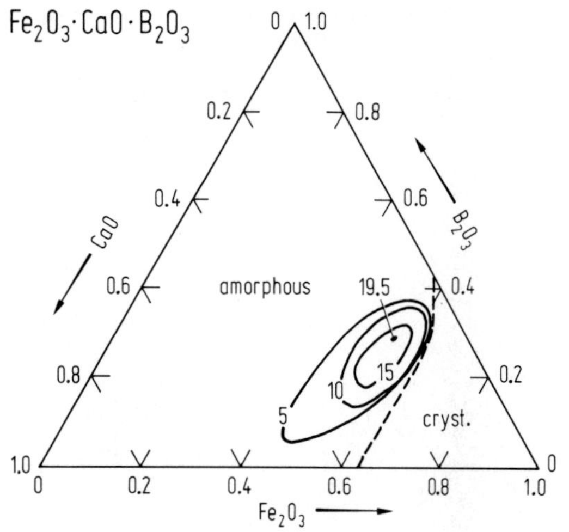

Fig. 29. Fe_2O_3·CaO·B_2O_3. Contours of constant room temperature magnetization (in G cm^3/g) for the amorphous ternary system [87N1].

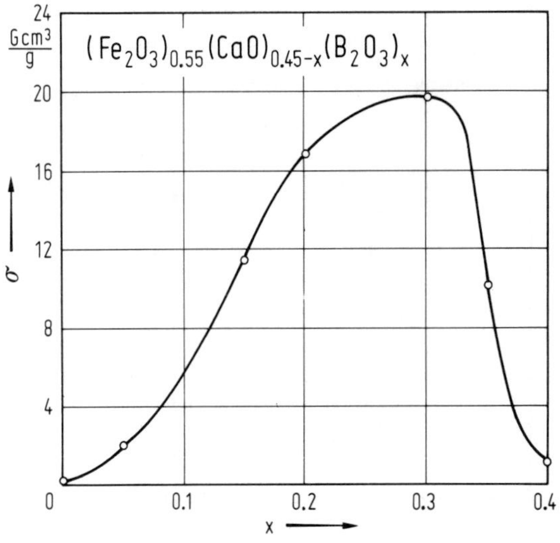

Fig. 30. $(Fe_2O_3)_{0.55}(CaO)_{0.45-x}(B_2O_3)_x$ glass. Dependence of the room temperature specific magnetization measured at 10 kOe vs. the content of B_2O_3 [87N1].

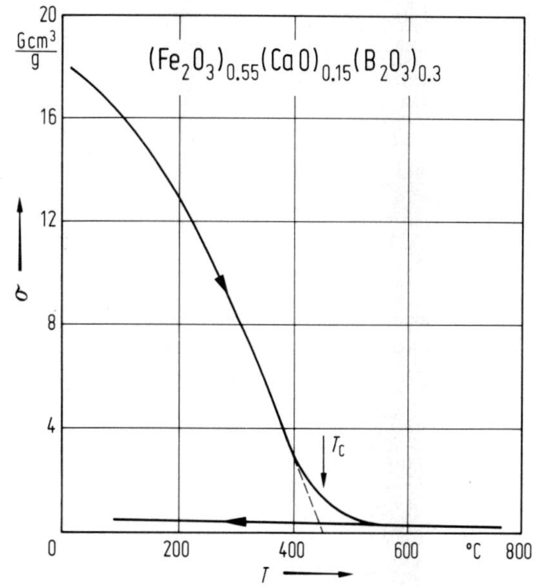

Fig. 31. $(Fe_2O_3)_{0.55}(CaO)_{0.15}(B_2O_3)_{0.3}$ glass. Specific magnetization vs. temperature at H = 10 kOe. The Curie temperature from the data in the figure is 450°C [87N1].

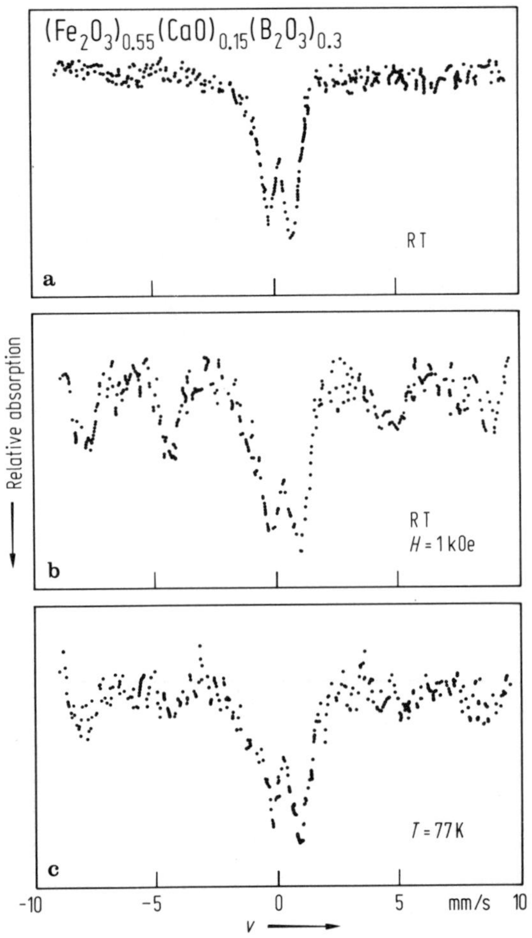

Fig. 32. $(Fe_2O_3)_{0.55}(CaO)_{0.15}(B_2O_3)_{0.3}$ glass. Mössbauer spectra at room temperature for $H = 0$ (a) and H = 1 kOe (b); and at 77 K, $H = 0$ (c) [87N1].

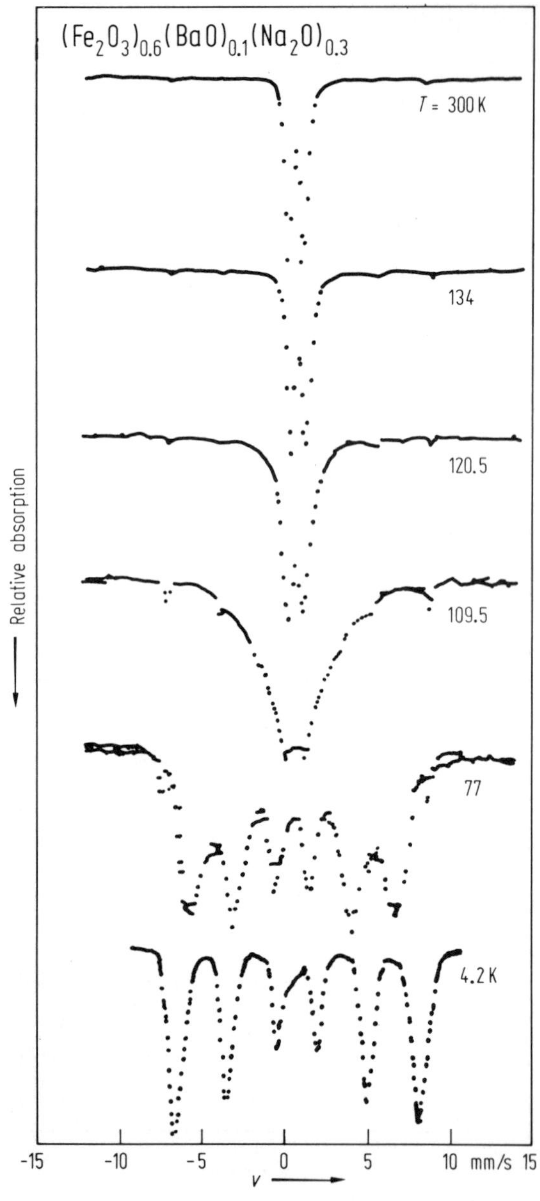

Fig. 34. $(Fe_2O_3)_{0.6}(BaO)_{0.1}(Na_2O)_{0.3}$ glass. Mössbauer spectra at a series of temperatures [78L].

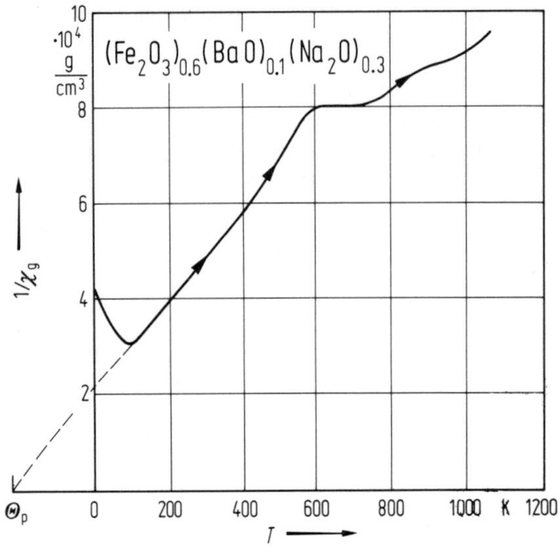

←
Fig. 33. $(Fe_2O_3)_{0.6}(BaO)_{0.1}(Na_2O)_{0.3}$ glass. Inverse susceptibility vs. temperature for increasing T. A magnetic transition at 100(10) K is evident from the data [78L].

5.7.2.3 Silicate glasses

5.7.2.3.0 General remarks

A long range magnetic order is reported for Mn^{2+}, Fe^{3+}- Fe^{2+}, Fe^{2+}, and Co^{2+} in aluminosilicate glasses containing a large amount of transition metal ions. Table 5 gives the type of order and the ordering temperature for such glasses. More information on crystalline and glassy silicates can be found in Landolt-Börnstein III/27i (to be published).

Table 5. Magnetic order and ordering temperature for transition metal oxide-silicate glasses (also refer to Tables 6···9).

3-d ion	Composition	Magnetic order	Ordering temp. [K]	Ref.	Figs.
Mn^{2+}	$MnO{\cdot}Al_2O_3{\cdot}SiO_2$ (Mn = 12.3···41.7 at%)	AF	3.3···6.5	75V	35···39
Fe^{3+} - Fe^{2+}	$[(Fe_3O_4)_{0.15}(CaO)_{0.425}(SiO_2)_{0.425}]$	F?	800 ?	85S	40···42
Fe^{2+}	$FeO{\cdot}Al_2O_3{\cdot}SiO_2$ (FeO = 43···51.8 mol%)	SG	10···15	82I	43,44
Co^{2+}	$CoO{\cdot}Al_2O_3{\cdot}SiO_2$ (Co = 11.8···33.5 at%)	AF	2.5···7.8	75V	45···50

5.7.2.3.1 Mn - aluminosilicate glasses

In concentrated manganese (12···41.7 at% Mn) aluminosilicate glasses, high temperature susceptibility (Table 6 and Fig. 35) show a Curie-Weiss behavior with a large paramagnetic Curie temperature, indicating the presence of a strong antiferromagnetic exchange. Low temperature susceptibility data (Fig. 36) indicate a magnetic transition at 3···7 K and is attributed to antiferromagnetic ordering of Mn-rich 5 nm clusters [75V]. The magnetic transition is also evident from ac susceptibility (Fig. 37) [78H2], [80R], [80F], specific heat [77W], and sound velocity (Fig. 38) [75M] studies. Optical spectra show a behavior similar to that of crystalline MnO (Fig. 39) [78W].

Table 6. $(MnO)_x(Al_2O_3)_y(SiO_2)_{1-x-y}$ glasses. The magnetic ordering temperature, paramagnetic Curie temperature Θ_p, and effective magnetic moment p_{eff} for manganese aluminosilicate glasses [75V].

Mn concentration [at%]	Magnetic ordering temperature [K]	Θ_p [K]	p_{eff} [μ_B]
12.3 (x = 0.35, y = 0.1)	3.3	−194	6.96
14.3 (x = 0.4, y = 0.1)	4.6	−194	6.86
19.2 (x = 0.5, y = 0.05)	5.55	−245	6.01
23.1 (x = 0.6, y = 0.1)	5.8	−226	4.69
29.2 (x = 0.7, y = 0.05)	6.15	−210	4.85
34.8 (x = 0.8, y = 0.05)	5.1	−236	5.2
41.7 (x = 0.9, y = 0.03)	6.5	−224	5.37

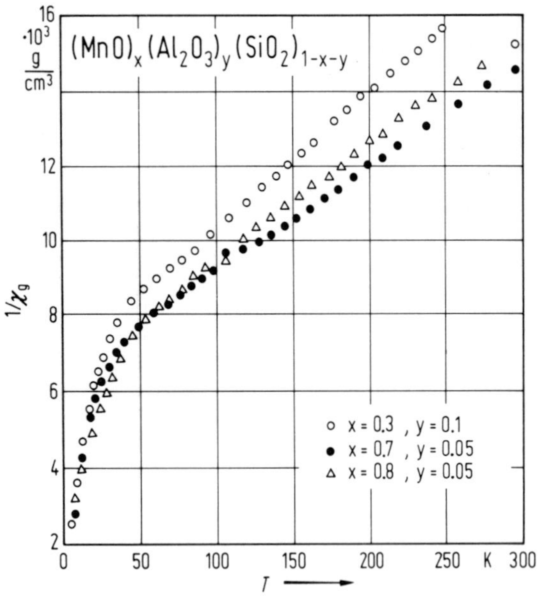

Fig. 35. $(MnO)_x(Al_2O_3)_y(SiO_2)_{1-x-y}$ glass. Reciprocal susceptibility vs. temperature for manganese aluminosilicate glasses with different composition [75V].

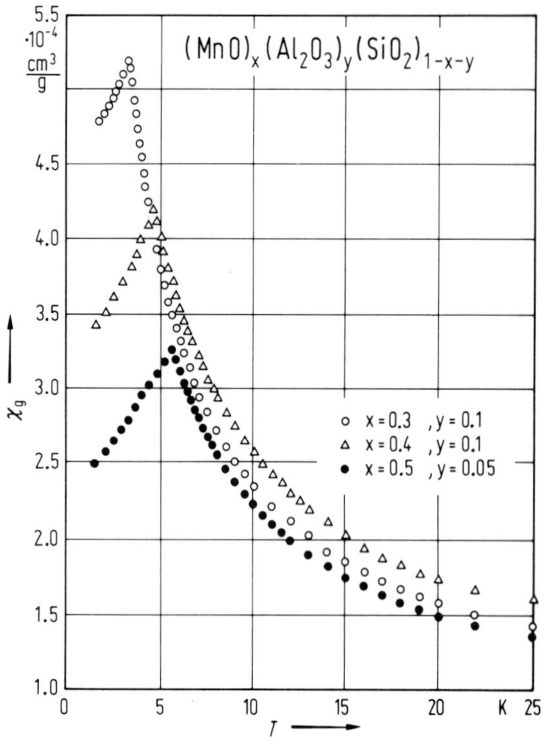

Fig. 36. $(MnO)_x(Al_2O_3)_y(SiO_2)_{1-x-y}$ glass. Susceptibility vs. temperature for aluminosilicate glasses of different composition. The magnetic transition is due to antiferromagnetic 5 nm clusters of Mn-rich regions [75V].

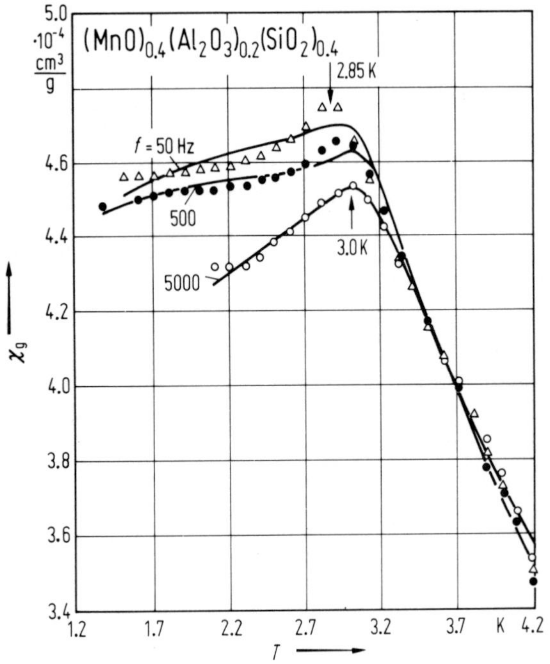

Fig. 37. $(MnO)_{0.4}(Al_2O_3)_{0.2}(SiO_2)_{0.4}$ glass. ac susceptibility vs. temperature at 50 Hz, 500 Hz, and 5000 Hz for a sample of 13.3 at% Mn-aluminosilicate glass. The solid lines are theoretical fits to the data, and arrows point to maxima [78H2].

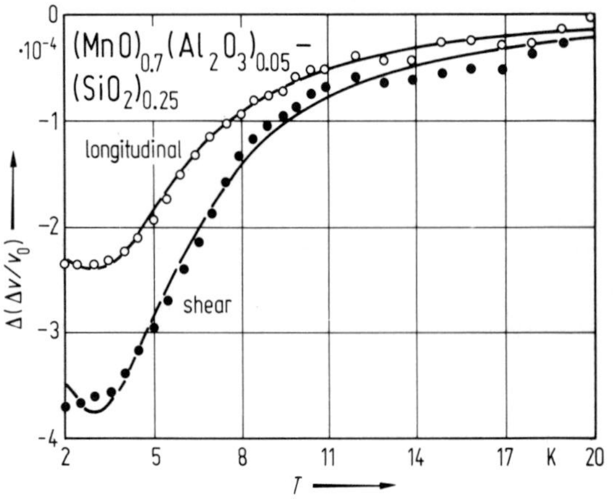

Fig. 38. $(MnO)_{0.7}(Al_2O_3)_{0.05}(SiO_2)_{0.25}$ glass. Longitudinal and shear sound velocity change vs. temperature. The experimental data were subtracted from a theoretical background. The solid lines are theoretical fits [75M].

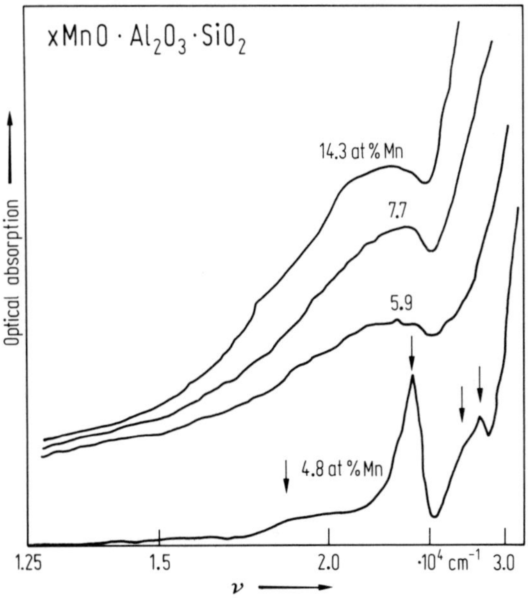

Fig. 39. $xMnO{\cdot}Al_2O_3{\cdot}SiO_2$ glass. Optical absorption spectra for a series of concentration in at% of Mn ions. The absorption peaks are identified as due to Mn^{2+} ions in a cubic field [78W].

5.7.2.3.2 Mixed - valence Fe silicate glasses

In silicate glasses containing iron, Mössbauer data indicate the presence of divalent and trivalent Fe in distorted tetrahedral and octahedral sites [77L2]. One observes in $Fe_2O_3{\cdot}SiO_2{\cdot}B_2O_3{\cdot}CaO{\cdot}Al_2O_3$ antiferromagnetic exchange between iron ions through susceptibility data for samples containing up to 12 mol% Fe_2O_3 (Fig. 40 and Table 7) [77O2]. However, the data do not indicate a magnetic transition. When the samples are heated, superparamagnetism is observed due to the growth of crystalline ferrite phases [68 O], [77O2].

Similar antiferromagnetic interactions between iron ions are evident from ESR and susceptibility studies in $(Fe_2O_3{\cdot}TiO_2){\cdot}BaO{\cdot}B_2O_3{\cdot}SiO_2$ [77T].

A long range magnetic order is reported in $Fe_3O_4{\cdot}CaO{\cdot}SiO_2$ glasses (Figs. 41 and 42) [85S]. A ferromagnetic character is seen for temperatures below 800 K for a-$(Fe_3O_4)_{0.15}(CaO)_{0.425}(SiO_2)_{0.425}$.

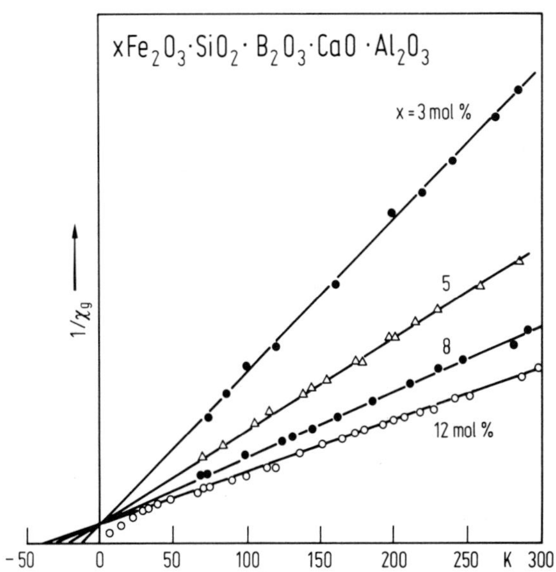

Fig. 40. $xFe_2O_3{\cdot}SiO_2{\cdot}B_2O_3{\cdot}CaO{\cdot}Al_2O_3$ glass. Reciprocal susceptibility vs. temperature for a series of Fe_2O_3 content in mol%. Refer to Table 7 for magnetic parameters estimated from the data and for detailed composition [77O2].

Table 7. $xFe_2O_3 \cdot SiO_2 \cdot B_2O_3 \cdot CaO \cdot Al_2O_3$ glasses. Magnetic parameters estimated from susceptibility data in Fig. 40 [77O2].

mol% Fe_2O_3 [1])	Θ_p [K]	p_{eff} [μ_B]
12	−42	4.7
8	−34	5.0
5	−24	5.1
3	−16	5.1

[1]) Base composition of glass system was (in mol%) $40SiO_2$, $30B_2O_3$, $20CaO$, $10Al_2O_3$. Iron cations were added in form of Fe_2O_3 such that the relative molar proportions of the base glass were maintained

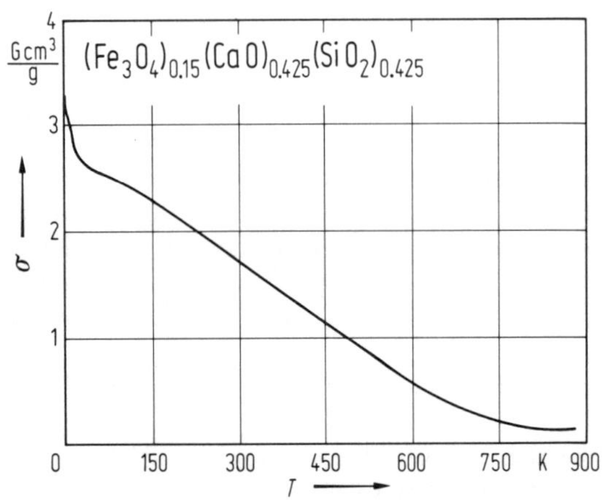

Fig. 41. $(Fe_3O_4)_{0.15}(CaO)_{0.425}(SiO_2)_{0.425}$ glass. Temperature dependence of the specific magnetization for H = 7 kOe. The indistinctiveness of a well defined Curie temperature is attributed to decomposition of the amorphous ordered phase [85S].

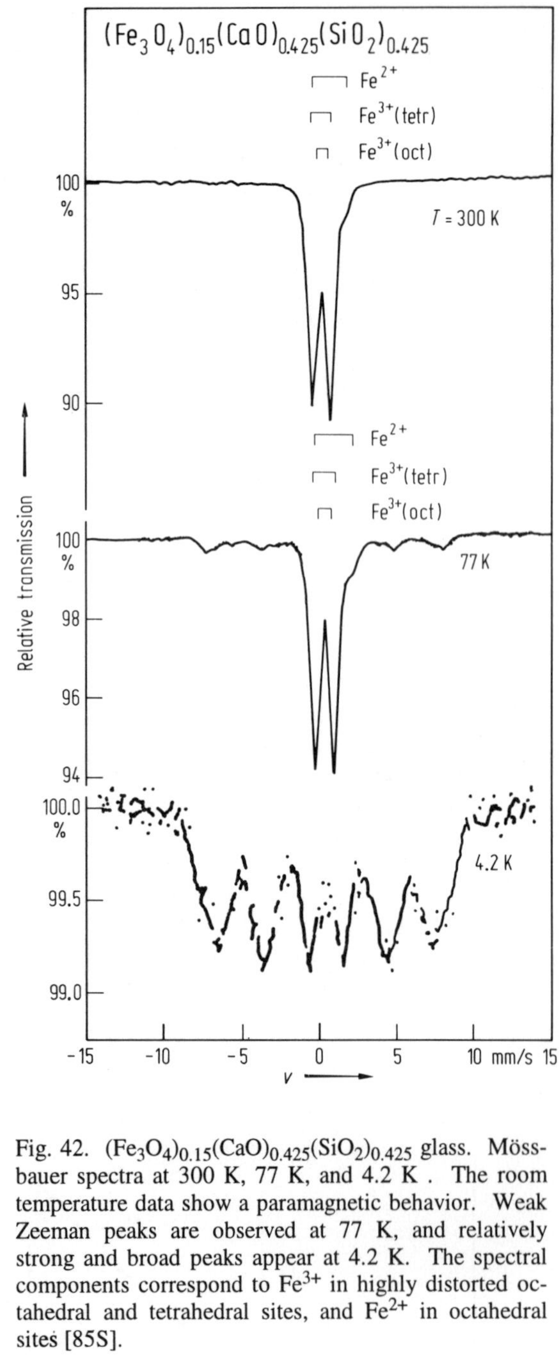

Fig. 42. $(Fe_3O_4)_{0.15}(CaO)_{0.425}(SiO_2)_{0.425}$ glass. Mössbauer spectra at 300 K, 77 K, and 4.2 K . The room temperature data show a paramagnetic behavior. Weak Zeeman peaks are observed at 77 K, and relatively strong and broad peaks appear at 4.2 K. The spectral components correspond to Fe^{3+} in highly distorted octahedral and tetrahedral sites, and Fe^{2+} in octahedral sites [85S].

5.7.2.3.3 Ferrous - silicate glasses

A spin glass behavior is reported for ferrous ions in aluminosilicate glasses [82I]. Samples of $FeO \cdot Al_2O_3 \cdot SiO_2$ with FeO concentration varying from 43 to 52 mol% show a transition to the spin glass state at 10···15 K (Table 8 and Fig. 43). Mössbauer spectra (Fig. 44) do not indicate the presence of Fe^{3+} in the samples.

Table 8. $(FeO)_x(Al_2O_3)_y(SiO_2)_{1-x-y}$ glasses. Composition, spin glass transition temperature T_f in zero magnetic field, effective magnetic moment p_{eff}, and the paramagnetic Curie temperature Θ_p for aluminosilicate glasses with ferrous ions [82I].

Chemical composition [mol%]			T_f [K]	p_{eff} [μ_B]	Θ_p [K]
FeO	Al_2O_3	SiO_2			
43.0	14.7	42.4	10.2	5.4	−82
45.2	8.4	46.5	12.3	5.83	−91
51.8	5.8	42.4	15.0	5.55	−84

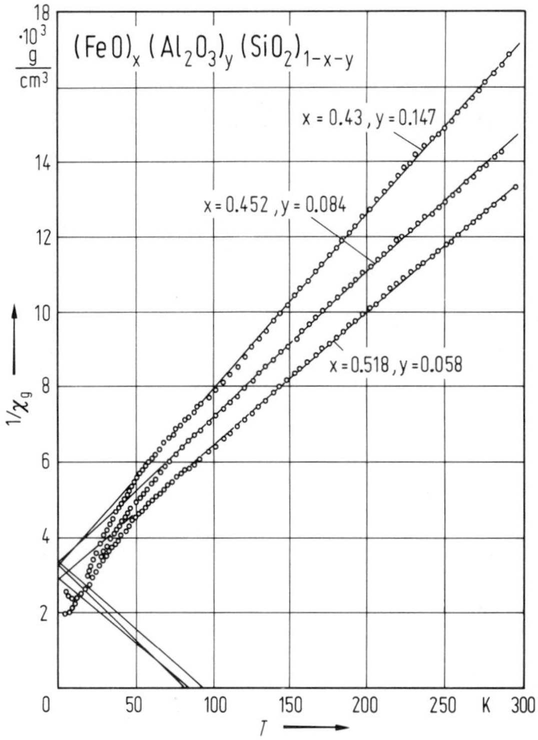

Fig. 43. $(FeO)_x(Al_2O_3)_y(SiO_2)_{1-x-y}$ glass. Temperature dependence of the inverse susceptibility for various aluminosilicate glasses. The samples show a spin glass transition at temperatures T_f given in Table 8 [82I].

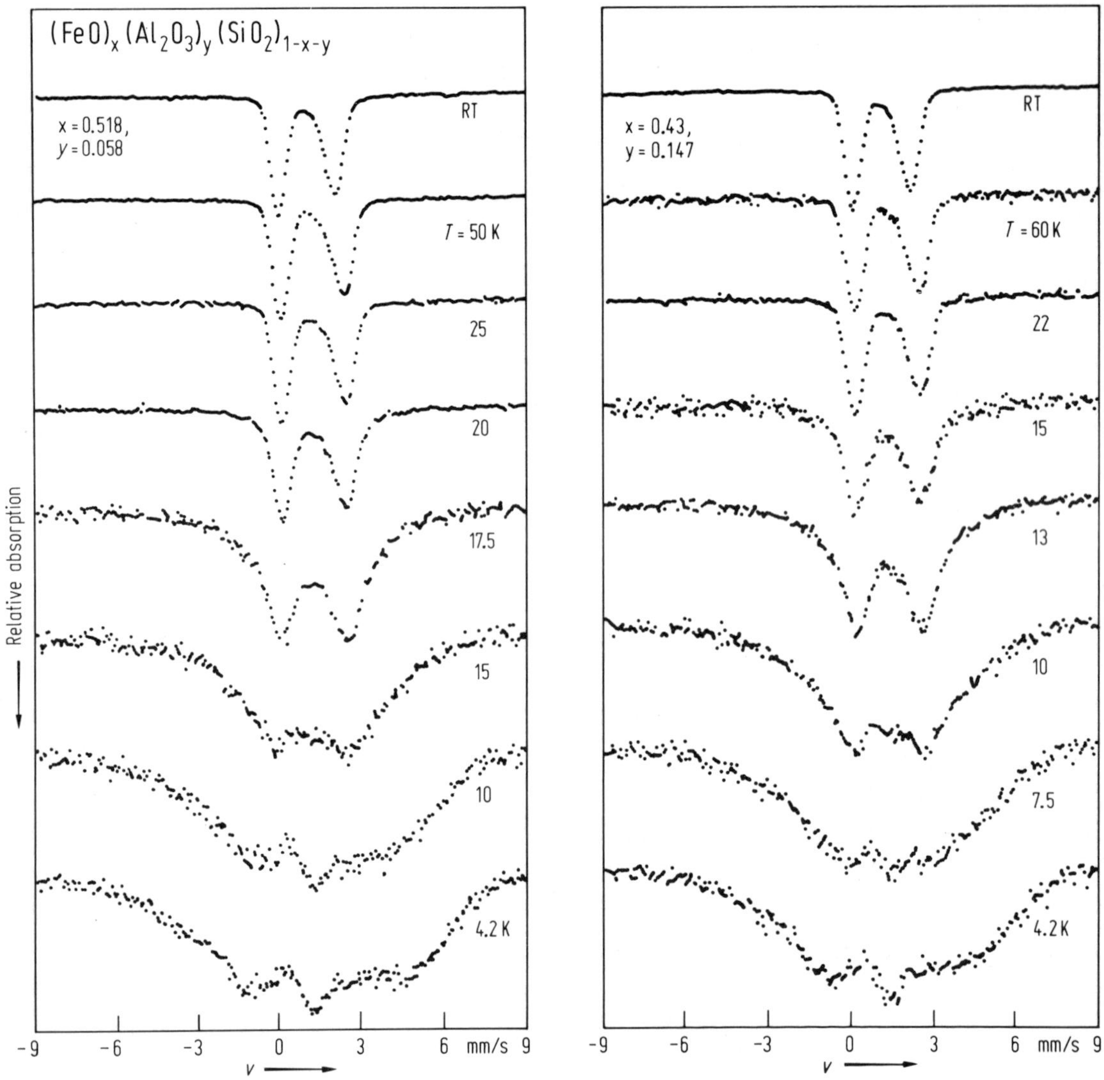

Fig. 44. $(FeO)_x(Al_2O_3)_y(SiO_2)_{1-x-y}$ glass. Mössbauer spectra for two aluminosilicate glasses. The samples undergo a transition to the spin glass state at 15 K and 10.2 K, respectively [82I].

5.7.2.3.4 Co - aluminosilicate glasses

In aluminosilicate glasses containing divalent cobalt, 5 nm clusters of cobalt rich regions order antiferromagnetically below 2.5···7.8 K for Co concentration ≈ 11···33 at% (Fig. 45) [75 V]. A negative exchange coupling between Co^{2+} ions is evident from high-temperature susceptibility data (Fig. 46 and Table 9) [75V]. ac susceptibility (Fig. 47) [78H2], Mössbauer (Fig. 48) [77B], specific heat (Fig. 49) [77W], and sound velocity (Fig. 50) [75M] measurements also show behaviors indicative of a long range ordering of cobalt moments in the system.

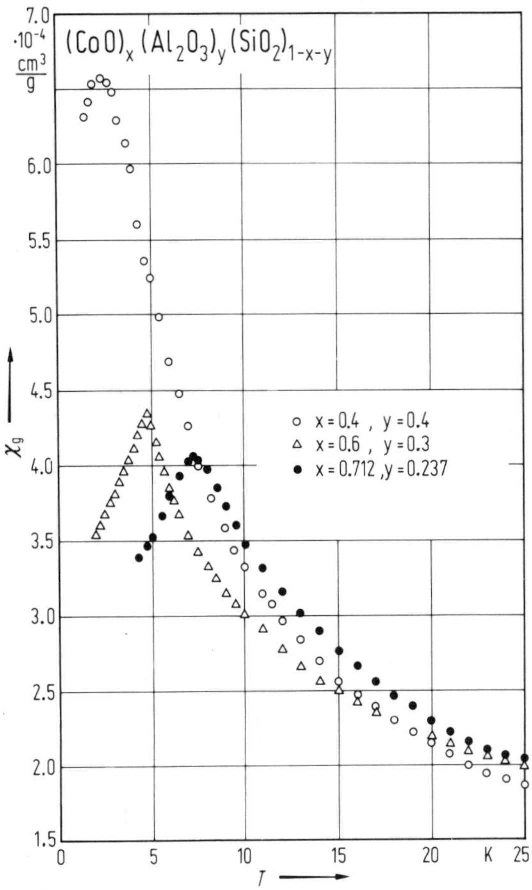

Fig. 45. $(CoO)_x(Al_2O_3)_y(SiO_2)_{1-x-y}$ glass. Susceptibility vs. temperature for various aluminosilicate glasses containing Co^{2+} ions. The data indicate an antiferromagnetic ordering of clusters of Co rich regions in the samples [75V].

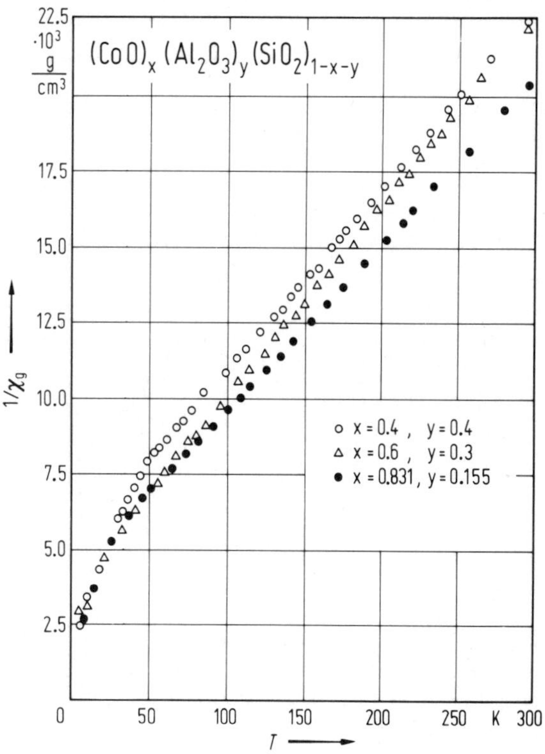

Fig. 46. $(CoO)_x(Al_2O_3)_y(SiO_2)_{1-x-y}$ glass. Inverse susceptibility vs. temperature data for various cobalt aluminosilicate glasses. Magnetic parameters obtained from the data are listed in Table 9 [75V].

Table 9. $(CoO)_x(Al_2O_3)_y(SiO_2)_{1-x-y}$ glasses. Magnetic transition temperature, paramagnetic Curie temperature, and effective magnetic moment obtained from susceptibility data as in Fig. 46 for cobalt aluminosilicate glasses [75V].

Co concentration [at%]	Magnetic transition temperature [K]	Θ_p [K]	p_{eff} [μ_B]
11.8 (x = 0.4, y = 0.4)	2.5	−95	6.7
16.1 (x = 0.5, y = 0.3)	2.3	−90	6.0
17.0 (x = 0.53, y = 0.324)	5.25	−88	5.5
20.0 (x = 0.6, y = 0.3)	4.75	−66	5.1
25.8 (x = 0.712, y = 0.237)	7.25	−77	4.9
33.5 (x = 0.831, y = 0.155)	7.8	−84	4.8

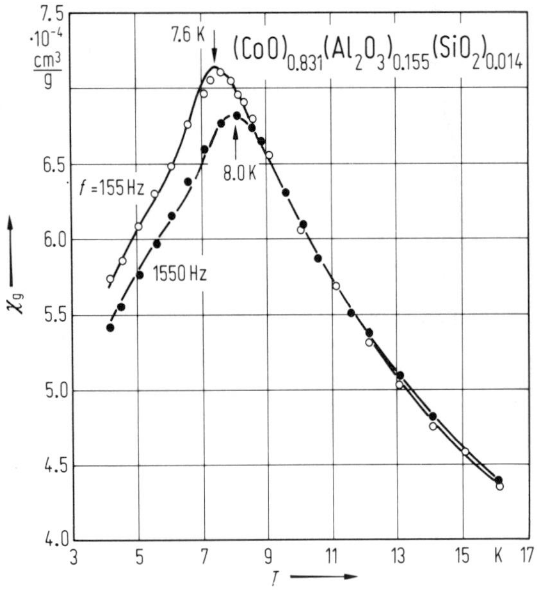

Fig. 47. $(CoO)_{0.831}(Al_2O_3)_{0.155}(SiO_2)_{0.014}$ glass. Variation of ac susceptibility at 155 Hz and 1550 Hz with temperature for a cobalt aluminosilicate glass with 33.5 at% Co [78H2].

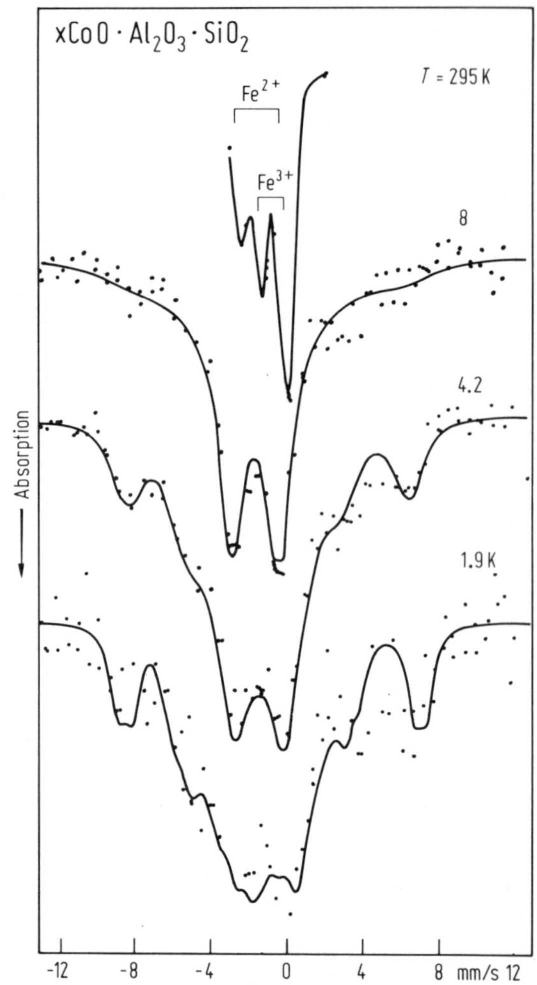

Fig. 48. $xCoO{\cdot}Al_2O_3{\cdot}SiO_2$ glass. Mössbauer spectra at various temperatures for a cobalt aluminosilicate sample (17 at% Co). Susceptibility data show a transition at 5.25 K. The solid curves are theoretical fits to the data [77B].

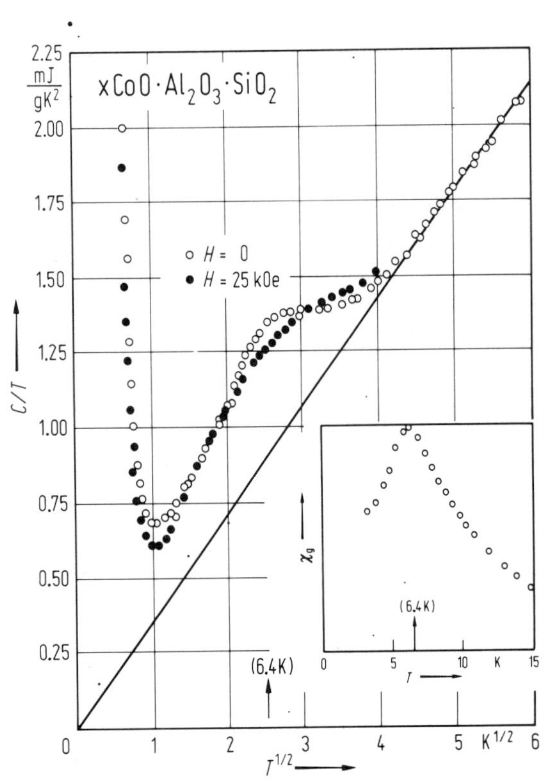

←
Fig. 49. $xCoO{\cdot}Al_2O_3{\cdot}SiO_2$ glass. Specific heat data for a sample of 14.3 at% cobalt aluminosilicate for static fields of 0 and 25 kOe. The solid line represents the $T^{3/2}$ contribution. Susceptibility data for the sample (inset) shows a peak at 6.4 K [77W]. Exact composition of glass is $(CoO)_{0.4}(Al_2O_3)_{0.1}(SiO_2)_{0.5}$.

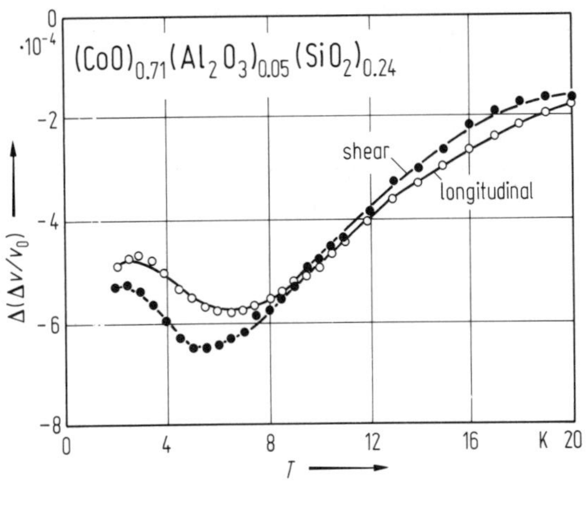

Fig. 50. $(CoO)_{0.71}(Al_2O_3)_{0.05}(SiO_2)_{0.24}$ glass. Longitudinal and shear sound velocity change vs. temperature. The experimental values of the velocities were subtracted from a calculated background. The solid curves are theoretical fits to the data [75M].

5.7.3 Amorphous oxides prepared by various techniques

5.7.3.1 Amorphous 3d oxides

5.7.3.1.0 General remarks

Investigations on noncrystalline oxides of chromium and iron show that antiferromagnetic interactions exist between 3d-ions. Iron oxide is reported to order antiferromagnetically in the range 48···225 K.

5.7.3.1.1 Cr - oxides

Amorphous CrO_2 films have been prepared by sputter deposition from a target of CrO_2 in an oxygen atmosphere [79H]. Use of an argon atmosphere or a Cr_2O_3 target always resulted in crystalline Cr_2O_3 films. a-CrO_2 films are brownish black in color and upon heating transform to a pale green crystalline Cr_2O_3 film. Susceptibility measurements from 4.2 K to 300 K do not show any evidence for magnetic ordering in the film [79H].

5.7.3.1.2 Fe (ferric) - oxides

a-Fe_2O_3 films prepared by sputter deposition from an Fe_2O_3 target or by thermal decomposition are reported to order antiferromagnetically at temperatures 48···225 K [76V], [78V], [79H], [79S1]. Susceptibility data (Fig.51) [78V] above the ordering temperature obey the Curie-Weiss law with an effective magnetization and paramagnetic Curie temperature given in Table 10. Mössbauer measurements at 4.2 K show a well resolved six-line hyperfine pattern which collapsed to a doublet above the ordering temperature (Figs 52 and 53).

Amorphous Fe_2O_3 samples sputtered from a target of Fe_3O_4, however, show a much higher ordering temperature and a complex temperature dependence for the susceptibility (Table 10, Fig. 54) [79M1]. Such films, when sputtered at high rf powers, were found to contain microcrystals of γ-Fe_2O_3 [79M1].

Table 10. a-Fe_2O_3. The magnetic order, ordering temperature T_N, effective magnetic moment p_{eff}, and paramagnetic Curie temperature Θ_p.

Magnetic order	T_N [K]	p_{eff} [μ_B]	Θ_p [K]	Remarks	Ref.	Figs.
AF	80	2.5	−200	Sus., Möss.	76V	
					78V	51
AF?	220	5.5	100	Sus.	79M1	54
	215			Möss.		
AF	100	3	–	Möss., Sus.	79H	52,53
AF	48···62	5.4	− 90	Sus., Möss.	79S1	

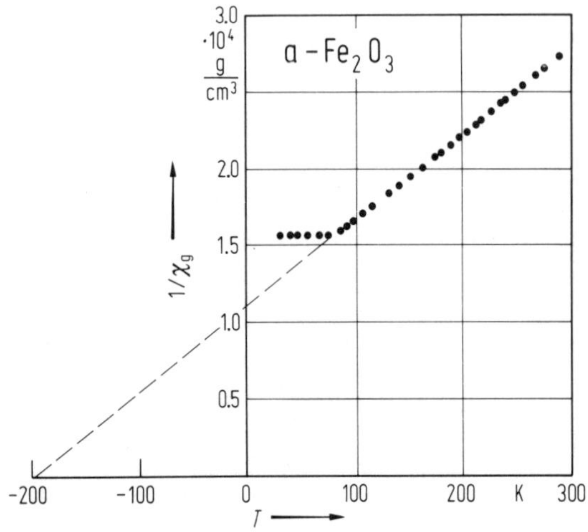

Fig. 51. a-Fe_2O_3 film. Inverse susceptibility measured at high fields vs. temperature. Sample was prepared by thermal decomposition method. The sample orders magnetically below 80 K [78V].

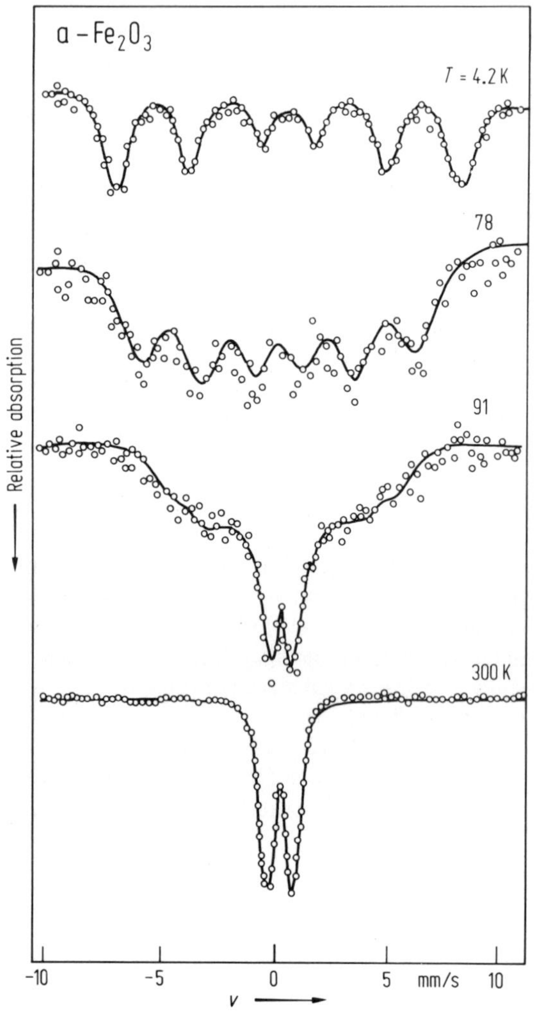

Fig. 52. a-Fe_2O_3 film. Mössbauer spectra at various temperatures. The spectral parameters indicate that the environment for iron is that of Fe^{3+} in a highly distorted octahedral site [79H].

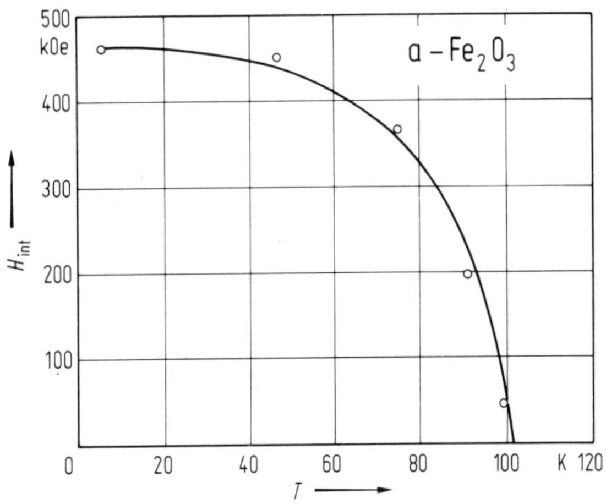

Fig. 53. a-Fe_2O_3 film. Temperature dependence of the hyperfine field H_{int} obtained from Mössbauer spectra as in Fig. 52. The ordering temperature obtained from the data is 100 K [79H].

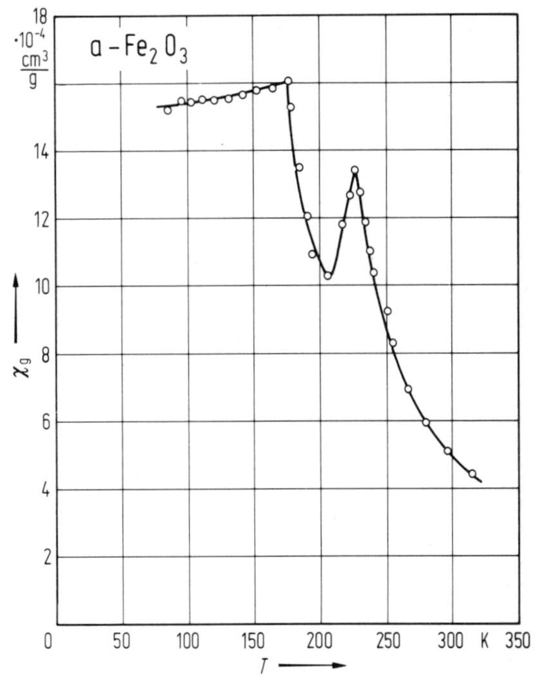

Fig. 54. a-Fe_2O_3 film. Magnetic susceptibility as a function of temperature for a sample sputtered from a target of Fe_3O_4. In the low temperature region ($T <$ 170 K), χ is nearly constant. A dip in χ is observed for T = 170···225 K. For $T >$ 225 K, χ decreases with T in accordance with the Curie-Weiss law. The data show a magnetic transition with T_N = 220 K. These films are reported to contain microcrystals of γ-Fe_2O_3 when sputtered at rf powers above 50 W [79M1].

5.7.3.2 Mixed binary oxides

5.7.3.2.0 General remarks

In this section, we provide data on magnetic parameters of noncrystalline compounds containing Fe_2O_3 and one of the following oxides: R_2O_3 (the composition corresponds to amorphous analogs of crystalline garnets $R_3Fe_5O_{12}$ for R = Y, Gd, Dy, Er, Tm, Yb, Lu, and Eu), Ga_2O_3, ZnO, SrO, Bi_2O_3 and CoO. Many of the compounds show evidence for a magnetically ordered state at low temperatures (refer to Tables 11 and 12). The ordering temperature is at least an order of magnitude smaller than for the corresponding crystalline oxide.

Data are also provided for Fe_2O_3·CuO and for CuO·Bi_2O_3.

5.7.3.2.1 Amorphous garnets

Amorphous analogs of crystalline ferrimagnetic garnets $R_3Fe_5O_{12}$ (RIG) prepared by flash-evaporation, pyrolysis, sputtering, or rapid-quenching order magnetically at very low temperatures. Details on the nature of magnetic order are provided in Table 11 and in subsequent figures (Figs. 55···64). In a-YIG, antiferromagnetic ordering occurs at temperatures ranging from 14 K to 40 K [79G], [86G]. However, a superparamagnetic behavior is observed for a-YIG prepared by pyrolysis [74P]. In amorphous rare earth iron garnets, the long range order is either antiferromagnetic or ferrimagnetic. The reported ordering temperature ranges from 4.2 K to 77 K [79L], [79M2], [86G].

Table 11(a). a-$R_3Fe_5O_{12}$. Magnetic order and ordering temperatures for amorphous garnets.

R	Magnetic order	T_N ,T_C [K]	Remarks	Ref.	Figs.
Y	AF	40	Sus.	79G	56···58
	AF	14···25	Möss.	86G	
Gd	AF ?	4.2···77	Möss.	79M2	63
	AF	14···25	Sus., Möss.	86G	62
Dy	Ferri.?	70	Möss.	79L	64
	AF	14···25	Möss.	86G	
Er	AF	14···25	Möss.	86G	
Tm	AF	14···25	Möss.	86G	
Yb	AF	14···25	Möss.	86G	
Lu	AF	14···25	Möss.	86G	
Eu	Ferri.?	62	Möss.	79L	64

Table 11 (b). a-$R_3Fe_5O_{12}$. Magnetic parameters obtained from susceptibility data for some amorphous garnets.

R	p_{eff} [μ_B]	Θ_p [K]	Ref.	Figs.
Y	2.9	−523	71S, 79G	55
	4.4	−500	79G	57
Gd	–	−170	79M2	
	–	− 10	86G	62
Dy	45(2)	−120	79L	
Eu	24(2)	−220	79L	

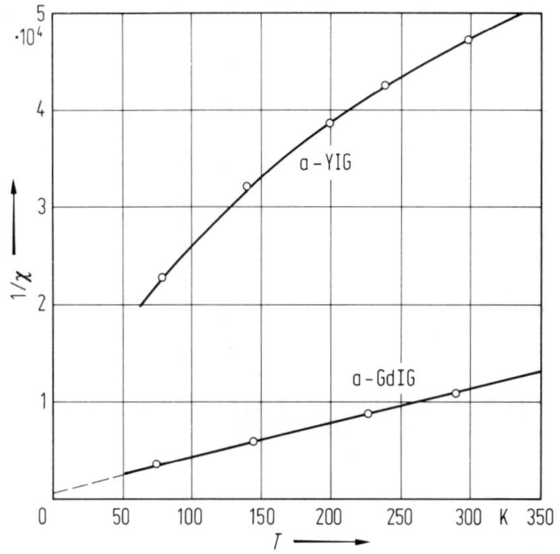

Fig. 55. a-YIG, a-GdIG. Inverse atomic susceptibility as a function of temperature for samples prepared by flash evaporation of metals in oxygen atmospheres. An antiferromagnetic exchange interaction between Fe ions in the compounds is evident for the data. The samples do not order magnetically down to 77 K [71S]. χ in cm^3g^{-1}.

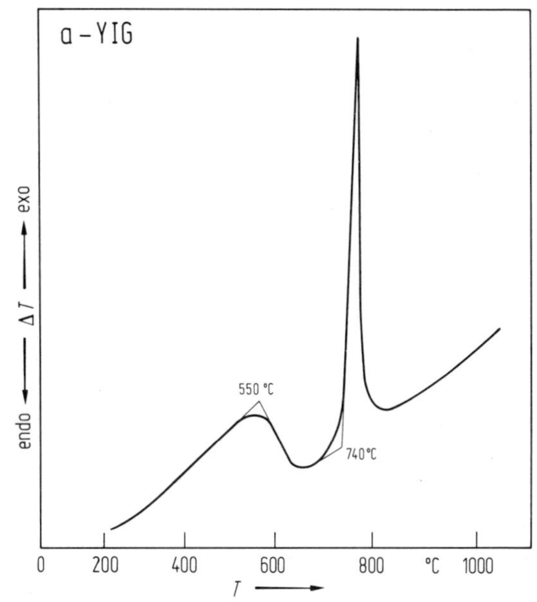

Fig. 56. a-YIG. Differential thermal analysis for a sample prepared by twin-roller quenching techniques. The data indicate a glass transition at 550°C and crystallization at 740°C. X-ray diffraction of the red colored transparent flakes gave weak diffused scattering with an estimated crystalline contamination of less than 3% [79G].

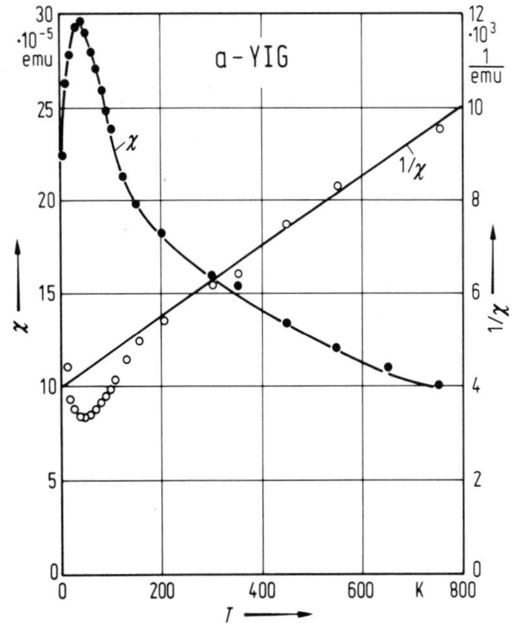

←

Fig. 57. a-YIG. Susceptibility χ and χ^{-1} vs. temperature for a sample prepared by twin-roller quenching techniques. The material is antiferromagnetic with a Néel temperature of 40 K [79G].

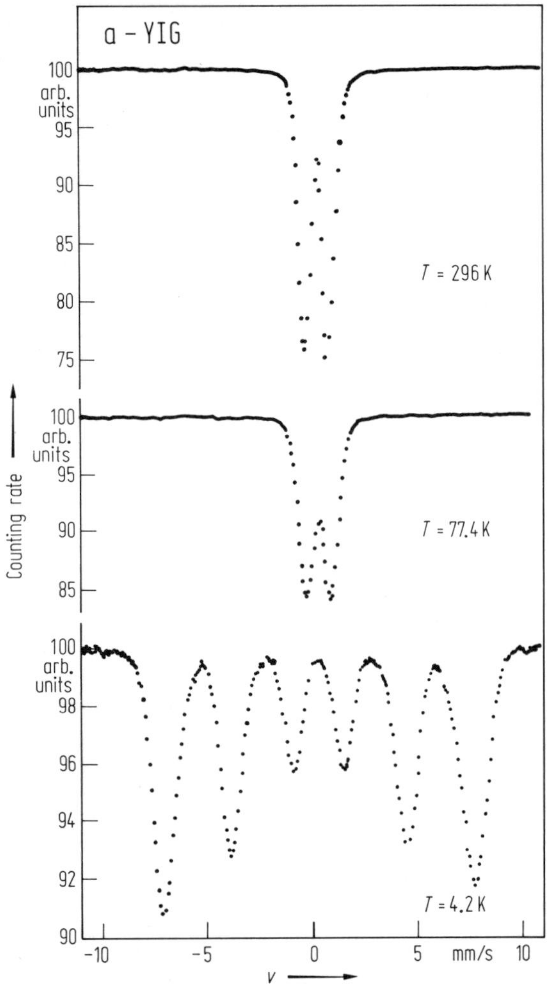

Fig. 58. a-YIG. Mössbauer spectra of ^{57}Fe in a sample prepared by twin-roller quenching. The isomer shift at room temperature, 0.31(1) mm/s with respect to iron foil, indicates that iron is in the Fe^{3+} valence state. The spectrum at 4.2 K corresponds to an average hyperfine field of 452 kOe [79G].

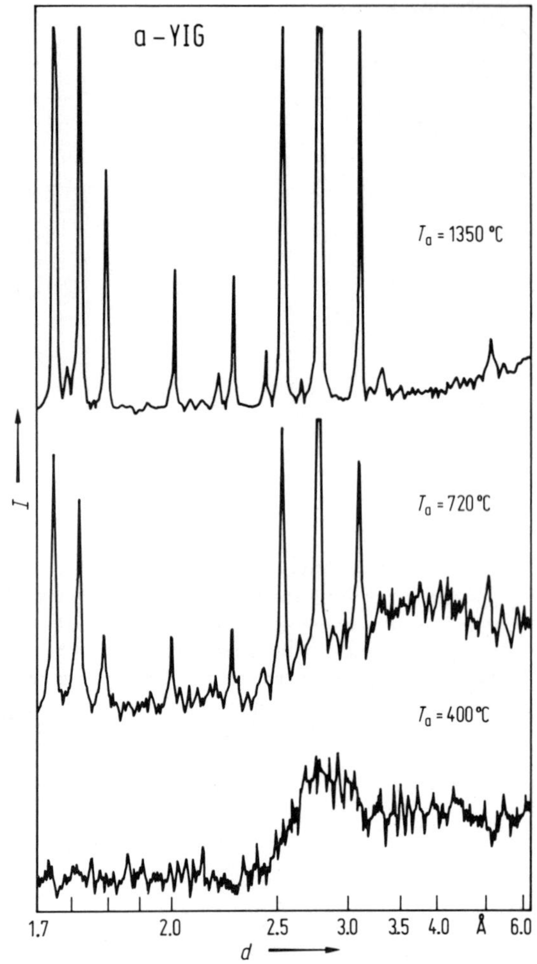

Fig. 59. a-YIG. X-ray diffractograms of a sample prepared by pyrolysis at 400°C. The sample consists of amorphous platelets made of 20 nm particles. The sample crystallizes when pyrolysis is done at above 680°C, as seen in the data for T_a = 720 and 1350°C [74P]. d: separation between lattice planes.

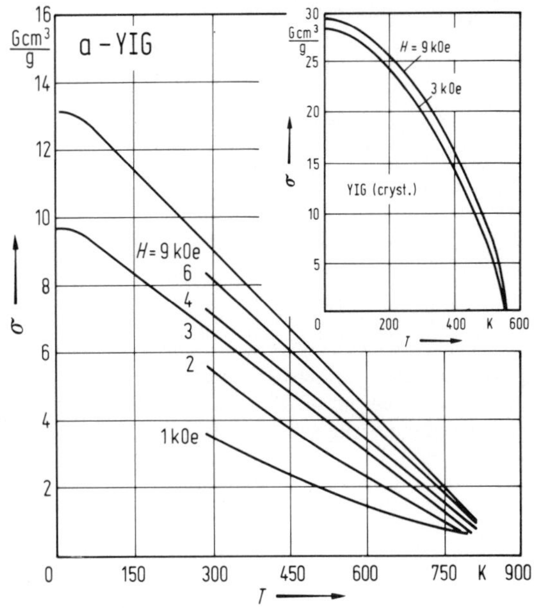

Fig. 60. a-YIG. Specific magnetization vs. temperature for a series of applied fields H for a sample prepared by pyrolysis at 400°C. The Curie temperature of about 850 K is much higher than T_C for the crystalline compound (560 K). The insert shows σ vs. T data for crystalline YIG obtained by heating the sample at 720°C [74P].

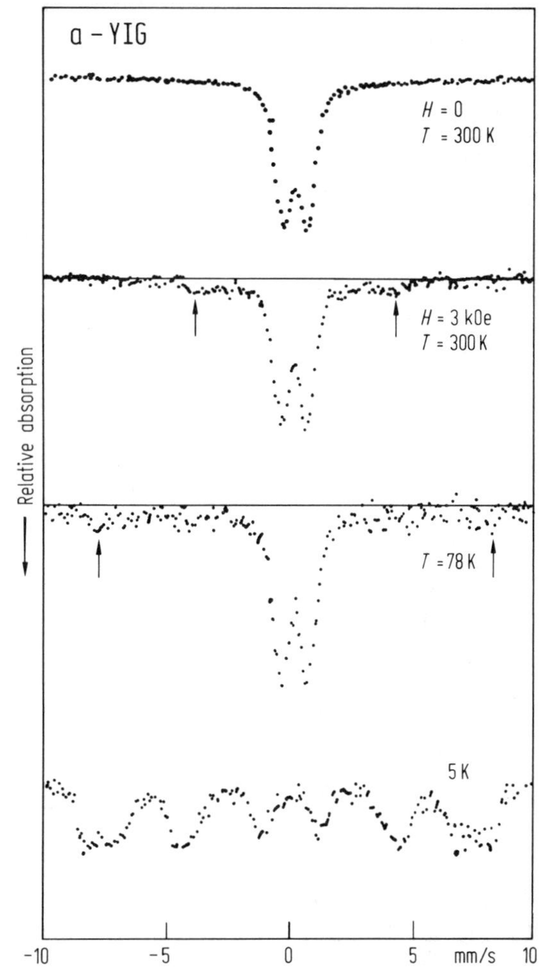

Fig. 61. a-YIG. Mössbauer spectra at 300, 78 and 5 K for a sample prepared by pyrolysis at 400°C. For the 300 K spectrum, the influence of an external magnetic field of 3 kOe is shown. The data are consistent with a superparamagnetic behavior for the sample [74P]. Arrows indicate weak absorption peaks.

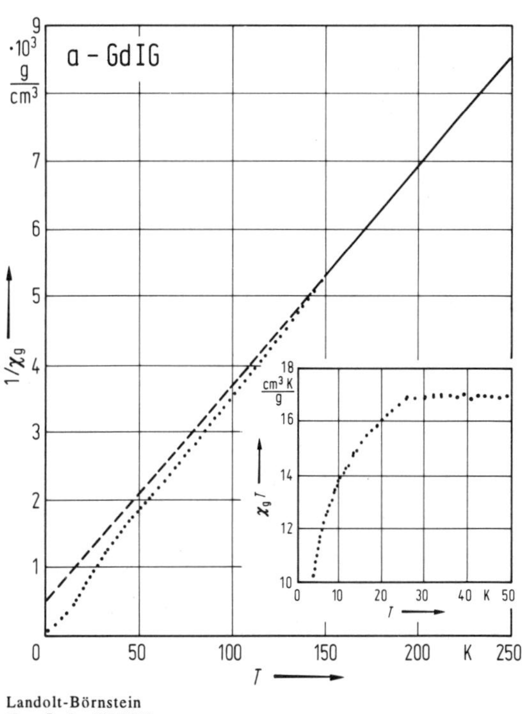

←
Fig. 62. a-GdIG. Inverse mass susceptibility vs. temperature. The insert shows χT vs. T. The sample orders magnetically below the temperature range 14···25 K [86G].

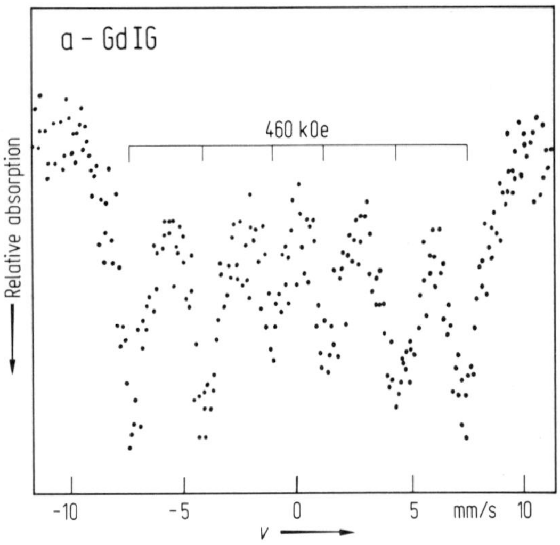

Fig. 63. a-GdIG. Mössbauer spectrum obtained with ^{57}Co in Cu source at 4.2 K for a sample prepared by rapid quenching. The sample orders magnetically below 77 K [79M2].

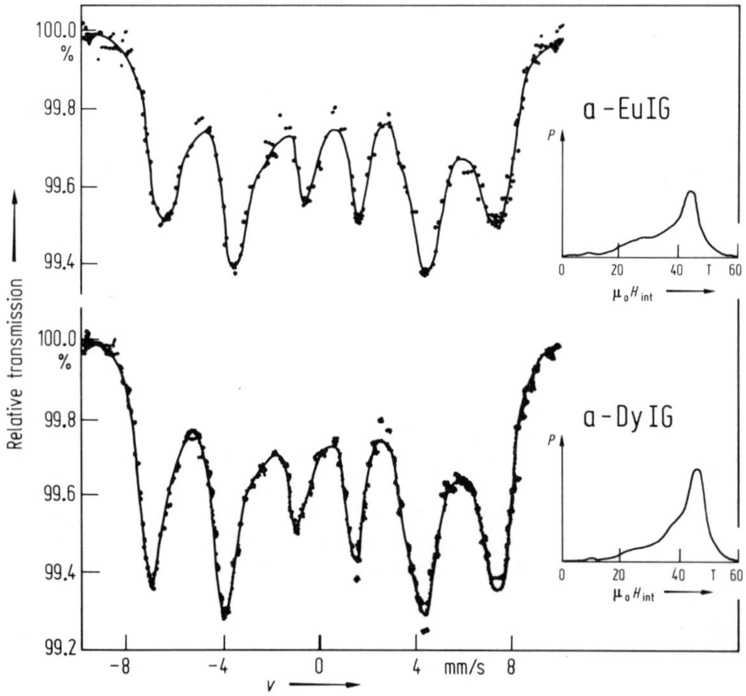

Fig. 64. a-EuIG, a-DyIG. ^{57} Fe Mössbauer absorption spectra at 4.2 K for samples prepared by dc sputtering. For spectral parameters, refer to [79L]. The samples order magnetically at 62 K and 70K, respectively. Inset shows probability distribution of magnetic hyperfine field.

5.7.3.2.2 Other mixed oxides

Table 12. Magnetic order in mixed amorphous oxides containing Fe or Cu ions.

Compound	Magnetic order	Magnetic ordering temperature [K]	Ref.	Figs.
$Fe_2O_3{\cdot}ZnO(ZnFe_2O_4)$	SG	60	91O	65
$Fe_{0.69}Ga_{1.37}O_3$	F	155	82C	66···70
$(Fe_2O_3)_{0.65}(SrO)_{0.35}$	Superpara.	–	91T1, 91T2	71···73
$(Fe_2O_3)_x(Bi_2O_3)_{1-x}$	Superpara.	–	89T, 91T1, 91T2	74···79
$Fe_2O_3{\cdot}CoO(CoFe_2O_4)$	?	284, 86	91O	80···82
$Fe_2O_3{\cdot}CuO(CuFe_2O_4)$	Ferri.	720	91S	83···86
$CuO{\cdot}Bi_2O_3(Bi_2CuO_4)$	?	100	91P	87···89

5.7.3.2.2.1 $ZnFe_2O_4$ [91O]

Films of a-$ZnFe_2O_4$ were deposited by the technique of rf sputtering. Static magnetization measured under zero-field-cooled and field-cooled conditions reveal a spin freezing transition at 60 K (Fig. 65). The magnetic transition is also evident from Mössbauer studies.

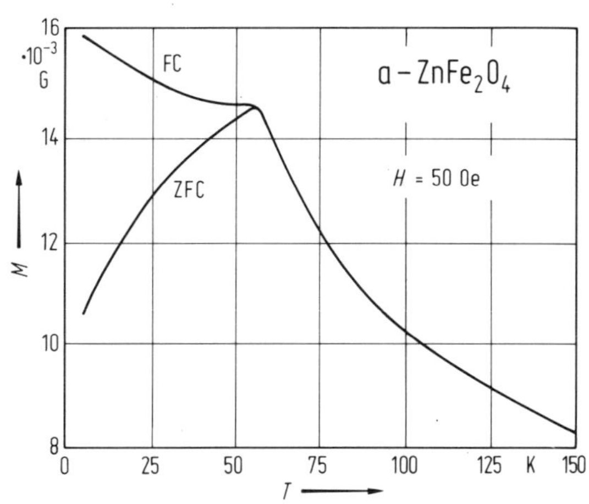

Fig. 65. a-$ZnFe_2O_4$. Zero-field-cooled (ZFC) and field-cooled (FC) magnetization for a sample prepared by rf sputtering. The data indicate spin freezing at 60 K [91O].

5.7.3.2.2.2 $Fe_{0.69}Ga_{1.37}O_3$ [82C]

Amorphous $Fe_{0.69}Ga_{1.37}O_3$ is produced by sputtering from a target of $FeGaO_3$ as black films of about 1 μm thick on Kapton and sapphire substrates. X-ray diffraction showed no sign of crystallinity. Mössbauer spectroscopy determined the composition to be $Fe^{3+}_{0.50}Fe^{2+}_{0.19}Ga_{1.37}O_3$. At room temperature the resistivity $\rho = 3\cdot10^4$ Ω cm but rises sharply with lowering temperature (Fig. 66) following the equation $\rho = \rho_0 \exp(E_a/kT)$ with $\rho_0 =$ 200 Ω cm and $E_a = 0.026(2)$ eV. The Mössbauer spectra are also consistent with the thermally activated $Fe^{2+}\Leftrightarrow Fe^{3+}$ electron transfer (Figs. 67 and 68). Hysteresis loop at 4.2 K (Fig. 69) shows ferromagnetism with 4.7 kOe coercivity and 0.46 reduced remanence. Temperature dependence of the remanence and Mössbauer count rate at zero velocity yields the Curie temperature T_C = 155(10) K (Fig. 70). The ferromagnetism is attributed to $Fe^{2+}\Leftrightarrow Fe^{3+}$ electron transfer.

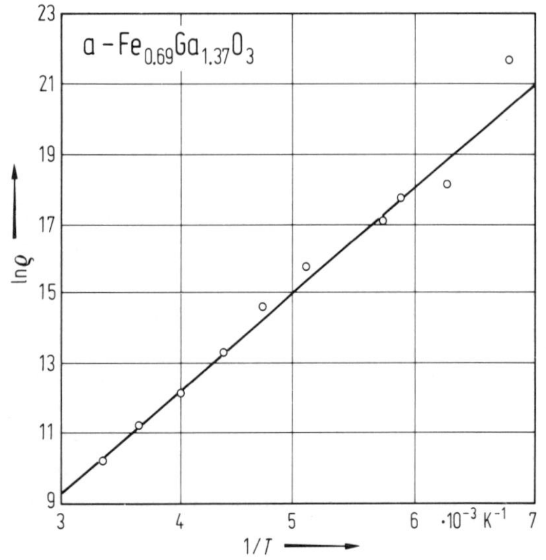

Fig. 66. a-$Fe_{0.69}Ga_{1.37}O_3$ films. Temperature dependence of the electrical resistivity. The straight line fit yields $E_a = 0.26(2)$ eV and $\rho_0 = 200$ Ω cm in the equation $\rho = \rho_0 \exp(E_a/kT)$ [82C].

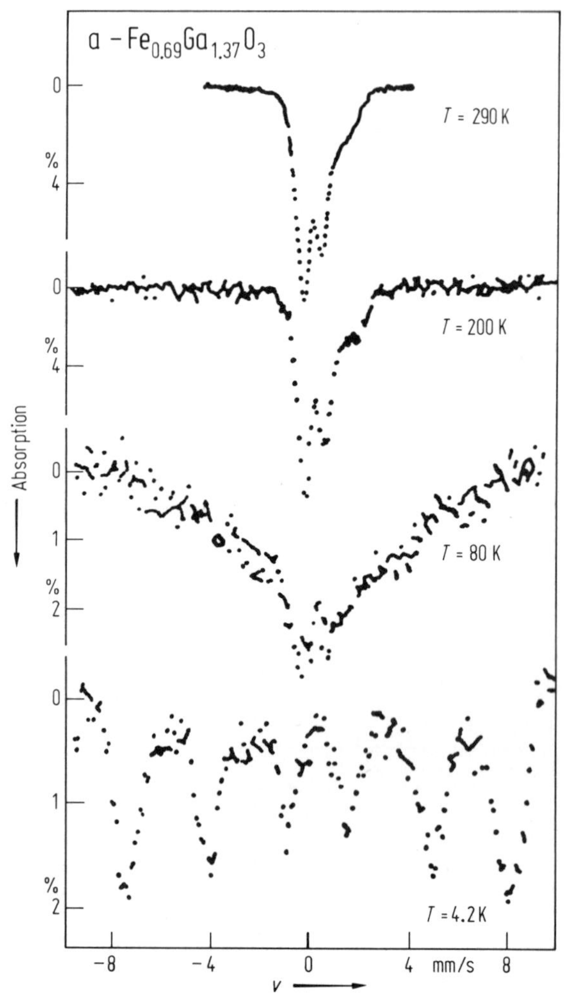

→

Fig. 67. a-$Fe_{0.69}Ga_{1.37}O_3$. Mössbauer spectra at several temperatures. The six line hyperfine splitting at 4.2 K indicates magnetic ordering and a ferric hyperfine field of 480 kOe. At high temperatures, both Fe^{2+} and Fe^{3+} quadrupole doublets are observed [82C].

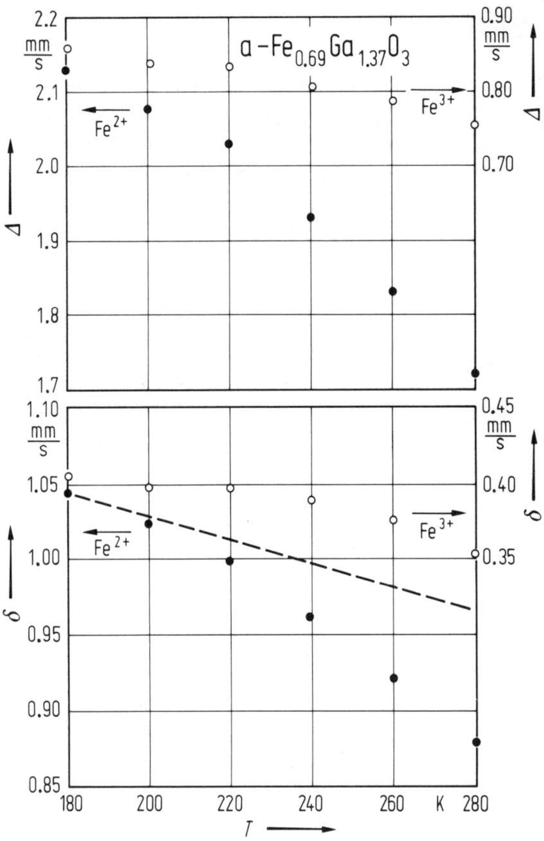

Fig. 68. a-$Fe_{0.69}Ga_{1.37}O_3$. Temperature dependence of the quadrupole splittings Δ and isomer shift δ (relative to iron metal) for Fe^{2+} and Fe^{3+} in the paramagnetic region. The dashed line shows the normal high-temperature variation of the isomer shift. The data are consistent with the $Fe^{2+} \Leftrightarrow Fe^{3+}$ relaxation process [82C].
←

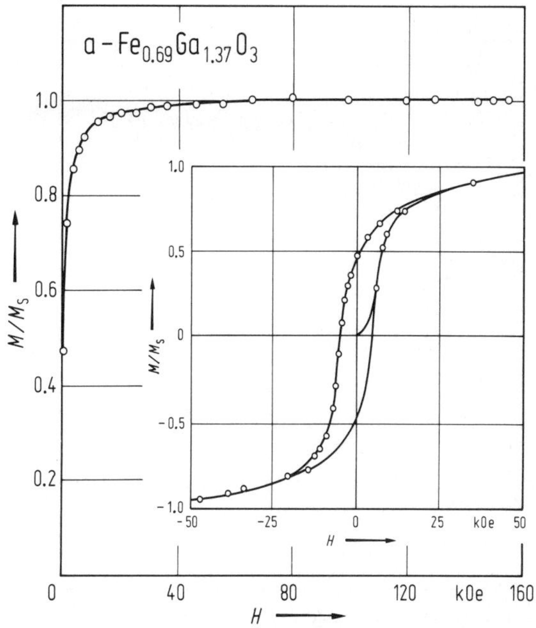

Fig. 69. a-$Fe_{0.69}Ga_{1.37}O_3$. Magnetization (hysteresis) curves at 4.2 K shows both coercivity (4.7 kOe) and remanence (0.46), typical of a hard ferromagnet [82C].

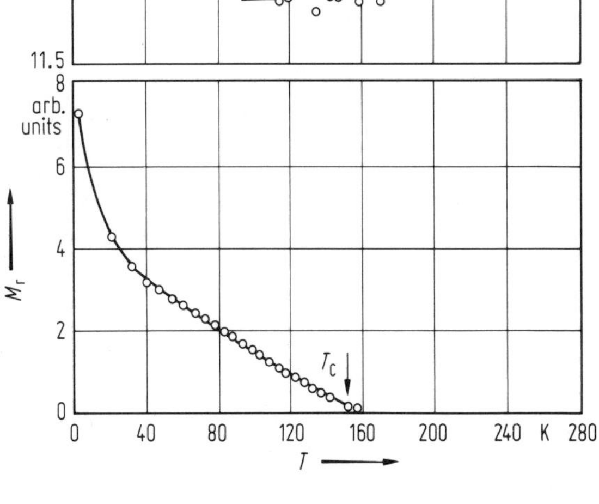

←
Fig. 70. a-$Fe_{0.69}Ga_{1.37}O_3$. Temperature dependence of the saturation remanence M_r and Mössbauer count rate at zero velocity are used to determine the Curie temperature T_C, yielding T_C = 156(6) K and 155(10) K, respectively [82C].

5.7.3.2.2.3 $(Fe_2O_3)_x (SrO)_{1-x}$ [91T1]

20 μm thick specimens of the composition $(Fe_2O_3)_{0.65}(SrO)_{0.35}$ were prepared by rapid quenching. Crystalline phases were not detected by X-ray diffraction studies (Fig. 71). A room temperature magnetization of 3.25 G cm^3/g at 10 kOe is obtained for the sample and M vs. H data show hysteresis and remanence (Fig. 72) with a coercive force of 250 Oe. Mössbauer spectra at room temperature consist of an asymmetrical paramagnetic quadrupole doublet. Spectra at low temperatures (Fig. 73) show the hyperfine structure. The magnetism is attributed to ordered clusters with a structure similar to that of $SrFe_{12}O_{19}$.

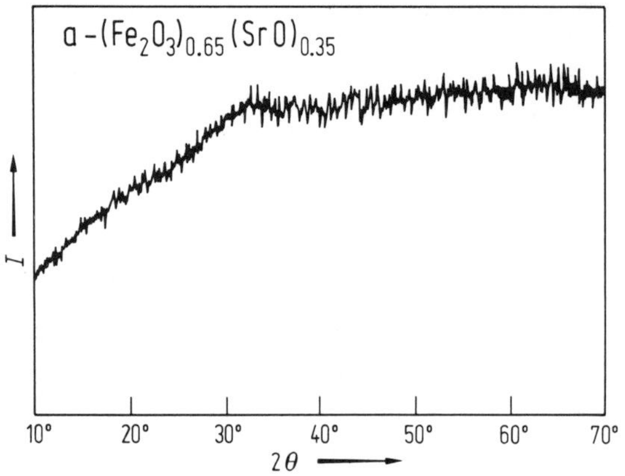

Fig. 71. a-$(Fe_2O_3)_{0.65}(SrO)_{0.35}$. X-ray diffraction pattern [91T1].

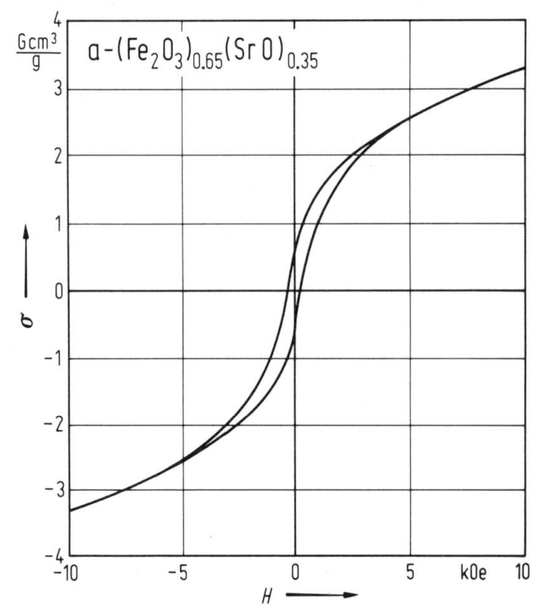

Fig. 72. a-$(Fe_2O_3)_{0.65}(SrO)_{0.35}$. Room temperature specific magnetization as a function of the static field. The spontaneous moment is suggested to be due to clusters of $SrFe_{12}O_{19}$ in the sample [91T1].

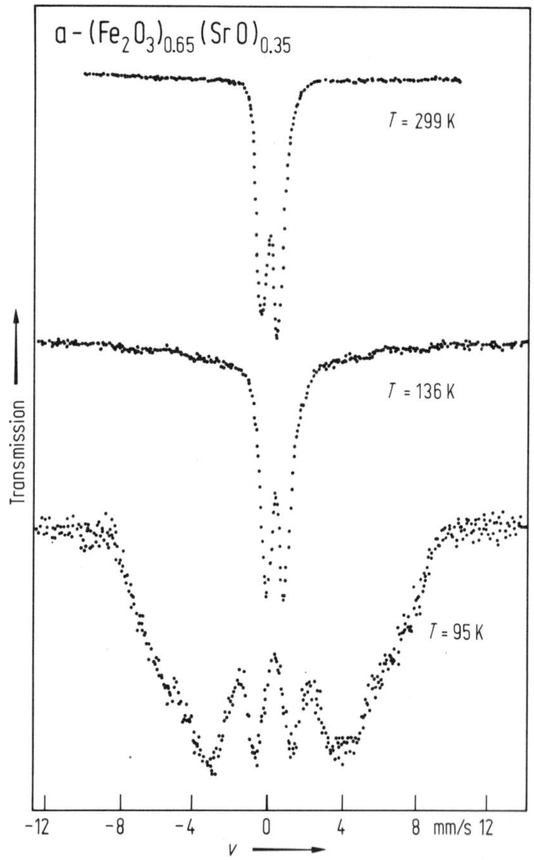

Fig. 73. a-$(Fe_2O_3)_{0.65}(SrO)_{0.35}$. Mössbauer spectra. Spectra at 299 K show a paramagnetic doublet with an isomer shift of 0.31 mm/s (rel. to α-Fe foil at RT) and a quadrupole splitting of 0.92 mm/s. The spectra show hyperfine structure at 136 K and 95 K. Ordered clusters with a spin structure similar to that of strontium hexaferrite are suggested to give rise to the observed spontaneous moment (Fig. 72) in the system [91 T1].

5.7.3.2.2.4 $(Fe_2O_3)_x (Bi_2O_3)_{1-x}$ [87T], [89T], [91T1], [91T2]

The system $Fe_2O_3{\cdot}Bi_2O_3$ is reported to have a wide glass forming region. Amorphous samples of $(Fe_2O_3)_x{\cdot}(Bi_2O_3)_{1-x}$, for x =0.1···0.625, were prepared by rapid quenching. Fig. 74 shows X-ray diffraction pattern for such samples. For samples with x = 0.1 to 0.6, a paramagnetic character at room temperature is observed by Mössbauer spectroscopy (Figs. 75 and 76) and electron spin resonance studies (g = 2.02···2.08) (Fig. 77). For x = 0.625, the amorphous bismuth iron garnet $Bi_3Fe_5O_{12}$, a ferromagnetic moment is observed at room temperature (Fig. 78). Mössbauer spectra for the sample show a weak hyperfine structure both at room temperature and at low temperatures (Fig. 79). The ferromagnetic character is suggested to arise from crystalline bismuth iron garnet-like clusters.

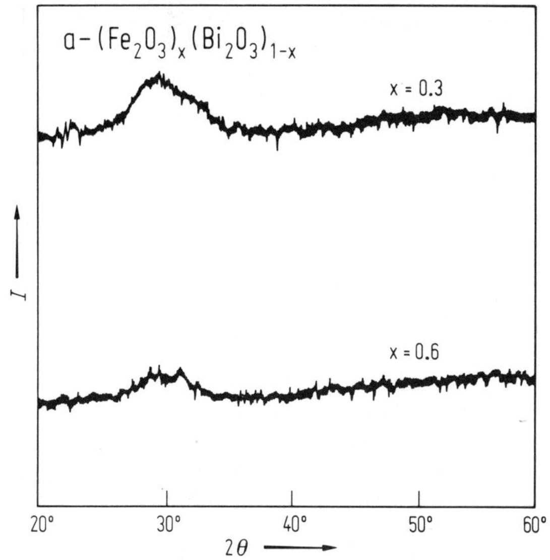

Fig. 74. a-$(Fe_2O_3)_x(Bi_2O_3)_{1-x}$. Room temperature X-ray diffraction data for samples with x = 0.3 and x = 0.6 prepared by rapid quenching. The data confirm the amorphous nature for the samples [89T].

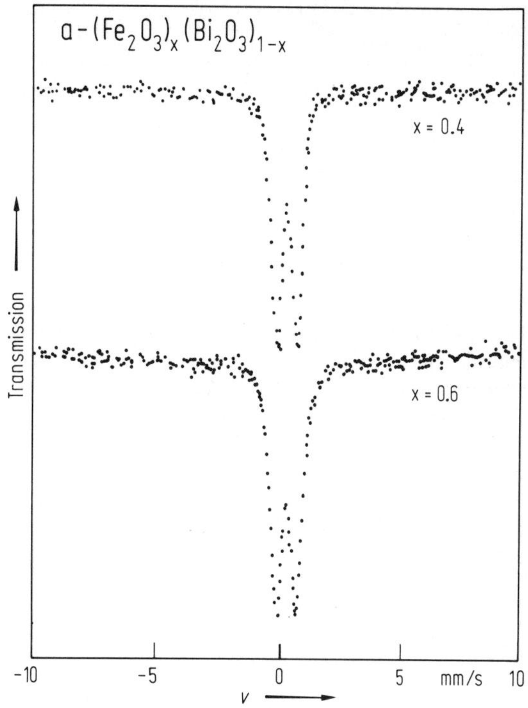

Fig. 75. a-$(Fe_2O_3)_x(Bi_2O_3)_{1-x}$. Mössbauer spectra at room temperature for amorphous samples with x = 0.4 and x = 0.6. The symmetric doublet is due to Fe^{3+} in tetrahedral sites [89T].

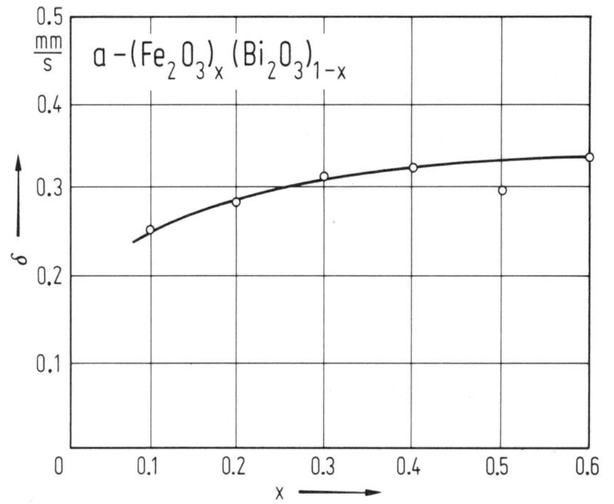

←
Fig. 76. a-$(Fe_2O_3)_x(Bi_2O_3)_{1-x}$. Dependence of the room temperature isomer shift on the content of Fe_2O_3.
δ is relative to metallic iron [89T].

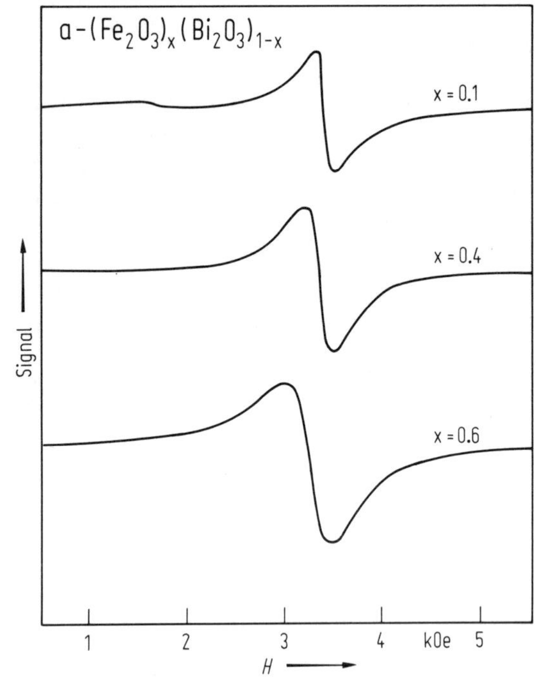

Fig. 77. a-$(Fe_2O_3)_x(Bi_2O_3)_{1-x}$. Room temperature electron spin resonance spectra at 9.5 GHz for amorphous samples with x = 0.1, 0.4 and 0.6. The *g*-value varies from 2.02 for x = 0.1 to 2.08 for x = 0.6 [89T].

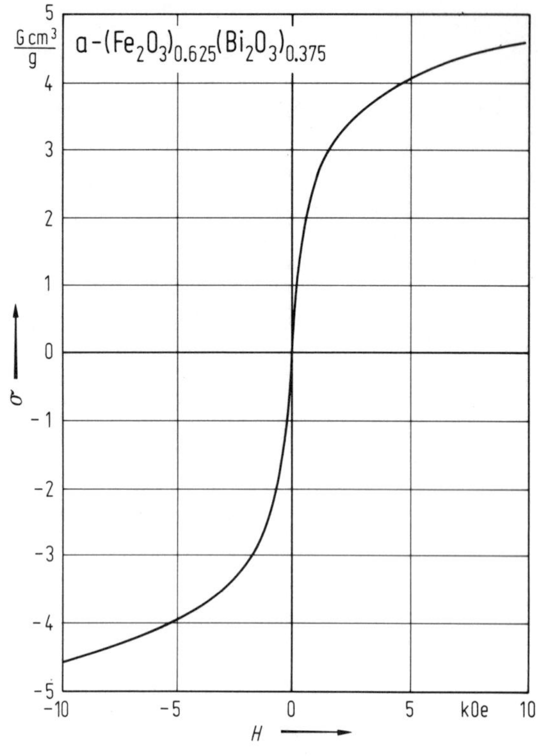

Fig. 78. a-$(Fe_2O_3)_{0.625}(Bi_2O_3)_{0.375}$ (= $Bi_3Fe_5O_{12}$). Specific magnetization vs. static field at room temperature. The observed ferromagnetic character was proposed to result from magnetically ordered clusters in the sample [91T1].

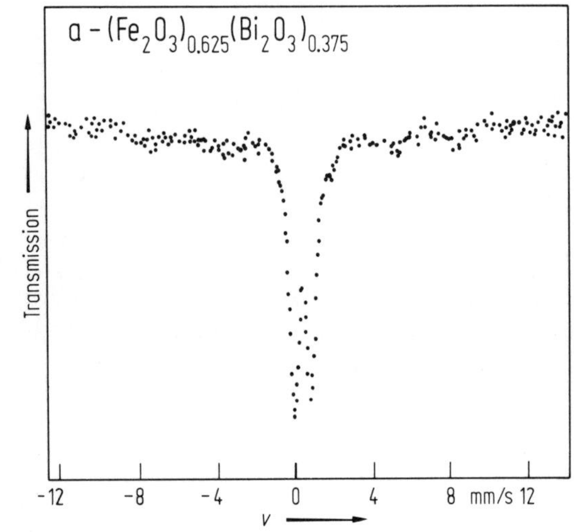

←
Fig. 79. a-$(Fe_2O_3)_{0.625}(Bi_2O_3)_{0.375}$ (= $Bi_3Fe_5O_{12}$). Mössbauer data at 98 K. The spectra show a weak hyperfine splitting due to the presence of ordered clusters. The room temperature spectrum also showed such a weak fine structure, and an isomer shift of 0.34 mm/s (rel. to α-Fe foil at RT) and a quadrupole splitting of 0.87 mm/s were estimated at room temperature [91T1].

5.7.3.2.2.5 $CoFe_2O_4$ [91O]

Films of a-$CoFe_2O_4$ were deposited on glass or mylar substrates by rf sputtering techniques using a sintered $CoFe_2O_4$ target. Amorphous nature of the samples was confirmed by X-ray diffraction, electron diffraction, and transmission electron microscopy. dc and ac susceptibility data show two spin freezing temperatures, T_{f1} = 284 K due to random anisotropy of Co ions and T_{f2} = 86 K due to frustration of antiferromagnetic interactions (Figs. 80 and 81). The Mössbauer spectrum at room temperature shows a typical paramagnetic doublet (Fig. 82). For $T < T_{f1}$, a sextuplet splitting appears and its intensity increases with decreasing T. The spin freezing at T_{f2} is evident from Mössbauer data.

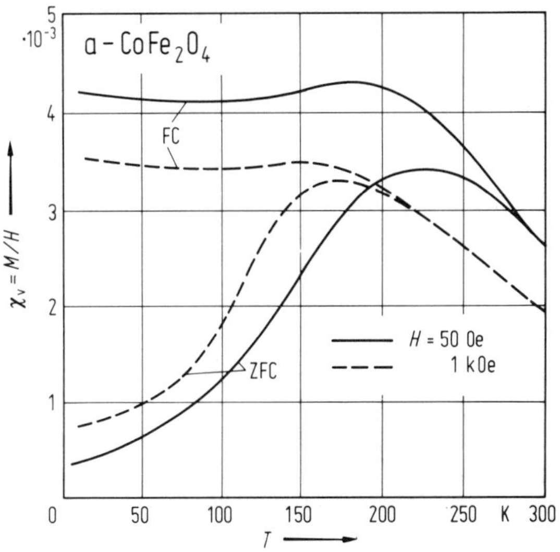

Fig. 80. a-$CoFe_2O_4$. Temperature dependence of zero-field-cooled (ZFC) and field-cooled (FC) susceptibility. The shape of the ZFC curves below irreversibility temperature T_{f1} resembles that of random anisotropy magnets. T_{f1} decreases with increasing static field H [91O]. χ in CGS-emu.

For Fig.81 see next page.

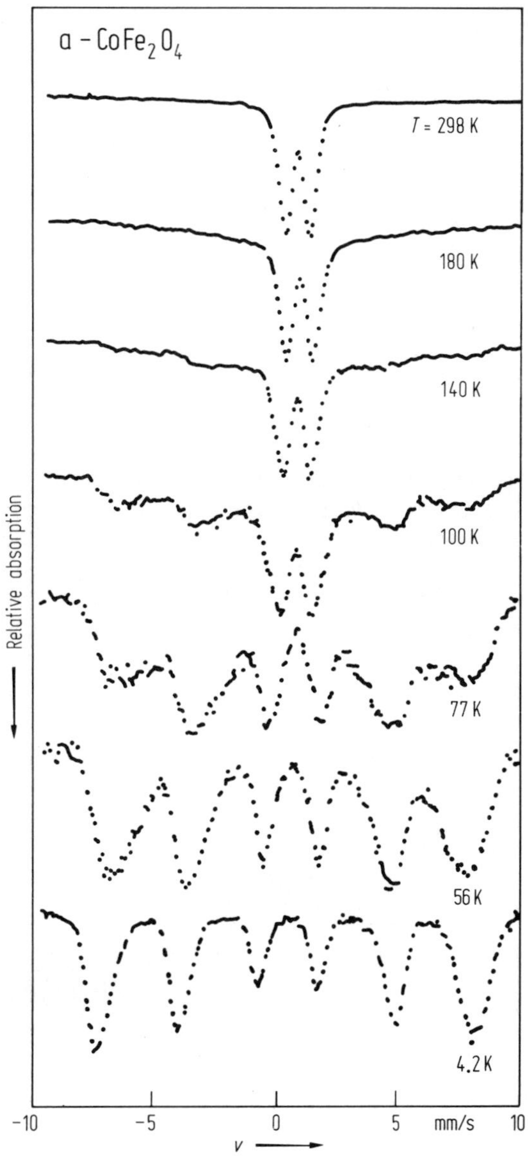

Fig. 82. a-$CoFe_2O_4$. Mössbauer spectra at various temperatures. Hyperfine splitting appears in the spectra for temperatures below T_{f1}. The separation between the two innermost lines begins to decrease below 100 K which is close to the spin freezing temperature T_{f2} = 86 K obtained from ac susceptibility data (Fig. 81) [91O].

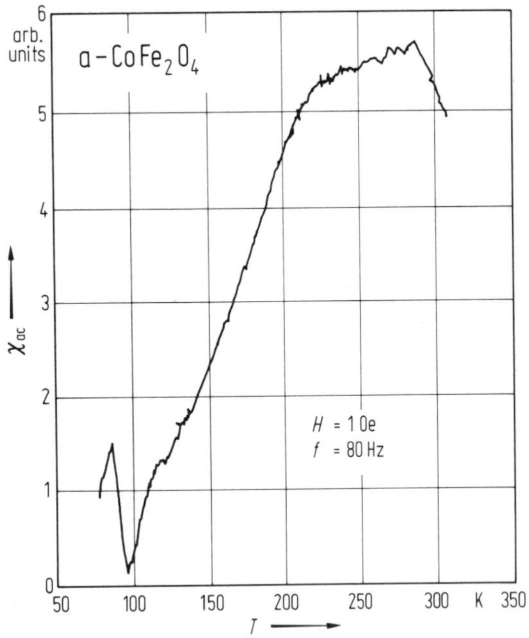

Fig. 81. a-$CoFe_2O_4$. ac susceptibility as a function of temperature. A cusp at 284 K is superimposed on a broad peak and it corresponds to the spin freezing at T_{f1}. A second spin freezing transition occurs at T_{f2} = 86 K [91O].

5.7.3.2.2.6 $CuFe_2O_4$ [91 S]

A long range magnetic order with a planar anisotropy is reported in as-deposited films of amorphous copper ferrite, $CuFe_2O_4$, prepared by rf sputtering. One micrometer thick films on glass substrates were sputtered in a mixed oxygen argon atmosphere. X-ray diffraction data indicate the absence of any crystalline phases (Fig. 83). The films show a large magnetization and anisotropy, and a paramagnetic-to-antiferromagnetic-like transition is evident at 50 K in the magnetization data (Figs. 84 and 85). The low temperature magnetization, Curie temperature, and gyromagnetic ratio (Fig. 86) are in agreement with values expected for crystalline copper ferrite. It is suggested that the film contains amorphous ordered clusters with a spin structure similar to that for $CuFe_2O_4$ and Fe_2O_3.

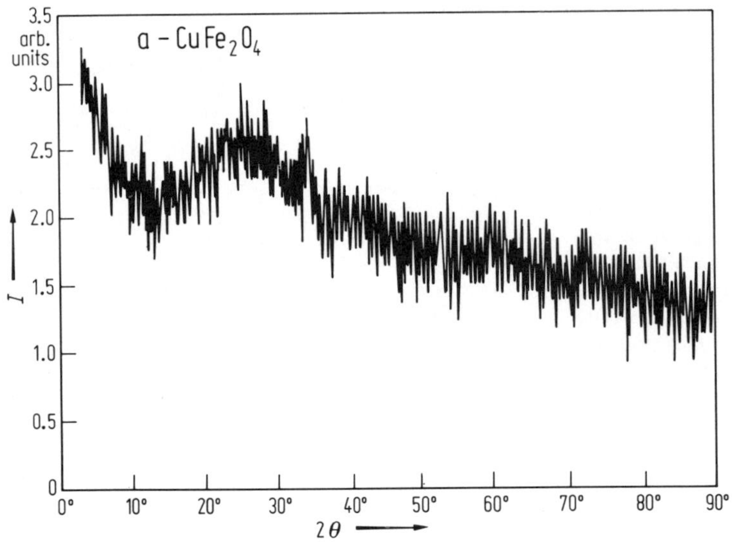

Fig. 83. a-$CuFe_2O_4$. X-ray diffraction pattern for a sample prepared by rf sputtering in a 50% oxygen and 50% argon atmosphere [91S].

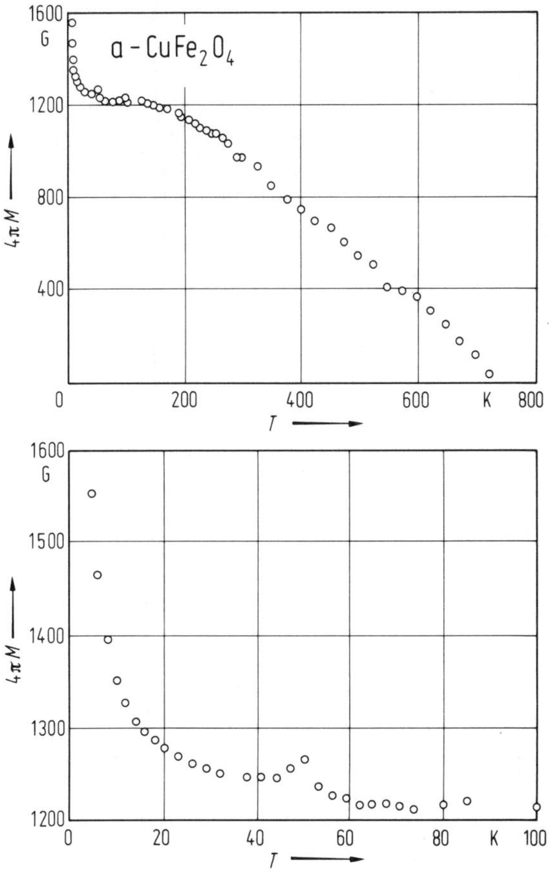

Fig. 84. a-$CuFe_2O_4$. Magnetization $4\pi M$ vs. temperature for an as-deposited film. The lower figure shows the data points in the low temperature region. The measurements were done for an in-plane static field of 5 kOe. The peak in the magnetization at 50 K is identified with a paramagnetic to antiferromagnetic transition for clusters of Fe_2O_3 in the film [91S].

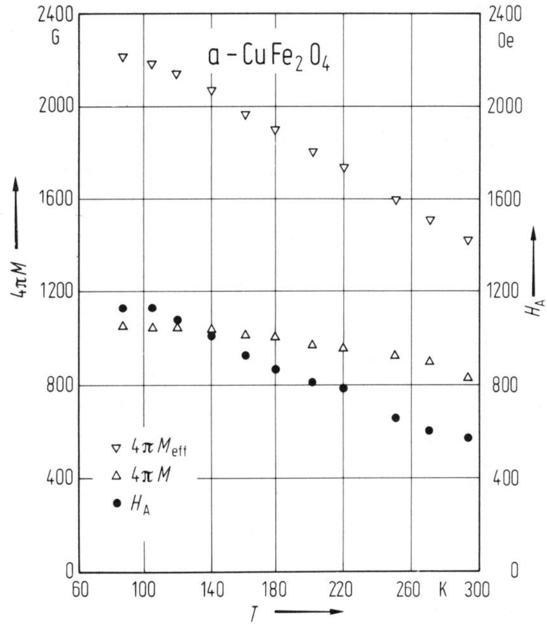

Fig. 85. a-$CuFe_2O_4$. Temperature dependence of the effective magnetization $4\pi M_{eff}$ obtained from ferromagnetic resonance at 9.1 GHz, magnetization shown in Fig. 84, and the in-plane anisotropy field $H_A = 4\pi M_{eff} - 4\pi M$ [91S].

Fig. 86. a-$CuFe_2O_4$ film. The effective gyromagnetic ratio γ vs. temperature obtained from FMR studies at 9.1 GHz [91S]. →

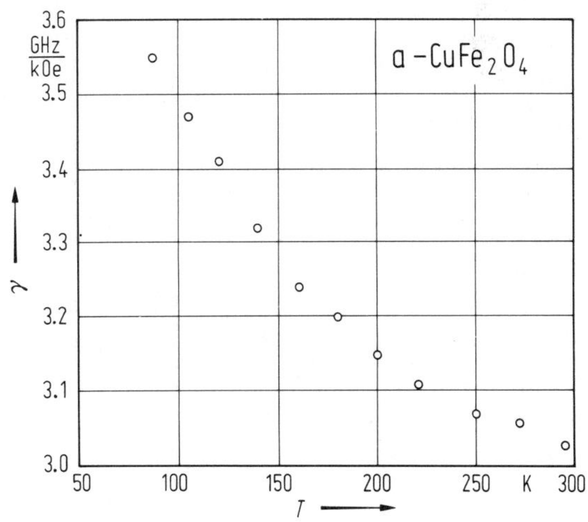

5.7.3.2.2.7 Bi_2CuO_4 [91 P]

Black colored powder samples of a-Bi_2CuO_4 with a crystallization temperature of 650 K (Fig. 87) were synthesized by melt quenching techniques. Susceptibility data show a strong dependence on the static field and temperature, with an effective moment of 1.5 μ_B for Cu^{2+} ions (Fig. 88). A magnetic transition of unknown nature is evident from data on ESR resonance field and linewidth (Fig. 89).

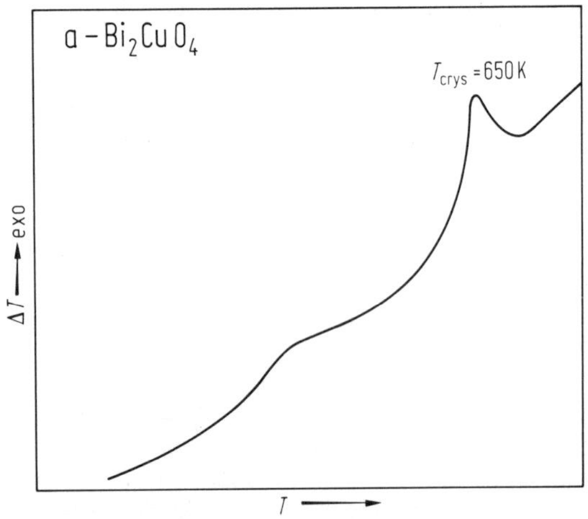

Fig. 87. a-Bi_2CuO_4. Differential thermal analysis data for a sample prepared by melt quenching. The crystallization temperature is 650 K [91P].

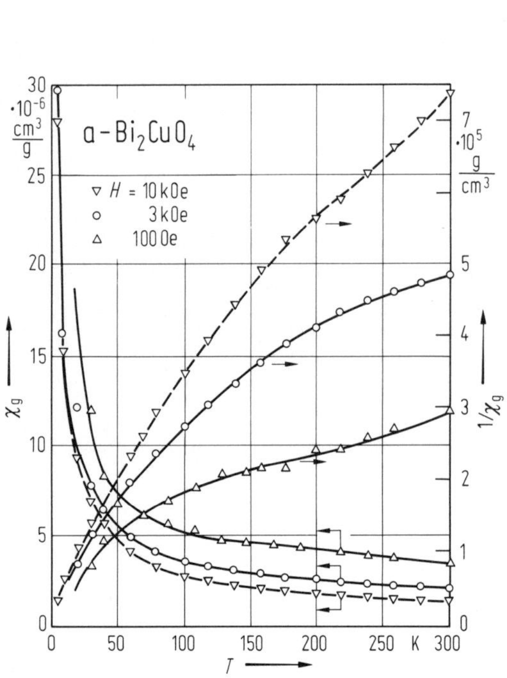

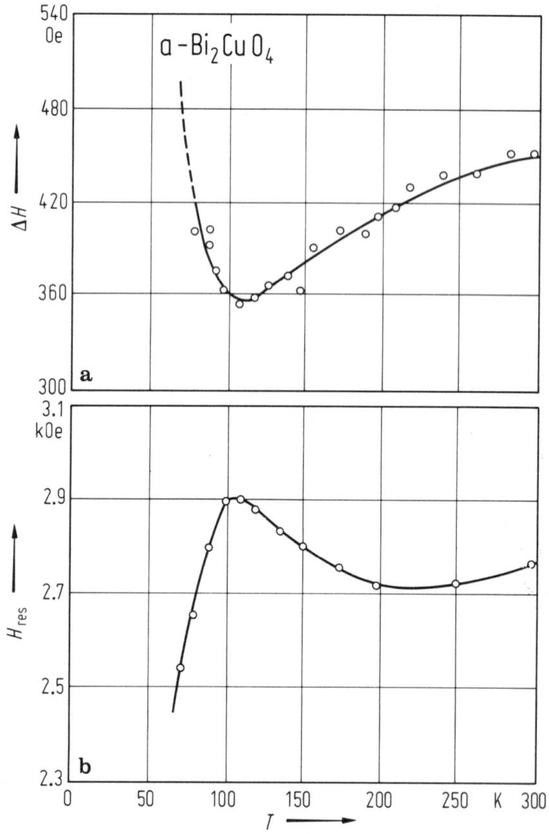

Fig. 89. a-Bi_2CuO_4. Variation with temperature of X-band ESR (a) linewidth and (b) resonance field. The data show evidence for a magnetic phase transition of unknown nature beginning at 100 K [91P].

←

Fig. 88. a-Bi_2CuO_4. Temperature dependence of the susceptibility. The data are for static fields H = 10 kOe, 3 kOe , and 100 Oe . An effective magnetic moment of 1.5 μ_B is obtained for Cu^{2+} ions from the data for H = 10 kOe [91P].

5.7.3.3 Fe_2O_3 - Bi_2O_3 - based ternary oxides

5.7.3.3.0 General remarks

A long range magnetic order is observed in a family of ternary amorphous oxides containing Fe_2O_3, Bi_2O_3 and one of the following compounds: Li_2O, K_2O, CaO, ZnO, CuO, and ABO_3 (AB = BaTi, PbTi, PbZr). The oxides show a ferromagnetic or ferrimagnetic character with a relatively high Curie temperature, of the order of 300···900 K. The samples are prepared by either rapid quenching or rf sputtering. A ferromagnetic moment is observed in as-quenched samples. However, as-sputtered samples are paramagnetic at room temperature and films annealed in air at high temperatures show a spontaneous moment. Refer to Table 13 for details on the nature of magnetic order and ordering temperatures.

Table 13. a-$(Fe_2O_3)_x(MO)_y(Bi_2O_3)_{1-x-y}$. Magnetic order and ordering temperatures for substituted amorphous compounds of Fe_2O_3·Bi_2O_3. For compounds consisting of two magnetic phases, data on ordering temperatures are provided for both phases.

Substituent	Magnetic order	Ordering temperature[K]	Ref.	Figs.
Li_2O	F	≈ 800 (x = 0.5, y = 0.25)	89N, 91N	90···96
	AF	200		
K_2O	F	673 (x = 0.734, y = 0.083)	92Y	97···100
CaO	F	760 (x = 0.65, y = 0.2)	87N2, 91N	101···105
	AF	80		
	Ferri.	600···750 (x = 0.5, y = 0.1···0.5)	92R1	106···109, 114
ZnO	S	460···520 (x = 0.5, y ≤ 0.2)	86O, 86S	115···121
	Ferri.	600 (x = 0.5, y < 0.5)	91R3	122...124, 129
CuO	F	650 (x = 0.5, y = 0.25)	90S	131···138
	S	200		
	Ferri.	700···750 (x = 0.5, y = 0···0.3)	93S2	143···147
ABO_3 (AB = BaTi, PbTi, PbZr)				
	F	473···773	88F, 91R1	150···153, 156 ,157

5.7.3.3.1 $Fe_2O_3 \cdot Li_2O \cdot Bi_2O_3$ oxides [89N], [91N]

Amorphous $Fe_2O_3{\cdot}Li_2O{\cdot}Bi_2O_3$ compounds were synthesized by rapid quenching of the molten mixture of the oxides by means of a twin-roller. Fig. 90 shows the phase diagram for the ternary system. As-quenched samples are X-ray amorphous. However, high resolution electron microscopy studies revealed the presence of a dense nanometer sized less-disordered phase. For $(Fe_2O_3)_{0.5}(Li_2O)_{0.25}$ -$(Bi_2O_3)_{0.25}$, the crystallization temperature is found to be 638 K (Fig. 91). A ferromagnetic character is observed for the samples (Figs. 92 and 93). However, the Curie temperature could not be determined accurately since T_C was higher than the crystallization temperature (Fig. 94). Mössbauer studies reveal the presence of two magnetic phases, a short-range antiferromagnetic phase and a ferromagnetic microcrystalline phase (Fig. 95). Samples show irreversible magnetization processes (Fig. 96) below 200 K in weak fields, most likely caused by the interaction between the above mentioned magnetic phases.

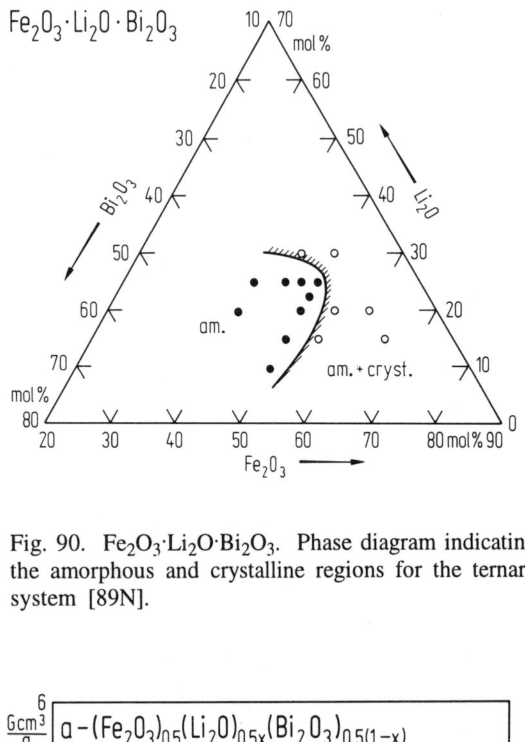

Fig. 90. $Fe_2O_3{\cdot}Li_2O{\cdot}Bi_2O_3$. Phase diagram indicating the amorphous and crystalline regions for the ternary system [89N].

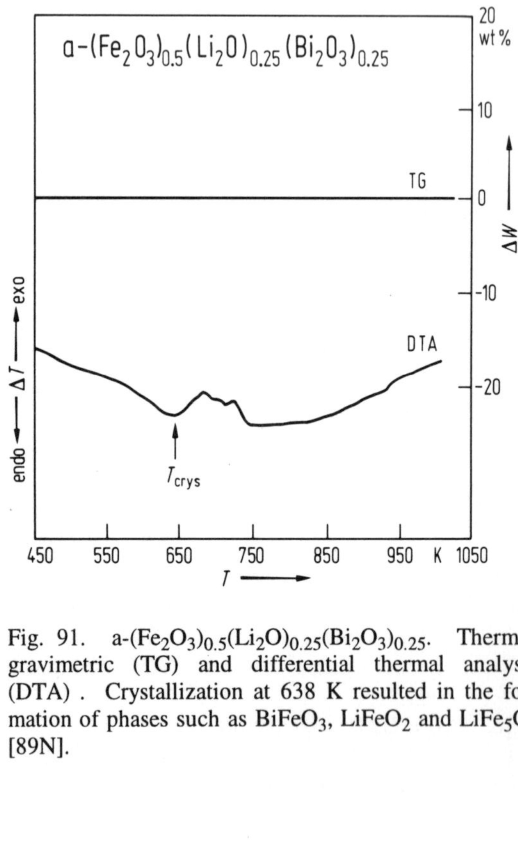

Fig. 91. a-$(Fe_2O_3)_{0.5}(Li_2O)_{0.25}(Bi_2O_3)_{0.25}$. Thermogravimetric (TG) and differential thermal analysis (DTA) . Crystallization at 638 K resulted in the formation of phases such as $BiFeO_3$, $LiFeO_2$ and $LiFe_5O_8$ [89N].

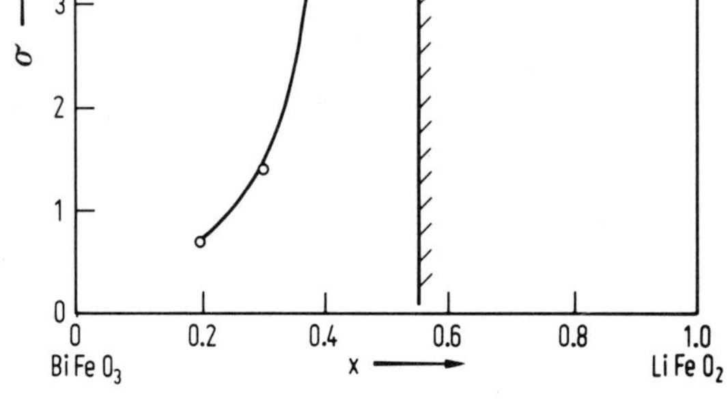

←
Fig. 92. a-$(Fe_2O_3)_{0.5}(Li_2O)_{0.5x}(Bi_2O_3)_{0.5(1-x)}$. Room temperature specific magnetization for samples in an applied static field of 15 kOe [89N]. am: amorphous; cryst: crystalline.

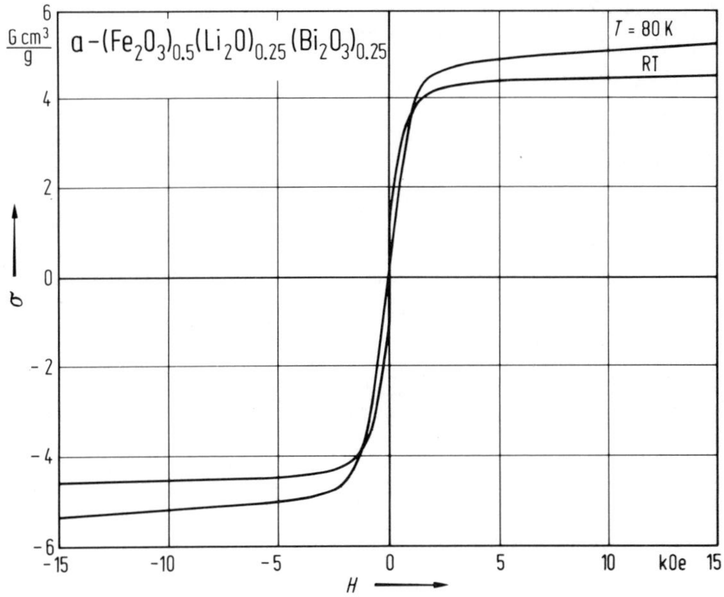

Fig. 93. a-$(Fe_2O_3)_{0.5}(Li_2O)_{0.25}(Bi_2O_3)_{0.25}$. Specific magnetization vs. static field at room temperature and 80 K. Saturation of the magnetization is not observed even at the largest field [89N].

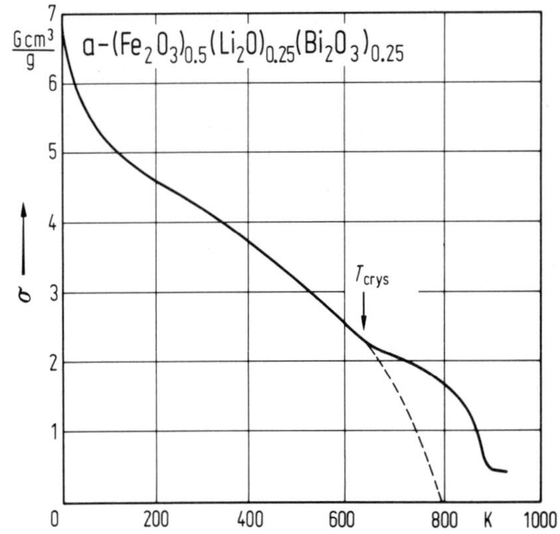

Fig. 94. a-$(Fe_2O_3)_{0.5}(Li_2O)_{0.25}(Bi_2O_3)_{0.25}$. Specific magnetization as a function of temperature for an applied static field of 11.4 kOe. T_C could not be determined since the sample crystallizes at a much lower temperature, at about 638 K [89N].

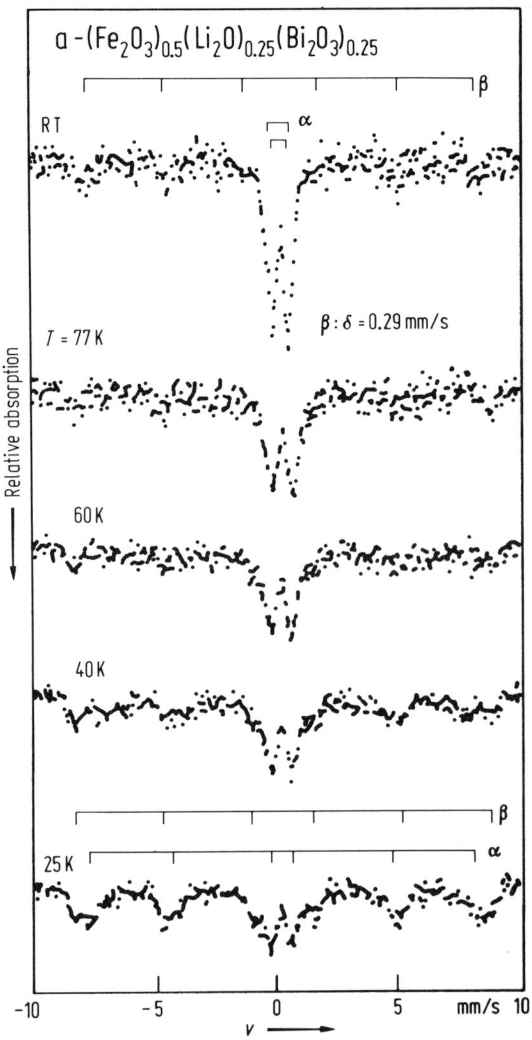

Fig. 95. a-$(Fe_2O_3)_{0.5}(Li_2O)_{0.25}(Bi_2O_3)_{0.25}$. Mössbauer spectra at a series of temperatures. The spectral components correspond to two magnetic phases, a short-range antiferromagnetic component (α) and a ferromagnetic microcrystalline phase (β). Values of isomer shifts (relative to metallic iron) indicate the absence of divalent iron ions [89N].

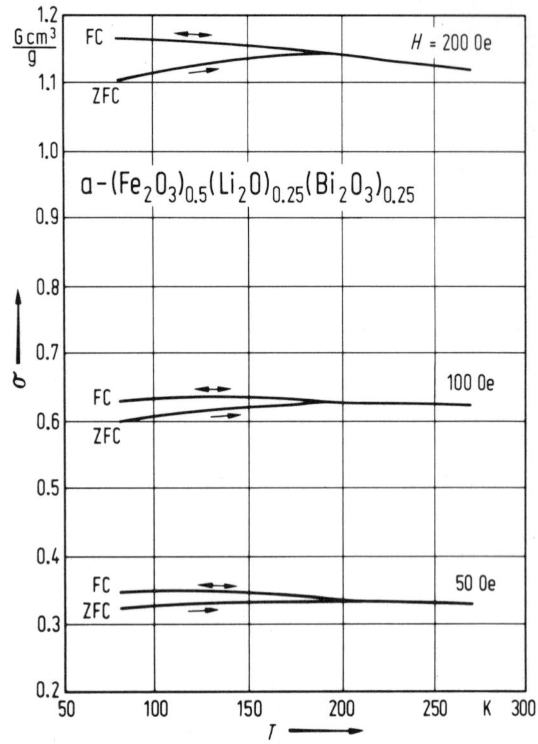

Fig. 96. a-$(Fe_2O_3)_{0.5}(Li_2O)_{0.25}(Bi_2O_3)_{0.25}$. Thermomagnetic curves showing σ vs. T for zero-field-cooled (ZFC) and field-cooled (FC) conditions. ZFC and FC curves separate at a finite temperature and this is indicative of frustration of exchange interactions [89N].

5.7.3.3.2 $Fe_2O_3 \cdot K_2O \cdot Bi_2O_3$ oxides [92Y]

Films of $Fe_2O_3 \cdot K_2O \cdot Bi_2O_3$ with a thickness of 0.5···1.3 μm were deposited on glass substrates by spin coating a gel solution of iron-, potassium-, and bismuth nitrates. The X-ray amorphous films are transparent from visible through near infrared. For $(Fe_2O_3)_{0.734}(K_2O)_{0.083}(Bi_2O_3)_{0.183}$, samples annealed at temperatures T_a below 500°C are paramagnetic at room temperature, and for higher T_a samples show a ferromagnetic character (Figs. 97 and 98). The magnetization of this sample showed irreversible changes upon heating and cooling in vacuum (Fig. 99). X-ray diffraction pattern for annealed ferromagnetic samples show intensity peaks corresponding to the crystalline antiferromagnetic $BiFeO_3$, but the films do not contain any crystalline ferromagnetic phases (Fig. 100).

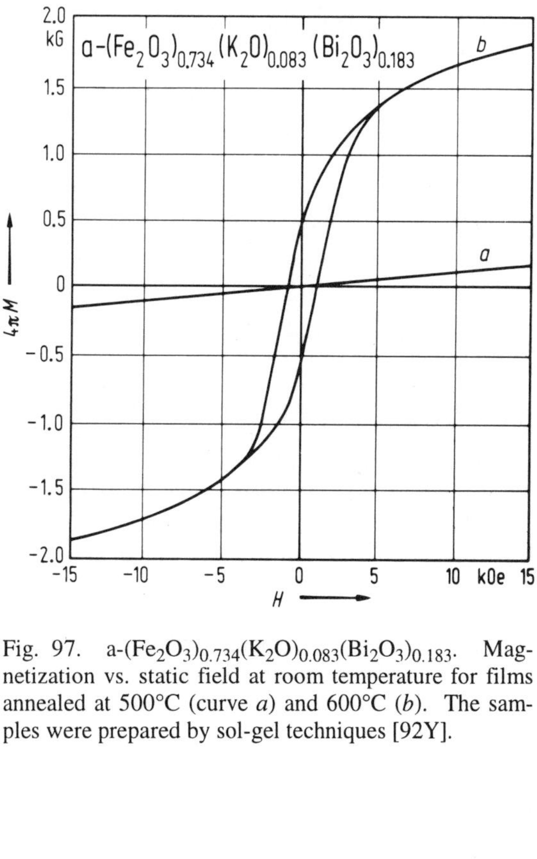

Fig. 97. a-$(Fe_2O_3)_{0.734}(K_2O)_{0.083}(Bi_2O_3)_{0.183}$. Magnetization vs. static field at room temperature for films annealed at 500°C (curve *a*) and 600°C (*b*). The samples were prepared by sol-gel techniques [92Y].

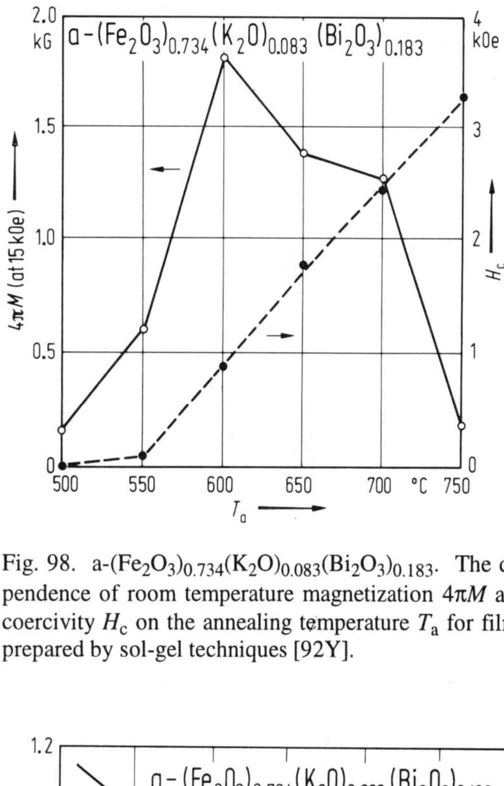

Fig. 98. a-$(Fe_2O_3)_{0.734}(K_2O)_{0.083}(Bi_2O_3)_{0.183}$. The dependence of room temperature magnetization $4\pi M$ and coercivity H_c on the annealing temperature T_a for films prepared by sol-gel techniques [92Y].

Fig. 99. a-$(Fe_2O_3)_{0.734}(K_2O)_{0.083}(Bi_2O_3)_{0.183}$. *M* vs. *T* for measurements in vacuum for a sample annealed at 600°C. Two magnetic phases, with Curie temperatures T_{C1} = 580°C and T_{C2} = 400°C, contribute to *M*. The phase with the Curie temperature T_{C1} was identified as Fe_3O_4 from X-ray diffraction and was formed by reduction in vacuum [92Y].

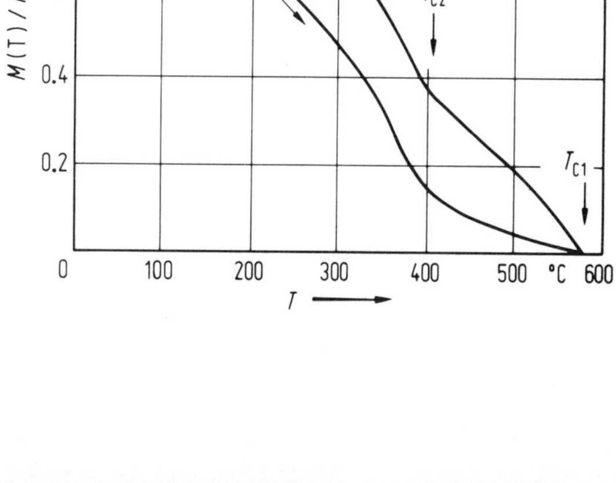

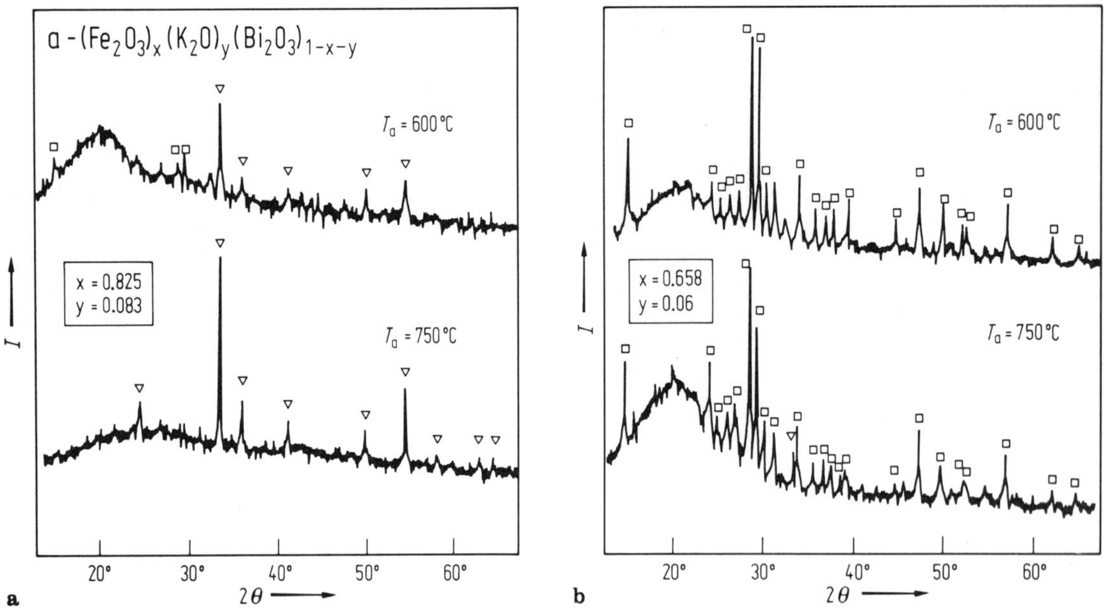

Fig. 100. a-$(Fe_2O_3)_x(K_2O)_y(Bi_2O_3)_{1-x-y}$. X-ray (Cu-$K_\alpha$) diffraction patterns for samples with (a) x = 0.825, y = 0.083 and (b) x = 0.658, y = 0.06 annealed at 600°C and 750°C. Crystalline antiferromagnetic phases $Bi_2Fe_4O_9$ (∇) and α-Fe_2O_3 (□) were identified in the films [92Y].

5.7.3.3.3 $Fe_2O_3 \cdot CaO \cdot Bi_2O_3$ oxides [87N2], [91T2], [92R1], [92S]

Amorphous Fe_2O_3·CaO·Bi_2O_3 prepared by either rapid quenching [87N2] or rf sputtering [92R1], [92S] show a long range magnetic order.

25 μm thick ribbons of the amorphous oxides were obtained by melting the oxides followed by quenching with a rotating twin roller. Fig. 101 shows the composition range over which the amorphous phase was achieved. DTA analysis indicates a two-step crystallization process in the samples (Fig. 102). A ferromagnetic character is observed at room temperature (Fig. 103 and 104) with a maximum magnetization of 13.2 G cm^3/g for $(Fe_2O_3)_{0.65}(CaO)_{0.2}(Bi_2O_3)_{0.15}$. Mössbauer spectra provide evidence for coexisting short-range antiferromagnetic and ferromagnetic phases (Fig. 105) [87N2].

About 1 μm thick samples of $(Fe_2O_3)_{0.5}(CaO)_x(Bi_2O_3)_{0.5-x}$ on glass or silicon substrates were deposited by rf sputtering in a pure oxygen or a mixed oxygen-argon atmosphere. As-sputtered films are X-ray amorphous and paramagnetic at room temperature. The films show a spontaneous moment when annealed in air at T_a above 400°C (Fig. 106). The room temperature magnetization decreases with increasing concentration of CaO (Fig. 107 and 108). FMR studies show a planar or uniaxial anisotropy in the films (Figs. 109···111). Magnetic and FMR parameters are found to depend sensitively on the oxygen partial pressure in the sputtering atmosphere (Figs. 112 and 113). X-ray diffraction studies do not indicate the precipitation of crystalline ferrimagnetic oxides in the films (Fig. 114), and the observed magnetization is attributed to ferrimagnetically ordered amorphous clusters [92R1], [92S].

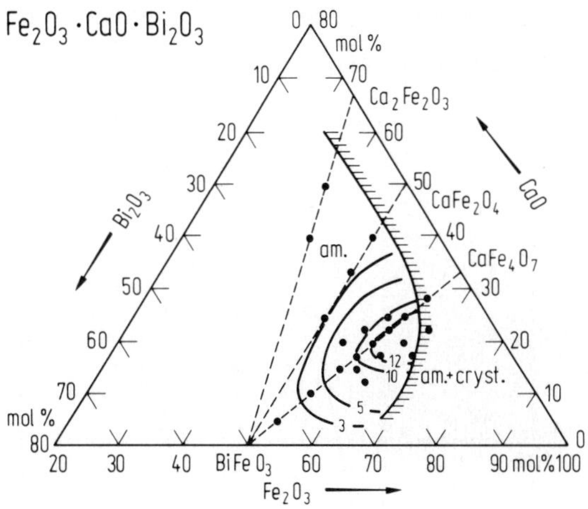

Fig. 101. $Fe_2O_3 \cdot CaO \cdot Bi_2O_3$. The phase diagram showing the composition range for amorphous (am) and crystalline (cryst) mixed oxides. The hatched side corresponds to the mixed phase region. Curves of constant room temperature magnetization (in G cm^3/g) for the amorphous compounds are also shown [87N2].

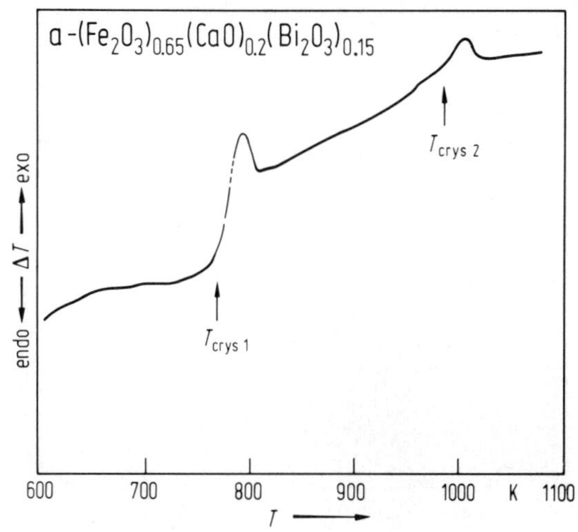

Fig. 102. a-$(Fe_2O_3)_{0.65}(CaO)_{0.2}(Bi_2O_3)_{0.15}$. DTA data. The crystallization occurs at two different temperatures, a cubic perovskite phase appears above $T_{crys\ 1}$ = 766 K and α-Fe_2O_3 phase is formed above $T_{crys\ 2}$ = 986 K [87N2].

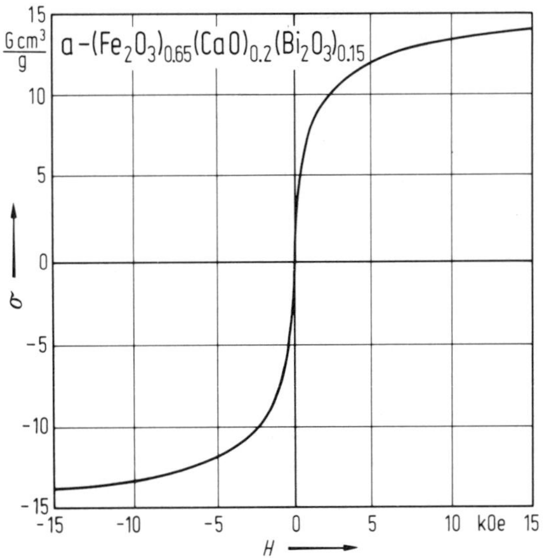

Fig. 103. a-$(Fe_2O_3)_{0.65}(CaO)_{0.2}(Bi_2O_3)_{0.15}$. The magnetic hysteresis curve at room temperature. The coercive field is about 80 Oe [87N2].

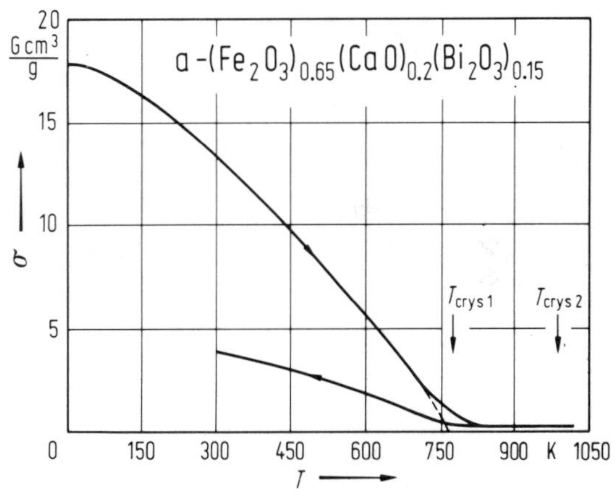

Fig. 104. a-$(Fe_2O_3)_{0.65}(CaO)_{0.2}(Bi_2O_3)_{0.15}$. σ vs. T data for increasing and decreasing temperatures for an applied static magnetic field of 10 kOe. σ is 17.8 G cm^3/g extrapolated to 0 K. Crystallization temperatures (Fig. 102) are indicated in the figure. The approximate Curie temperature is 760 K [87N2].

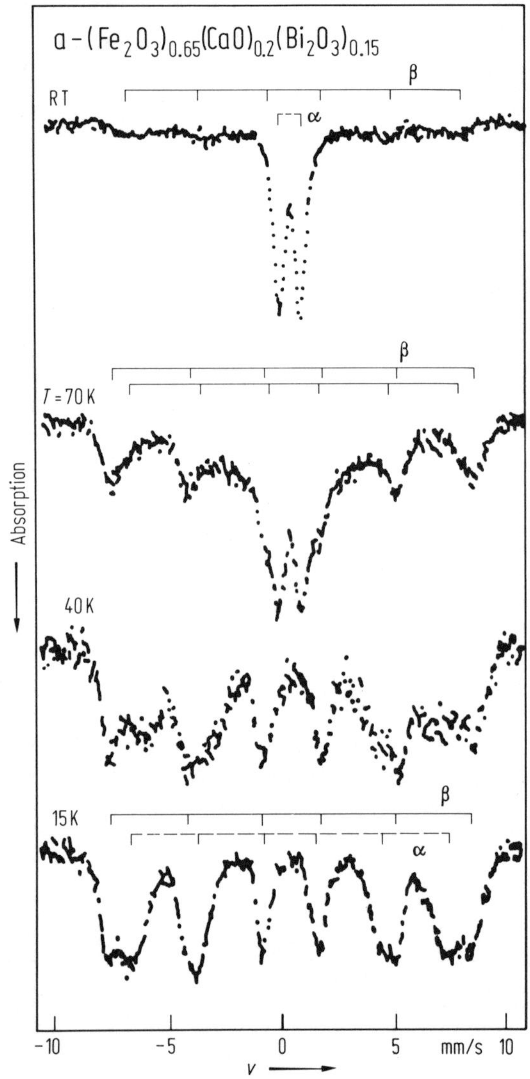

Fig. 105. a-$(Fe_2O_3)_{0.65}(CaO)_{0.2}(Bi_2O_3)_{0.15}$. Spectra obtained by Mössbauer studies at several temperatures. α and β indicate the spectral components for a short-range antiferromagnetic phase ($T_N \approx 80$ K) and a ferromagnetic phase ($T_C \approx 760$ K), respectively. At 15 K, the hyperfine fields are 435.7 kOe (for α) and 502.9 kOe (for β) [87N2].

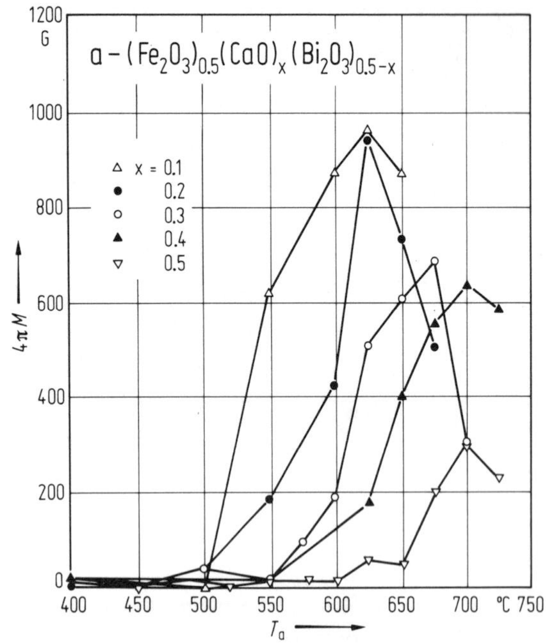

Fig. 106. a-$(Fe_2O_3)_{0.5}(CaO)_x(Bi_2O_3)_{0.5-x}$. Room temperature magnetization measured at 5 kOe vs. annealing temperature for rf sputtered films for x = 0.1···0.5. The films were sputtered in a pure oxygen atmosphere and the annealing was done for 60 min in air. X-ray diffraction data do not show the precipitation of crystalline ferro- or ferrimagnetic phases in annealed sample [92R1].

→

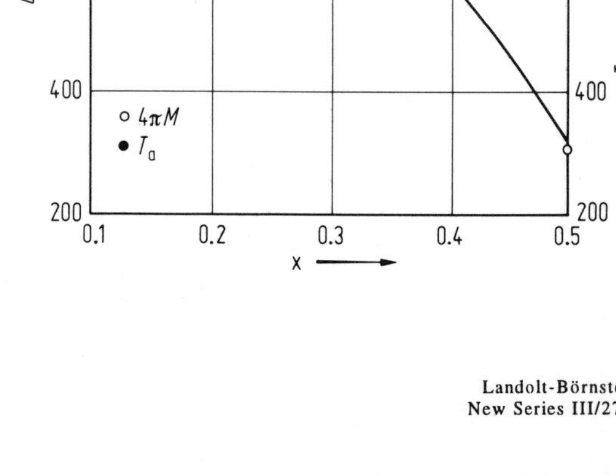

Fig. 107. a-$(Fe_2O_3)_{0.5}(CaO)_x(Bi_2O_3)_{0.5-x}$. The variation with the concentration of CaO, x, of the maximum room temperature $4\pi M$ at 5 kOe in annealed samples. The variation with x of the annealing temperature T_a corresponding to the maximum $4\pi M$ is also shown. The samples were sputtered in oxygen atmospheres [92R1].

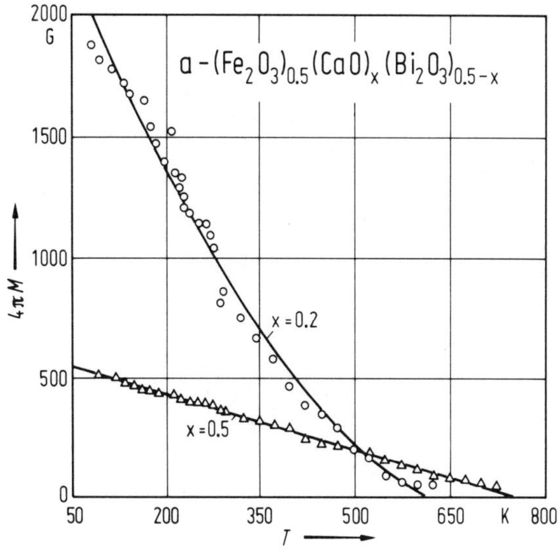

Fig. 108. a-$(Fe_2O_3)_{0.5}(CaO)_x(Bi_2O_3)_{0.5-x}$. The dependence of $4\pi M$ (measured at 5 kOe) on temperature for annealed samples sputtered in oxygen. Annealing conditions correspond to maximum in $4\pi M$ (see Fig. 106). T_C varies from a minimum of 600(10) K for x = 0.2 to a maximum of 750(10) K for x = 0.5 [92R1].

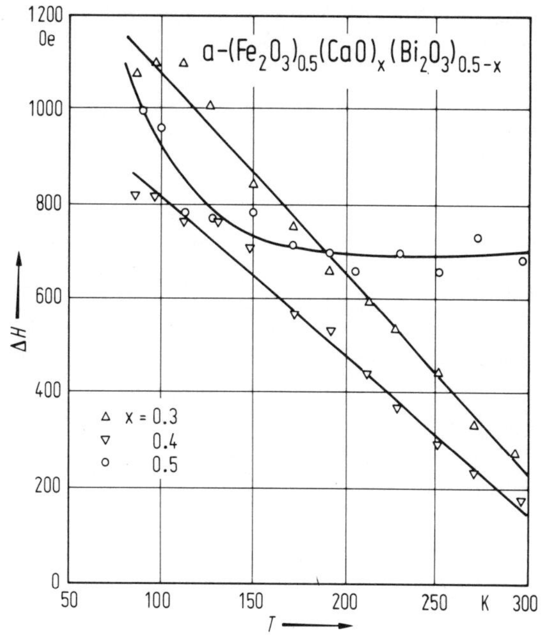

Fig. 110. a-$(Fe_2O_3)_{0.5}(CaO)_x(Bi_2O_3)_{0.5-x}$. Data on FMR linewidth ΔH as a function of temperature for annealed films (T_a = 600···650°C in air) sputtered in a 50% O_2 + 50% Ar atmosphere. The linewidth was measured for static fields parallel to the film plane [92S].

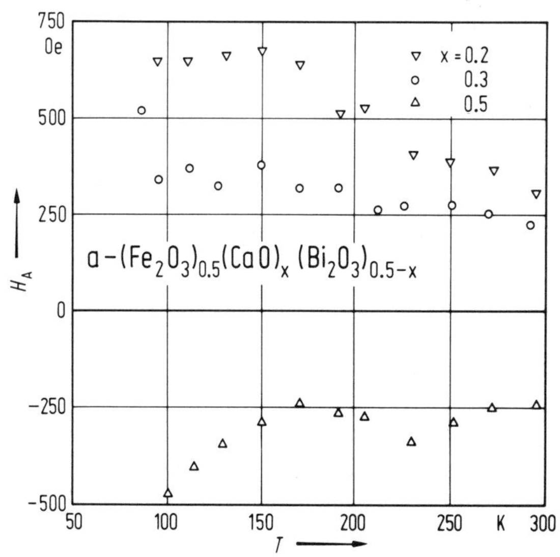

Fig. 109. a-$(Fe_2O_3)_{0.5}(CaO)_x(Bi_2O_3)_{0.5-x}$. Anisotropy field vs. temperature for annealed samples (T_a = 600···650°C in air) sputtered in a 50% oxygen + 50% argon atmosphere. Positive (negative) values for H_A indicate uniaxial (planar) anisotropy for the sample. H_A-values were estimated from data on $4\pi M$ and FMR values for the effective magnetization [92R1].

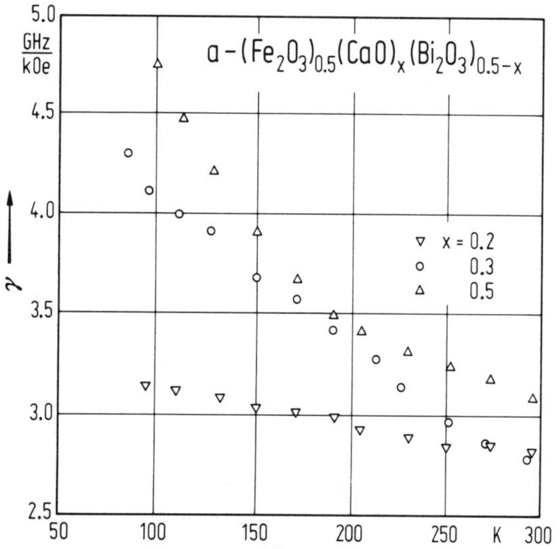

Fig. 111. a-$(Fe_2O_3)_{0.5}(CaO)_x(Bi_2O_3)_{0.5-x}$. The gyromagnetic ratio γ vs. T for annealed samples (T_a = 600···650°C in air) sputtered in a 50% O_2 + 50% Ar atmosphere [92S].

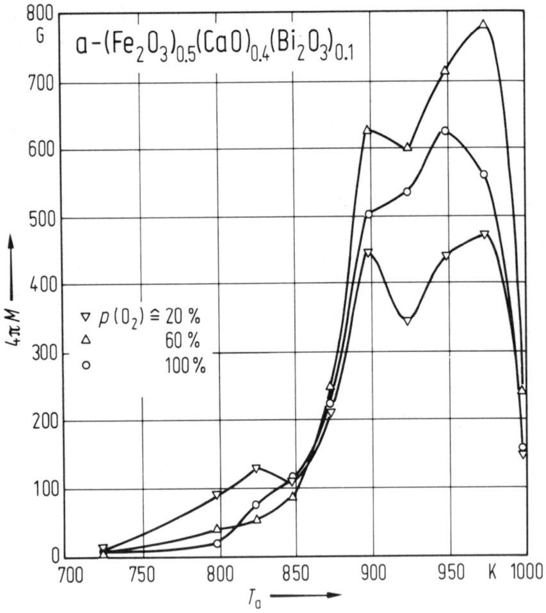

Fig. 112. a-$(Fe_2O_3)_{0.5}(CaO)_{0.4}(Bi_2O_3)_{0.1}$. Variation of room temperature $4\pi M$ (measured at 5 kOe) with T_a at oxygen partial pressures $p(O_2)$ corresponding to 20, 60, and 100% of total Ar + O_2 sputtering atmosphere [92S].

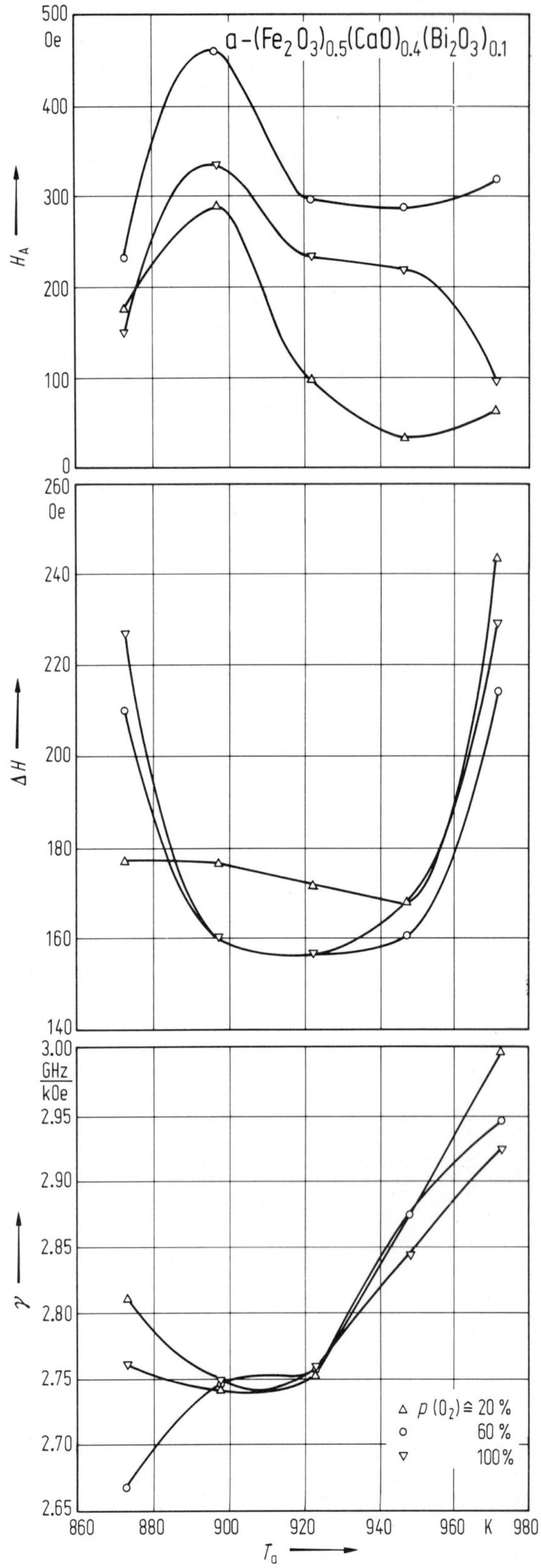

Fig. 113. a-$(Fe_2O_3)_{0.5}(CaO)_{0.4}(Bi_2O_3)_{0.1}$. The annealing temperature T_a dependence of the room temperature anisotropy field H_A, FMR linewidth ΔH, and the gyromagnetic ratio γ for samples sputtered with $p(O_2)$ corresponding to 20, 60, and 100% of the total atmosphere. The samples were annealed in air [92S].

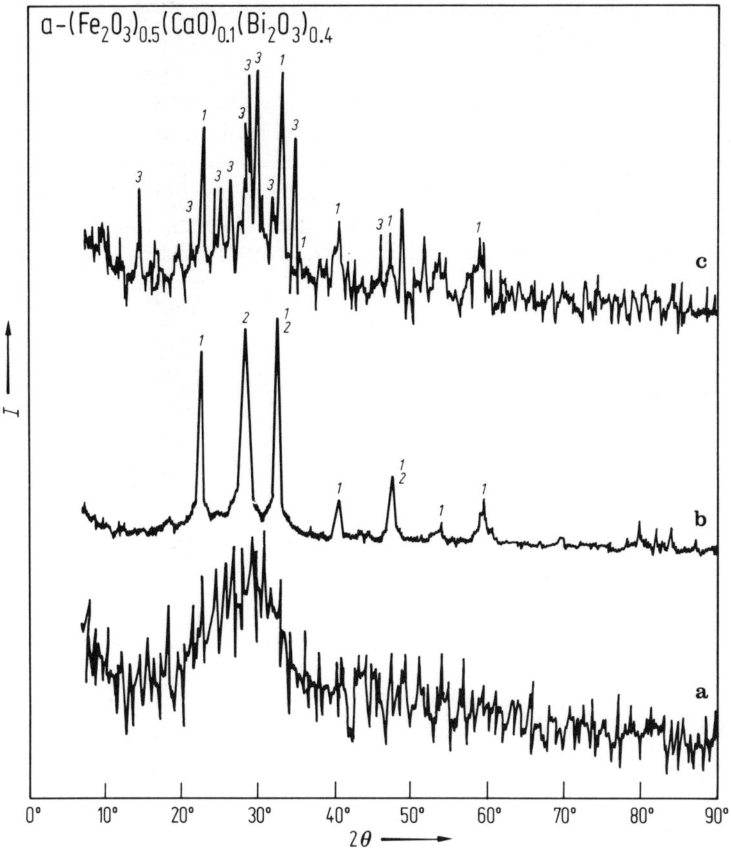

Fig. 114. a-$(Fe_2O_3)_{0.5}(CaO)_{0.1}(Bi_2O_3)_{0.4}$. X-ray diffraction pattern for (a) as-deposited samples and for films annealed in air at (b) 550°C and (c) 650°C. Intensity peaks for crystalline $BiFeO_3$, $Bi_{3.11}Ca_{0.89}O_{5.56}$ and $Bi_2Fe_4O_9$ are indicated by numbers 1, 2 and 3, respectively. The data do not indicate the presence of crystalline ferrimagnetic phases such as γ-Fe_2O_3, Fe_3O_4, or $Bi_3Fe_5O_{12}$ [92R1].

5.7.3.3.4 $Fe_2O_3 \cdot ZnO \cdot Bi_2O_3$ oxides [86S], [86O], [92T], [91R3], [92R2], [93S1]

A long range order is observed in a-Fe_2O_3·ZnO·Bi_2O_3 synthesized by melt quenching, cluster beam deposition [86O] , or rf sputtering [91R3], [92R2], [93S1]. Properties of rapidly quenched samples are discussed first. Ribbons of 10···20 μm in thickness were obtained by melt quenching with a rotating copper wheel and a cooling rate of 10^5 K/s. Samples of $(Fe_2O_3)_{0.5}(ZnO)_x(Bi_2O_3)_{0.5-x}$ are amorphous for $x \leq 0.2$ (Figs. 115···117). A ferromagnetic character with a Curie temperature of 460···520 K is observed in these oxides (Figs. 118···120). Mössbauer spectra show a single doublet at room temperature and a hyperfine pattern at 4.2 K (Fig. 121). Based on these results, it is concluded that the samples are either short-range antiferromagnets [86S], [86O], or superparamagnets due to magnetically ordered clusters [92T]. Samples prepared by cluster beam deposition techniques showed a much higher magnetization than for rapidly quenched ribbons [86S].

Samples prepared by radio frequency sputtering showed a similar ferromagnetic behavior. As-sputtered samples are paramagnetic at room temperature and annealing in air at T_a = 400···775°C resulted in a ferromagnetic moment in the films (Figs. 122 and 123). The Curie temperature is independent of x and is about 600 K (Fig. 124) [91R3]. Magnetic parameters such as the magnetization (Figs. 125 and 126), uniaxial anisotropy (Fig. 127), FMR linewidth, and gyromagnetic ratio (Fig. 128) are found to depend on the nature of the sputtering atmosphere [92R2]. X-ray diffraction studies show evidence for the precipitation of either nonmagnetic or antiferromagnetic crystalline phases in the films (Fig. 129) [91R3]. The magnetization, therefore, is proposed to arise from ferrimagnetically ordered clusters. Samples sputtered at high temperatures on heated substrates show a spontaneous magnetic moment (Fig. 130) possibly due to the formation of such ordered clusters [93S1].

For Fig. 115 see next page.

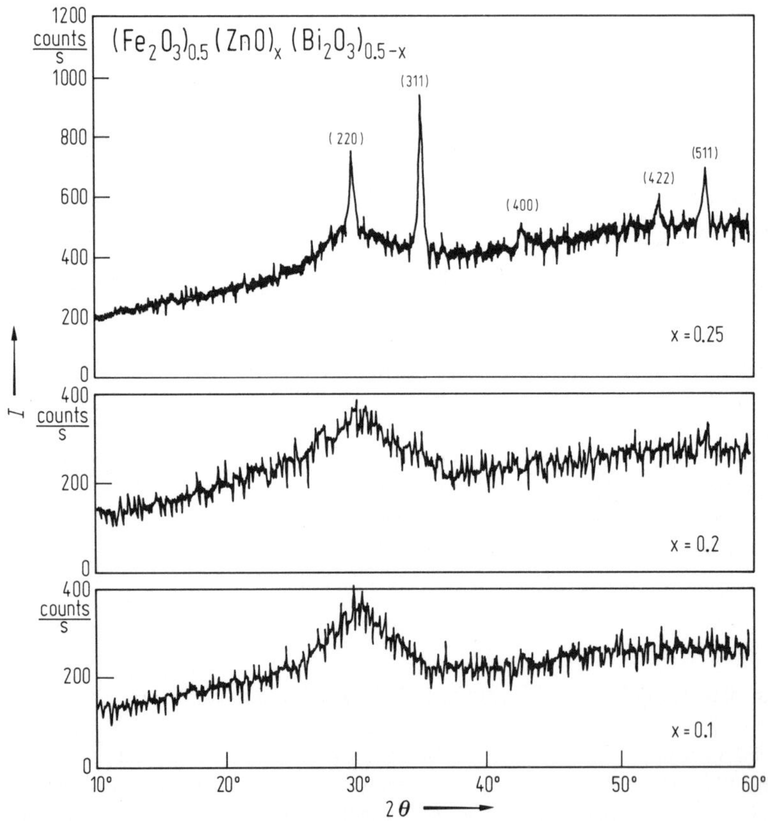

Fig. 116. $(Fe_2O_3)_{0.5}(ZnO)_x(Bi_2O_3)_{0.5-x}$. X-ray diffraction data for rapidly quenched samples. The ribbons remain amorphous for x ≤ 0.2. Data for x = 0.25 show intensity peaks due to crystalline $ZnFe_2O_4$ [86S].

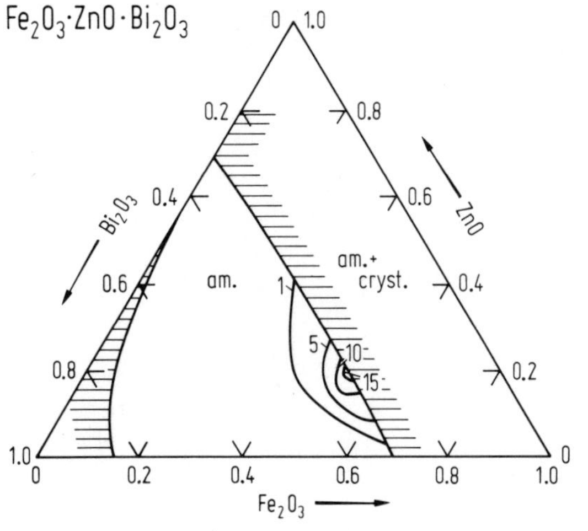

Fig. 115. Fe_2O_3·ZnO·Bi_2O_3. Diagram showing the composition range for the formation of amorphous (am) oxide [86O]. Contours of constant room temperature magnetization (in G cm^3/g) for the amorphous compounds are also shown.

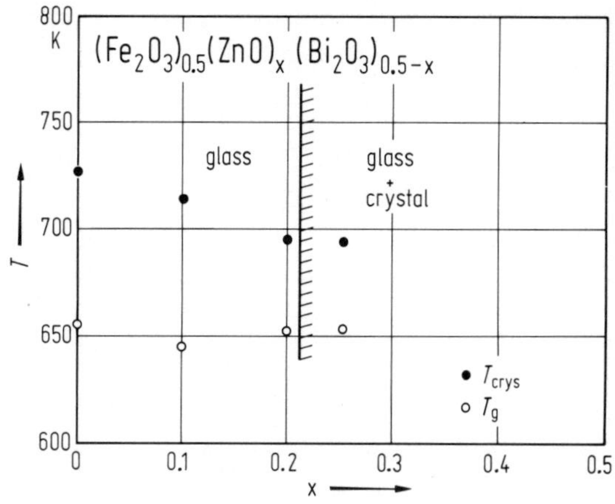

Fig. 117. $(Fe_2O_3)_{0.5}(ZnO)_x(Bi_2O_3)_{0.5-x}$. Glass transition temperature T_g and crystallization temperature T_{crys} obtained by DTA for rapidly quenched samples [86S].

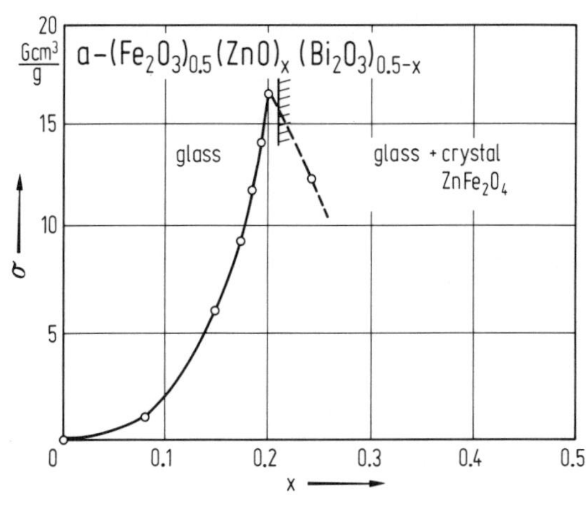

Fig. 118. a-$(Fe_2O_3)_{0.5}(ZnO)_x(Bi_2O_3)_{0.5-x}$. Room temperature magnetization as a function of the concentration of ZnO, x. The data are for a static field of 10 kOe [86S].

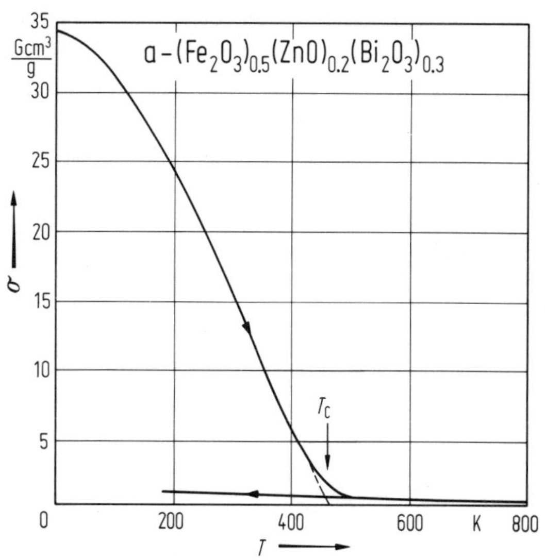

Fig. 119. a-$(Fe_2O_3)_{0.5}(ZnO)_{0.2}(Bi_2O_3)_{0.3}$. σ vs. T for a sample prepared by rapid quenching. The data are for a static field of 10 kOe. The Curie temperature is 480 K. The magnetization decreases drastically, as shown, when the sample is heated to the crystallization temperature of 700 K and cooled down to room temperature [86S].

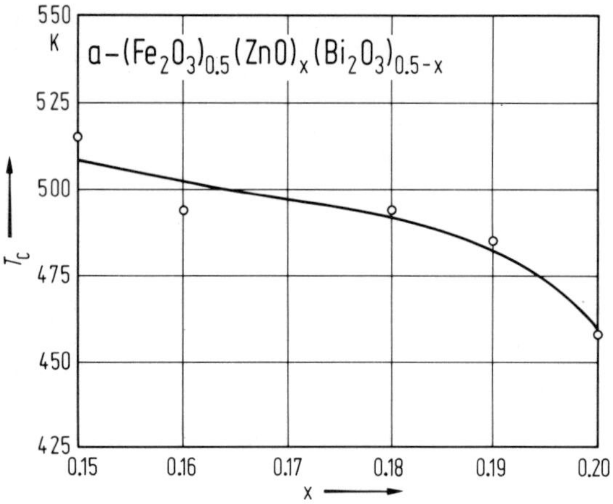

Fig. 120. a-$(Fe_2O_3)_{0.5}(ZnO)_x(Bi_2O_3)_{0.5-x}$. Curie temperature vs. concentration x of ZnO for rapidly quenched samples [86S].

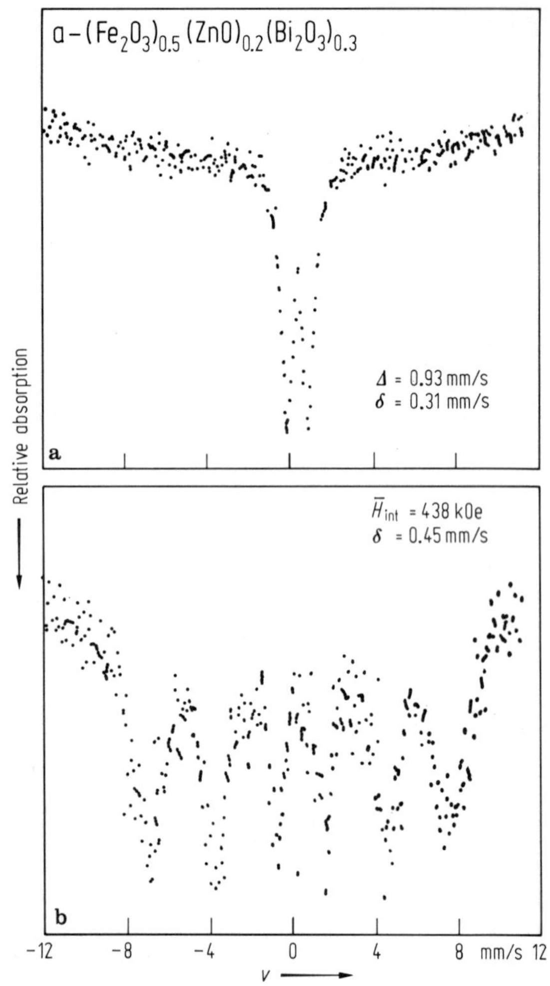

Fig. 121. a-$(Fe_2O_3)_{0.5}(ZnO)_{0.2}(Bi_2O_3)_{0.3}$. Mössbauer data at room temperature (a) and at 4.2 K (b) [86S]. Isomer shift δ is relative to metallic iron.

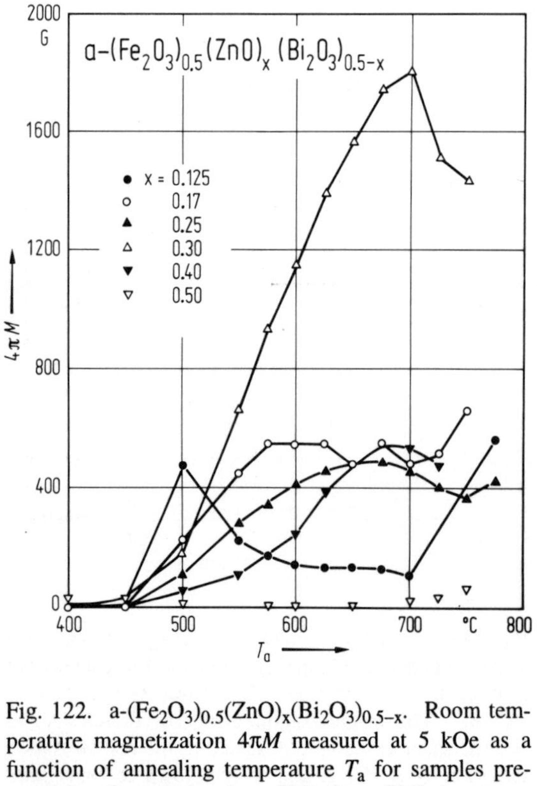

Fig. 122. a-$(Fe_2O_3)_{0.5}(ZnO)_x(Bi_2O_3)_{0.5-x}$. Room temperature magnetization $4\pi M$ measured at 5 kOe as a function of annealing temperature T_a for samples prepared by rf sputtering in a 50% O_2 + 50% Ar atmosphere. The annealing was done in air at T_a for 60 min [91R3].

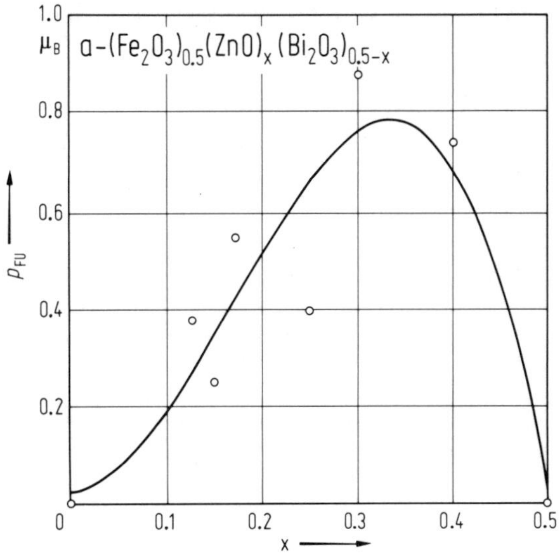

Fig. 123. a-$(Fe_2O_3)_{0.5}(ZnO)_x(Bi_2O_3)_{0.5-x}$. Magnetic moment (measured at 5 kOe) per formula unit p_{FU} at 80 K as a function of concentration of ZnO in samples prepared by rf sputtering with $p(O_2)$ corresponding to 50% of the total Ar + O_2 atmosphere. The samples were annealed in air at T_a corresponding to maximum in M (Fig. 122). The solid line is guide to eye [91R3].

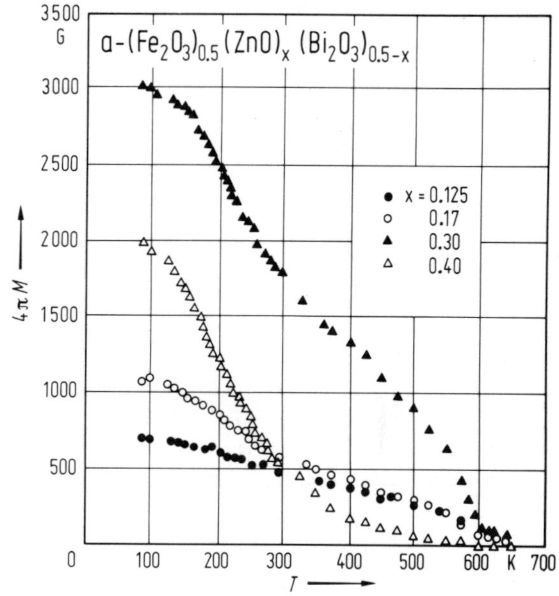

Fig. 124. a-$(Fe_2O_3)_{0.5}(ZnO)_x(Bi_2O_3)_{0.5-x}$. $4\pi M$ (measured with H = 5 kOe) vs. T for rf sputtered annealed (T_a = 650···700°C in air) samples. The Curie temperature is 600 K [91R3].

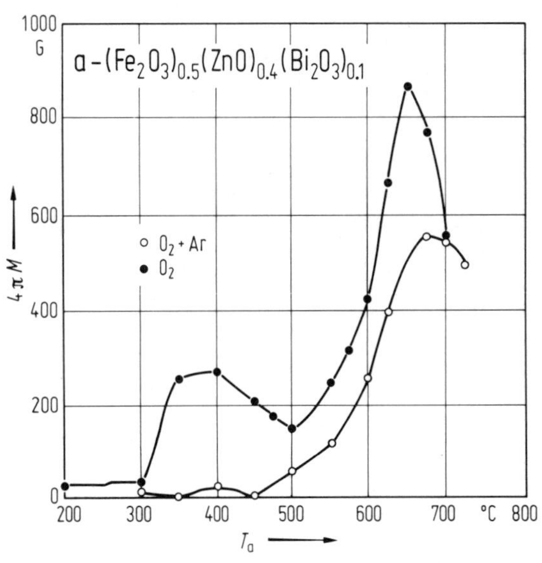

Fig. 125. a-$(Fe_2O_3)_{0.5}(ZnO)_{0.4}(Bi_2O_3)_{0.1}$. Room temperature $4\pi M$ (for H = 5 kOe) vs. T_a for a sample sputtered in pure oxygen and in mixed oxygen and argon atmospheres [92R2].

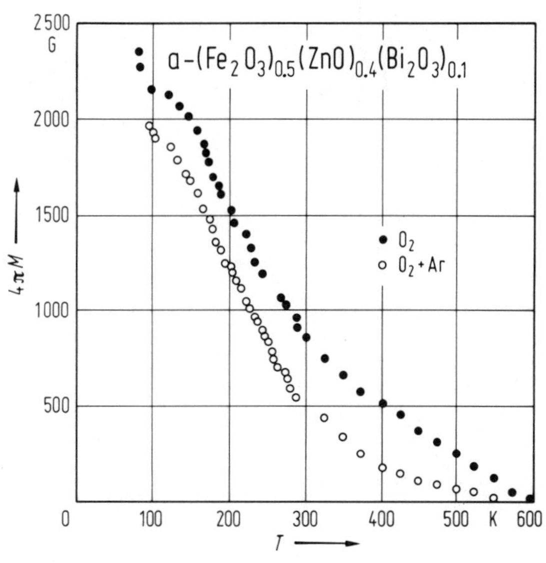

Fig. 126. a-$(Fe_2O_3)_{0.5}(ZnO)_{0.4}(Bi_2O_3)_{0.1}$. $4\pi M$ (for H = 5 kOe) vs. T for a sample annealed in air at 650°C. The films were rf sputtered in oxygen or oxygen + argon [92R2].

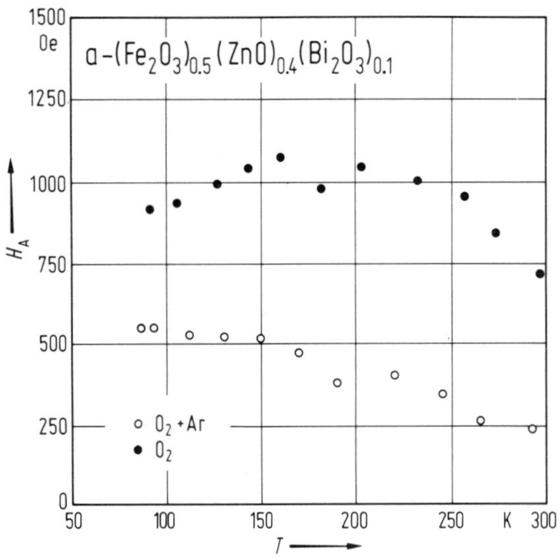

Fig. 127. a-$(Fe_2O_3)_{0.5}(ZnO)_{0.4}(Bi_2O_3)_{0.1}$. Uniaxial anisotropy field obtained from FMR and static magnetization measurements for annealed (in air at T_a = 650°C) films sputtered in oxygen and oxygen + argon [92R2].

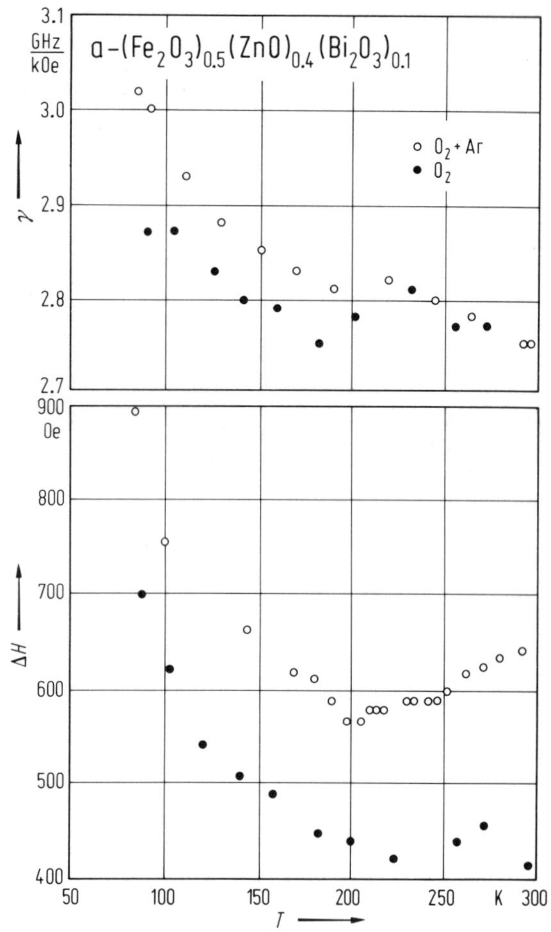

Fig. 128. a-$(Fe_2O_3)_{0.5}(ZnO)_{0.4}(Bi_2O_3)_{0.1}$. Variation of the FMR linewidth ΔH at 9.2 GHz and the gyromagnetic ratio γ with temperature for oxygen and oxygen + argon sputtered samples annealed at 650°C [92R2].

For Fig. 129 see next page.

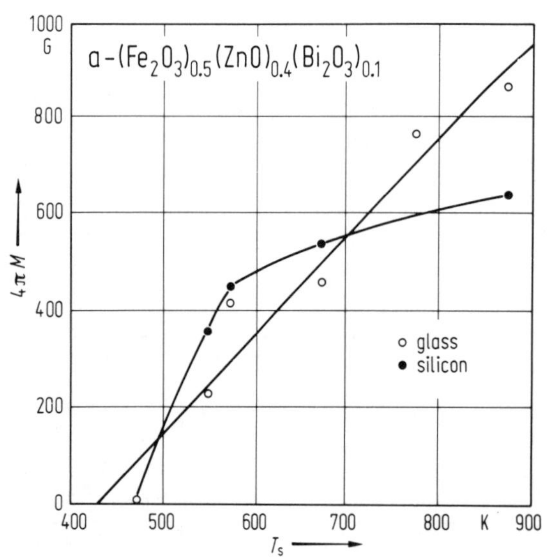

Fig. 130. a-$(Fe_2O_3)_{0.5}(ZnO)_{0.4}(Bi_2O_3)_{0.1}$. Room temperature $4\pi M$ as a function of substrate temperature T_s for samples sputtered on glass or silicon substrates in oxygen-argon atmospheres. The films retain the amorphous structure for $T_s \leq 575$ K. At higher T_s, nonmagnetic crystalline phases are formed. The spontaneous moment measured for H = 5 kOe was suggested to arise from ordered clusters formed *in situ* in the film [93S1].

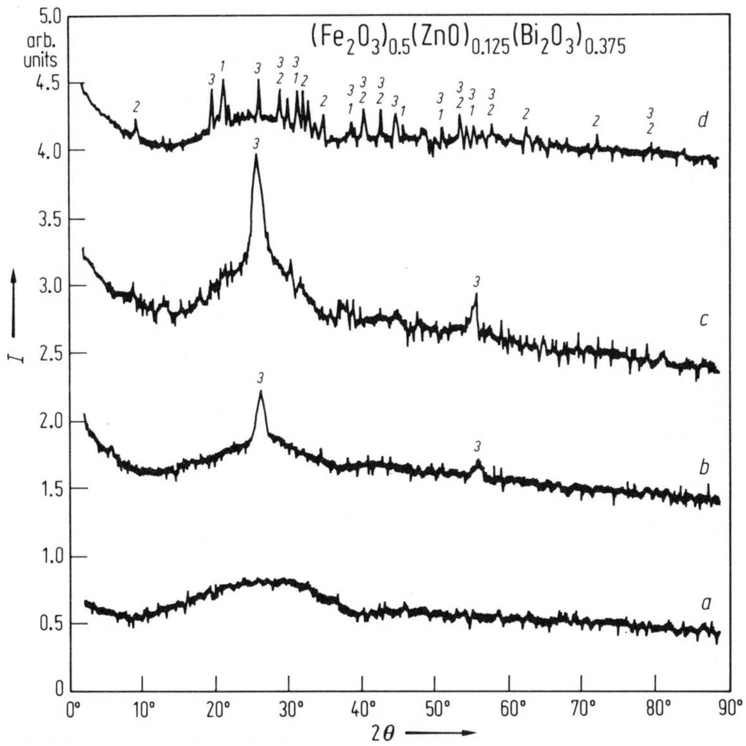

Fig. 129. $(Fe_2O_3)_{0.5}(ZnO)_{0.125}(Bi_2O_3)_{0.375}$. X-ray diffraction at room temperature for (*a*) as-sputtered films and for films annealed at (*b*) 475°C, (*c*) 525°C, and (*d*) 775°C. The high temperature annealing leads to the precipitation of crystalline (1) $BiFeO_3$, (2) γ-Fe_2O_3, and (3) $24Bi_2O_3{\cdot}ZnO$ [91R3].

5.7.3.3.5 $Fe_2O_3 \cdot CuO \cdot Bi_2O_3$ oxides [90 S], [91R2], [92R3], [93S2]

The phase diagram for the existence of amorphous and crystalline phases in the $Fe_2O_3{\cdot}CuO{\cdot}Bi_2O_3$ systems is shown in Fig. 131.

Rapidly quenched amorphous samples of the composition $(Fe_2O_3)_{0.5}(CuO)_x(Bi_2O_3)_{0.5-x}$ (x = 0···0.3) are studied (Fig. 131). Small angle X-ray scattering studies reveal the presence of 10 nm clusters. The crystallization temperature obtained from DTA is 633 K (Fig. 132). Amorphous samples with x > 0.1 show a spontaneous moment at room temperature which increases with increasing x. The magnetization for x = 0.25 originates from two magnetic phases, a ferromagnetic phase with T_C = 650 K and with a spin structure similar to the one for crystalline copper ferrite, and a random antiferromagnetic phase with T_N = 200 K (Figs. 133···137). Mössbauer absorption spectra could be understood in terms of the two-phase model, and the spectral parameters indicate Fe^{3+} in distorted octahedral and tetrahedral sites (Fig. 138).

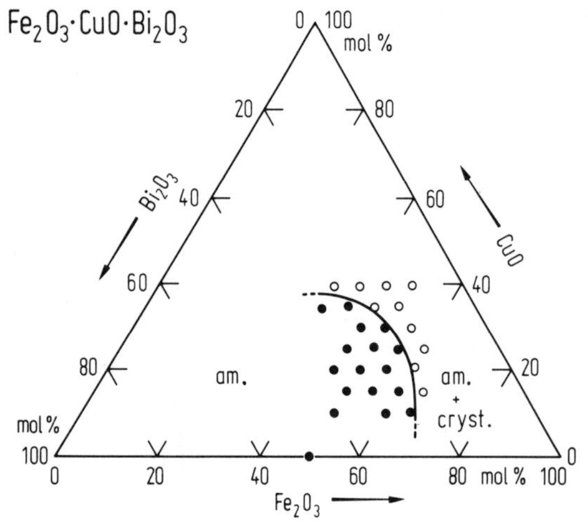

Fig. 131. Fe_2O_3·CuO·Bi_2O_3. Phase diagram for the rapidly quenched ternary oxide system [90S].

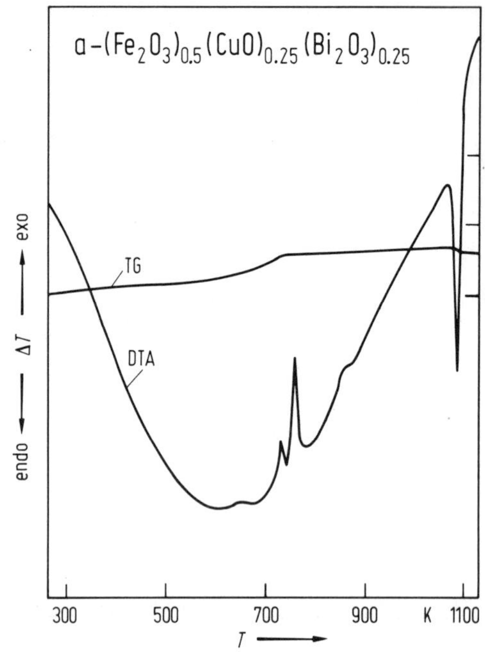

Fig. 132. a-$(Fe_2O_3)_{0.5}(CuO)_{0.25}(Bi_2O_3)_{0.25}$. DTA and TG data for a rapidly quenched sample. The data were for a heating rate of 10 K/min. The DTA curve shows four exothermic peaks at 633, 723, 748, and 848 K. At higher temperatures crystalline phases such as $CuFe_2O_4$, Bi_2CuO_4, $BiFeO_3$, and $Bi_2Fe_4O_9$ were formed. In the TG curve, the weight gain observed up to 723 K is attributed to the oxidation of Cu^{1+} to Cu^{2+}. The ratio of Cu^{1+} to Cu^{2+} in as-quenched sample is 3:1 [90S].

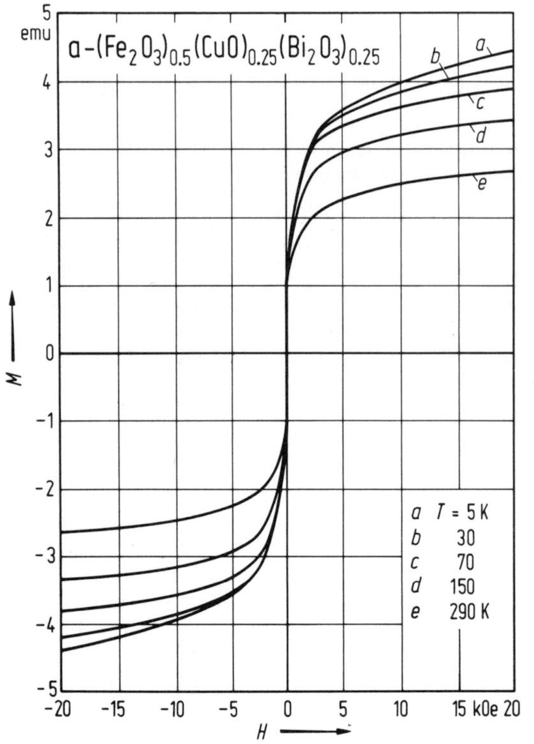

←
Fig. 133. a-$(Fe_2O_3)_{0.5}(CuO)_{0.25}(Bi_2O_3)_{0.25}$. Magnetic hysteresis curves at several temperatures. Notice the increase in the slope of the curves in the high field region with decreasing temperature. The magnetization M can be expressed as $M = M_0 + \chi H$, where M_0 is the magnetization due to a ferromagnetic phase and χH is the magnetization due to a random antiferromagnetic phase [90S].

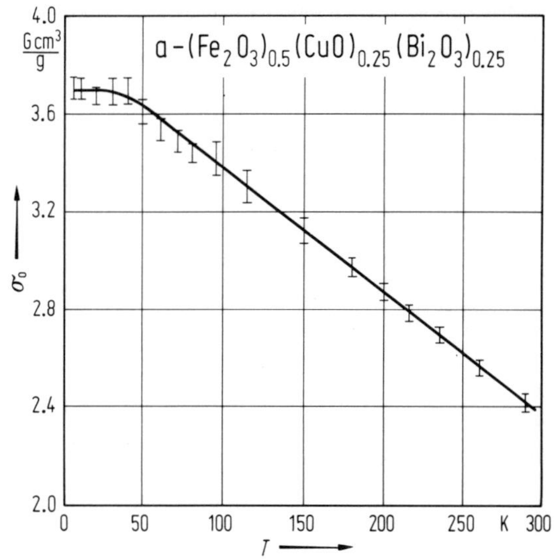

Fig. 134. a-$(Fe_2O_3)_{0.5}(CuO)_{0.25}(Bi_2O_3)_{0.25}$. Temperature dependence of the specific magnetization σ_0 for the ferromagnetic component. σ_0 values were estimated from data as in Fig. 133 [90S].

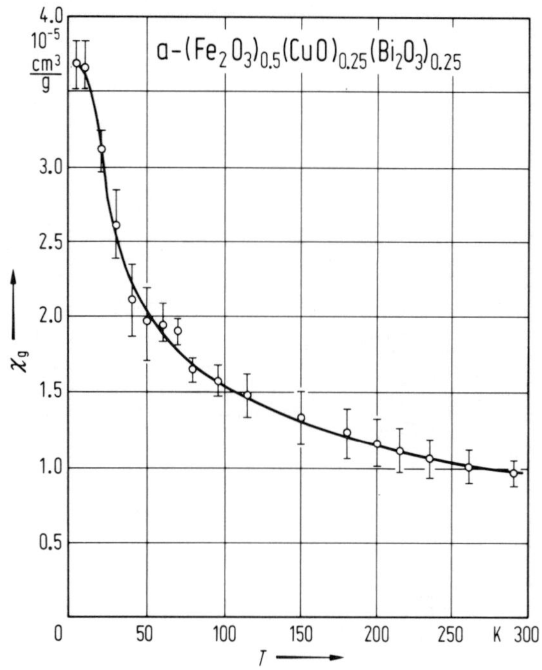

Fig. 135. a-$(Fe_2O_3)_{0.5}(CuO)_{0.25}(Bi_2O_3)_{0.25}$. Susceptibility χ_g for the random antiferromagnetic component vs. temperature [90S].

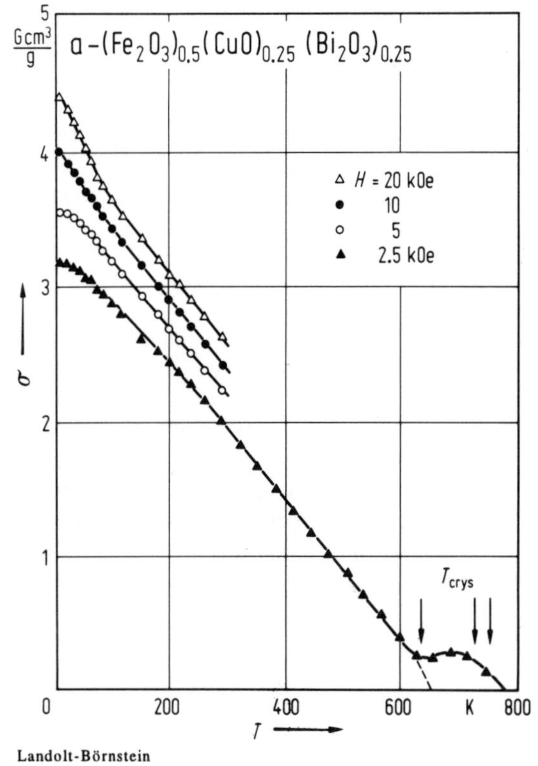

Fig. 136. a-$(Fe_2O_3)_{0.5}(CuO)_{0.25}(Bi_2O_3)_{0.25}$. σ vs. T data. The Curie temperature (≈ 650 K) could not be accurately determined since the crystallization occurred below T_C [90S].

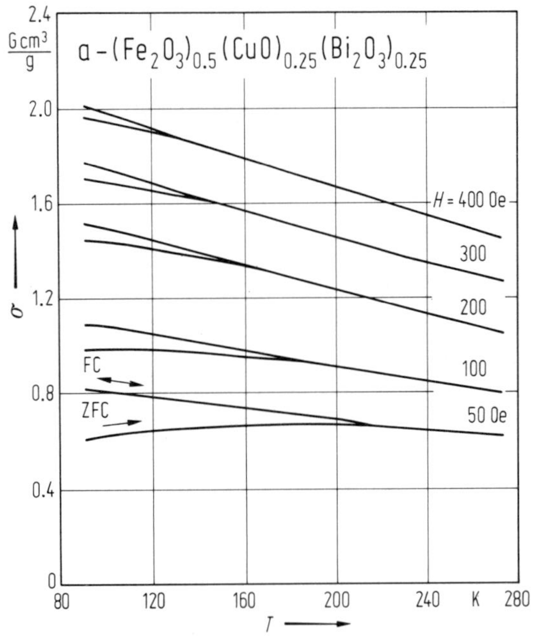

Fig. 137. a-$(Fe_2O_3)_{0.5}(CuO)_{0.25}(Bi_2O_3)_{0.25}$. Thermomagnetic curves. ZFC and FC denote data obtained for zero-field-cooled and field-cooled conditions, respectively. In weak fields, the data indicate irreversible thermomagnetization below 200 K. This temperature is identified with a transition from paramagnetic to random antiferromagnetic phase [90S].

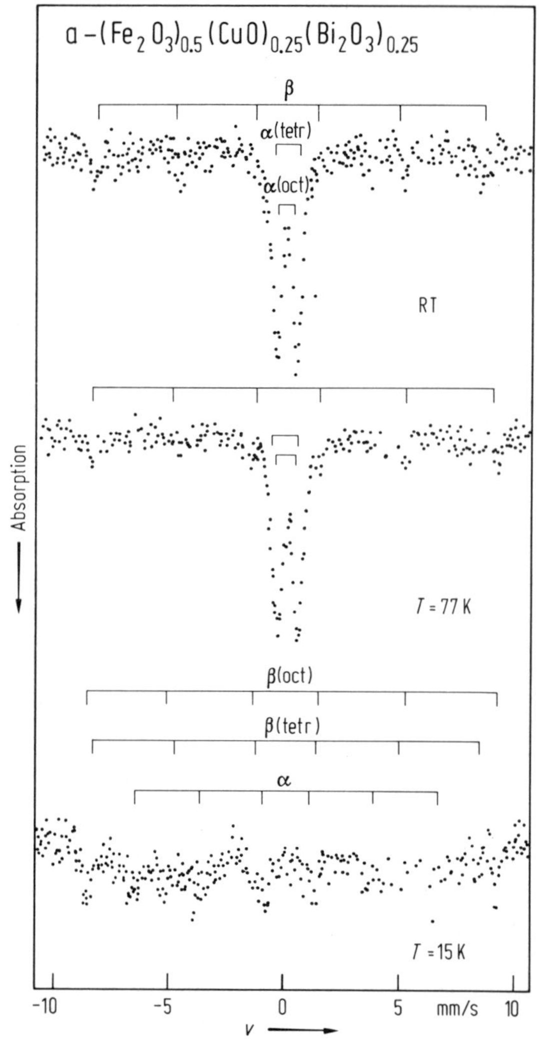

Fig. 138. a-$(Fe_2O_3)_{0.5}(CuO)_{0.25}(Bi_2O_3)_{0.25}$. Mössbauer spectra at a series of temperatures. α and β denote spectral components for the random antiferromagnetic and ferromagnetic phases, respectively. The spectral parameters correspond to Fe^{3+} in octahedral and tetrahedral sites [90S].

In rf sputtered samples of $(Fe_2O_3)_{0.5}(CuO)_x(Bi_2O_3)_{0.5-x}$ (x = 0···0.3), a ferrimagnetic character with a uniaxial anisotropy was observed through static and dynamic magnetic measurements [91R2], [92R3], [93S2]. Samples with thickness in the order of 1 μm were prepared in a mixed oxygen-argon atmosphere and were X-ray amorphous (Fig. 139). As-sputtered films are paramagnetic and films annealed at T_a > 450°C show a spontaneous moment (Figs. 140 and 141). The Curie temperature, 460···480°C, is smaller than the crystallization temperature (Fig. 142). Studies on the dependence of magnetic parameters on sputtering conditions show that the magnetization, Curie temperature, FMR linewidth, and gyromagnetic ratio increase with increasing oxygen partial pressure $p(O_2)$ whereas the uniaxial anisotropy field decreases with increasing $p(O_2)$ (Figs. 143···147). The spontaneous moment in annealed films is suggested to result from the formation of ordered copper ferrite-like clusters.

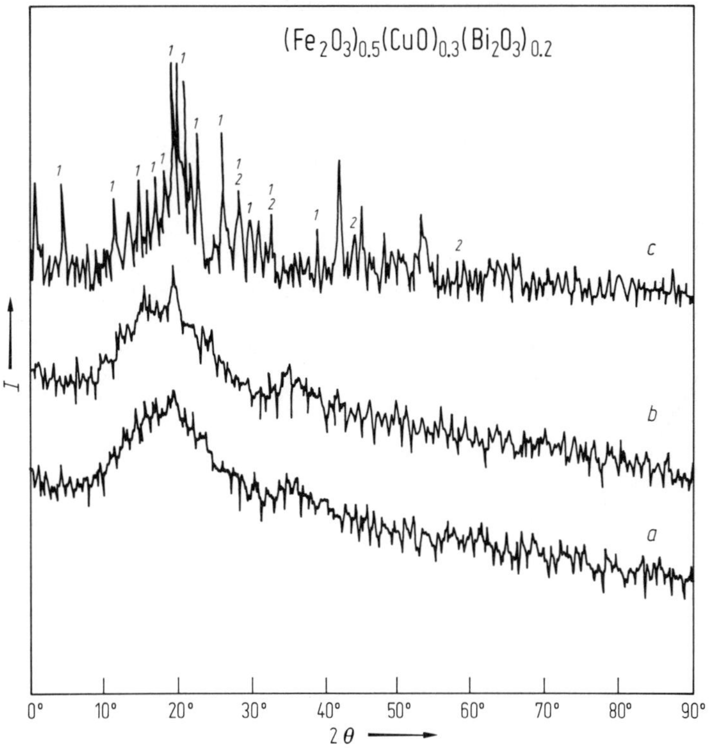

Fig. 139. $(Fe_2O_3)_{0.5}(CuO)_{0.3}(Bi_2O_3)_{0.2}$. X-ray diffraction pattern for an rf sputtered sample (50% Ar, 50% O_2 atmosphere). The data are for films annealed in air at (*a*) 500°C, (*b*) 550°C, and (*c*) 600°C for 60 min. The films are X-ray amorphous for annealing at temperatures below 600°C. For T_a = 600°C, diffraction peaks corresponding to the crystalline phases $Bi_2Fe_4O_9$ and CuO, indicated by 1 and 2 respectively in the figure, are seen [92R3].

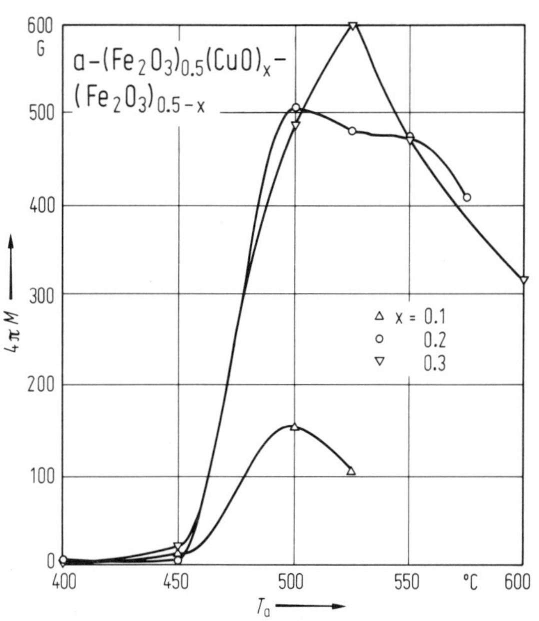

Fig. 140. a-$(Fe_2O_3)_{0.5}(CuO)_x(Fe_2O_3)_{0.5-x}$. Annealing temperature T_a vs. room temperature magnetization $4\pi M$ for rf sputtered films. The films were sputtered in a 50% O_2 + 50% Ar atmosphere and annealed in air for 60 min. The magnetization was measured with an in-plane static field of 5 kOe [92R3].

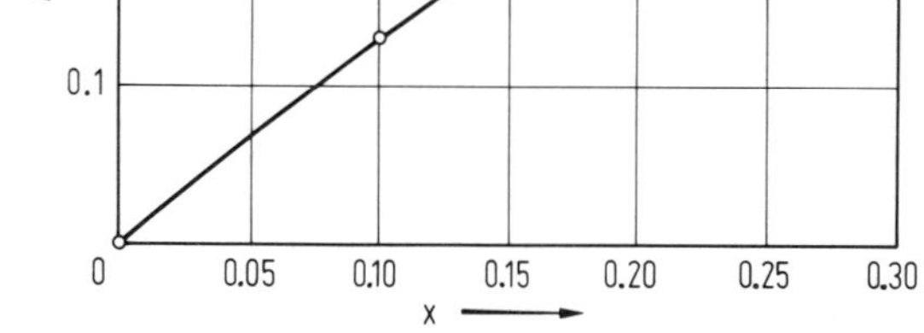

Fig. 141. a-$(Fe_2O_3)_{0.5}(CuO)_x(Bi_2O_3)_{0.5-x}$. Variation of the magnetic moment (at H = 5 kOe) per formula unit p_{FU} at 80 K as a function of the concentration x of CuO in annealed rf sputtered films (annealing conditions: T_a = 500···600°C in air; sputtering in 50% Ar, 50% O_2 atmosphere). The data indicate a linear increase in magnetization with x, and the magnetization is proposed to arise from copper ferrite like clusters [92R3].

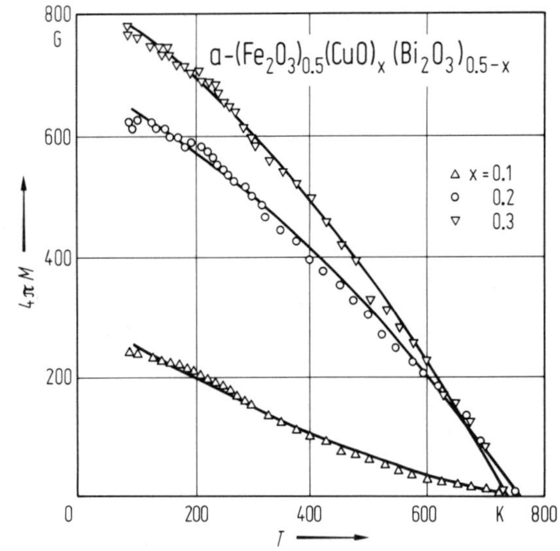

Fig. 142. a-$(Fe_2O_3)_{0.5}(CuO)_x(Bi_2O_3)_{0.5-x}$. $4\pi M$ measured at 5 kOe as a function of temperature for samples sputtered in 50% O_2 + 50% Ar and annealed at 500···600°C in air. The Curie temperatures for the films, 460···480°C, are in agreement with T_C for crystalline copper ferrite [92R3].

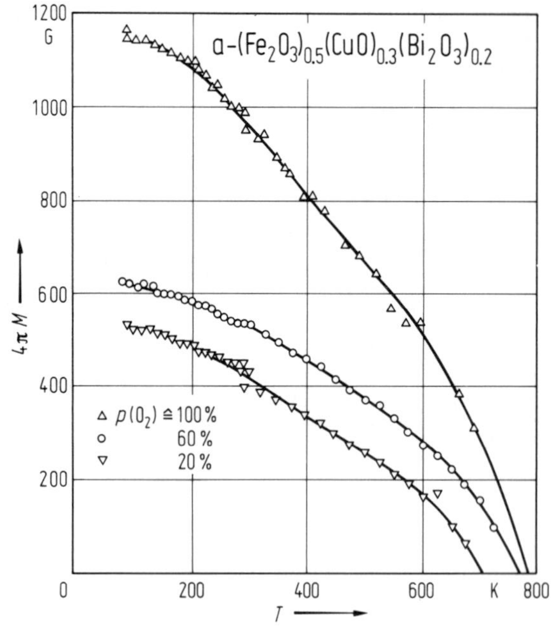

Fig. 143. a-$(Fe_2O_3)_{0.5}(CuO)_x(Bi_2O_3)_{0.5-x}$, x = 0.3. $4\pi M$ (for H = 5 kOe) vs. T for samples sputtered in an oxygen-argon atmosphere with the oxygen partial pressure $p(O_2)$ corresponding to 100, 60, 20% of the total atmosphere and annealed in air at T_a =700···800 K. The anticipated presence of Fe^{2+} in films sputtered in an argon rich atmosphere may account for the observed decrease of M and T_C with decreasing $p(O_2)$ [93S2].

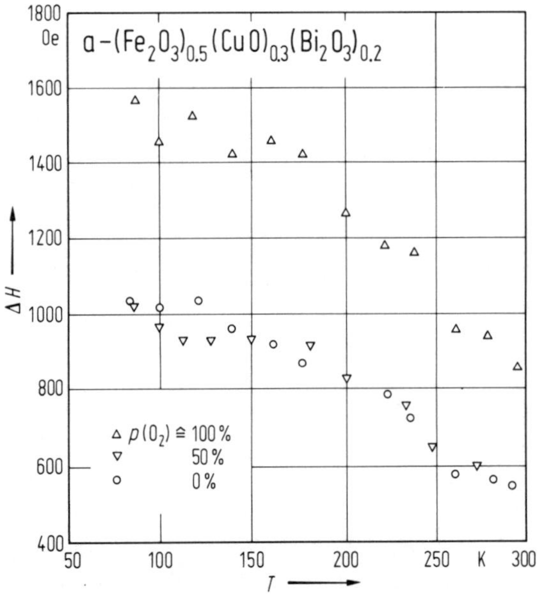

Fig. 145. a-$(Fe_2O_3)_{0.5}(CuO)_x(Bi_2O_3)_{0.5-x}$, x = 0.3. Ferromagnetic resonance linewidth ΔH at 9.2 GHz as a function of temperature for annealed (in air at 750···800 K) films sputtered with $p(O_2)$ corresponding to 0, 50, and 100% of the total atmosphere [93S2].

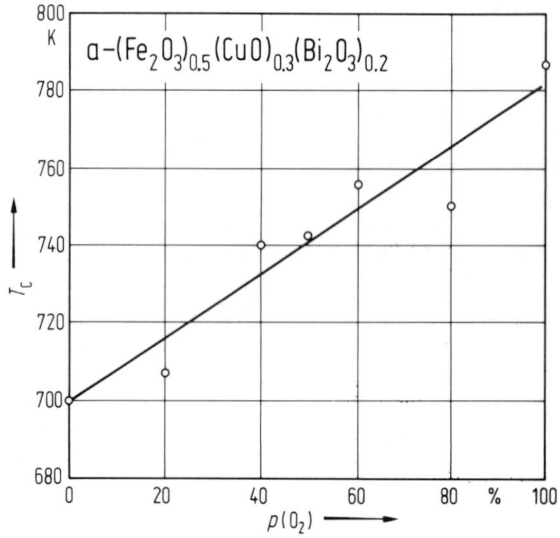

Fig. 144. a-$(Fe_2O_3)_{0.5}(CuO)_x(Bi_2O_3)_{0.5-x}$, x = 0.3. The dependence of the Curie temperature of annealed (in air at 750···800 K) films on the percentage of oxygen partial pressure in the sputtering atmosphere. T_C-values were determined from $4\pi M$ vs. T data as in Fig. 143 [93S2].

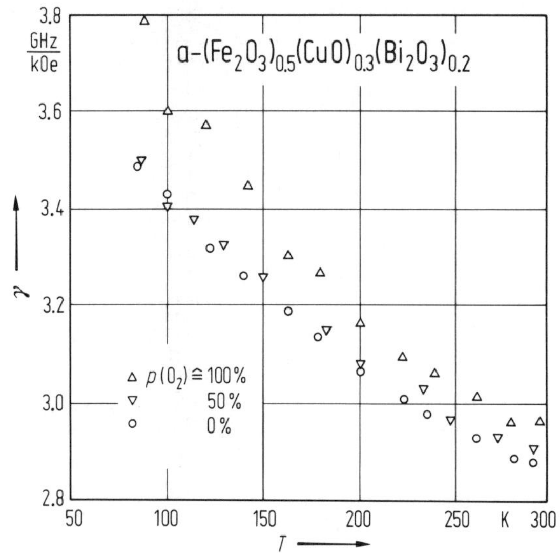

Fig. 146. a-$(Fe_2O_3)_{0.5}(CuO)_x(Bi_2O_3)_{0.5-x}$, x = 0.3. Variation of the gyromagnetic ratio γ with T for films sputtered with $p(O_2)$ corresponding to 0, 50, and 100% of the total atmosphere and annealed in air at 750···800 K [93S2].

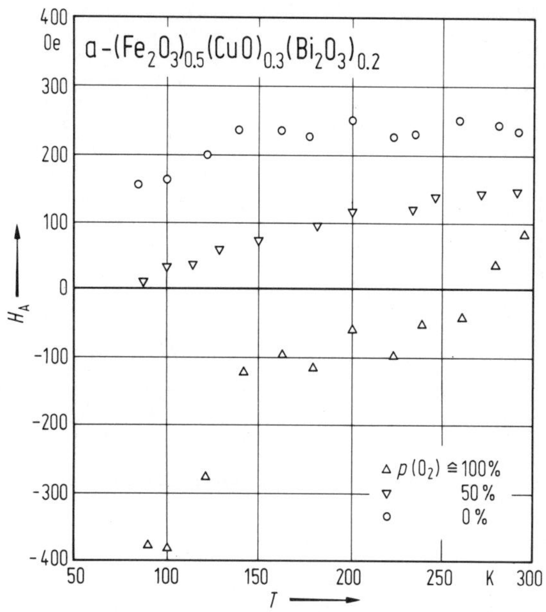

Fig. 147. a-$(Fe_2O_3)_{0.5}(CuO)_x(Bi_2O_3)_{0.5-x}$, x = 0.3. Temperature dependence of the anisotropy field H_A for annealed (in air at 750···800 K) films sputtered with $p(O_2)$ corresponding to 0, 50, and 100% of the total atmosphere. The H_A-values were estimated from static magnetization and effective magnetization obtained by FMR at 9.2 GHz. Positive (negative) values for H_A indicate out-of-plane (in-plane) anisotropy [93S2].

5.7.3.3.6 $Fe_2O_3 \cdot Bi_2O_3 \cdot ABO_3$ oxides [88F], [91K1], [91K2], [91R1], [91R4]

Amorphous compounds of the composition $(Fe_2O_3)_{0.5-x/2}(Bi_2O_3)_{0.5-x/2}(ABO_3\ )_x$ (AB = BaTi, PbTi, PbZr) show coexisting ferromagnetism and ferroelectricity at ambient temperatures. Such compounds prepared by rf sputtering are paramagnetic in the as-deposited state and show a ferromagnetic moment when annealed at high temperatures. The magnetization for the films depend sensitively on the annealing temperature, the nature and amount of ABO_3 compounds, and the nature of the sputtering atmosphere (Figs. 148···153). The films are hard magnets with hysteresis and remanence (H_c at RT = 200 Oe···8 kOe) (Figs. 154 and 155). The Curie temperature for the samples varies from 200°C to 500°C (Figs. 156 and 157). X-ray diffraction data show the precipitation of collinear or canted antiferromagnetic crystalline phases in samples annealed at very high temperatures (Fig. 158). However, crystallization occurs preferentially on the film surface and the magnetization increases when such magnetically dead surface layers are removed (Fig. 159 and 160). A long range order is evident from Mössbauer studies (Figs. 161). The films are transparent in the visible through the near-infrared region and show a large magneto-optic coefficient (Fig. 162 and 163). In addition, the films are ferroelectric at room temperature (Fig. 164).

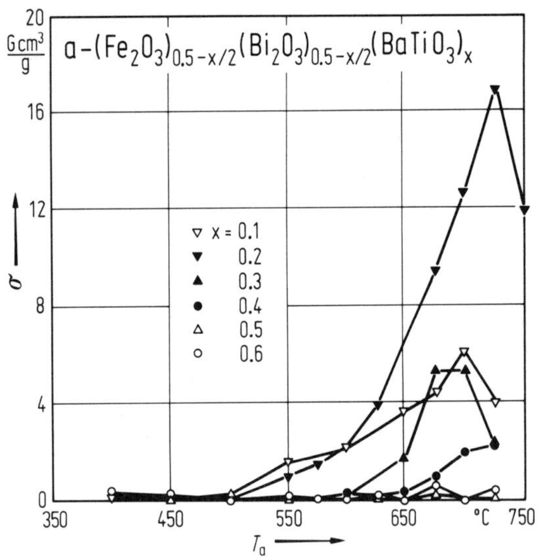

Fig. 148. a-$(Fe_2O_3)_{0.5-x/2}(Bi_2O_3)_{0.5-x/2}(BaTiO_3)_x$. Room temperature specific magnetization measured at 6 kOe vs. annealing temperature T_a for films prepared by rf sputtering in a 50% O_2 + 50% Ar atmosphere. Samples were annealed for 60 min in air at T_a. X-ray diffraction pattern do not indicate the precipitation of any ferromagnetic crystalline phases in the films. The sample magnetization is due to the formation of ordered clusters upon annealing at T_a. The decrease in magnetization at very high T_a is due to the loss of Fe from the ordered clusters [91R4].

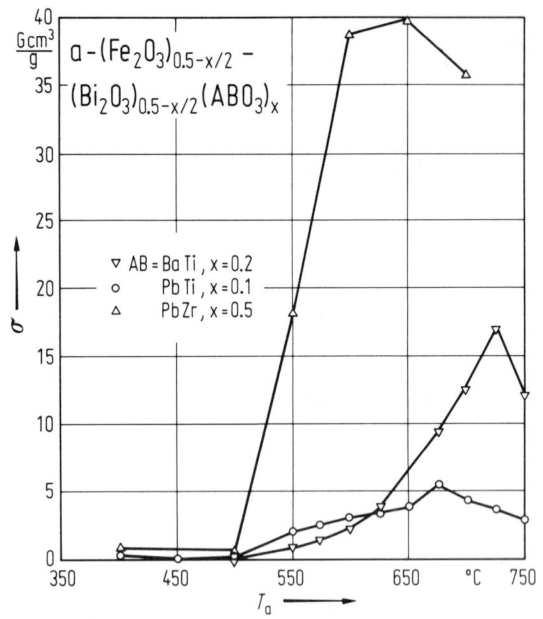

Fig. 149. a-$(Fe_2O_3)_{0.5-x/2}(Bi_2O_3)_{0.5-x/2}(ABO_3)_x$ for $(ABO_3)_x$ = $(BaTiO_3)_{0.2}$, $(PbTiO_3)_{0.1}$, and $(PbZrO_3)_{0.5}$. Data as in Fig. 148. The significance of these three specific samples is that the room temperature magnetization is the highest for the films in their respective series. The film with $(PbZrO_3\ )_{0.5}$ shows a $4\pi M$ of 3.8 kG, the largest value reported for any amorphous oxide [91R4].

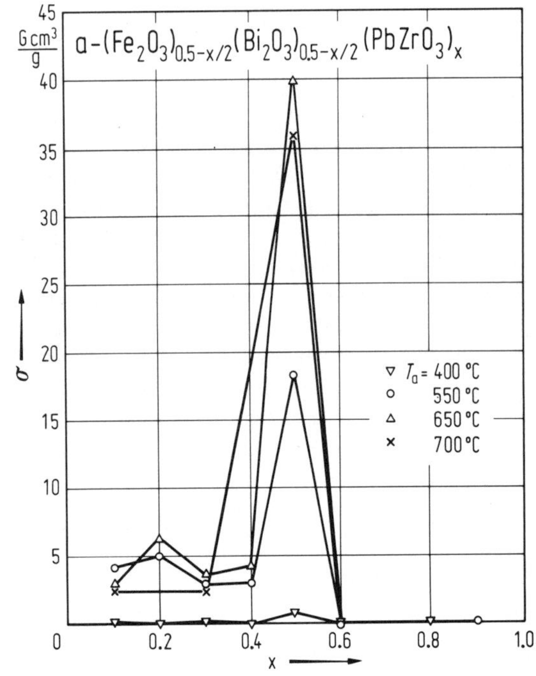

Fig. 150. a-$(Fe_2O_3)_{0.5-x/2}(Bi_2O_3)_{0.5-x/2}(PbZrO_3)_x$. Room temperature σ vs. the concentration x of $PbZrO_3$ for a series of annealing temperatures T_a. The films were deposited in 50% O_2 + 50% Ar and annealed in air [91R1].

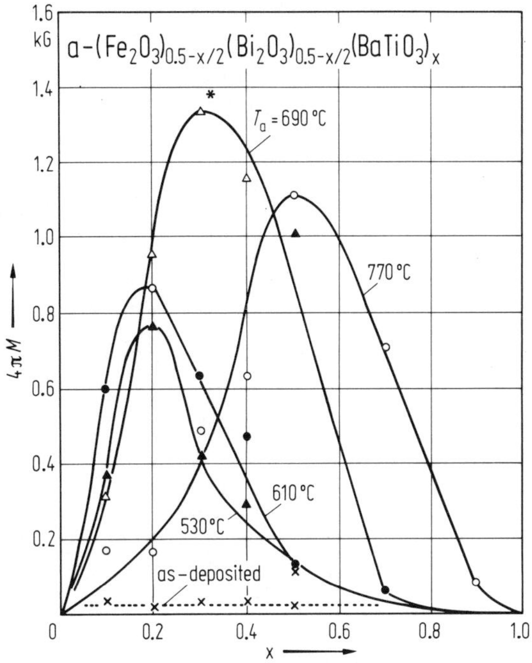

Fig. 151. a-$(Fe_2O_3)_{0.5-x/2}(Bi_2O_3)_{0.5-x/2}(BaTiO_3)_x$. Data as in Fig. 150 for films sputtered in 25% O_2 + 75% Ar atmospheres and annealed in air [88F].

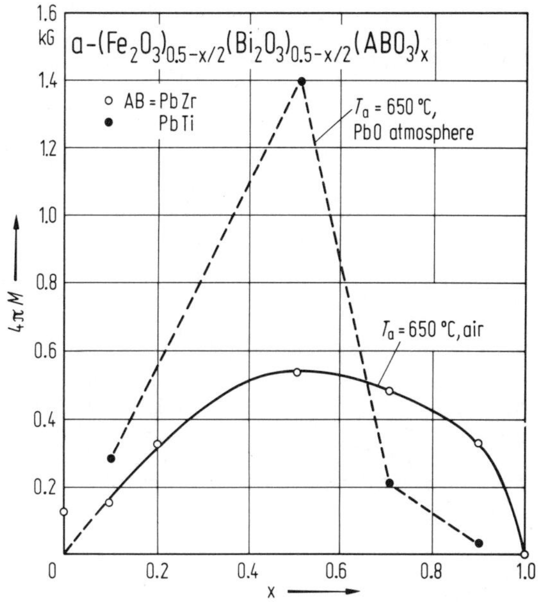

Fig. 152. a-$(Fe_2O_3)_{0.5-x/2}(Bi_2O_3)_{0.5-x/2}(ABO_3)_x$. Similar data as in Fig. 151 for amorphous films containing $PbTiO_3$ or $PbZrO_3$ at different T_a and annealing atmosphere. The films were sputtered in $4{\cdot}10^{-2}$ Torr of total pressure with O_2 : Ar = 1:3 [88F].

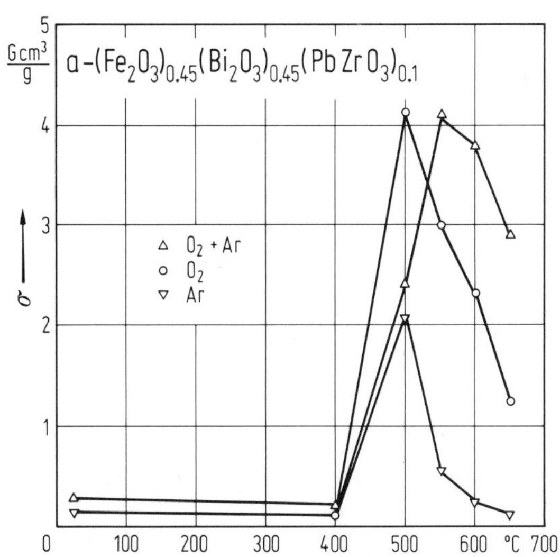

Fig. 153. a-$(Fe_2O_3)_{0.45}(Bi_2O_3)_{0.45}(PbZrO_3)_{0.1}$. Variation of the room temperature σ at 6 kOe with T_a for samples sputtered in oxygen, argon, and oxygen + argon atmospheres and annealed in air [91R1].

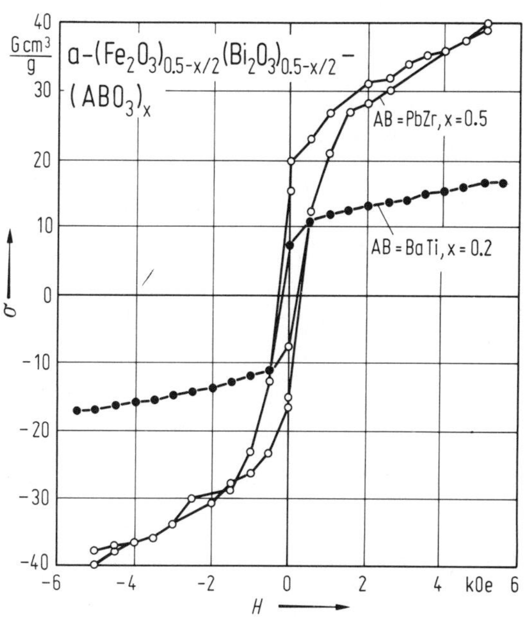

Fig. 154. a-$(Fe_2O_3)_{0.4}(Bi_2O_3)_{0.4}(BaTiO_3)_{0.2}$, a-$(Fe_2O_3)_{0.25}(Bi_2O_3)_{0.25}(PbZrO_3)_{0.5}$. In-plane static field dependence of the room temperature specific magnetization for samples annealed in air at T_a = 725°C (AB = BaTi) and 650°C (AB = PbZr). The coercive field for these 50% O_2 + 50% Ar sputtered samples is about 200···300 Oe [91R4].

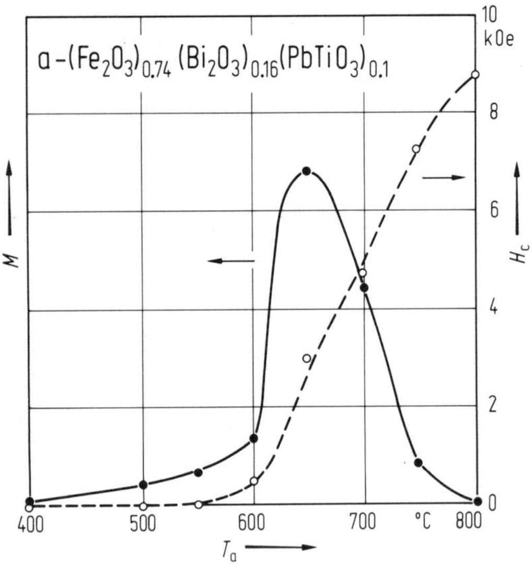

Fig. 155. a-$(Fe_2O_3)_{0.74}(Bi_2O_3)_{0.16}(PbTiO_3)_{0.1}$. Variation of M and H_c at room temperature for rf sputtered sample with annealing temperature T_a. The films were sputtered in an argon rich atmosphere and showed a much larger H_c than for films deposited in 50% O_2 + 50% Ar (refer to Fig. 154) [91K2].

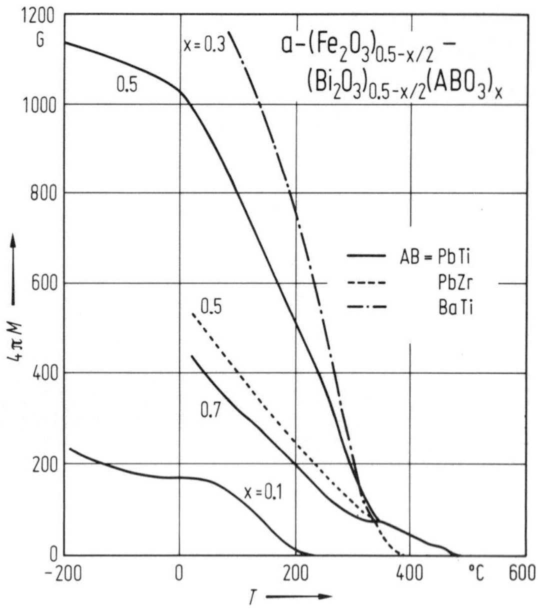

Fig. 156. a-$(Fe_2O_3)_{0.5-x/2}(Bi_2O_3)_{0.5-x/2}(ABO_3)_x$ (AB = BaTi, PbTi, PbZr). $4\pi M$ vs. T data for films sputtered in argon rich atmospheres. T_C ranges from 200°C to 400°C [88F].

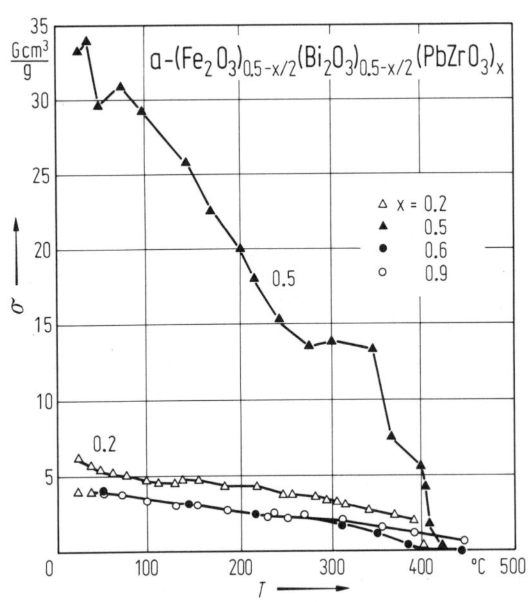

Fig. 157. a-$(Fe_2O_3)_{0.5-x/2}(Bi_2O_3)_{0.5-x/2}(PbZrO_3)_x$. Specific magnetization vs. temperature for samples annealed at T_a = 650°C in air. These films sputtered in 50% O_2 + 50% Ar atmospheres have T_C = 400···500°C [91R1].

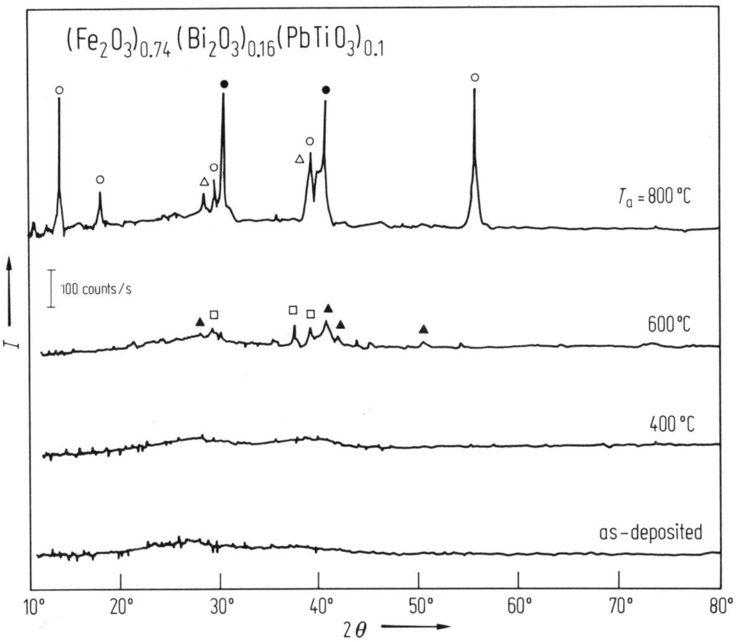

Fig. 158. $(Fe_2O_3)_{0.74}(Bi_2O_3)_{0.16}(PbTiO_3)_{0.1}$. X-ray ($FeK_\alpha$) diffraction data for as-deposited and annealed films. Crystalline collinear or canted antiferromagnetic phases such as $BiFeO_3$-$PbTiO_3$ (full triangle), $PbFe_4O_7$ (square), α-Fe_2O_3 (full circle), $Pb_2Bi_4Ti_5O_8$ (open circle), or $BiFeO_3$ (open triangle) are formed in the film at high T_a. The crystallization is shown to occur at the film surface (Fig. 159 and 160) and the film magnetization (Figs. 148···153) is shown to arise from the amorphous interior [91K2].

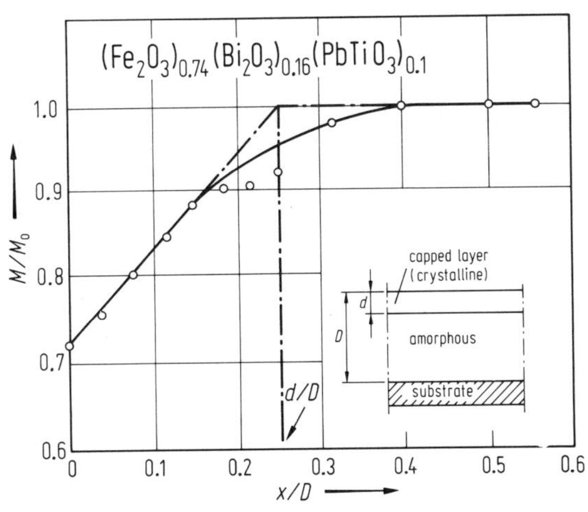

Fig. 159. a-$(Fe_2O_3)_{0.74}(Bi_2O_3)_{0.16}(PbTiO_3)_{0.1}$. Variation of the room temperature magnetization with the film thickness for annealed films. x represents the thickness of the surface layer removed by etching the film. M increases with x and levels off when $x = d =$ 145 nm, i.e., when the magnetically dead crystalline top layer of the film is removed [91K2].

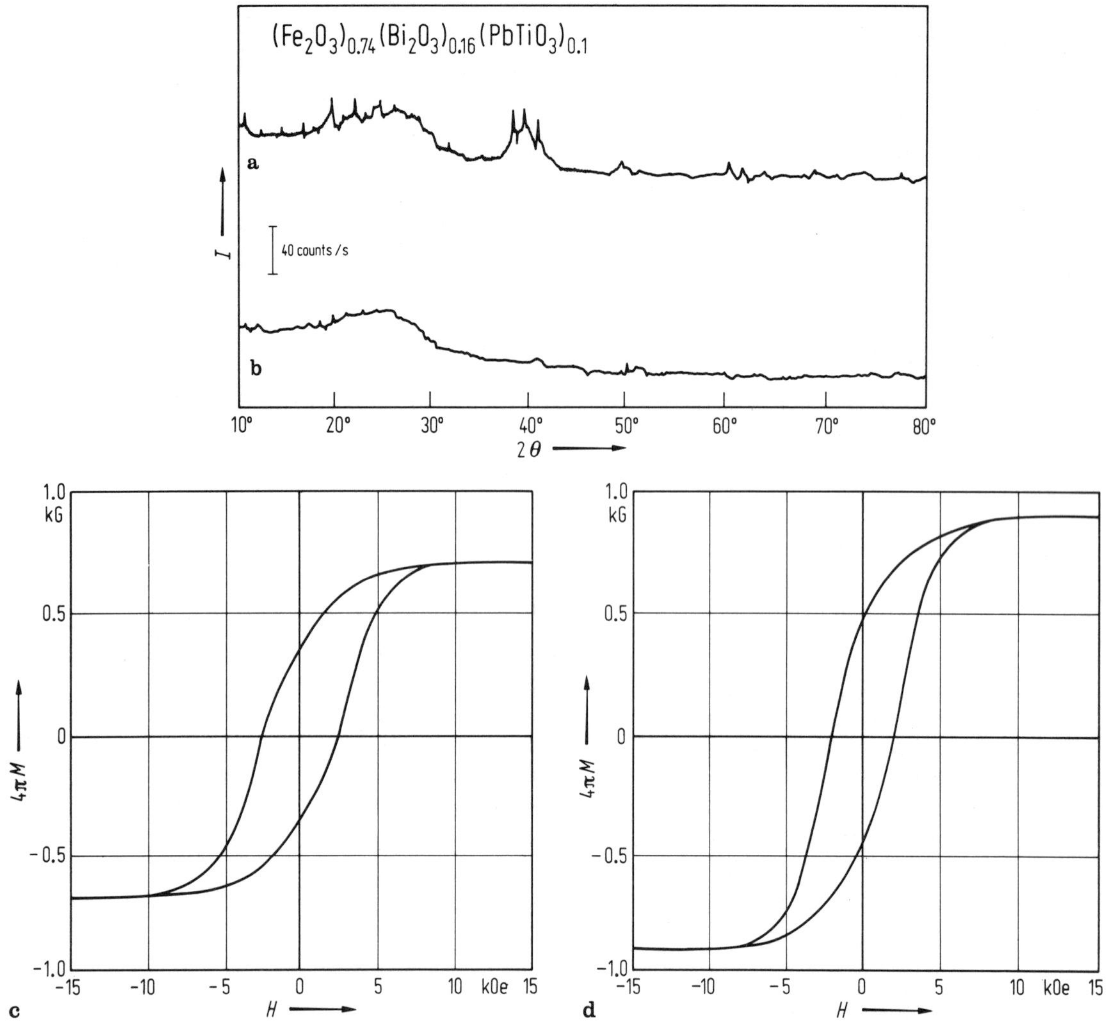

Fig. 160. $(Fe_2O_3)_{0.74}(Bi_2O_3)_{0.16}(PbTiO_3)_{0.1}$. Changes in the X-ray ($FeK_\alpha$) diffraction pattern, (a, b), and in the magnetic hysteresis loop, (c, d) before (a, c) and after (b, d) a 145 nm thick magnetically dead surface layer is removed from the sample (also refer to Fig. 159). It is evident from the data that the magnetization arises from the amorphous interior of the film [91K2].

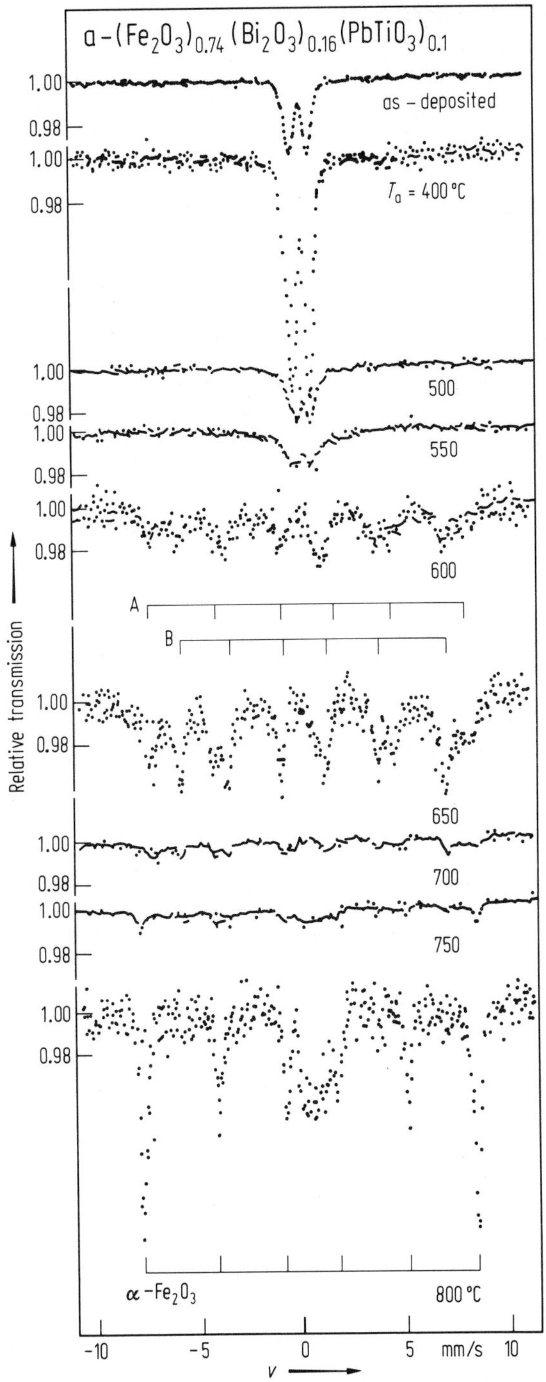

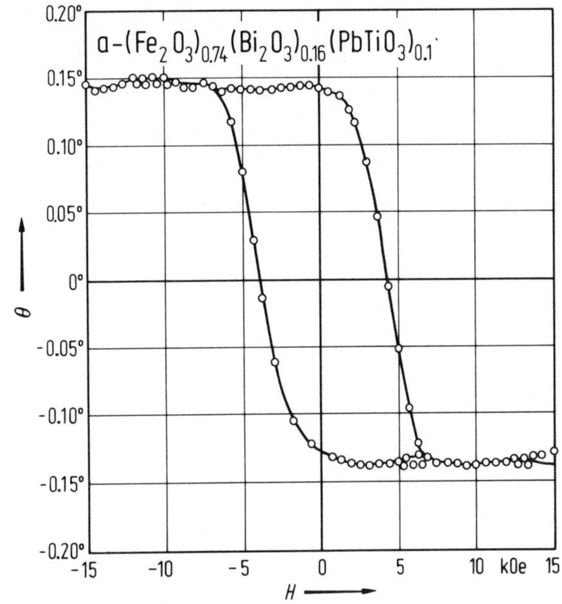

Fig. 163. a-$(Fe_2O_3)_{0.74}(Bi_2O_3)_{0.16}(PbTiO_3)_{0.1}$. The magneto-optic Faraday effect hysteresis loop for an annealed sample having the maximum magnetization of this oxide system for a wavelength of 633 nm and for static magnetic fields applied normal to the film plane. The saturation Faraday rotation $2\theta_F$ = − 1200°/cm [91K1].

For Fig. 162 see next page.

←

Fig. 161. $(Fe_2O_3)_{0.74}(Bi_2O_3)_{0.16}(PbTiO_3)_{0.1}$. Mössbauer spectra at RT for as-deposited and air-annealed films. Spectra for α-Fe_2O_3 are shown for comparison. A hyperfine structure, indicative of a long range magnetic order, appears for T_a = 600°C. Symbols A and B denote the spectral components for Fe^{3+} ions in crystallographic sites similar to $4f_1$ and 12k sites, respectively, in crystalline PbO·$6Fe_2O_3$. Spectra at T_a > 700°C resemble the structure shown for α-Fe_2O_3 [91K2].

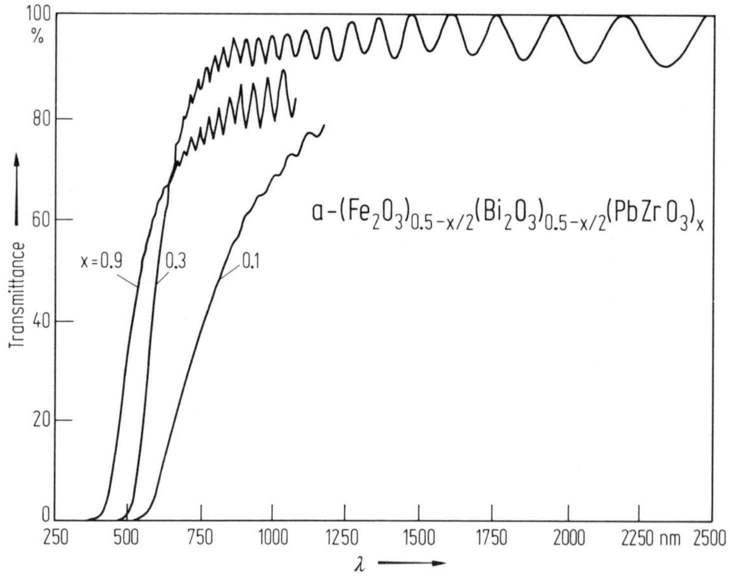

Fig. 162. a-$(Fe_2O_3)_{0.05}(Bi_2O_3)_{0.05}(PbTiO_3)_{0.9}$ (2.67μm thick), a-$(Fe_2O_3)_{0.35}(Bi_2O_3)_{0.35}(PbTiO_3)_{0.3}$ (5.88 μm thick), a-$(Fe_2O_3)_{0.45}(Bi_2O_3)_{0.45}(PbTiO_3)_{0.1}$ (5.25μm thick). Transmittance vs. wavelength for amorphous films. The oscillations are due to interference resulting from multiple reflections in the film. The films have a refractive index of 3.1 for a wavelength of 1.5μm. The samples are transparent in the visible to near infrared region of the electromagnetic spectrum [91K1].

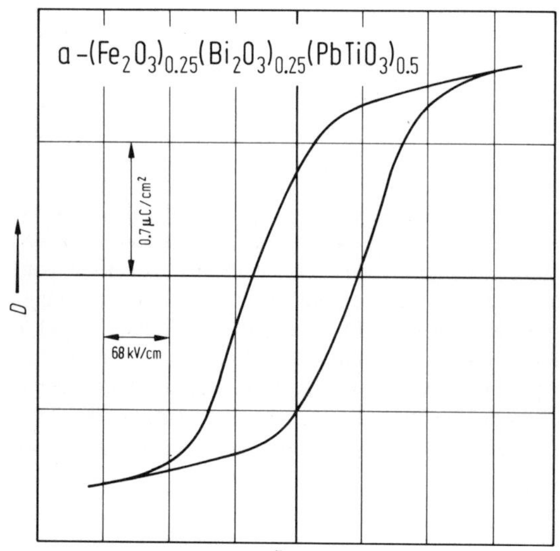

←
Fig. 164. a-$(Fe_2O_3)_{0.25}(Bi_2O_3)_{0.25}(PbTiO_3)_{0.5}$. Ferroelectric hysteresis loop at 60 Hz [91K1].

5.7.4 Theoretical status

Here we summarize the theoretical research on the question of whether amorphous materials should order magnetically since they lack the periodic crystalline order and consequently there is an inherent frustration in the exchange coupling between the neighboring spins. Equally important is the nature of magnetic ordering in such systems. A ferromagnetic order with a finite Curie temperature in amorphous materials was first predicted by Gubanov [60G]. Handrich [69H] estimated the spontaneous moment for such structures with a specific type of magnetic atoms and showed that the magnetization would depend sensitively on the randomness (or fluctuations) in the structure and that the magnetization would vanish for a certain critical value of the structural fluctuations. However, the model predicted a paramagnetic Curie temperature and T_C independent of the structural fluctuation parameter.

In amorphous oxides, the superexchange interactions between nearest neighbors and next nearest neighbors are expected to be short range. Based on an effective field model, Simpson showed that an antiferromagnetic order would exist in amorphous oxides with a T_N proportional to the square root of the average number of interacting nearest neighbors [70S], [70K]. Results of this model were in general agreement with magnetic parameters for amorphous yttrium and gadolinium iron garnets [71S]. Hasegawa, based on a molecular field model of amorphous antiferromagnets, obtained several characteristic features in the temperature dependence of the susceptibility, depending on the sign and magnitude of fluctuations in the exchange constants [70H].

Kaneyoshi in a series of theoretical works in recent years showed that ferrimagnetism could exist in a two dimensional amorphous structure [87K], [89K], [92K1], [92K2]. In this model, magnetic parameters were estimated for a two dimensional network consisting of spin-1/2 and spin-1 ions. The calculated susceptibility showed features expected for a ferrimagnet [87K]. The theory predicted a collinear or canted ferrimagnetic structure, with a random distribution for the canting angle, depending on the strength and fluctuation in the exchange constants [89K]. Important features such as a compensation point and hysteresis were also expected in the temperature dependence of the magnetization in such systems [92K1], [92K2].

In our estimation, the current theoretical understanding of the nature of magnetism and magnetic order in amorphous oxides is quite inadequate. The kind of ordering that are possible in amorphous oxides is shown in Fig. 165 [84M]. However, it is not clear whether speromagnetism, asperomagnetism, and sperimagnetism can be established by neutron scattering experiments. Also, it needs to be established whether the observed magnetism is truly long order or due to ordering of some microclusters. Therefore, there is ample room for additional theoretical and experimental studies in amorphous oxides.

(Fig. 165 see next page)

Acknowledgement

One of us (G.S) gratefully acknowledges the support received from the Petroleum Research Fund, administered by the American Chemical Society, for this project.

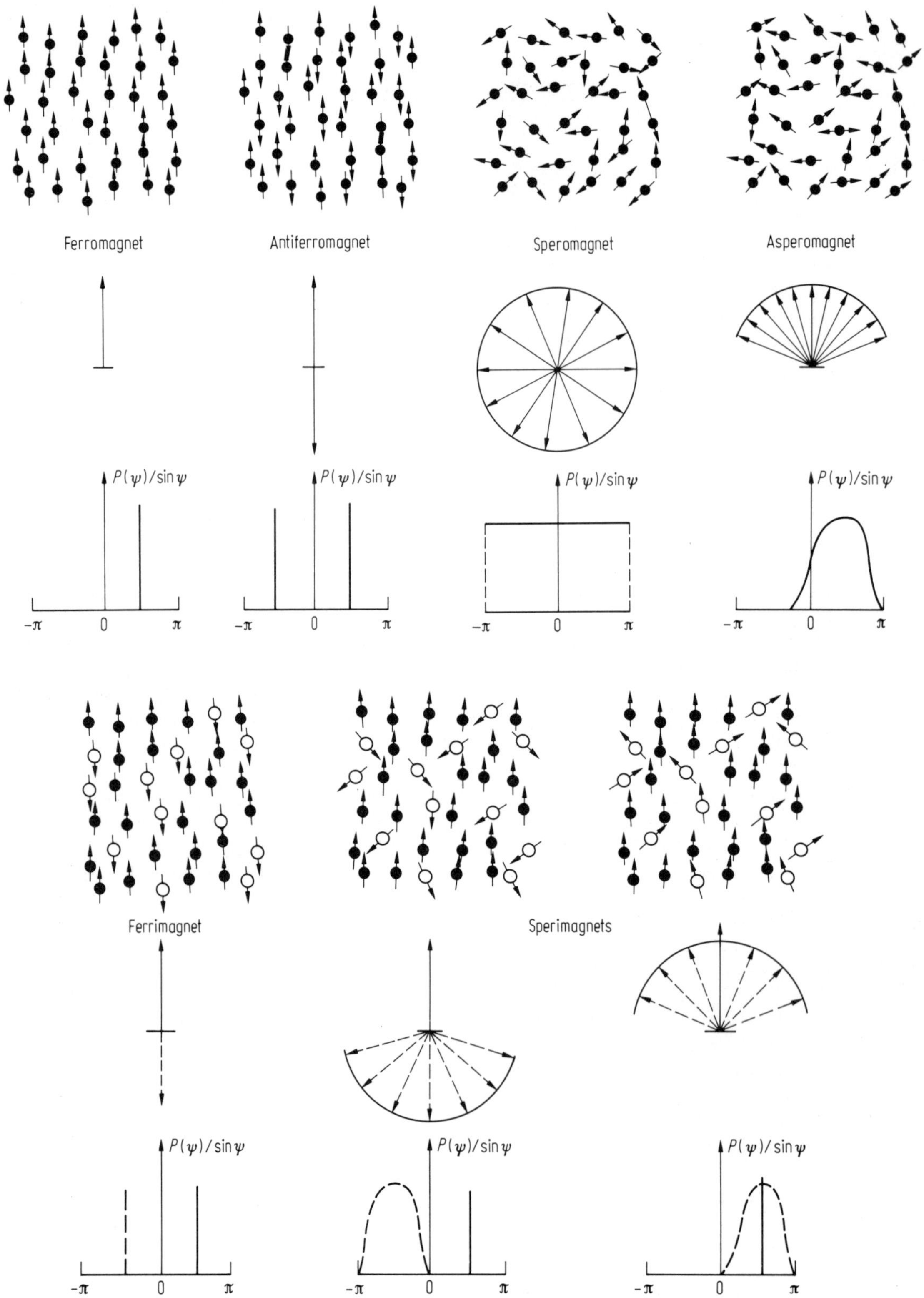

Fig. 165. Possible two-subnetwork magnetic structures in amorphous solids showing schematically the spatial distribution of moment directions and the corresponding normalized angular probability distributions $P(\psi)/\sin\,\psi$. Solid and dashed lines represent moments of two chemical species [84 M].

5.7.5 References for 5.7

60 G Gubanov, G.: Fiz. Tverd. Tela. **2** (1960) 502.
62 Y Yafaev, N. R., Yablokov, Yu. V.: Sov. Phys. Solid State **4** (1962) 1123.
67 L Landry, R. J., Fournier, J. T., Young, C. G.: J. Chem. Phys. **46** (1967) 1285.
69 B Burkey, R. R., Kenealy, P. F., Beard, G. B., Hooper, H. O.: J. Appl. Phys. **40** (1969) 4289.
69 H Handrich, K.: Phys. Status Solidi **32** (1969) K55.
70 H Hasegawa, R.: Phys. Status Solidi (b) **44** (1970) 613.
70 K Kobe, S., Handrich, K.: Phys. Status Solidi **42** (1970) K69.
70 S Simpson, A. W.: Phys. Status Solidi **40** (1970) 207.
71 F Fournier, J. T., Landry, R. J., Bartram, R. H.: J. Chem. Phys. **55** (1971) 2522.
71 S Simpson, A. W., Lucas, J. M.: J. Appl. Phys. **42** (1971) 2181.
72 E Egami, T., Sacli, O. A., Simpson, A. W., Terry, A. L., Wedgwood, F. A.: J. Phys.C: Solid State Phys. **5** (1972) L261.
72 F Friebele, E. J., Wilson, L. K., Kinser, D. L.: J. Am. Ceram. Soc. **55** (1972) 164.
73 E Egami, T., Sacli, O. A., Simpson, A. W., Terry, A. L., Wedgwood, F. A.: Amorphous Magnetism, Hooper, H. O., de Graaf, A. M.(eds.), New York: Plenum Press 1973.
73 H1 Hooper, H. O., de Graaf, A. M.,(eds.): Amorphous Magnetism, New York: Plenum Press 1973.
73 H2 Hooper, H. O., Beard, G. B., Catchings, R. M., Burkey, R. R., Forrest, M., Kenealy, P. F., Kline, R. W., Moran, jr., T. J., O'Keefe, J. G., Thomas, R. L., Verhelst, R. A.: Amorphous Magnetism, Hooper , H. O., de Graaf, A. M.(eds.), New York: Plenum Press 1973.
73 M MacCrone, R. K.: Amorphous Magnetism, Hooper, H. O., de Graaf, A. M.(eds.), New York: Plenum Press 1973.
73 W Wilson, L. K., Friebele, E. J., Kinser, D. L.:Amorphous Magnetism, Hooper, H. O., de Graaf, A. M.(eds.), New York: Plenum Press 1973.
74 F1 Friebele, E. J., Koon, N. C.: Solid State Commun. **14** (1974) 1247.
74 F2 Friebele, E. J., Koon, N. C., Wilson, L.K., Kinser, D. L.: J. Am. Ceram. Soc. **57** (1974) 237.
74 P Popma, Th. J. A., van Diepen, A. M.: Mater. Res. Bull. **9** (1974) 1119.
75 C Cargill, G. S.: Structure of metallic alloy glasses, Solid State Physics **30** (1975) 227, Seitz, F., Turnbull, D., Ehrenreich. S. (eds.), New York: Academic Press 1975.
75 M Moran, T. J., Batra, N. K., Verhelst, R. A., de Graaf, A. M.: Phys. Rev. B**11** (1975) 4436.
75 V Verhelst, R. A., Kline, R. W., de Graaf, A. M., Hooper, H. O.: Phys. Rev. B**11** (1975) 4427.
76 V van Diepen, A. M., Popma, Th. J. A.: J. Phys. Colloq. C**6** (1976) 755.
77 B Bieman, L. H., Kenealy, P. F., de Graaf, A. M.: Amorphous Magnetism II, Levy, R. A., Hasegawa, R. (eds.), New York: Plenum Press 1977.
77 K Kim, K. J., Maley, M. P., MacCrone, R. R.: Amorphous Magnetism II, Levy, R. A., Hasegawa, R. (eds.), New York: Plenum Press 1977.
77 L1 Levy, R. A., Hasegawa, R.(eds.): Amorphous Magnetism II, New York: Plenum Press 1977.
77 L2 Levy, R. A.: Amorphous Magnetism II, Levy, R. A., Hasegawa, R. (eds.), New York: Plenum Press 1977.
77 O1 O'Horo, M., Steinitz, R.: Amorphous Magnetism II, Levy, R. A., Hasegawa, R. (eds.), New York: Plenum Press 1977.
77 O2 O'Horo, M. P., O'Neil, J. F.: Amorphous Magnetism II, Levy, R. A., Hasegawa, R. (eds.), New York: Plenum Press 1977.
77 P Perry, C. H., Kinser, D. L., Wilson, L. K.: Amorphous Magnetism II, Levy, R. A., Hasegawa, R. (eds.), New York: Plenum Press 1977.
77 T Trombetta, L., Williams, J., MacCrone, R. K.: Amorphous Magnetism II, Levy, R. A., Hasegawa, R. (eds.), New York: Plenum Press 1977.
77 W Wenger, L. E., Keesom, P. H.: Amorphous Magnetism II, Levy, R. A., Hasegawa, R., (eds.), New York: Plenum Press 1977.

78 C Coey, J. M. D.: J. Appl. Phys. **49** (1978) 1646.
78 H1 Horie, O., Syono, Y., Nakagawa, Y., Ito, A., Okamura, K., Yajima, S.: Solid State Commun. **25** (1978) 423.
78 H2 Huang, F. S., Bieman, L. H., de Graaf, A. M., Rechenberg, H. R.: J. Phys. C: Solid State Phys. **11** (1978) L271.
78 L Laville, H., Bernier, J. C., Sanchez, J. P.: Solid State Commun. **27** (1978) 259.
78 V van Diepen, A. M., Popma, Th. J. A.: Solid State Commun. **27** (1978) 121.
78 W Wong, Y. H., Thomas, D., Thomas, R. L.: J. Appl. Phys. **49** (1978) 1640.
79 G Gyorgy, E. M., Nassau, K., Eibschutz, M., Waszczak, J. V., Wang, C. A., Shelton, J. C.: J. Appl. Phys. **50** (1979) 2883.
79 H Heiman, N., Kazuma, N. S.: J. Appl. Phys. **50** (1979) 7633.
79 L Litterst, F. J., Tejada, J., Kalvius, G. M.: J. Appl. Phys. **50** (1979) 7636.
79 M1 Marusak, L. A., Messier, R.: J. Appl. Phys. **50** (1979) 7630.
79 M2 Matsuzawa, T., Okamura, K., Shishido, T., Yajima, S.: J. Phys. Colloq. C**2** (1979) 149.
79 P Perry, C. H., Kinser, D. L. , Wilson, L. K., Vaughn, J. G.: J. Appl. Phys. **50** (1979) 1601.
79 S1 Shigematsu, T., Bando, Y., Takada, T.: J. Phys. Colloq. C**2** (1979) 153.
79 S2 Syono, Y., Ito, A.: J. Phys. Colloq. C**2** (1979) 115.
80 C Chen, H. S.: Rep. Prog. Phys. **43** (1980) 353.
80 F Ferre, J., Pommier, J., Renard, J. P., Knorr, K.: J. Phys. C: Solid State Phys. **13** (1980) 3697.
80 I Imshennik, V. K., Suzdalev, I. P., Maksimov, Yu. V., Litterst, F. J.: J. Phys. Colloq. C**1** (1980) 267.
80 L Laville, H., Bernier, J. C.: J. Mater. Sci. **15** (1980) 73.
80 R Rechenberg, H. R., de Graaf, A. M.: J. Phys. C: Solid State Phys. **13** (1980) L397.
82 B Bonnenfant, A., Friedt, J. M., Maurer, M., Sanchez, J. P.: J. Phys. (Paris) **43** (1982) 1475.
82 C Coey, J. M. D., Devlin, E., Gambino, R. J.: J. Appl. Phys. **53** (1982) 7810.
82 I Ito, A., Torikai, E., Yamaguchi, H., Syono, Y.: J. Phys. C: Solid State Phys. **15** (1982) 2759.
82 S Sugimoto, M., Hiratsuka, N.: Jpn. J. Appl. Phys. **21** (1982) 197.
83 H Hasegawa, R., (ed.): Glassy metals, New York: CRC Press, 1983.
83 S Sugimoto, M., Hiratsuka, N.: J. Magn. Magn. Mater. **31-34** (1983) 1533.
84 C Coey, J. M. D., Ryan, D. H.: IEEE Trans. Magn. MAG-**20** (1984) 1278.
84 K Kaneyoshi, T.: Amorphous Magnetism, Boca Raton (Florida): CRC Press 1984.
84 M Moorjani, K., Coey, J. M. D.:Magnetic Glasses, Amsterdam: Elsevier 1984. (See chapter 2 on the preparation and characterization of glasses and chapter 5 on insulating glasses.)
85 S Sugimoto, M., Takahashi, T., Naya, K., Soeno, Y., Hiratsuka, N.: IEEE Trans. Magn. TJMJ-**1** (1985) 783.
86 G Girnus, W., Beuthien, H., Priess, R., Gunsser, W.: J. Magn. Magn. Mater. **54-57** (1986) 225.
86 O Ota, N., Okubo, M., Masuda, S., Suzuki, K.: J. Magn. Magn. Mater. **54-57** (1986) 293.
86 S Suzuki, K., Onodera, H., Sakurai, M., Masuda, S., Matsumoto, A., Sadamura, H.: IEEE Trans. Magn. MAG-**22** (1986) 1090.
87 K Kaneyoshi, T.: Philos. Mag. Lett. **55** (1987) 69.
87 N1 Nagata, Y., Inoh, T., Ohta, K.: IEEE Trans. Magn. MAG-**23** (1987) 2317.
87 N2 Nakamura, S., Ichinose, N.: J. Non-Cryst. Solids **95-96** (1987) 849.
87 T Tanaka, K., Soga, N.: J. Non-Cryst. Solids **95-96** (1987) 255.
88 F Fujii, T., Jinzenji, S., Asahara, Y., Kajima, A.: J. Appl. Phys. **64** (1988) 5434.
88 I Inomata, K., Hashimoto, S., Nakamura, S.: Jpn. J. Appl. Phys. **27** (1988) L883.
89 K Kaneyoshi, T.: Solid State Commun. **69** (1989) 91.
89 N Nakamura, S., Ichinose, N.: Jpn. J. Appl. Phys. **28** (1989) 984.
89 T Tanaka, K., Kajima, K., Yoko, T.: J. Non-Cryst. Solids **109** (1989) 289.
90 I Inomata, K., Hashimoto, S, Okuno, S., Matsubara, E., Waseda, Y.: J. Appl. Phys. **67** (1990) 5796.
90 M Matsubara, E., Waseda, Y., Hashimoto, S., Inomata, K.: Phys. Status Solidi (a) **117** (1990) K1.
90 S Soeya, S., Nakamura, S., Ichinose, N.: J. Appl. Phys. **68** (1990) 2875.

91 K1 Kajima, A., Kaneda, T., Ito, H., Fujii, T., Okamoto, I., Kimura, T., Ohdan, K.: J. Appl. Phys. **69** (1991) 3663.
91 K2 Kajima, A., Kaneda, T., Ito, H., Fujii, T., Okamoto, I.: J. Appl. Phys. **70** (1991) 3760.
91 N Nakamura, S., Hirotsu, V., Ichinose, N.: Jpn. J. Appl. Phys. **30** (1991) L844.
91 O Okuno, S. N., Hashimoto, S., Inomata, K., Morimoto, S., Ito, A.: J. Appl. Phys. **69** (1991) 5072.
91 P Petrakovskii, G. A., Sablina, K. A., Vorotinov, A. M., Vasiliev, V. N., Kruglik, A. I., Balaev, A. D., Velikanov, D. A., Kiselev, N. I.: Solid State Commun. **79** (1991) 317.
91 R1 Rao, B. U. M., Srinivasan, G., Babu, V. S., Seehra, M. S.: J. Appl. Phys. **69** (1991) 5463.
91 R2 Rao, B. U. M., Srinivasan, G.: J. Appl. Phys. **70** (1991) 6317.
91 R3 Rao, B. U. M., Srinivasan, G.: Phys. Rev. B **44** (1991) 395.
91 R4 Rao, B. U. M., Srinivasan, G.: Appl. Phys. Lett. **58** (1991) 2441.
91 S Srinivasan, G., Rao, B. U. M., Zhao, J., Seehra, M. S.: Appl. Phys. Lett. **59** (1991) 372.
91 T1 Tanaka, K., Hirao, K., Soga, N.: J. Appl. Phys. **69** (1991) 7752.
91 T2 Tanaka, K., Hirao, K., Soga, N., Mori, H.: Jpn. J. Appl. Phys. **30** (1991) L2095.
92 K1 Kaneyoshi, T., Jascur, M.: J. Magn. Magn. Mater. **115** (1992) 217.
92 K2 Kaneyoshi, T., Jascur, M.: J. Phys.: Condens. Matter **4** (1992) L281.
92 R1 Rao, B. U. M., Srinivasan, G.: IEEE Trans. Magn. **28** (1992) 2444.
92 R2 Rao, B. U. M., Srinivasan, G.: J. Magn. Magn. Mater. **103** (1992) L228.
92 R3 Rao, B. U. M., Srinivasan, G.: J. Magn. Magn. Mater. **111** (1992) 249.
92 S Srinivasan, G., Rao, B. U. M., Saligram, U.: Appl. Phys. A **55** (1992) 549.
92 T Tanaka, K., Hirao, H., Soga, N.: The Physics of non-crystalline solids, London: Taylor and Francis, 1992, p. 296.
92 Y Yamaguchi, K., Matsumoto, K., Fujii, T.: IEEE Trans. Magn. **28** (1992) 2447.
93 S1 Srinivasan, G., Rao, B. U. M.: J. Appl. Phys. **73** (1993) 5713.
93 S2 Srinivasan, G., Rao, B. U. M.: J. Magn. Magn. Mater. **123** (1993) 304.
93 S3 Srinivasan, G., Rao, B. U. M.: Chen, J.: IEEE. Trans. Magn. **29** (1993) 3078.